Carolyn Eberhard

CORNELL UNIVERSITY

BIOLOGY LABORATORY
A Manual to Accompany

Second Edition

Karen Arms
Pamela S. Camp

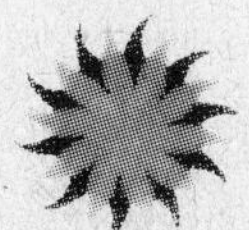

SAUNDERS COLLEGE PUBLISHING

Philadelphia New York Chicago
San Francisco Montreal Toronto
London Sydney Tokyo Mexico City
Rio de Janeiro Madrid

Address orders to:
383 Madison Avenue
New York, NY 10017

Address editorial correspondence to:
West Washington Square
Philadelphia, PA 19105

This book was set in Souvenir by Intergraphic Technologies.
The editors were Michael Brown, Lee Walters, Michael Fisher, and Don Reisman.
The art & design director was Richard L. Moore.
The text design was done by Phoenix Studio, Inc.
The cover design was done by Richard L. Moore.
The artwork was drawn by Fine Line Illustrations, Inc. and Monica Howland.
The production manager was Tom O'Connor.
This book was printed by Connecticut Printers.
Cover credit: Scarlet Ibis in Trinidad Swamp, ©Towsend Dickinson

BIOLOGY LABORATORY — A manual to accompany BIOLOGY 2/e ISBN 0-03-059963-6

Library of Congress catalog card number 81-53080.

345 019 987654

CBS COLLEGE PUBLISHING
Saunders College Publishing
Holt, Rinehart and Winston
The Dryden Press

PREFACE

This manual is designed to be used in the laboratory section accompanying a general biology course. The order of topics is similar to that in the accompanying textbook *Biology*. However, the descriptive material in this laboratory manual is sufficiently comprehensive and detailed that it can be used with other texts.

Since the content and format of the first edition have successfully met the needs of many instructors, the second edition is a modest revision rather than a drastically altered manual. About one-third of the figures have been redrawn, and many have been enlarged. A description of the metric system is given in Appendix D, and an index has also been included. A short reference list for students appears at the end of each topic in the laboratory manual, as well as in the instructor's manual. Finally, there is an added project dealing with the effect of nutritional deficiencies on plant seedlings. The order of topics has been changed, with Population Genetics following Mendelian genetics and Animal Development appearing near the end of the manual.

To accommodate the varying requirements of instructors teaching biology laboratory, 30 topics have been included — more than would ordinarily be covered in a two-semester course. Among these topics is introductory material important to both majors and nonmajors taking laboratory: seven topics on the diversity of living organisms, four topics emphasizing vertebrate anatomy, and two topics discussing the angiosperms. Other topics in this manual, such as those on genetics, muscle physiology, and animal behavior, help students learn how to gather and analyze data and to design experiments. We believe it is necessary that a general biology course provide a thorough introduction to the basic biology of plants and animals because many students, including biology majors, do not pursue studies in these areas. Moreover, those who do take more advanced courses often find that the instructor assumes they are familiar with material that in fact was not emphasized or was eliminated altogether from their introductory biology courses.

Three projects for independent study are given at the end of the manual in which groups of students are asked to design or conduct, and analyze experiments with plant hormones, plant nutrition, and pond microecosystems. The projects emphasize input from the students rather than the use of elaborate equipment, and they have been used successfully with students who had varying levels of preparation.

Each topic begins with Objectives that describe what knowledge or skills the student should be able to demonstrate after completing the exercise. General references to relevant subjects and specific references to the Arms and Camp text are given at the beginning of each topic. The exercises contain questions concerning information recall, concepts, and analysis of laboratory material to stimulate student discussion during the laboratory period. Student worksheets are provided at the end of many topics for students to use to record data and observations of particular experiments. These sheets can then be collected for grading at the end of the laboratory period. This edition contains an index for the convenience of both the student and instructor in using the manual.

The exercises in this manual present techniques such as pipeting, chromatography, and the use of the spectrophotometer and kymograph. We have tried to minimize the use of complicated procedures and elaborate equipment wherever possible so that each experiment is feasible in some form at most colleges and universities. The major recurrent budget items for these exercises are preserved fetal pigs and live material. Although we encourage the use of live material, we have often provided for the use of preserved material as well in the event that live material is unavailable or too expensive.

The separate *Instructor's Manual,* which may be obtained from the publisher free of charge, suggests how many topics may be combined, shortened, or expanded. It also provides lists of materials and equipment needed for these exercises, sources for specific supplies, directions for preparing solutions and handling live material, and an estimate of the cost of each exercise. A list of reference books, films, and film loops is also included.

We are indebted to our colleagues at Cornell University and Ithaca College for their support, suggestions, and contributions to this manual: Renee Alexander, Jeanne Appling, Alan Bennett, Antonie W. Blackler, Everette Busbee, Judith Byrnes, Neil Campbell, Gary Creason, Lin Davidson, Jan Factor, Gail Ferstandig, Sue Giffen, Larry Greenberg, Donna Hill, Monica Howland, Barry Ingber, Leanne Ketcham, Debra Kirchhof-Glazier, Marie Knowlton, Joshua Korman, Martha Lackey, Carolyn Louie, Martha Lyon, Carol Mapes, Michael Meredith, Elizabeth Mort, Sharman O'Neill, Donald Rhodes, Helen Roland, William Schrader, Buffy Silverman, Paul Sisco, Daniel Solomon, Adrian Srb, George Swenson, Craig Tendler, Eileen Whalen, Richard Wodzinski, Nancy Wolf, and Hal Yingling.

We are extremely grateful to Monica Howland for preparing many new and revised figures at short notice and under somewhat arduous conditions.

We also wish to acknowledge the following individuals who offered instructive comments on the original manuscript: Betty D. Allamong, Ball State University; Russell C. Hollingsworth, Tarrant County Junior College, Northeast; Carmita Love, Community College of Philadelphia; Roger Sawyer, University of South Carolina; and Joseph Wood, University of Missouri.

We are indebted to the following for suggestions and comments that were extremely valuable during the preparation of the second edition: Robert Amy, Southwestern at Memphis; Paul Davidson, Rollins College; John Erickson, Western Washington University; Cesar Fermin, Florida Institute of Technology; Eileen Gregory, Rollins College; Terry William Hill, Southwestern at Memphis; Pius Horner, San Bernadino Valley College; Virginia Keville, Salem State College; Stephen Klein, Schenendehowa Central High School; William Leonard, University of Nebraska; Karen Loftin, University of Houston; W. Wallace Martin, Randolph Macon College; Allen Schroeder, Gettysburg College; Robert Smith, San Bernadino Valley College; and Gail Stetten, Johns Hopkins Hospital.

We are grateful to Mary Ahl, Susan Boardman, Muriel Milks, and Mary Wright for typing the original manuscript, and to the editorial staff of Holt, Rinehart and Winston for its help in producing the first edition of the manual. The Saunders editorial staff, Michael Brown, Michael Fisher, Don Reisman, their assistants, and Margaret Mary Kerrigan, were most helpful in bringing the second edition to fruition.

Finally, we wish to express our gratitude to the literary executor of the late Sir Ronald Fisher, F.R.S., Dr. Frank Yates, F.R.S., and Longman Group, Ltd. in London for permission to reprint material from their book, *Statistical Tables for Biological, Agricultural, and Medical Research* (6th edition 1974).

C.E.
Ithaca, New York K.A.
October, 1981 P.S.C.

CONTENTS

	To the Student vii
TOPIC 1	Biological Molecules 1
TOPIC 2	Enzymatic Activity 9
TOPIC 3	Introduction to the Microscope 19
TOPIC 4	Structure of Cells and Cell Types 25
TOPIC 5	Cell Membranes 35
TOPIC 6	Cellular Respiration 41
TOPIC 7	Photosynthesis 49
TOPIC 8	Cell Reproduction 61
TOPIC 9	Mendelian Genetics and the Analysis of Data 73
TOPIC 10	Human Genetics 85
TOPIC 11	Population Genetics 95
TOPIC 12	Classification of Organisms 103
	Introduction to Living Organisms (Topics 13-19) 111
TOPIC 13	Introduction to Monera and Protista 113
TOPIC 14	Introduction to Fungi and the Lower Plants 121
TOPIC 15	Introduction to the Higher Plants: Embryophytes 129
TOPIC 16	Introduction to the Lower Invertebrates 137
TOPIC 17	Introduction to the Higher Invertebrates: Annelids and Molluscs 143
TOPIC 18	Introduction to the Arthropods 149

TOPIC 19 Introduction to the Vertebrates 157
TOPIC 20 Digestion 169
TOPIC 21 Gas Exchange 177
TOPIC 22 The Transport System in Vertebrates 183
TOPIC 23 Function of the Transport System 191
TOPIC 24 Excretion and Reproduction 203
TOPIC 25 Animal Development 213
TOPIC 26 The Nervous System 225
TOPIC 27 Muscles and Skeletons 239
TOPIC 28 Behavior 251
TOPIC 29 Plant Structure and Function 265
TOPIC 30 Plant Reproduction 279
Introduction to the Projects 291
Project A Plant Hormones 293
Project B Plant Nutrition 297
Project C The Pond Ecosystems 305
Appendix A Writing a Lab Report 311
Appendix B Biological Experiments and Statistics 314
Appendix C Laboratory Equipment 320
Appendix D The Metric System 322
Index 325

TO THE STUDENT

OBJECTIVES

At the beginning of each topic in this manual is a list of specific Objectives describing the subject areas and skills that you should master upon completion of the topic. In addition, there are other, broader objectives that you should try to achieve in the course of using this manual. We hope that you will:

1. Interact with other students and your instructor by asking questions about the laboratory material, by comparing your findings, and, when possible, by presenting your results in class discussions at the end of the laboratory session.
2. Acquire knowledge that will provide a foundation for further scientific study.
3. Apply what you have learned in this course to your daily life, and try to make more informed decisions about such matters as nutrition, medicine, natural resources, endangered species, environmental hazards, and the like.

SAFETY

Experiments in biology laboratories sometimes involve working with potentially dangerous chemicals and expensive equipment that must be used properly. To ensure your personal safety, follow these simple rules:

1. Do not eat in the laboratory.

2. Do not drink in the laboratory.

3. Never smoke in the laboratory.

4. Protect your face and eyes by wearing safety glasses and contact lenses. Tie back long loose hair.

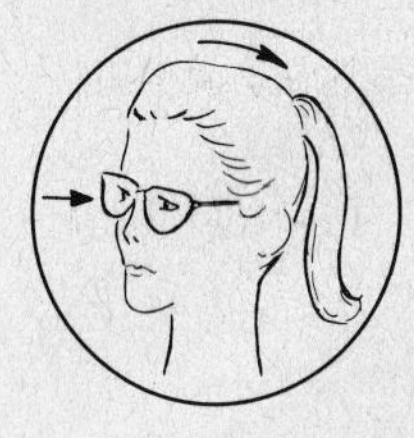

5. Roll up long sleeves when using equipment or open flames.

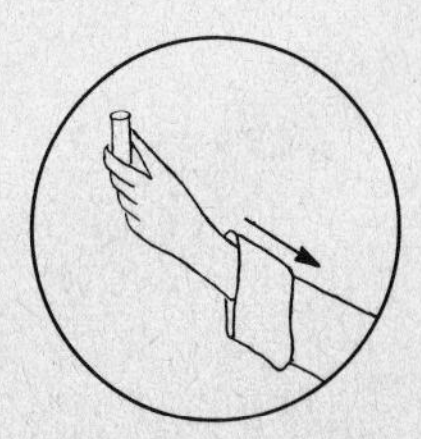

6. Keep open flames away from flammable chemicals or materials and *yourself!*

7. Report any injury to yourself or another student, no matter how minor, to the instructor.

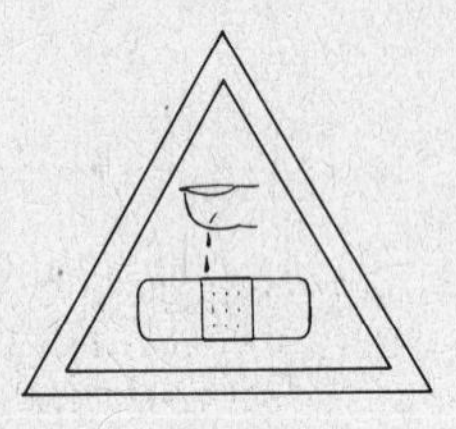

8. Report any spills to your instructor, and clean up the spills immediately.

9. Determine the shortest exit route from the building, and be prepared to leave the room immediately in case of a fire.

PREPARATION

1. Read the assigned exercise before the laboratory session. Otherwise your time will be spent less productively than it should be.
2. Use your textbook to complement each lab by reading the topics or text sections listed at the beginning of each topic while the laboratory work is still fresh in your mind. If the material presented in the topic has not yet been discussed in lecture, read the relevant material *before* the laboratory session.

METRIC SYSTEM AND SCIENTIFIC NOTATION

If you are not familiar with the metric system, consult Appendix D at the end of this manual. You will be expected to use the metric system in this course.

Some measurements in biology have very small or very large units, so it is awkward to write them out as decimals. Instead, we use scientific notation in which the decimal point is placed after the first digit and the number is expressed as a multiple of the power of 10. For example, 506 kilograms is expressed as 5.06×10^2 kg, and 1/200 (.005) of a second is expressed as 5×10^{-3} sec. Scientific notation makes it easy to use the metric system. In our second example, 5×10^{-3} sec equals 5 msec because 10^{-3} sec equals 1 msec.

Introductory biology laboratory may be one of your few opportunities to explore first-hand many of the topics covered in this manual. We hope that you will take advantage of this time to interact with your classmates and your instructor as much as possible so that this experience will be both rewarding and enjoyable.

C.E.
K.A.
P.S.C.

Ithaca, New York
October, 1981

TOPIC 1

Biological Molecules

OBJECTIVES

When you have completed this topic, you should be able to:

1. Define the term functional group and give the functional group or groups characteristic of an acid, an amino acid, a carbohydrate, a fatty acid, and an alcohol.
2. Describe how to carry out a simple laboratory test for a monosaccharide, a polysaccharide, a protein, and a fat.
3. Describe how a protein can be broken down into its building blocks—amino acids.
4. Show by thin-layer chromatography that different amino acids have different chemical properties.
5. Explain how chromatography works to separate chemically different substances.

TEXT REFERENCES

Chapter 2; Chapter 3 (A–E).
Atoms, bonds, compounds, chemical reactions, water, carbon chemistry, biological molecules.

INTRODUCTION

An **element** is a particular type of atom that is different from the atoms of all the other elements. The elements most abundant in living material are:

Carbon	= C	Nitrogen	= N
Hydrogen	= H	Sulfur	= S
Oxygen	= O	Phosphorus	= P

These are by far the most important elements in biological systems: they can be combined in enormous variety to make up the chemical molecules on which life is based.

In most biological molecules the atoms are held together by **covalent bonds** in which electrons are shared between adjacent atoms. Ionic bonds, hydrogen bonds, and disulfide bonds are also important. If you are not sure how each of these bonds is formed, refer back to your text and review them.

Fortunately for us, covalent bonds are not very easily broken down under ordinary conditions. Some small molecules are especially stable and are used as the building blocks or **subunits** of very large molecules called **polymers.** A polymer often consists of many similar subunits strung together in a long chain. It is easier to break down a polymer into its subunits than to break apart the subunits themselves.

Biological molecules can be categorized on the basis of similar parts called **functional groups:**

Carboxyl	Hydroxyl	Amino
—C(=O)—O—H	—OH	—N(—H)—H

Aldehyde	Ketone	Sulfhydryl
—C(=O)—H	—C(=O)—	—S—H

For example, any molecule that has the carboxyl group —C(=O)—O—H is an acid. (Remember that one line represents one covalent bond and two lines represent a double bond.) The term ***fatty acid*** refers to the group of acids that have the carboxyl group and several H—C—H groups besides. An ***amino acid*** molecule has an amino group and a carboxyl group. A carbohydrate such as a monosaccharide or simple sugar has an aldehyde or ketone group and some hydroxyl groups. (Aldehyde and ketone groups are specific examples of carbonyl compounds which always contain —C=O.)

Amino acid molecules may have additional functional groups (sulfhydryl, additional carboxyl groups, additional hydroxyl groups, and so on), but they all share the functional groups characteristic of an amino acid.

IDENTIFICATION OF BIOLOGICAL MOLECULES

Molecules of a certain class have similar chemical properties because they have the same functional groups. A chemical test that is sensitive to these groups can be used to identify molecules that are in that class. Practice the following tests by using them to see what various foods contain.

Iodine Test for Starch

Starch is a polysaccharide composed of hundreds of glucose subunits joined in long chains,. Iodine solution is yellow but it turns dark or black in the presence of starch.

□ Use a razor blade to cut a small slice of potato and a small square of paper for comparison; add a drop of Lugol's solution to each sample and note your results in Table 1–1 at the end of the exercise.

Starch and cellulose (paper) are both glucose polymers. Does the iodine test distinguish one from the other?

____

□ Obtain five test tubes and mark each one 5 cm from the bottom, using a ruler to measure the distance; label the tubes #1 through #5.

IF YOU ARE USING PIPETS IN THIS LAB, USE 2 mL OF SOLUTION PER CENTIMETER AND OMIT MARKING THE TUBES.

□ Add 5 drops (0.3 mL) of soluble starch to #1, 10 drops (0.6 mL) to #2, and 15 drops (0.9 mL) to #3; add 15 drops (0.9 mL) of glucose to #4 and 15 drops (0.9 mL) of water to #5.
□ Fill the tubes to the 5-cm mark with water (9.7, 9.4, or 9.1 mL), mix, and add 5 drops (0.3 mL) of iodine solution to each tube.
□ Mix once more and record the color of each of the solutions in Table 1–1. Draw your conclusion about the starch content of each tube as indicated by this test.

Benedict's Test for Reducing Sugar (Monosaccharides and Some Disaccharides)

Benedict's reagent tests for the presence of a free aldehyde group. Monosaccharides (simple sugars like glucose) usually have a free aldehyde or ketone group, as do some disaccharides (double sugars like maltose or lactose). These sugars are called **reducing sugars** and will give a positive test with Benedict's reagent.

□ Take four test tubes and use a ruler to mark them at 1 cm and 3 cm from the bottom; label them #1, #2, #3, and #4.
□ Fill tube #1 to the first mark with distilled water (2 mL); fill #2 to the first mark with egg white solution (2 mL); fill #3 to the first mark with honey solution (2 mL); fill #4 to the first mark with glucose solution.
□ Add Benedict's reagent (6 mL) to all four tubes up to the second mark and mix the tubes well.
□ Place all four tubes in a boiling water bath, being careful not to scald yourself, and wait exactly 3 min.
□ Use a test tube clamp to remove the tubes, place them in your rack, and record the color of each one in Table 1–1.

If there is no sugar at all, the blue color will not change; a small amount of sugar will change the color to green or yellow green; a large amount of sugar will change the color to orange or even brick red.

You knew before the experiment that one tube would not change. Which one? ______________ You also know that one *would* change.

Which one? ______________

Why did you test these tubes anyway?

____

____

What are these tubes called? ______________

Which of the other two test substances had more reducing sugar? ______________

Biuret Test for Protein

When amino groups are joined in the **peptide bonds** that connect the amino acids into a protein, they react with copper ions under the right conditions to give an attractive blue color.

□ Mark three test tubes, which you should have thoroughly washed and rinsed, with a line 3 cm from the bottom and another line 5 cm from the bottom.
□ Fill tube #1 to the first mark with water (6 mL); fill #2 to the first mark with egg white solution (6 mL); and fill #3 to the first mark with honey solution (6 mL).
□ Fill all the tubes to the second mark with 10% NaOH (sodium hydroxide) (4mL).

CAUTION: NaOH IS CAUSTIC—USE CAREFULLY. WASH YOUR HANDS IMMEDIATELY IF YOU GET IT ON YOU. ASK YOUR INSTRUCTOR TO HELP YOU CLEAN UP ANY SPILLS IMMEDIATELY.

□ Mix each tube well and add 5 drops (0.3 mL) of 1% $CuSO_4$ (copper sulfate).
□ Mix again and record the results in Table 1–1.

Circle one or more of the substances in the following list that would be easy to test for protein, using a blender and the Biuret reaction: hamburger, milk, beer, cheese, chicken soup.

Sudan III Test for Fat

A fat molecule is made up of three long fatty acid chains joined to a glycerol molecule:

```
 H       O  H  H    /H\       H
 |       ‖  |  |    | |       |
H—C—O—C—C—C—  ( C )  —C—H
 |          |  |    | |       |
 |          H  H    \H/ n*    H
 |       O  H  H    /H\       H
 |       ‖  |  |    | |       |
H—C—O—C—C—C—  ( C )  —C—H
 |          |  |    | |       |
 |          H  H    \H/ n*    H
 |       O  H  H    /H\       H
 |       ‖  |  |    | |       |
H—C—O—C—C—C—  ( C )  —C—H
 |          |  |    | |       |
 H          H  H    \H/ n*    H
```

*Group *n* repeats *n* (usually 12 to 14) times.

Sudan III is a red dye that stains the long $-\underset{H}{\overset{H}{C}}-$ chain of the fatty acid red orange. Because fats are not soluble in water (hydrophobic interactions cause them to stick together and stay completely undissolved), it is necessary to dissolve them in another solvent, ethanol (ethyl alcohol = EtOH).

□ Mark your clean test tubes 1 cm from the bottom and again at 3 cm from the bottom; label them #1, #2, and #3.
□ Fill tube #1 to the first mark with water (2 mL), #2 to the first mark with white flour, and #3 to the first mark with oil or liquid margarine.
□ Fill all the tubes to the second mark with 95% ethanol (4 mL) and shake each one until it is well mixed.
□ After the flour has had a chance to settle, mark on a piece of filter paper in pencil your initials and three dots labeled #1, #2, and #3.
□ Use a Pasteur pipet to transfer a small drop from tube #1 to dot #1. The solution will spread out very rapidly, so try to just touch the pipet to the paper.
□ Next transfer a drop from tube #2 to dot #2, and finally another drop from tube #3 to dot #3; use the pipet in the order 1-2-3.
□ Dry the spots by blowing on the paper for a minute, or by holding it in front of a drier.
□ Submerge the paper in a dish of Sudan III solution for exactly 1 min.
□ Remove the paper with forceps and soak it in a container of distilled water for 1 min to remove the excess dye. Continue longer if necessary to distinguish the spots from the background color.
□ Observe the color of each spot and record your results in Table 1–1.

Unknown

Your instructor may provide additional or unidentified foods as unknowns and ask you to perform one or more of the tests you have learned on them. Or, you may select one of the foods used for learning a test and perform the other tests on it.

SUGGESTED TESTS

Iodine: egg white, honey, gelatin, sucrose, onion juice, beer
Benedict's: soluble starch, sucrose, glucose, gelatin, onion juice, beer
Biuret: gelatin, skim milk, onion juice, beer
Sudan III: ground coffee, instant coffee, coconut, ground peanuts

SUGGESTED UNKNOWNS

Group A: soy flour, glucose, powdered skim milk, enriched flour
Group B: corn starch, honey, table sugar, egg white
Group C: ground coffee, instant coffee
Group D: Vermont Maid syrup, maple syrup
Group E: glucose, table salt, potato starch, gelatin

Record the unknown code number or food used and your results in Table 1–1.

AMINO ACID CHROMATOGRAPHY

If you touch a pen containing washable black ink to a piece of tissue, the ink will spread out on the paper and will separate into red, green, and blue bands. As the liquid in which the colored pigments are dissolved is absorbed by the cellulose fibers of the tissue, the pigments can't keep up: some will move almost as fast as the solvent but some will move very slowly. This is because each pigment has a tendency to stick or **adsorb** to the cellulose fibers, and those that stick more strongly will be slowed down the most. As a result, each substance will have its own characteristic rate of movement and the pigments will separate from each other. This is the essential principle of **chromatography.**

Chromatography has been adapted to separate all kinds of mixtures into their individual components. The substances do not have to be colored, so long as they can be identified somehow when the chromatography is finished. To make the chromatography reproducible, the mixture is applied in a tiny spot to the **matrix,** or stationary phase, which might be cellulose, silica gel, alumina, or some other inert substance. The **solvent** or mobile phase is then allowed to be **absorbed** by the matrix. As it moves along, it dissolves the substances in the mixture, and allows them to migrate along the matrix. Different substances will migrate different distances. The solvent is carefully formulated so that the substances are not completely soluble in it, or they would move as rapidly as the solvent itself. They must also not be completely insoluble or they would not migrate at all. As they move they tend to adsorb or stick to the matrix to different degrees, so adsorption also helps to separate them. When the solvent has almost reached the edge of the matrix, the chromatography is stopped, and the finished **chromatogram** is dried after noting the exact position of the **solvent front** (the leading edge of the area wet by the solvent). Finally the individual substances are identified, and the distance moved by each one from the origin is carefully measured. This distance is compared with the distance traveled by the solvent itself: the ratio of the two is the R_f and is always constant for a given substance in a particular solvent system:

$$R_f = \frac{\text{distance traveled by substance (cm)}}{\text{distance traveled by solvent (cm)}}$$

When a protein is broken down by hydrolysis into its subunits—amino acids—the result is a mixture of 20 or more individual amino acids.

$$\text{protein} + H_2O \xrightarrow{\text{hydrolysis}} \text{20 amino acids}$$

How are proteins broken down during digestion in the body?

____

____

____

The problem is to determine that these subunits are indeed amino acids and identify the ones present. Although each amino acid has its characteristic R_f, when 20 amino acids are involved it isn't practical to try to separate them all in a single chromatography. Instead you will try to separate a mixture of two or three at a time to see how chromatography works. The matrix you will be using is silica gel and the solvent contains butanol, acetic acid, and water. You will be given pure amino acid solutions to use as **standards** from which to calculate their correct R_f values. Then you will also chromatograph unknown solutions, calculate the R_f values of the components, and identify them by comparison with the standard R_f's. You will need three slides to run your chromatograms and will chromatograph two solutions on each slide.

□ To prepare a slide for thin layer chromatography, first stir the slurry of silica gel and acetone with the glass rod to mix it well.
□ Then use a pair of forceps to dip the clean slide into the slurry to within 5 mm of the top, gently move it back and forth to make sure it is evenly coated, and carefully withdraw it, holding it in a vertical position as you do so.
□ Let the slide dry in air, choose the most smoothly coated side, and wipe the silica gel off the *other* side with a paper towel. Touch the slide as little as possible so that you do not contaminate it with finger grease.
□ Use a teasing needle to mark the origin: make a small scratch on each edge of the silica gel layer about 1 cm from the bottom so that your slide looks like the one shown in Figure 1–1.
□ Using a separate capillary pipet for each solution as you spot, apply two standard amino acid solutions between the origin marks: touch the gel very quickly with only the end of the pipet or your spot will be too large.

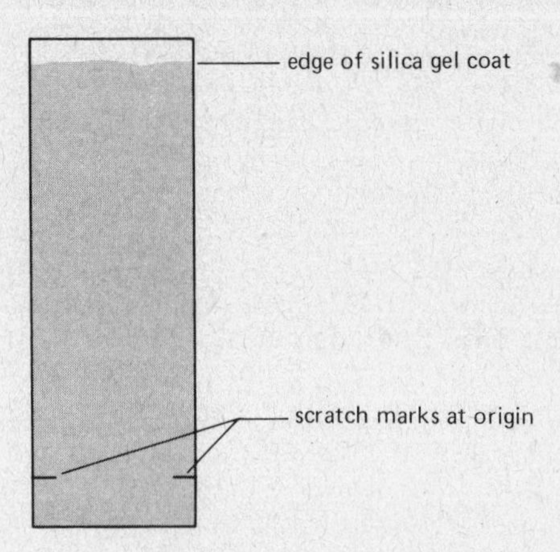

Figure 1–1. Chromatography slide.

After the first drop dries, apply a second drop beside it. The two amino acid spots should be at least 1 cm apart, and the spots must not overlap. (See Figure 1–2.)
□ Let the spots dry thoroughly in air.
□ While your first slide is drying, prepare your second and third slides in the same way: spot two more standard amino acid solutions and two different mixtures of unknown amino acids, unknown mixtures #1 and #2. Your instructor may ask you to substitute unknown amino acids for some of the standards.
□ When you are ready to chromatograph, use forceps to set the slide *with the amino acid spots at the bottom* into a Coplin jar containing the solvent. The level of solvent should be low enough so that your spots are above, not in, the solvent at the start. (Your spots will dissolve in the solvent rather than migrate if they are in the solvent.)
□ Let the solvent move up the gel until it is within 1 cm of the top of the silica gel layer. This takes about 30 min. (See Figure 1–3.)
□ Remove the slide with forceps and immediately make a scratch mark on each edge of the silica gel to show the position of the solvent, the **solvent front.**
□ Allow nothing to touch the silica gel while you let it dry in air.
□ Use forceps to transfer the slide to the box provided for ninhydrin spraying. Ninhydrin reacts with amino acid spots to produce identifiable visible spots.

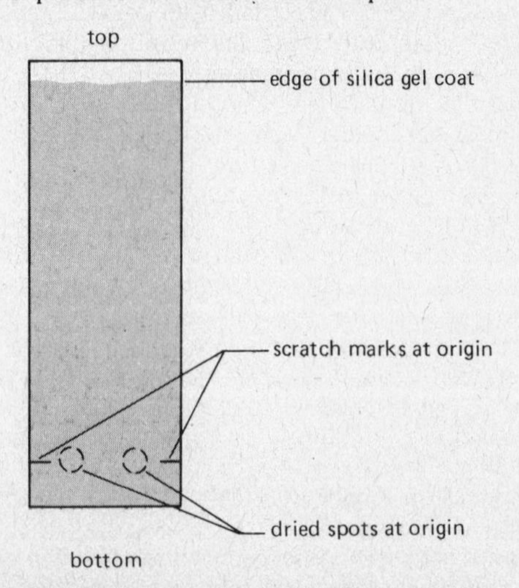

Figure 1–2. Chromatography slide after spotting samples.

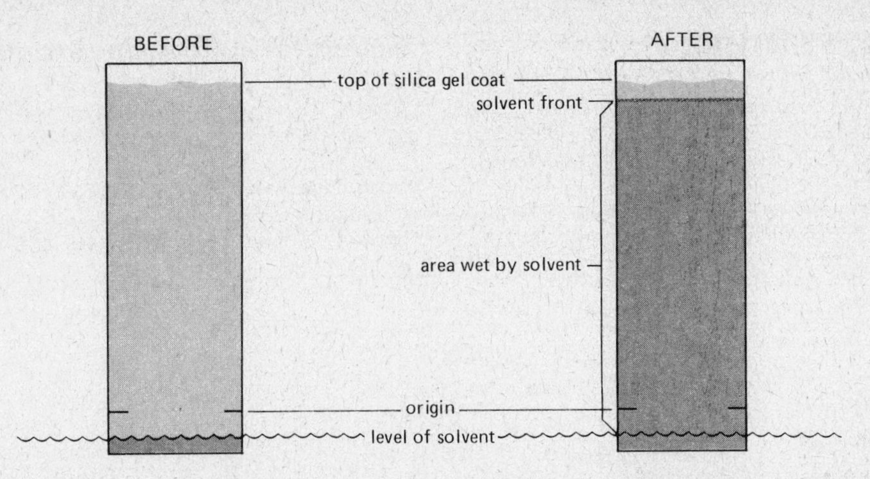

Figure 1–3. Running a chromatogram. Stop the chromatography before the solvent reaches the top of the silica gel.

BE VERY CAREFUL WHEN USING TOXIC NINHYDRIN SPRAY AND WORK IN A FUME HOOD IF AT ALL POSSIBLE.

□ Spray the slide as demonstrated by your instructor.
□ Allow the slides to dry until colored spots appear.

Did your amino acids migrate away from the origin all right? ______________

□ For each spot, record the color in Table 1–2 and measure the distance traveled. (The exact distance is the distance from the center of the original spot at the origin to the center of the colored spot.)
□ Measure the exact distance traveled by the solvent in each case, so that you can calculate the R_f values for each amino acid. (See Figure 1–4.)
□ Finally you can now identify the amino acids in your unknown amino acid mixtures by comparing the R_f of each spot with the R_f values for the standard amino acids. Since you determined the R_f values for only four amino acids, you may have to obtain R_f's from other members of the class to identify some of your unknown spots.

Return all materials and solutions and clean up your work area before leaving the laboratory. Hand in your results according to the instructions given.

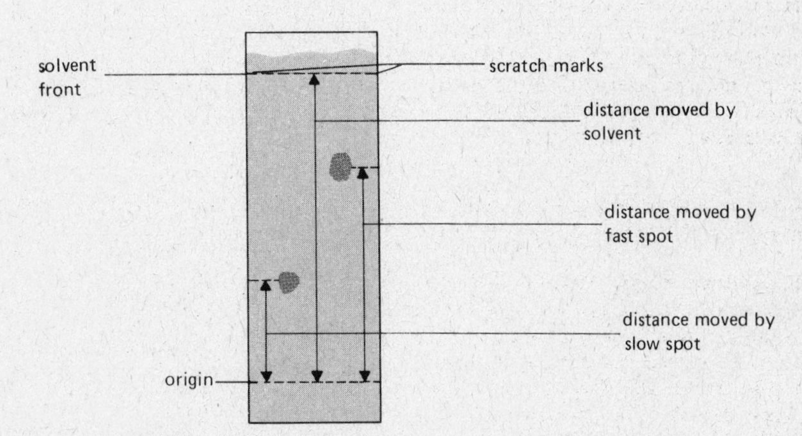

Figure 1–4. Measuring a chromatogram. Measure the distance from the origin to the center of each spot.

EXPLORING FURTHER

Baker, Jeffrey J. W., and Garland E. Allen. *Matter, Energy and Life: An Introduction to Chemical Concepts.* 4th ed. Reading, Mass.: Addison-Wesley, 1981.

Scientific American Articles

Calvin, M. *Organic Chemistry of Life: Readings from Scientific American.* San Francisco: W. H. Freeman, 1973.

Scrimshaw, Nevin S., and Lance Taylor. "Food." September 1980.

Sharon, Nathan. "Carbohydrates." November 1980.

Student Name ______________________ **Date** ______________

TABLE 1–1. RESULTS FROM CHEMICAL TESTS

1.

Sample	Color	Starch Present
Potato		
Paper		

Tube #	Test Substance	Color	Starch Present
1	Starch		
2	Starch		
3	Starch		
4	Glucose		
5	Water		

2.

Tube #	Test Substance	Color	Reducing Sugar Present
1	Water		
2	Egg white		
3	Honey		
4	Glucose		

3.

Tube #	Test Substance	Color	Protein Present
1	Water		
2	Egg white		
3	Honey		

4.

Tube #	Test Substance	Color of Spot	Fat Present
1	Water		
2	Flour		
3	Oil (margarine)		

5.

Unknown # or Food Tested	Test Performed	Results	Substance Present

TABLE 1–2. RESULTS FROM AMINO ACID CHROMATOGRAPHY

Standard Amino Acids	**Color**	**Distance Moved by Amino Acid**	**Distance Moved by Solvent**	R_f
Alanine				
Arginine				
Asparagine				
Aspartic acid				
Cysteine				
Glutamic acid				
Glutamine				
Glycine				
Histidine				
Isoleucine				
Leucine				
Lysine				
Methionine				
Phenylalanine				
Proline				
Serine				
Threonine				
Tryptophan				
Tyrosine				
Valine				
Unknown Amino Acids				
#1 =				
#2 =				
#3 =				
#4 =				
Unknown Mixture #1				
Unknown Mixture #2				

What amino acids did unknown mixture #1 contain? ____________________

What amino acids did unknown mixture #2 contain? ____________________

TOPIC 2

Enzymatic Activity

OBJECTIVES

When you have completed this topic, you should be able to:

1. Give the class of macromolecules to which peroxidase belongs and the subunits that make it up.
2. Name the substrate and products of the peroxidase catalyzed reaction.
3. Explain the role of guaiacol in this experiment.
4. Describe how temperature, pH, enzyme concentration, and substrate concentration affect the reaction rate.
5. Explain why peroxidase is a necessary enzyme for all aerobic or oxygen-utilizing cells.

TEXT REFERENCES

Chapter 2 (E-G); Chapter 3 (E,G).
Reactions, water, pH, proteins, enzymes, enzyme-substrate interactions.

INTRODUCTION

Enzymes are **biological catalysts** that carry out the thousands of chemical reactions that occur in living cells. They are generally large proteins made up of several hundred amino acids, and often contain a special structure called the **prosthetic group** that is important in the actual catalysis.

In an enzyme-catalyzed reaction, the substance to be acted upon, or **substrate,** binds to the **active site,** or business end, of the enzyme. The enzyme and substrate are held together in an **enzyme-substrate complex** by hydrophobic interactions, hydrogen bonds, and ionic bonds.

The enzyme then converts the substrate to the reaction products in a process that often requires several chemical steps, and may involve covalent bonds. Finally, the products are released into solution and the enzyme is ready to form another enzyme-substrate complex. As is true of any catalyst, the enzyme is not used up as it carries out the reaction but is recycled over and over. One enzyme molecule can carry out thousands of reaction cycles every minute.

Each enzyme is specific for a certain reaction because its amino acid sequence is unique and causes it to have a unique three-dimensional structure. The active site also has a specific shape so that only one or a few of the thousands of compounds present in the cell can interact with it. If there is a prosthetic group on the enzyme, it will form part of the active site. Any substance that blocks or changes the shape of the active site will interfere with the activity and efficiency of the enzyme.

If these changes are large enough, the enzyme can no longer act at all, and is said to be **denatured.** There are several factors that are especially important in determining the enzyme's shape, and these are closely regulated both in the living organism and in laboratory experiments to give the **optimum** or most efficient enzyme activity:

1. **Salt concentration.** If the salt concentration is very low or zero, the charged amino acid side chains of the enzyme molecules will stick together. The enzyme will denature and form an inactive precipitate. If, on the other hand, the sale concentration is very high, normal interaction of charged groups will be blocked, new interactions will occur, and again the enzyme will precipitate. An intermediate salt concentration such as that of blood (0.9%) or cytoplasm is the optimum for most enzymes.
2. **pH.** pH is a logarithmic scale that measures the acidity or H^+ concentration in a solution. The scale runs from 0 to 14 with 0 being highest in acidity and 14 lowest. When the pH is in the rage of 0–7, a solution is said to be acidic; if the pH is around 7, the solution is neutral; and if the pH is in the range of 7–14, the solution is basic. Amino acid side chains contain groups such as —COOH and $—NH_2$ that readily gain or lose H^+ ions. As the pH is

lowered, an enzyme will tend to gain H^+ ions, and eventually enough side chains will be affected so that the enzyme's shape is disrupted. Likewise, as the pH is raised, the enzyme will lose H^+ ions and eventually lose its active shape. Many enzymes have an optimum in the neutral pH range and are denatured at either extremely high or low pH. Some enzymes, such as those which act in the human stomach where the pH is very low, will have an appropriately low pH optimum. A **buffer** is a compound that will gain or lose H^+ ions so that the pH changes very little.

3. **Temperature.** All chemical reactions speed up as the temperature is raised. As the temperature increases, more of the reacting molecules have enough kinetic energy to undergo the reaction. Since enzymes are catalysts for chemical reactions, enzyme reactions also tend to go faster with increasing temperature. However, if the temperature of an enzyme-catalyzed reaction is raised still further, a **temperature optimum** is reached: above this point the kinetic energy of the enzyme and water molecules is so great that the structure of the enzyme molecules starts to be disrupted. The positive effect of speeding up the reaction is now more than offset by the negative effect of denaturing more and more enzyme molecules. Many proteins are denatured by temperatures around 40–50°C, but some are still active at 70–80°C, and a few even withstand being boiled.
4. **Small molecules.** Many molecules other than the substrate may interact with an enzyme. If such a molecule increases the rate of the reaction it is an **activator,** and if it decreases the reaction rate it is an **inhibitor.** The cell can use these molecules to regulate how fast the enzyme acts. Any substance that tends to unfold the enzyme, such as an organic solvent or detergent, will act as an inhibitor. Some inhibitors act by reducing the —S—S— bridges that stabilize the enzyme's structure. Many inhibitors act by reacting with side chains in or near the active site to change or block it. Others may damage or remove the prosthetic group. Many well-known poisons such as potassium cyanide and curare are enzyme inhibitors that interfere with the active site of a critical enzyme.

TURNIP PEROXIDASE

In this experiment you will study the enzyme **peroxidase** from turnips. Peroxidases are widely distributed in plant and animal cells and catalyze the oxidation of organic compounds by hydrogen peroxide as follows:

$$RH + H_2O_2 \xrightarrow{\text{peroxidase}} ROH + H_2O$$

What are the substrates in this reaction?

____

What are the products?

____

Any cell using molecular oxygen in its metabolism will produce small amounts of H_2O_2 as a highly toxic byproduct. It is critical that the H_2O_2 be quickly removed by enzymes such as peroxidase before it can do damage to the cell.

You will use a reducing agent, guaiacol, that changes color when it is oxidized. This change can be easily measured in the spectrophotometer.

$$\underset{\text{(colorless)}}{4\text{ guaiacol}} + 2H_2O_2 \xrightarrow{\text{peroxidase}} \underset{\text{(brown)}}{\text{tetraguaiacol}} + 8H_2O$$

THE SPECTROPHOTOMETER

A colored solution such as oxidized guaiacol solution appears that way because some of the light entering the solution is absorbed by the colored substance. A clear solution will allow almost all of the light to pass through. The amount of **absorbance** can be determined by using a spectrophotometer, which measures quantitatively what fraction of the light passes through a given solution, and indicates on the absorbance scale the amount of light absorbed compared to that absorbed by a clear solution. The darker the solution, the greater its absorbance.

Inside the machine there is a light that shines through a filter (which can be adjusted to control the color, or wavelength, of light), then through the sample and onto a light-sensitive phototube. The phototube produces an electric current proportional to the amount of light striking it. The absorbance meter measures how much light has been blocked by the sample and thereby prevented from striking the phototube. A clear tube of water or other solvent is the **blank** and has zero absorbance. A solution that contains a small amount of a colored substance might show an absorbance of 0.1, a solution with a moderate amount might show an absorbance of 0.4, and so forth. In fact, in the lower portion of the absorbance scale, the amount of substance in solution is directly proportional to the absorbance reading so that a graph of absorbance versus concentration will give a straight line. This very useful relationship is known as **Beer's law.**

When you are ready to read a sample, use Figure 2–1 to follow the steps below.

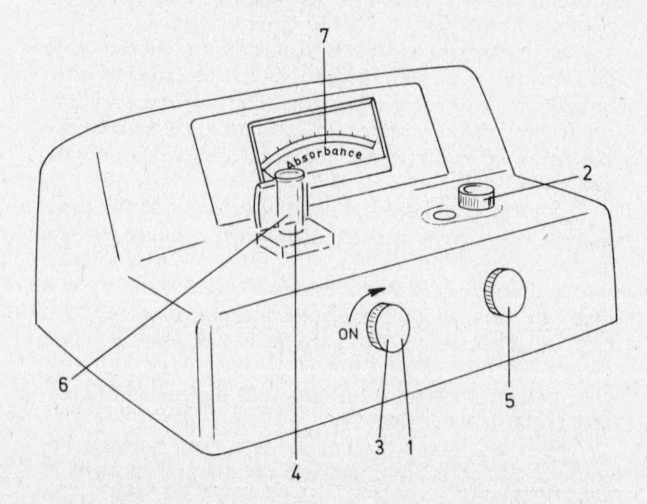

Figure 2–1. Spectronic 20 spectrophotometer. The numbered controls correspond to steps in the operating instructions (see text).

Here are the instructions for reading a sample:

1. Turn on the instrument with the power switch knob. Allow 5 min for warm-up time.
2. Adjust to the desired wavelength (in this case use 500 nm).
3. With the sample chamber *empty* and the cover closed, use the power switch knob to set the meter needle to read **infinity absorbance.** (When the chamber is empty, a shutter blocks all light from the phototube.)

ALWAYS READ FROM THE ABSORBANCE SCALE, NOT THE TRANSMISSION SCALE.

---

4. Fill a spectrophotometer tube or cuvette halfway with distilled water (or other solvent when water is not the solvent in your experiment) to serve as the blank. Wipe it free of moisture or fingerprints with a lint-free tissue (Kimwipe), and insert it into the sample holder. Line up the etched mark with the raised line on the front of the sample holder, and close the cover.
5. With the right-hand knob, adjust the meter to read **zero absorbance.** Remove the tube, empty it, and shake it as dry as possible. This precaution is essential for accuracy.
6. Fill and insert the sample cuvette.
7. Read the absorbance directly from the meter. Rinse the cuvette with clean water and shake it as dry as possible.
8. It is necessary to reset the machine to infinity absorbance and zero absorbance before each set of readings because the settings drift a bit. Whenever the wavelength is changed, the infinity and zero absorbance also must be reset.

EXPERIMENTAL PROCEDURE

Preparation of the Turnip Extract

□ Weigh out 1 g of turnip (the peeled, inner portion).
□ Place it in a blender with 200 mL of distilled water.
□ Blend it thoroughly at the high setting for about 1 min.

This suspension is the **turnip extract** (abbreviated t.e.), and contains the enzyme peroxidase. The activity of the turnip extract will vary from day to day, depending on the size and age of the turnip and the extent of blending.

Kinetics of the Peroxidase Reaction

Pipets will be used to measure accurately the solutions used in this experiment. Your instructor will demonstrate how to use a pipet correctly.

BE SURE TO USE A DIFFERENT PIPET FOR EACH SOLUTION SO THAT THE REAGENTS ARE NOT RUINED BY CROSS-CONTAMINATION.

□ Label the pipets with a tape or marking pencil so each one can be reused with the proper solution.
□ Obtain a spectrophotometer tube and label it #1.
□ Obtain two test tubes and label them #2 and #3. (#1 will contain a blank reaction without H_2O_2. The contents of #2 and #3 will be mixed to start the reaction.)
□ Set up the three tubes as follows, and make a record of these additions in Table 2–1 (run 1, base line) at the end of this topic.

Tube #1 (blank tube without H_2O_2). Add 0.1 mL of guaiacol, 1.0 mL of turnip extract, and 8.9 mL of distilled water; mix well.
Tube #2. Add 0.1 mL of guaiacol, 0.2 mL of 1% H_2O_2, and 4.7 mL of distilled water.
Tube #3. Add 1.0 mL of turnip extract and 4.0 mL of distilled water.

□ Adjust the Spectronic 20 to zero absorbance using tube #1 and following steps 1–5 in the instructions for its operation.

You have now set up the instrument so that any difference in the meter reading with a change in the sample will reflect a difference in oxidized guaiacol concentration.

□ Obtain a stopwatch, wind it up, and be sure that you understand how to use it correctly.
□ Prepare the sample: obtain a clean spectrophotometer tube, tissue, and tubes #2 and #3 filled with the solutions given above. Label the clean tube #4.

YOU WILL HAVE 20 SEC TO MIX THE CONTENTS OF TUBES #2 and #3, POUR THE CONTENTS INTO THE CLEAN SPECTROPHOTOMETER TUBE, WIPE THE TUBE, AND TAKE YOUR FIRST READING.

□ When you are completely ready, mix the contents of tubes #2 and #3, pour the contents back and forth two times, and then pour quickly into tube #4; start the stopwatch when the tubes are mixed. ($t = 0$ when the tubes are mixed.)
□ Wipe the outside of the tube and place it in the spectrophotometer.
□ Take your first reading 20 sec after the tubes were mixed ($t = 20$ sec).
□ Continue to read the absorbance every 20 sec for 2 min.

Record the readings in Table 2–2, and graph the absorbance versus time in Figure 2–2 (at end of topic).

This curve will represent the **base line** of enzyme activity with which the enzyme activity under varying conditions will be compared. In the following experiments you will vary one condition at a time and compare the results with the base line. Be sure you have already recorded the setup for your base line experiment under run 1 in Table 2–1. Check your graph with your instructor before proceeding with the rest of the experiment.

Since this exercise contains too many projects to be performed by one pair of students, the instructor will specify how the remaining work will be divided up. Use the experimental setup space in Table 2–1 to record your experiment number and what you put into your tubes; record your data in Table 2–2, and graph your results.

1. Effect of Enzyme Concentration

What happens when you use:

(a) TWICE THE AMOUNT OF ENZYME

Tube #1. Add 0.1 mL of guaiacol, 2.0 mL of turnip extract, and 7.9 mL of distilled water.
Tube #2. Add 0.1 mL of guaiacol, 0.2 mL of 1% H_2O_2, and 4.7 mL of distilled water.
Tube #3. Add 2.0 mL of turnip extract and 3.0 mL of distilled water.

(b) HALF THE AMOUNT OF ENZYME

Tube #1. Add 0.1 mL of guaiacol, 0.5 mL of turnip extract, and 9.4 mL of distilled water.
Tube #2. Add 0.1 mL of guaiacol, 0.2 mL of 1% H_2O_2, and 4.7 mL of distilled water.
Tube #3. Add 0.5 mL of turnip extract and 4.5 mL of distilled water.

Notice that tube #1 always contains 10 mL and tubes #2 and #3 together total 10 mL. Why is this important?

____

____

Repeat the procedure for preparing the sample, read it, record the data, and graph your results. How does changing the concentration of enzyme affect the rate of the reaction?

____

2. Effect of Varying the Substrate Concentration

Design and carry out an experiment to determine whether varying the concentration of the substrate H_2O_2 affects the rate of reaction.
Conclusion:

____

____

3. Effect of Temperature

What is the effect of temperature on enzyme activity? Set up an experiment to show the difference in activity found between 0° and 37°C. Use the conditions of your base line experiment, but run the reaction in the water baths at different temperatures that are available to you, such as 4°, 15°, room temperature = about 22°, 30°, and 37°C. Record the exact temperature for each run and graph your results.

At what temperature do you see maximal activity?

____

This temperature is called the enzyme's temperature

__________.

4. Effect of pH

Measure the effect of pH by using the base line conditions but substituting buffers at pH 3, 7, and 11 for distilled water. Measure the pH of the distilled water so that you can compare the base line and buffer results.

At which pH is the enzyme most effective?

____

Why is this important biologically?

____

____

____

5. Effect of Heat

Heat 5 mL of turnip extract in a boiling water bath for 10 min and test its activity as in Run 1.

Is the boiled enzyme active? __________

Explain what happened during boiling:

____

____

____

____

____

6. Effect of Inhibitors

Hydroxylamine is a small molecule whose structure is similar enough to that of H_2O_2 that it attaches to the iron atom that is part of the peroxidase enzyme. To test how this substance affects enzyme activity, add 5 drops of 10% hydroxylamine to the enzyme extract and let it stand for 1 min. Then measure the activity of the peroxidase.

Is the peroxidase still active? __________

Explain the action of hydroxylamine on the enzyme's activity:

____

____

____

____

____

When you are finished with your work, wash and thoroughly rinse your glassware, return the reagents, and discard any debris left from your experiment.

EXPLORING FURTHER

Barker, R. *Organic Chemistry of Biological Compounds.* Englewood Cliffs, N.J.: Prentice-Hall, 1971.

Dickerson, R. E., and I. Geis. *The Structure and Action of Proteins.* New York: Harper and Row, 1969.

Scientific American Articles

Allison, Anthony. "Lysosomes and disease." November 1967 (#1085).

Calvin, M. *Organic Chemistry of Life: Readings from Scientific American.* San Francisco: W. H. Freeman, 1973.

Changeaux, Jean Pierre. "The control of biochemical reactions." April 1965 (#1008).

Koshland, D. E., Jr. "Protein shape and biological control." October 1973 (#1280).

Levinthal, Cyrus. "Molecular model-building by computer." June 1966 (#1043).

Mosbach, Klaus. "Enzymes bound to artificial matrixes." March 1971 (#1216).

Student Name ______________________ **Date** ______________

TABLE 2–1. EXPERIMENTAL DESIGN FOR PROJECT # _____

	Tube #1	**Tube #2**	**Tube #3**
Run 1 (Base line)			
Run 2			
Run 3			
Run 4			
Run 5			
Run 6			
Run 7			

TABLE 2–2. RESULTS FROM TURNIP PEROXIDASE EXPERIMENT

Time (sec)	Run 1 (Base line)					
20						
40						
60						
80						
100						
120						

Time (sec)						
20						
40						
60						
80						
100						
120						

Student Name ______________________ **Date** ______________

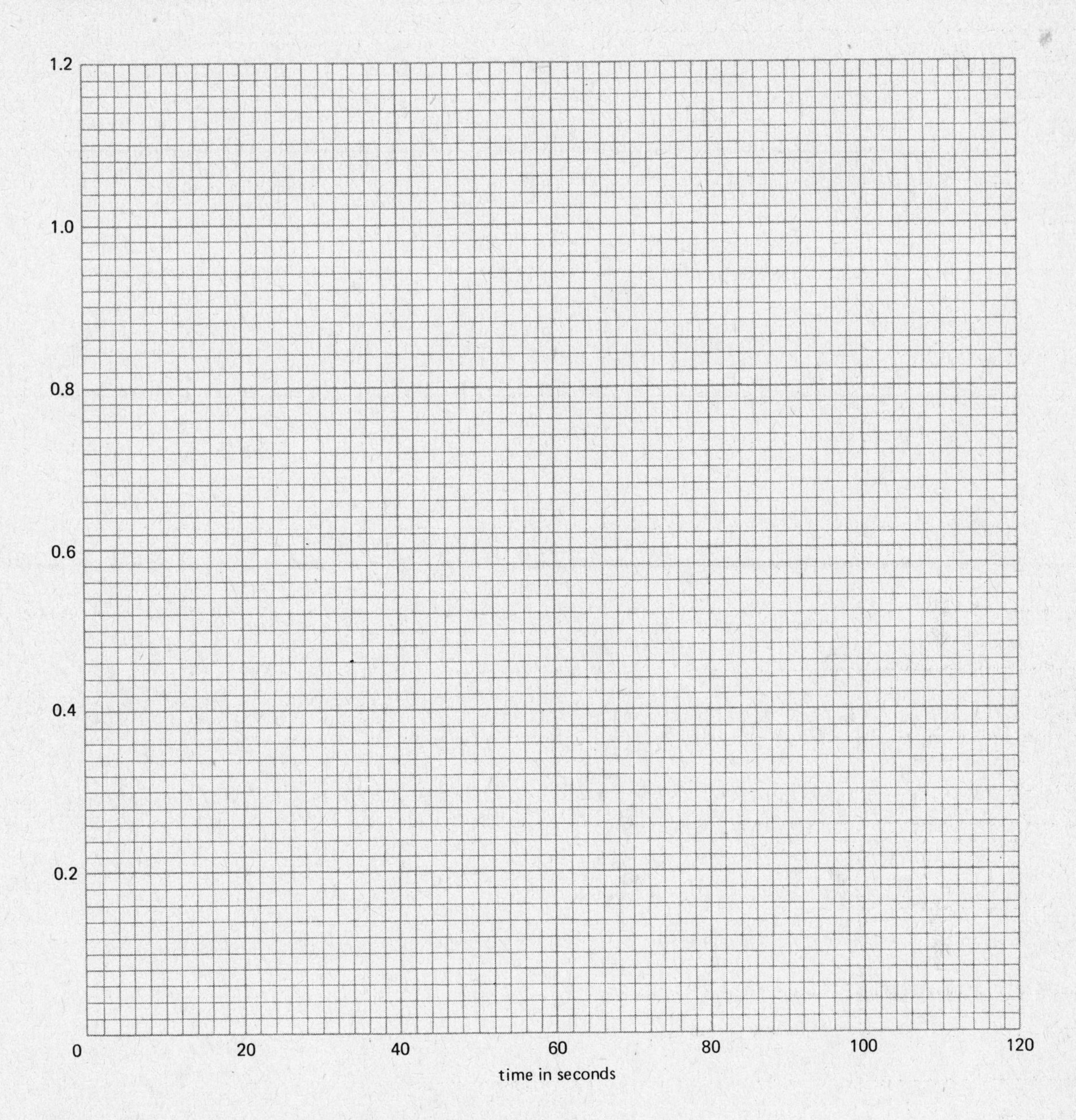

Figure 2–2. Kinetics of turnip peroxidase activity.

Results of experiment # ______
Conclusions:

____

____

____

____

TOPIC 3

Introduction to the Microscope

OBJECTIVES

When you have completed this topic, you should be able to:

1. Carry a microscope correctly and place it correctly on your lab desk.
2. Explain the function of each part of the microscope as shown in Figures 3–1 and 3–3.
3. Clean the microscope lenses correctly using lens paper.
4. Prepare the microscope for examining a ready-made (prepared or permanent) slide.
5. Locate the object of interest and focus on it.
6. Switch from low to high power and refocus.
7. View a slide under oil immersion, and remove the oil from the slide and lens afterwards.
8. Determine the magnification of the microscope.
9. Measure a tiny object by viewing it in the microscope field at a known magnification.
10. Prepare a wet mount.
11. State the advantages and disadvantages of a compound microscope and stereomicroscope.
12. Store the microscopes properly.

TEXT REFERENCES

Chapter 4 (A).
Metric units, light microscope, specimen preparation.

INTRODUCTION

The purpose of a microscope is to see inside of organisms and cells—to see what is invisible to the naked eye. The eye can be aided with a **simple microscope** which is nothing more than a magnifying glass with one lens, or with a **compound microscope** which has two lenses at opposite ends of a tube. The **ocular lens** is the one nearest the eye, and the **objective lens** is nearest the object or specimen. Both of the microscopes that you will be using today are compound and quite a bit more sophisticated than the exquisite simple microscopes with which Anthony van Leeuwenhoek made his discovery of the microbes. A microscope is a precision instrument, and easily damaged. You should not attempt to use it until you have grasped the function of each vital part. Study Figure 3–1 as you locate each of the following.

Arm: Supports the body tube and is the part that you can grasp to carry the microscope. Pick up your microscope by its arm, keeping it upright, and supporting it underneath with your free hand. Set it gently on your lab desk.

Base: Gives the microscope a firm, steady support.

Ocular lens: Magnifies ten times (10×). This lens is often unattached, and thus it may fall out unless the microscope is kept upright.

Objective lens: Magnifies the object by the factor marked on the particular lens. Low power (10×) gives the smallest image, high power (sometimes called high dry) gives a large image (40×), and oil immersion gives the largest image (100×). Sometimes a very low power scan objective (4×) replaces the oil immersion lens. Objective lenses are always used in the order: low, high, oil immersion.

Nosepiece: The revolving part to which objectives are attached. It must be firmly clicked into position when the objective is changed. Rough treatment can cause it to snap off.

Body tube: Joins the nosepiece to the ocular lens.

Stage: Supports the slide that is held onto it by **stage clips,** and has a hole so that light can shine up through the specimen. Always center the specimen over this hole.

Coarse adjustment: Moves the body tube or stage up and down, depending on the design of the microscope, to approximately the right position so that the specimen is in focus. This knob is used only with the **low power** or **scan** (focus finder) **objective.**

Fine adjustment: Moves the body tube (or stage) up and down to precisely the right position so that the specimen is perfectly in focus. Use it to achieve fine focus with the low power objective and for *all focusing* with the high power and oil immersion objectives.

Light source: Usually a small electric light beneath the stage that is controlled by a push-button light switch. Sometimes a mirror is used to reflect light from another source into the microscope.

Iris diaphragm: Regulates how much light and lamp heat go through the specimen. It is controlled by a lever that is moved back and forth.

Condenser: A lens (substage condenser or light director), located above the diaphragm, which concentrates the light before it passes through the specimen. Its position is controlled by a knob on microscopes in which it is adjustable.

Observer: The microscope is useless when the observer *looks* but does not see.

HOW TO USE THE COMPOUND MICROSCOPE

Cleaning the Lenses

Clean lenses are necessary for a clear view of your specimen.

NOTHING SHOULD TOUCH THE MICROSCOPE LENSES EXCEPT THE SPECIAL LENS PAPER AND APPROPRIATE SOLVENT.

Use a clean piece of lens paper to wipe the ocular and objective lenses. Sometimes you may need to moisten the lens paper with alcohol or xylene to remove grease

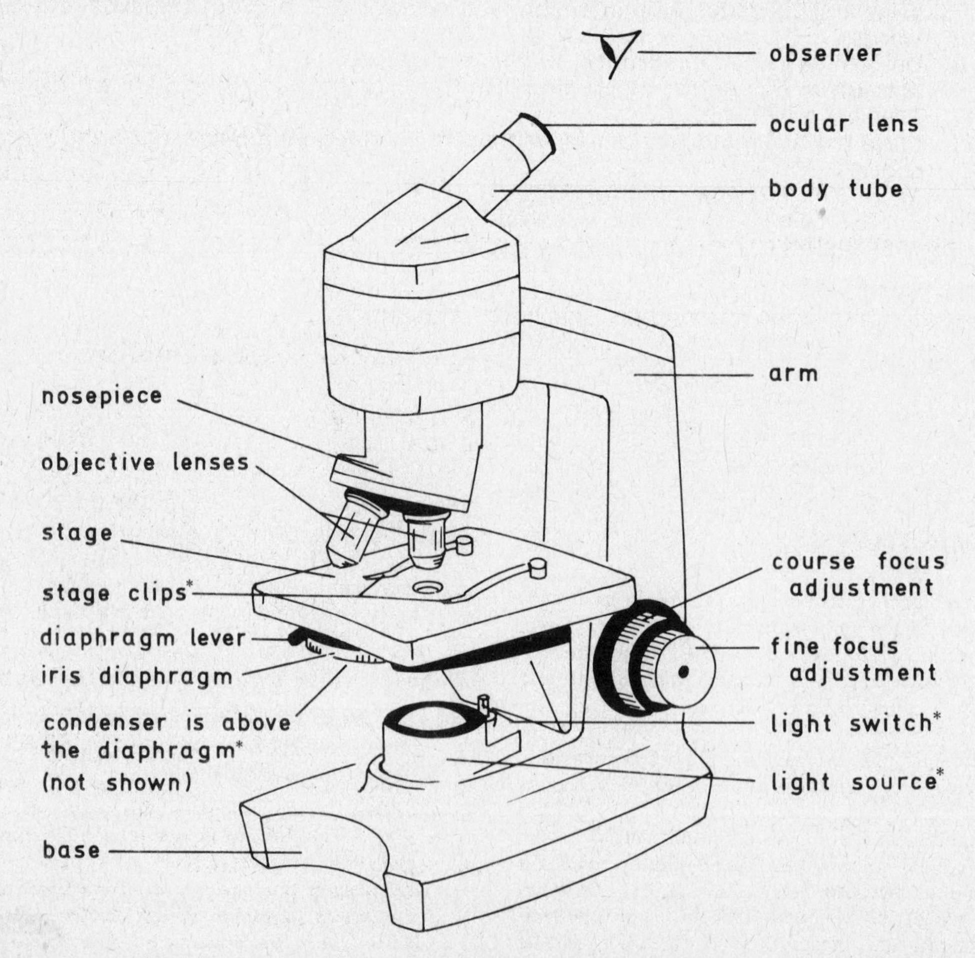

Figure 3–1. Compound microscope. Your microscope is similar to the one shown in this diagram. Some of the parts that may not be present on your microscope are marked with an asterisk (*). Your instructor may be able to provide a diagram that represents your microscope exactly.

or oil which will blur the image. Follow the directions of your instructor regarding the use of solvent.

If there's dirt in your microscope scene,
Use only lens paper to clean,
If you do it just right,
Your field will be bright,
If not, you can try some xylene.

Viewing a Prepared Slide

Develop familiarity with your microscope by looking at a prepared slide of a letter ("e," "k," and so on) or some other object.

□ First click the lowest power objective into position.
□ Then use the coarse adjustment to raise the ocular (or lower the stage) as far as possible.
□ Place the slide (oriented so that you can read the letter "e" properly) on the stage and position the letter over the hole in the stage.
□ Now look through the microscope and adjust the diaphragm lever (and condenser knob) so that enough light is coming through to make the field bright.

In which direction do you move the lever to increase the amount of light? right

□ View the objective lens *from the side* and turn the coarse adjustment until the lens *almost* touches the slide.

Did you have to turn the knob toward you or away from you? away

□ Look into the microscope and slowly turn the coarse adjustment in the opposite direction so that the objective lens is moving *away* from the slide.
□ Continue turning until the letter "e" comes into focus.

Remember that the coarse adjustment is used only on low power, so you should not have to use it again with this slide.

□ Finish focusing with the fine adjustment knob.

Does the object appear normal or upside down?

normal

□ Move the slide a tiny bit away from you while observing the object in the microscope.

Does it move away from you or toward you?

toward

□ Move the slide back and forth and note that the inverted image always moves in the direction opposite to the direction in which the slide is moved.
□ Center the image in the middle of the field of view.
□ Move the high power lens into position, making sure that it clicks in correctly; you may have to focus a little with the fine focus adjustment knob but do *not* touch the coarse focus adjustment; you may also have to increase the amount of light in order to have a clear view.

If you do not see anything in the field, you probably did not have the image properly centered. Return to low power, center the image, and then switch to high power. Only the central portion of the low power field is visible under high power. Ask your instructor for help if you are still having trouble.

□ When you have finished viewing the letter and are ready to go on, return to low power, turn the coarse adjustment knob to move the objective away from the slide, and return the slide to the front desk.

Using Oil Immersion

Learn to use oil immersion by viewing one of the designated slides. In this technique the resolution of the image is increased by filling the space between the oil immersion lens and the slide with oil. It works because oil has a much higher index of refraction than air and contributes to the magnification of the image.

OIL IS USED ONLY WITH THE OIL IMMERSION LENS. DO NOT USE OIL WITH ANY OF THE OTHER LENSES.

□ With your low power lens in position, locate the object and center it in the field.
□ Switch to high power and refocus.
□ Swing the high power lens away from the slide, and place a drop of immersion oil in the center of the cover slip and over the specimen.
□ Swing the oil immersion lens into position and refocus using the fine focus only. Note the improved resolution obtained by using immersion oil.
□ When you are finished viewing, swing the immersion lens away, but do not put any other lens into position. Raise the body tube with the coarse adjustment and remove the slide. First wipe most of the oil off with lens paper, then use a piece of lens paper wet with xylene to wipe the rest of the oil off the lens and the slide. Finally, wipe the lens and slide with dry lens paper.
□ Click the low power lens into position and you are ready to go on.

Computing Magnification

Magnification is a measure of how big an object looks to your eye compared to "life size". Life size images are specified as 1. Magnification is usually written by a number followed by "×," which stands for "times life size" (for example, 10× means 10 times life size). A simple lens like a magnifying glass magnifies 3× to 10× or so. A compound light microscope can magnify things up to 1000×. **Resolution,** on the other hand, measures how clearly you can see details in the microscope, and is usually given as the distance between two objects that can just barely be resolved. It depends in part on the quality of the lenses used. Resolution increases with magnification up to a theoretical limit: the maximum resolution in an ordinary light microscope is $\frac{1}{2}\lambda$ where λ is the wavelength of light used. If $\lambda = 500$ nm, as in green light, the resolution is $\frac{1}{2} \times 500$ nm $= 250$ nm (0.25 μm).* The use of immersion oil will increase the resolving power somewhat to about 0.18 μm.

The ocular lens on your microscope gives 10× magnification of the image made by the objective lens.

---

*Refer to the description of metric units in Appendix D if these terms are unfamiliar to you.

The objective lens magnifies the object 4× or more, so total magnification is the magnification of the first lens *times* the magnification of the second lens. The magnification of the objective lens may be 4×, 10×, 40×, or 100×. Fill in the magnifications of the objective lenses of your microscope and the total magnification:

Ocular	Objective	Total Magnification
10×	10× ×	100× ×
10×	43 ×	430 ×
10×	____ ×	____ ×
10×	____ ×	____ ×

Measuring Objects in the Microscope

If the field of view is a certain size under the lowest power or magnification, it will be correspondingly *smaller* as the power is increased. If the magnification is known, the relative sizes of the two fields can be calculated.

□ Place a ruler on your stage and measure the size of the field under the lowest magnification:

field = 1.5 mm (a)

lowest total magnification = 100 × (b)

□ Switch to another magnification:

total magnification = ____ × (d)

□ Calculate the field size of the second magnification where:

$$\underset{\text{(a)}}{\text{field 1}} \times \underset{\text{(b)}}{\text{magnification 1}} = \underset{\text{(c)}}{\text{field 2}} \times \underset{\text{(d)}}{\text{magnification 2}}$$

$$\text{field 2} = c = \frac{a \times b}{d} = \_\_\_\_ \text{ mm}$$

□ Calculate the approximate field sizes in μm for your microscope:

Objective	Total Magnification	Field size in μm
scan	____ ×	____ μm
low power	____ ×	____ μm
high power	____ ×	____ μm
oil immersion	____ ×	____ μm

Once the size of the field in micrometers is known, it is possible to estimate the size of an object by comparing its size to the size of the field. For example, suppose an object is one-half the diameter of the field under low power. The size of the field is 1500 μm under low power. The size of the object is

$$\tfrac{1}{2} \times 1500 \ \mu\text{m} = 750 \ \mu\text{m}$$

□ Obtain a prepared slide containing small organisms.

□ View the slide under low power and determine whether higher magnification is needed to determine the size of the organisms.
□ Switch to higher powers if necessary.
□ Note the size of the organism relative to the diameter of the field, and record the objective lens used, the magnification, and the size of the field in the chart below.
□ Calculate the approximate size of the organism in micrometers.

Organism measured	Volvox
Objective lens	430 10×
Total magnification	430
Size of field (μm)	350
Relative size of organism	
Size of organism (μm)	750 μm

Preparing a Wet Mount

Wet mounts are used to study fresh, living material. They can be used only for a little while because they will soon dry out, but they are useful for observing qualities such as color, movement, or behavior that cannot be observed on dead, stained material. Follow the steps shown in Figure 3–2 to prepare your wet mount.

□ Obtain a slide, coverslip, and teasing needle from your instructor.
□ Place a drop of the sample on your slide; it will contain some interesting living organisms.
□ Touch the coverslip to one edge of the drop, and gently lower it with the teasing needle, as shown.

If you have been careful, the slide will not have any bubbles. If not, you will see them as circles of various sizes with very dark edges when you look at your sample.

□ Locate your organism under low power and then switch to high power.
□ Make a sketch of what you see and record the total magnification:

Total magnification = 430 ×

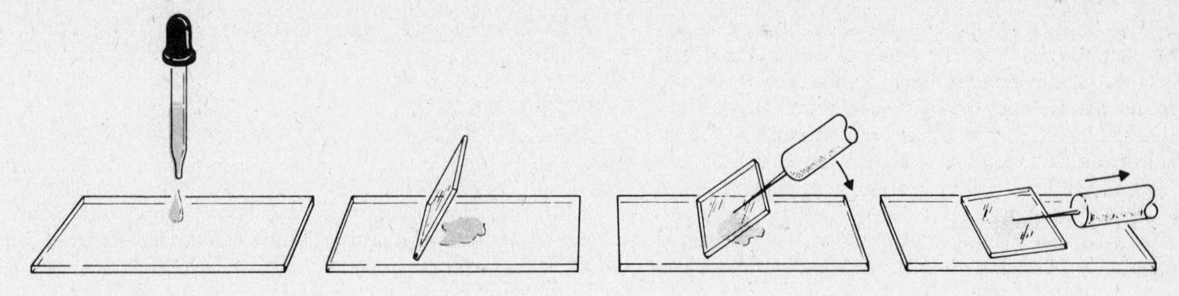

Figure 3–2. Wet mount. Mount your living specimen as shown to avoid trapping air bubbles under the cover slip.

Identify the organism if you can:

____

____

____

□ When you are finished, discard the coverslip and clean the slide.

Studying Biological Organisms

Your instructor will provide cultures of living organisms or prepared slides for you to study in the microscope. Always view the organisms under low power first, then increase the magnification as necessary. You may be asked to sketch and estimate the size of some of the following organisms:

yeast	*Paramecium caudatum*	*Amoeba*
Volvox	*Paramecium bursaria*	*Euplotes*
Spirogyra	*Chlamydomonas*	*Stentor*

Storing the Microscope

□ Return to the low power objective.
□ Raise the body tube or lower the stage with the coarse adjustment so the lenses can't strike the stage accidentally.
□ Clean all lenses with lens paper and (if you have not already done so) clean the oil immersion lens with xylene.
□ Turn off the light.
□ Cover the microscope, if there is a cover for it.
□ Pick up the microscope by its arm, supporting it under the base, and return it to its storage box or cabinet.

HOW TO USE THE STEREOMICROSCOPE

This microscope has relatively low magnification, usually 4× to 30×, and is used for viewing and manipulating larger objects. It is really two microscopes with two ocu-

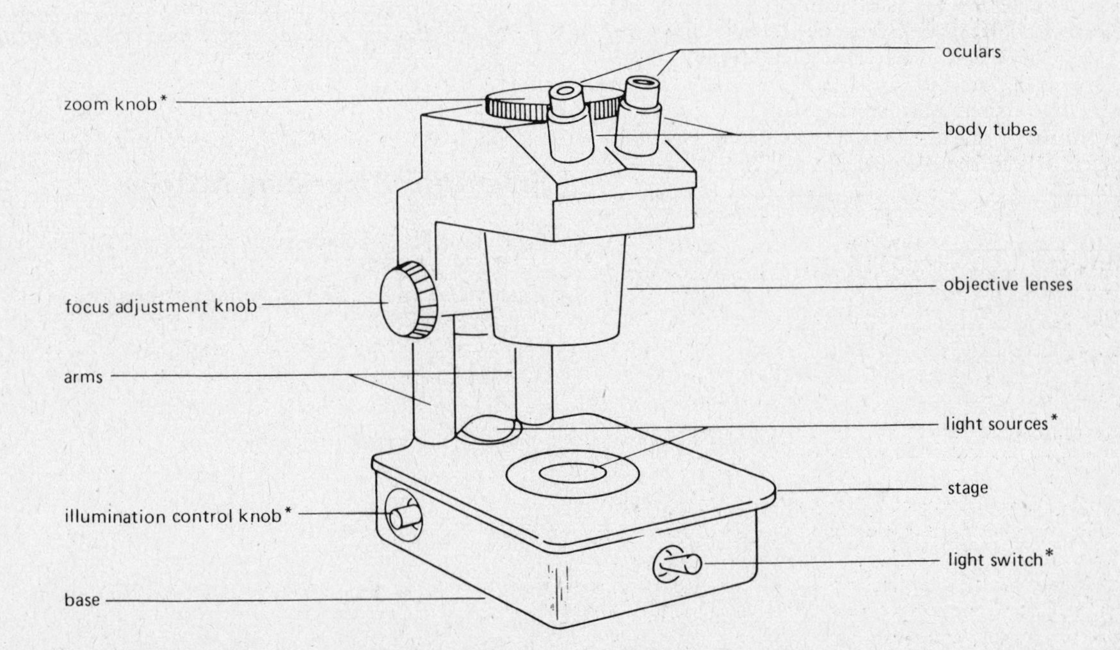

Figure 3–3. Stereomicroscope. Your microscope may be similar to the one shown in this diagram. Some of the parts that may not be present are marked with an asterisk (*), and some microscopes will have a rotating nosepiece to change magnification. Your instructor may be able to provide a diagram that represents your microscope exactly.

lars so that objects can be viewed with binocular vision in three dimensions. Also, the depth of the field that is in focus at the same time is much greater than with the compound microscope. The light souce can be directed down onto the object, in addition to or instead of up through the object, so that you can view objects too thick to transmit light. Otherwise the stereomicroscope is similar to the compound microscope that you have already used except that the magnification may be varied continuously by rotating the "zoom" knob rather than by rotating the objectives.

□ Carrying the stereomicroscope by its arm, place it on the table, and locate the following parts using Figure 3–3 as your guide.

Base with built-in illuminator

Light switch: Turns on the illuminator. Push the switch on the front of the base to turn on the light. If your microscope has a separate light source, turn it on and aim it down onto the stage.

Illumination control knob: Rotates to provide reflection or transmission or, in some microscopes, to combine transmission and reflection.

1. **Transmission:** Light travels up through specimen.
2. **Reflection:** Light is directed down onto the speciman from which it is reflected up into the microscope.
3. **Transmission-reflection:** Combines (1) and (2). Rotate the knob to get the best possible illumination for a particular sample.

Oculars: Move the oculars together or apart until you feel comfortable looking into the microscope. The two fields of view should overlap completely so that you see a single circle of light.

"Zoom" knob: Controls magnification. It is located next to the oculars on the body of the microscope. Turn it until the indicator mark is on 4× or the lowest magnification. Some microscopes have revolving objectives much like a compound microscope.

Stage: Place a small ruler on the stage.

Focus knob: A single knob that you should turn until the ruler is in sharp focus.

How big is the field under the lowest possible magnification? ______________ mm

Turn the zoom knob to the highest magnification. How big is the field? ______________ mm

Now look at your finger or some other handy object under low power. Is it right side up or upside down?

____

Move your finger to the right; which way does the image move? ______________

Your instructor will provide objects or cultures of biological interest for you to view in the stereomicroscope such as:

insect	*Volvox*
leaf	*Planaria*
fungus	*Hydra*

When you are finished, remove the sample, turn off the light, clean the lenses with lens paper, cover the stereomicroscope, and store it in its box or cabinet.

You can see that this microscope is extremely useful when a small amount of magnification is needed, or when you need to manipulate or dissect things too small to be easily seen with the naked eye. One task that is greatly facilitated under a stereoscope is removing scratches from a stereo record with a fine needle: the grooves that have been crushed can be easily straightened out again so that the needle no longer skips. Similar microscopes are used in human microsurgery.

EXPLORING FURTHER

Corrington, Julian D. *Getting Acquainted with the Microscope.* Rochester: Bausch and Lomb, 1969.

Headstrom, Richard. *Adventures with a Microscope.* New York: Dover Publications, 1977.

———. *Adventures with a Hand Lens.* New York: Dover Publications, 1976.

Leeuwenhoek, Antony van. *Antony van Leeuwenhoek and his Little Animals.* New York: Dover Publications, 1960.

Scientific American Articles

Fein, Jack M. "Microvascular surgery for stroke." April 1978 (#1385).

Quate, Calvin F. "The acoustic microscope." October 1979 (#3061).

TOPIC 4

Structure of Cells and Cell Types

OBJECTIVES

When you have completed this topic, you should be able to:

1. Describe the structures that you can expect to see in a typical plant cell with the light microscope.
2. Describe the structures that you can expect to see in a typical animal cell with the light microscope.
3. Describe how a bacterial cell looks under the light microscope.
4. Explain how a prokaryotic cell such as a bacterium basically differs from a typical eukaryotic cell.
5. List structures found in unicellular organisms that are an adaptation to their free-living life, and are not usually found in the cells of multicellular organisms.
6. List the advantages and disadvantages to an individual cell of being part of a multicellular organism.
7. Describe specialized adaptations found in cells in multicellular organisms and in tissues of higher organisms.

TEXT REFERENCES

Chapter 4; Chapter 22 (A,D); Chapter 23 (A,B,H,J,K); Chapter 25 (F).
Light microscopy, staining, cell structure, prokaryotic cells, protists, green algae, animal tissues, plant tissues.

INTRODUCTION

In today's laboratory you will study biology from the viewpoint of the individual cell. Although the cell is considered to be the building block of all organisms, cells differ enormously in shape, size, and capability. Prokaryotic cells are less complex, are usually found only in unicellular organisms, and have more limited capabilities than eukaryotic cells. Higher organisms are made up of highly integrated aggregations of specialized eukaryotic cells, but some sophisticated organisms consist of a single eukaryotic cell. The purpose of your observations should be, first, to review the basic organization of cells and, second, to study how the cell has become adapted to a variety of biological roles through the process of evolution.

A "TYPICAL" CELL

There is really no "typical" cell in which you can see all the features that a cell might have, so you will instead look at three ordinary cells to see what they are like. These cells have a true nucleus with a nuclear membrane and other characteristics not found in prokaryotes.

Onion Epidermis (Eukaryotic Plant Cell)

☐ Break or cut off with a razor blade a piece of a single layer of onion from within the onion (the outermost layer may contain only dead cells). See Figure 4–1.

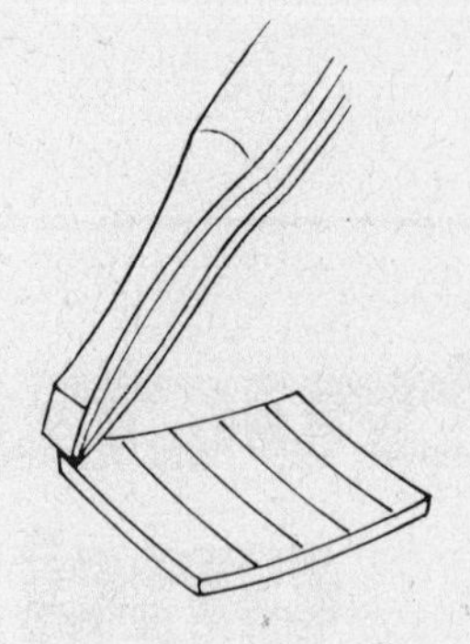
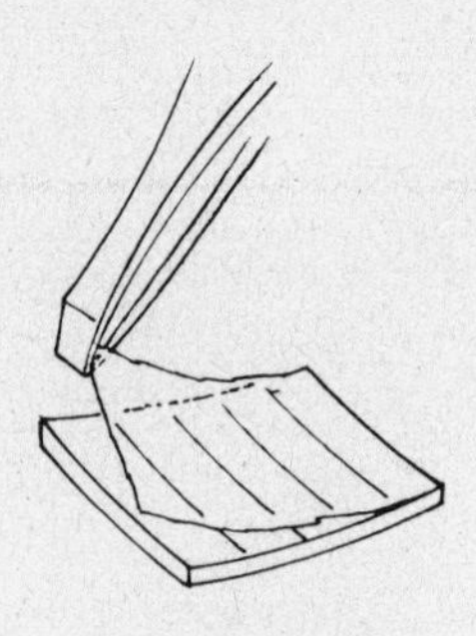

Figure 4–1. Onion epidermis. The desired tissue is extremely thin and transparent.

□ Return the onion to its storage container.
□ Snap the piece in half and then use forceps to peel off a bit of tissuelike transparent epidermis from the inner layer as shown in Figure 4–1.
□ Mount the epidermis in tap water on a slide so that it is flat and not doubled over on itself. Add a coverslip.
□ View the slide under low power.

What is the shape of the cells?

brick shaped

What is the thick layer that surrounds each of the cells?

cell wall

What does the clear part of each cell contain?

cytoplasm

If you are looking at a confusing mix of overlapping cell parts, you probably do not have a good piece of epidermis. Ask your instructor for help. Some cells will not have a **vacuole,** if they are young. Also, the **plasma membrane,** which lines the inner surface of the cell wall, is too thin to be seen in the light microscope, but you can see exactly where it must be.

□ Raise the coverslip and add a drop of 45% acetocarmine dye; this will stain the **cytoplasm** and **nucleus** of your cells pink or red.
□ Distinguish between the vacuoles and the cytoplasm. The cytoplasm is within the cell membrane and is "living," whereas the contents of the vacuoles are often inert storage materials.
□ Locate the nucleus, which now should be red.

What kinds of molecules are located in the nucleus?

nucleolus DNA & Chromosomes

What is the function of the nucleus?

Reproduction

□ Make a sketch of the onion cell to show the parts that you saw.

SKETCHES FOR THIS LAB SHOULD BE MADE IN THE SPACES PROVIDED AT THE END OF THE EXERCISE. YOUR INSTRUCTOR MAY ASK THAT THEY BE HANDED IN WHEN YOU LEAVE.

□ When you are finished with each specimen, discard the coverslip and clean the slide for the next specimen. Your instructor will tell you whether you should discard, rather than wash, a slide on which you used stain.

Human Cheek Epithelium (Eukaryotic Animal Cell)

□ Use a clean toothpick to scrape the inside of your cheek gently.
□ Stir the scrapings into a drop of tap water and add a drop of methylene blue dye to stain the cheek cells. Add a coverslip.
□ View your cells under low power and locate some that are spread out rather than bunched up in a big clump.

The cells you are looking for are pale blue and have a very dark blue nucleus. You may see some that are folded over so that you can see how flat they are.

□ Switch to high power and take a closer look at one of the cells.

What shape is the cell?

round

Do you see a vacuole or cell wall? NO

Again, you cannot see the plasma membrane but can assume it surrounds the cytoplasm.

□ Look closely at the nucleus and locate the nucleolus within it (try another cell if necessary).

What is the function of the nucleolus?

Synthesis of ribosomes & ribosomal materials

This cell is highly specialized and has a protective function. As a result, it cannot carry out certain functions.

What are some of these functions?

Bacterium (Prokaryotic Cell)

□ Use a toothpick to place a tiny dab of yogurt on your slide. Add a small drop of tap water and stir the toothpick around to spread out the cells. Add a coverslip.

□ Focus on the slide under low power and then switch to high power.

The swarms of rod-shaped organisms that you see are cells of yogurt bacterium *Lactobacillus.* These bacteria are adapted to live on milk sugar (lactose) and are used by human beings to convert ordinary milk into yogurt, which is acidic and keeps much longer than milk.

□ Add a drop of immersion oil to your slide and view the bacteria under oil immersion if your microscope has an oil immersion lens.

Can you see any structure within the cells?

____

Bacteria have tough cell walls but do not have any of the membrane-bound organelles that are found in eukaryotic cells.

What organelle is found in both eukaryotic and prokaryotic cells? ______________________________

If time permits, look at some of your own oral bacteria:

□ Use a clean toothpick to scrape some of the tartar from between your teeth.
□ Smear out the scum on a slide and pass the slide over the small flame of an alcohol lamp or Bunsen burner to fix the cells onto the slide.
□ Add a drop of crystal violet stain and wait 1 min.
□ Wash off the excess dye with a gentle stream of water and blot off the excess water with a tissue, being careful not to touch the spot where the bacteria are.
□ If you have an oil immersion lens, let the slide dry and then add a drop of oil and view under oil immersion.
□ If you do not have oil immersion, add a drop of water and coverslip and view under high power.

The bacteria are tiny and darkly stained. They may be rod-shaped, round, or spiral-shaped, and some may stick together in clumps. Choose some different types and draw what you see in the space provided at the end of this topic.

□ If a demonstration slide is available, view the three main types of bacteria using oil immersion.

What shapes can you see? ______________________

____

____

SPECIALIZED ANIMAL CELLS

Amoeba

This organism is a single cell with an irregular shape and many internal organelles. Cytoplasmic streaming allows it to move by changing shape so it actively seeks prey and can engulf food by trapping it within arms called **pseudopods.** *Amoeba* digests the food within its membrane-bound **food vacuoles.** It has a **contractile vacuole** to take up excess water and eject it to the exterior so that the *Amoeba* doesn't swell up and burst. The **nucleus** controls the activity of the cell and is duplicated when the cell divides by simply balling up and splitting in half.

□ Use a dropping pipet to take a sample from the bottom of the *Amoeba* culture. Try to get your sample near some debris.

***AMOEBA* IS SOMETIMES LARGE ENOUGH TO SEE WITH THE NAKED EYE. IF NOT, YOU MAY USE A STEREOMICROSCOPE WHEN LOCATING AN *AMOEBA* AND TRANSFERRING IT TO YOUR SLIDE.**

□ Place the drop on your slide and view it under low power to make sure that you have at least one *Amoeba*: if not, try again using a dissecting microscope to make sure that your dropper sucks one up.

SMALL, FAST-MOVING ORGANISMS ARE FOOD, NOT *AMOEBAE*!

□ Add a coverslip and view the *Amoeba* under high power.

A healthy *Amoeba* puts on quite an active performance, extending pseudopods in all directions, creeping along the slide and perhaps even encircling some of the food organisms in the culture fluid. This organism works in slow motion, so you may have to watch carefully to see exactly what your *Amoeba* is doing.

Keep in mind that the *Amoeba* is very sensitive to heat from the microscope lamp, so you should turn off the lamp frequently to let it cool, and always turn off the lamp when you are not looking through the microscope.

What is the approximate diameter of your *Amoeba,* in micrometres? (Make this simple calculation by comparing the size of the *Amoeba* to the known diameter of your microscope field at high power.) ______________ μm

Note that the inner part of the cytoplasm is doing most of the streaming. This is the **sol** part of the cytoplasm whereas the more rigid outer part is called the **gel.** Cytoplasm can be interconverted from sol to gel. The interconversion is important in cytoplasmic streaming and in the **amoeboid movement** that results.

□ Locate the contractile vacuole that periodically ejects its contents to the outside of the cell.

About how long does each cycle of contraction take?

____

Some of the white blood cells in your own body are very similar to this *Amoeba.* They circulate between the cells of your tissues and engulf harmful bacteria and cell debris that would otherwise accumulate.

Paramecium

This organism is also a single cell with some of the same features of the *Amoeba*; it moves, ingests particles of food, digests the food in vacuoles, removes water with a contractile vacuole, and reproduces by dividing in half. Its life style in doing these things is so different, however, that it is hard to see how *Amoeba* and *Paramecium* are similar at all. The *Paramecium* is in constant motion, thanks to its **cilia,** which cover the whole body and beat in a coordinated sequence to propel the organism along. The body has a definite shape which allows it to move efficiently, and there is an **oral groove** that tends to accumulate possible food items for efficient ingestion. Cytoplasmic streaming helps circulate vacuoles around the cytoplasm, and wastes tend to be eliminated from old food vacuoles at the same site on the membrane.

□ Place a small drop of the *Paramecium* culture on a clean slide and add a drop of methyl cellulose (a viscous solution such as Protoslo that slows the speed of the organisms). Stir the drop with a toothpick and add a coverslip.
□ View the slide under low power so that you can see what *Paramecium* looks like, and then switch to high power.

Approximately how large is your specimen?

_______________ μm

□ Make an outline drawing of *Paramecium* in the space provided at the end of this topic, showing the oral groove and labeling the anterior end.

Does a *Paramecium* always swim in the same direction?

____

□ Locate the large nucleus, which is the macronucleus.
□ Locate the active contractile vacuole(s). Observe the cycle through which it slowly fills with excess water and then suddenly contracts to expel water.

Yeast cells are a good food source for *Paramecium.* When they are heated and stained with Congo red dye, they can be easily seen even after they have been ingested and are enclosed in food vacuoles.

□ Add a small drop of *Paramecium,* a drop of yeast cells stained with Congo red dye, and a drop of Protoslo to a clean slide and stir with a toothpick. Add a coverslip and wait a few minutes for the yeast to be taken up.
□ Locate a *Paramecium* under high power.

Where are the yeast cells found inside the cell?

____

____

____

What color are the yeast cells when they are inside the *Paramecium*? _______________

What factor affected the color of the dye?

____

Stentor (Alias the "Blue Vortex")

Stentor is an enormous unicellular organism that is related to the *Paramecium* because it also moves by means of cilia. As in *Paramecium,* cilia are important in circulating food items and directing them toward a spot where they can be ingested. A swimming *Stentor* looks very different from a sessile, or stationary, *Stentor.* (See Figure 4–2.)

□ Add a drop of the *Stentor* culture to a clean slide or depression slide and observe under the dissecting microscope. Do not add a coverslip while this organism is swimming unless you are using a depression slide.
□ Draw an outline of a *Stentor* specimen and label it either swimming or sessile; if it is swimming, label the anterior end, and if it is sessile, label the top.
□ Add a drop of Protoslo and a coverslip and locate the *Stentor* under low power of your compound microscope.

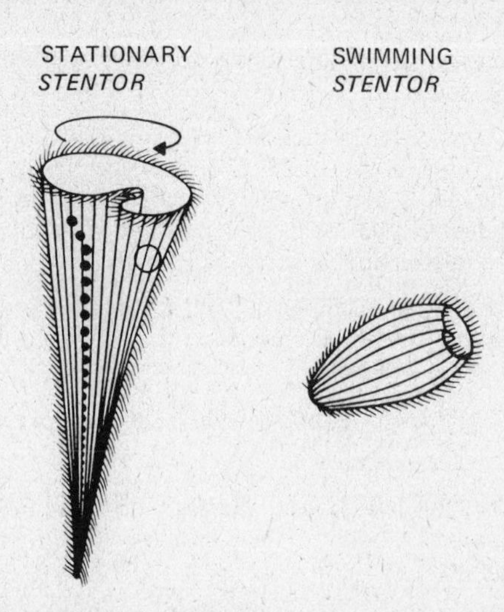

Figure 4–2. Stentor. A unicellular eukaryote. (From Jahn, Theodore L., Eugene C. Bovee, and Frances F. Jahn, *How to Know the Protozoa,* 2nd Ed. © 1977 Wm. C. Brown Company Publishers, Dubuque, Iowa. Reprinted by Permission.)

Note the circular band of cilia around the "top" of the organism. You may be able to see the circulation of food particles near the cilia if your *Stentor* is stationary and adjusted to being observed on the slide.

□ If your *Stentor* is holding still, observe it under high power and locate the nucleus: it is completely different from other nuclei that you have seen and looks like a string of beads in the center of the organism.

How big is *Stentor* in its largest dimension?

_______________ μm

Intestinal Cells

In more complex multicellular organisms, the cells are actually much simpler than some of the cells that you have already seen; the complexity is created by *combinations* of cells rather than the cells themselves. Each individual cell tends to lose some functions, such as motility or protection, and to specialize in other functions, such as support, secretion, or contraction.

□ Take a prepared slide of the cross section of intestine and view it under low power.

Note the hollow inner space which is the **lumen.** There are five or more layers of different tissues surrounding the lumen. They are organized in sequence from the inner side into the **mucosa, submucosa, muscle,** and the **peritoneum.** Study the cell types in each layer in turn, using Figure 4–3 as your guide. Make a sketch of each cell type in the space provided at the end of this topic.

COLUMNAR EPITHELIAL CELLS

These cells make up the inner layer of the mucosa. They follow its highly folded inner surface and secrete mucus, enzymes, and fluid into the lumen of the gut. Their specialty is protection and secretion.

GOBLET CELLS

These are large cuplike spaces facing the lumen of the gut and are filled with mucus secreted by the goblet cells. These cells are highly specialized for secretion.

CONNECTIVE TISSUE CELLS

These cells make up the bulk of the structure of the submucosa. The dense connective tissue cells secrete collagen fibers that support the other cells and the many blood vessels and lymph vessels that pass through this layer. Since blood is a fluid connective tissue itself, the blood cells within the vessels are another type of connective tissue cell specialized for oxygen transport or defense against harmful organisms.

MUSCLE CELLS

Muscle cells are in two layers: the first is oriented around the lumen so that the cells are cut lengthwise and appear spindle-shaped; the second is oriented so that its cells are cut crosswise and look like circles. You can see the cell nuclei, which are darkly stained. These cells are smooth muscle cells that are specialized for contraction. The two layers of muscle cells counteract each other: if the inner layer contracts, the section of gut will become long and thin, but if the outer layer contracts, the section will become short and fat.

SQUAMOUS EPITHELIAL CELLS

The peritoneum is the outermost layer of cells and is only one cell layer thick. The cells are flattened and serve to protect the outer surface of the intestine. The nuclei may be seen as small elevations. The cells lining the blood vessels are also squamous epithelium.

What type of squamous epithelial cells did you see earlier in this exercise?

cheek cells
___

SPECIALIZED PLANT CELLS

Chlamydomonas

Chlamydomonas is a tiny plant that belongs to the green algae. It is an **autotroph** because it can carry out photosynthesis but is otherwise very different from the plants to which we are accustomed.

□ Take a drop from the *Chlamydomonas* culture, add a drop of Protoslo, stir, and add a coverslip.

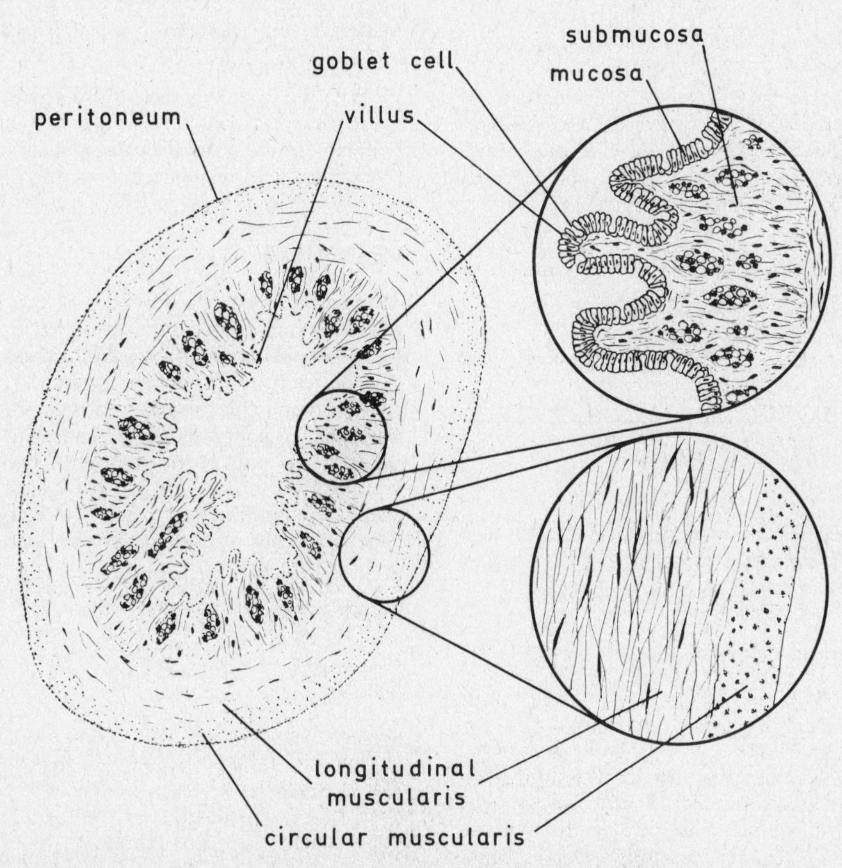

Figure 4–3. Intestine in cross section. The lumen is the hollow interior of the intestine where you would find undigested food. The projections into the lumen are the villi.

□ Observe under low power to locate the organism, which is tiny and, being related to the green plants, quite green.

What "animal"-like feature of these cells do you notice right away?

____

□ When you have found a cell that is not moving too much, switch to high power and take a closer look.

How big is *Chlamydomonas*?

______________ μm

□ Find the green organelle which contains the green pigment, chlorophyll.

What is this organelle called? ______________

What basic life process goes on here? ______________

What type of nutrition does *Chlamydomonas* use?

____

□ Adjust the illumination until you can see the pair of **flagella** that allow *Chlamydomonas* to move.

The flagella are like cilia only much, much longer. They are also found in animal cells such as sperm and some protozoans.

Why would a plantlike organism that can synthesize its own food need to move?

____

____

Volvox

This organism consists of a colony of 50–50,000 algal cells, each very much like the one you just saw, joined together in a large hollow ball. An active *Volvox* is an aesthetic delight and a healthy culture dish of them is spectacular.

□ View the culture of *Volvox* under the stereoscope to see the whole organisms in motion.

How does the *Volvox* look as it moves?

____

____

What are the structures called that propel it?

____

□ Place a drop of the culture on your clean slide and put a bit of broken coverslip on either side of the drop to protect it. Then add a coverslip.
□ View a colony under low power.

Does your colony have daughter colonies inside of it?

____

How many? ______________

□ The *Volvox* colony does not split in half during reproduction; instead, specialized reproductive cells are released into the interior of the colony where they join to start new colonies. When the daughter colonies have grown large enough, the parent colony bursts open to release them.

Other cells in the colony are specialized for photoreception, and help direct the colony to swim toward the light. This is a basic response of photosynthetic organisms.

□ Examine the colony under high power and adjust the illumination so that you can see the flagella; use Protoslo if you cannot find a quiet colony.

The action of the flagella of the individual cells can be coordinated because there are thin strands of cytoplasm between all of the cells, permitting a primitive form of coordination.

Leaf Cells

□ Take a prepared slide of the cross section of a leaf and study the types of cells that it contains.

Within the leaf the cells are organized into three tissues: **surface tissue, fundamental tissue,** and **vascular tissue.** Each tissue has its own characteristic types of cells specialized to carry out certain functions. Surface tissue is **epidermis** with cells specialized to protect the leaf. Fundamental tissue includes **mesophyll** cells specialized for photosynthesis and gas exchange. Vascular tissue includes **xylem** and **phloem** tissue for transport.

EPIDERMIS

Cells of the epidermis form a single protective layer on both the upper and lower surface of the leaf. They secrete a waxy **cuticle** that waterproofs the leaf surface and helps protect it.

□ Note the openings, called **stomata,** in the lower leaf epidermis. The two small cells on either side of each opening are the **guard cells,** specialized epidermal cells that control the passage of gases into and out of the leaf.

MESOPHYLL

These cells make up the bulk of the inner part of the leaf. The densely packed cells below the upper epidermis form the **palisade mesophyll.** These cells are specialized to carry out photosynthesis very efficiently: they contain many chloroplasts and their shape lets them trap much of the light striking the leaf. The **spongy mesophyll** is the part of the mesophyll beneath the palisade layer. Here the cells are loosely packed and irregular in shape so that the gases can be freely exchanged between the cells and air within the leaf.

□ Note that the spaces between the cells connect with the openings in the lower epidermis.

XYLEM

Vascular tissue occurs in vascular bundles of xylem and phloem and transports materials into and out of the leaf. Xylem is one type of vascular tissue that is so specialized to carry out its function of transport that it has lost all other functions, and is dead.

□ Locate oval or round **vascular bundles** in your leaf. They may be stained red.

□ Locate the xylem vessels. They are the large empty-looking structures within the vascular bundles.
□ Find the spiral thickenings in the walls of xylem vessels. These are easy to see where the vessels run parallel to the cut through the leaf and usually stain red.

In the stems of woody plants, many xylem vessels are bunched together to make up the woody part. The cell walls are made of cellulose and are greatly thickened to give support to the stems of these large plants and to conduct water and minerals over great distances without collapsing.

PHLOEM

Phloem tissue is also specialized for transport, but has only living cells and controls the substances that travel in them. Phloem conducts food from the leaves to the stem and roots for storage, and then from storage back to the next crop of leaves.

□ Locate the phloem cells alongside the xylem vessels; they are smaller, more lightly stained and may lack a nucleus.
□ Label the important cell types in the drawing of the leaf cross section at the end of this topic.

When you are finished with your observations, return the materials and store your microscopes properly.

EXPLORING FURTHER

DeRobertis, E. D. P., F. A. Saez, and E. M. F. DeRobertis. *Cell Biology.* 7th ed. Philadelphia: W. B. Saunders, 1979.
Jahn, T. L., and F. Jahn. *How to Know the Protozoa.* 2nd ed. Dubuque: W. C. Brown, 1979.
Prescott, G. W. *How to Know the Freshwater Algae.* Dubuque: W. C. Brown, 1954.

Scientific American Articles

Albrecht-Buehler, Guenter. "The tracks of moving cells." April 1978 (#1386).
Dustin, Pierre. "Microtubules." August 1980 (#1477).
Lake, James A. "The ribosome." August 1981.
Lazarides, Elias, and Jean Paul Revel. "The molecular basis of cell movement." May 1979 (#1427).
Porter, Keith R., and Jonathan B. Tucker. "The ground substance of the living cell." March 1981.

Student Name ______________________________ **Date** ______________

onion cell

type of bacteria

***Paramecium* (size = ____________ μm)**

***Stentor* (size = ____________ μm)**

columnar epithelium (intestine)

goblet cells (intestine)

connective tissue cells (intestine)

muscle cells (intestine)

squamous epithelium (intestine)

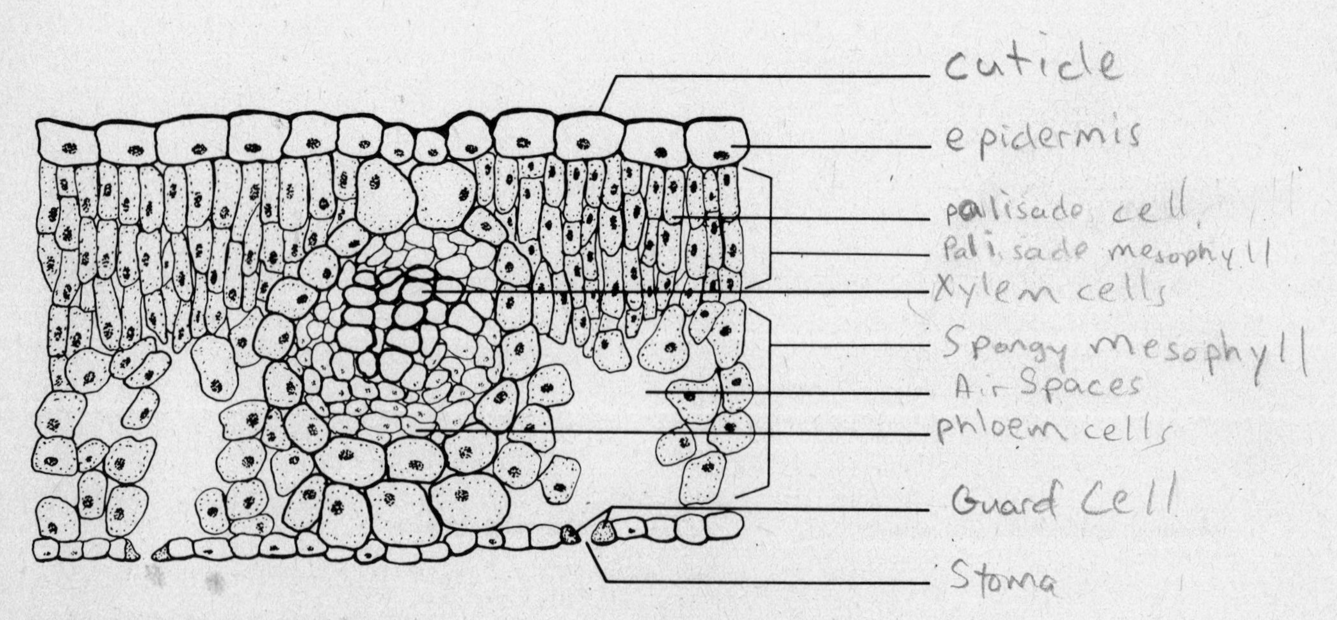

Figure 4–4. Leaf in cross section.

TOPIC 5

Cell Membranes

OBJECTIVES

When you have completed this topic, you should be able to:

1. Define solute, solvent, and chemical gradient.
2. Define the following processes and identify the characteristics that distinguish them from one another: diffusion, osmosis, active transport.
3. Give three factors that affect the rate of diffusion and state whether they increase or decrease the rate.
4. Define and use correctly the following terms: osmotic potential, permeable, hypertonic, isotonic, hypotonic.
5. Describe what happens when an animal cell and a plant cell are put into distilled water; and then describe what happens when they are put into a concentrated salt solution; use the following terms correctly in your answer: turgor, plasmolysis, lysis, wall pressure, hypertonic, hypotonic, osmosis.
6. State the function of the contractile vacuole in protozoans like *Paramecium* and *Amoeba.*
7. State the function of the central vacuole and tonoplast in plant cells.
8. Given any two solutions of different osmotic potentials that are separated by a semipermeable membrane, state which solution is hypertonic and in which direction there will be a net flow of water.

TEXT REFERENCES

Chapter 5.
Diffusion, currents, osmosis, permeability, active transport, cell membrane.

Know bold face words

INTRODUCTION

In this exercise you will study two ways in which substances move into and out of cells: diffusion and active transport.

What kinds of substances must enter a typical human cell?

____

____

What kinds of substances must move out of such a cell?

____

____

Cell membranes are semipermeable in that only certain substances may pass through them.

If you put a high concentration of a solid substance, the **solute,** into a liquid, the **solvent,** the randomly moving molecules of the dissolving solute will spread throughout the liquid. This spontaneous or "downhill" process is called **diffusion** and it does not require any input of energy. As they spread out, the molecules form a **concentration gradient** or **chemical gradient** with the highest concentration near the dissolving solute and the lowest concentration farthest away. The diffusing molecules will tend to move down the concentration gradient from higher to lower concentration until they reach equilibrium. Then they will be spread out evenly and there will be no further *net* movement. The free energy of the solution will be *lower* and the entropy (disorder) will be *higher* after diffusion has occurred.

Diffusion takes place rapidly in a gas, but more slowly in a liquid or solid. You are experiencing both types of diffusion when you smell odors. Diffusion in air is followed by diffusion in a liquid, as the chemical odor reaches your nasal membranes, where it must dissolve before exciting the sensory cells.

When diffusion occurs across a cell membrane, solutes may pass into or out of the cell in the process called **dialysis.** Each solute will move from the side where its concentration is higher to the side with a lower concentration, because the direction of movement is always down the concentration gradient. Its movement will be independent of the movement of the other solutes.

If the substance diffusing across the membrane is water, the process is called **osmosis.** During osmosis water will flow from the side with the most water, and hence the least negative (higher) osmotic potential and lowest solute concentration, to the side with the lower water concentration, most negative (lower) osmotic potential, and higher solute concentration.

PURE WATER HAS AN OSMOTIC POTENTIAL OF ZERO AND ANY SOLUTION WILL HAVE A NEGATIVE OSMOTIC POTENTIAL OF LESS THAN ZERO.

Cells cannot control the process of osmosis directly since cell membranes are freely permeable to water, and since osmosis is a spontaneous downhill process requiring no energy input. Cells can and must control their water content, however, in order to keep their solutes at the correct concentrations required for life processes to function. They can do this by removing excess water with special structures such as the contractile vacuole of protozoans, or they can increase their solute concentration, permitting water to enter the cell by osmosis.

Many substances are brought into the cell or removed from the cell *against* their chemical gradients in the process of **active transport.** Since this is an uphill process, a biologically active membrane that can utilize energy from cell respiration is required. The process of active transport is highly specific in that each particular substance to be carried across the membrane must be transported by a specific carrier molecule. The carriers are membrane proteins and their transport function must always be coupled with cell respiration. If there is no cell respiration, or if the integrity of the membrane is destroyed, active transport cannot occur.

DIFFUSION

Diffusion of a Liquid in a Liquid

Since liquids are highly sensitive to convection currents, the success of the experiment depends on not allowing your setup to be disturbed or jiggled once the experiment is under way.

□ Fill a small beaker with water and set it in a protected place. Wait 10 min.
□ Add a drop of red dye to the surface of the solution as gently as possible. Try not to disturb the surface, but hold the dropper as close to it as you can.
□ Watch the dye spread through the solution from time to time during the lab period.

At the end of period answer the questions given below:

Is the **concentration** of the colored part of the solution greater at the beginning or end of the experiment?
beginning

Is the **free energy** of the solution greater at the beginning or at the end? beginning

Is the **entropy** of the solution greater at the beginning or at the end? the end

What difference would you have seen if you had originally set up the beaker in the refrigerator?
Diffusion would have been slower

Diffusion in a Solid

The solid you will use is agar, which is clear like jello so you can see what is happening. The substances that will be diffusing are solids that react with each other to give colored compounds. The position of the colored bands shows where the two substances first come in contact. The band will be closer to the substance that diffuses more slowly. The setup for this experiment is shown in Figure 5–1.

□ Take a Petri dish of 1.5% agar and turn it upside down. Make a dot near one edge and label it A. Use a ruler to measure 15 mm from A and make a second dot B. Make dots C and D also exactly 15 mm from A and as evenly spaced as you can.
□ Use a cork borer to make holes in the agar at A, B, C, and D. Work carefully and slowly so you don't damage the agar.
□ Use a Pasteur pipet or dropper to add these solutions to the corresponding holes. Be careful *not to overfill the holes* and try to fill them all to about the same level.

A:	1 M $AgNO_3$	silver nitrate
B:	1 M NaCl	sodium chloride
C:	1 M KBr	potassium bromide
D:	1 M $K_3Fe(CN)_6$	potassium ferricyanide

□ Place the Petri dish in the dark and check it from time to time until you can see colored bands forming where the solutions meet solution A.

DIFFUSION IN AGAR

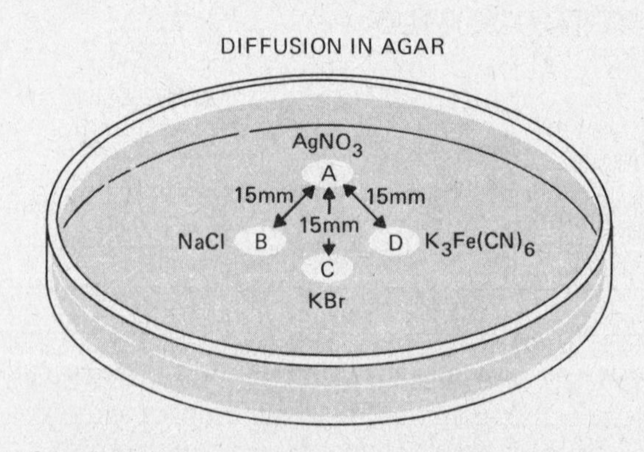

Figure 5–1. Diffusion in a solid. A Petri dish of agar has four holes filled with the following solutions. A contains 1 M $AgNO_3$; B contains 1 M NaCl; C contains 1 M KBr; and D contains 1 M $K_3Fe(CN)_6$. Run the experiment in the dark.

Ag^+ will react with negative ions from the other solutions to give colored substances. Record what colors appear in Table 5–1 at the end of this topic.

□ Measure the distance between each hole and the colored band, and record it in the table. Then calculate the ratio of the distance each ion moved from holes B, C, and D versus the distance moved by the Ag^+ ion: Cl^- distance/Ag^+ distance, and so on. Record these ratios in the table. Since time is constant, the ratio of distances traveled is equal to the ratio of rates.

$$\frac{D_1}{D_2} = \frac{R_1}{R_2}$$

Rank the relative rates of the Cl^-, Br^-, and $Fe(CN)_6^{\equiv}$ ion as slow, faster, or fastest.

As you might expect, the larger an ion is, the slower it will diffuse through the agar. Given that the molecular weights (MW) of the ions in holes B, C, and D can be ranked $Cl^- < Br^- < Fe(CN)_6^{\equiv}$, which ion should move fastest?

____

Which should move slowest? _______________

Did your results agree with your expectations?

____

Calculating Diffusion Rates

The diffusion rate is inversely related to the square root of the molecular weight, so the ratio of rates will be

$$\frac{\text{rate}_1 = \dfrac{1}{\sqrt{MW_1}}}{\text{rate}_2 = \dfrac{1}{\sqrt{MW_2}}} \text{ or } \frac{R_1}{R_2} = \sqrt{\frac{MW_2}{MW_1}}$$

□ Given the molecular weights of the ions, calculate the expected ratio of their diffusion rates, and enter your results in Table 5–1.
□ Enter your observed ratios next to your calculated ratios for comparison.

OSMOSIS

Osmosis in a Model Cell

You will use the following system to become familiar with osmosis: Two solutions will be separated by a **semipermeable membrane** that is permeable to water, glucose, and small molecules, but not to larger molecules. Water will diffuse from one to the other in the process of osmosis, and small molecules will diffuse across the membrane in the process of **dialysis.** This setup is shown in Figure 5–2.

□ Obtain a short piece of dialysis tubing (12–15 cm), and soften it by soaking it in distilled water.
□ Fold one end over and tie it tightly with a piece of thin string or strong thread.
□ Fill the bag with solution A, a liquid "meal" containing:

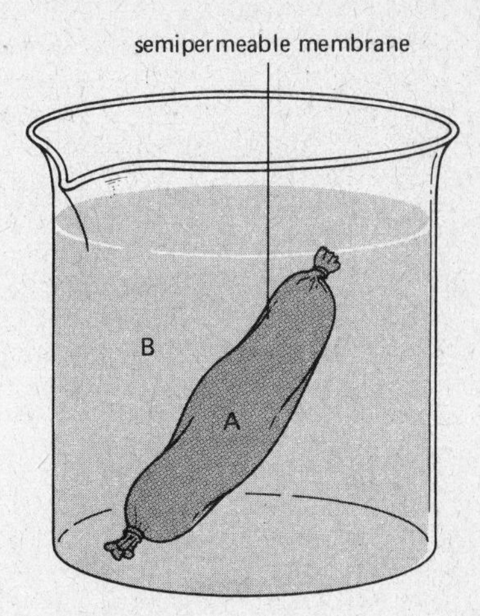

Figure 5–2. Model cell. A semipermeable membrane filled with solution A represents the cell. It is placed in solution B to start the experiment.

25% glucose
0.5% egg albumin
1% starch

□ Fold over the top, squeeze out all the air, and tie tightly. The bag should be limp.
□ Place the bag in a small beaker and add enough of solution B (distilled water containing a very small amount of iodine) to cover it. Let it stand for an hour or so while you go on.

At the end of your experiment answer the questions in Table 5–2, remembering that the bag is permeable to glucose, iodine, and water but not to starch or protein.

Osmosis in a Living Cell

To watch osmosis across a living membrane, see what happens when a plant cell comes in contact with a hypertonic solution. Note that the plant is kept in an aquarium which contains fresh water that is hypotonic to the plant cell. The cell is prevented from swelling and undergoing **lysis** (bursting) because its rigid cellulose walls exert wall pressure on the cell contents. Many animal cells have no such protection against lysis.

□ Make a wet mount of an *Elodea* leaf and focus on it under high power.
□ Add a drop or two of 10% NaCl solution to the edge of the coverslip.
□ Touch a piece of tissue to the opposite side of the coverslip to pull the solution through and observe.

Figure 5–3 shows the cell *before*; sketch the cell *after* NaCl in the space provided.

The cell is now (turgid, flaccid) flaccid_______.

The cell has undergone (lysis/plasmolysis/death)

lysis_______________.

□ Now add a few drops of distilled water to the edge of the coverslip and pull it through with tissue while watching the cell.

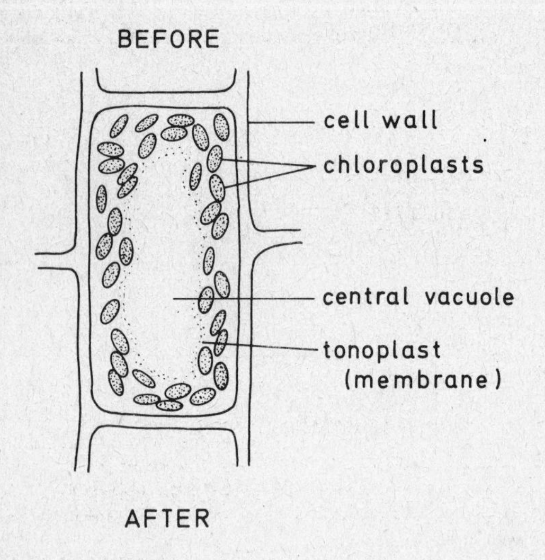

Figure 5–3. ***Elodea* cell.**

Is the process you have observed reversible?

Yes

ACTIVE TRANSPORT

Work in pairs to study active transport in yeast, a fungus. The substance to be transported is neutral red, a dye that is red in a slightly acidic solution and yellow in a basic solution. Sodium carbonate (Na_2CO_3) will be used to make the dye solution basic initially. Record your results in Table 5–3.

□ Obtain two 1-g samples of baker's yeast.
□ In a 250-mL Erlenmeyer flask (Flask A) place 1 g of yeast, 25 mL of 0.75% Na_2CO_3 (sodium carbonate), and mix well. Boil this flask gently for 2 min and cool. The cells are now dead.
□ Add 25 mL of 0.02% neutral red. Record the color of the suspension.
□ In a second flask (Flask B) place 25 mL of 0.75% Na_2CO_3 and 25 mL of 0.02% neutral red. Record the color of the solution.
□ Add 1 g of yeast cells, swirl the flask, and watch for a color difference compared to Flask A. Record the final color of each flask.

Either the cells in Flask B put out something to change the solution or the dye entered the cells and was changed in the cytoplasm. To test this compare cells in Flasks A and B:

□ Filter a portion of the contents of Flask A into a test tube. Record the color of the yeast cells and the solution.
□ Filter a portion of the contents of Flask B into a test tube. Record the color of the cells and the solution.

Think about your observations, then answer the questions in Table 5–3 about Flask B.

□ Add an equal volume of 0.75% acetic acid to the remaining contents of Flask A to acidify the suspension. Record the color of the suspension in Flask A.
□ Filter a portion of the suspension. Record the color of the cells and complete Table 5–3.

When you are finished, wash the glassware that you have used and clean up the lab desk for the next student.

EXPLORING FURTHER

See also Topic 4 references.

Giese, A. C. *Cell Physiology*. 5th ed. Philadelphia: W. B. Saunders, 1979.

Scientific American Articles

Kennedy, D., ed. *Cellular and Organismal Biology*. San Francisco: W. H. Freeman, 1974.
Lodish, Harvey F., and James E. Rothman. "The assembly of cell membranes." January 1979 (#1415).
Rubenstein, Edward. "Diseases caused by impaired communication among cells." March 1980 (#1463).
Staehelin, L. Andrew, and Barbara E. Hull. "Junctions between living cells." May 1978 (#1388).

Student Name ______________________ **Date** ______________

TABLE 5–1. RESULTS FROM THE DIFFUSION EXPERIMENT

Band	Reaction	Color
AB	$Ag^+ \rightarrow Cl^- = AgCl$	
AC	$Ag^+ \rightarrow Br^- = AgBr$	
AD	$Ag^+ \rightarrow Fe(CN)_6^{\equiv} = Ag_3Fe(CN)_6$	

Ion	Measurement	Distance (mm)	Ratio $D_1/D_2 = R_1/R_2$	Relative Rate
Cl^- Ag^+	B → AB band A → AB band	D_1 = D_2 =	Cl^-/Ag^+ =	Cl^- moved ______
Br^- Ag^+	C → AC band A → AC band	D_1 = D_2 =	Br^-/Ag^+ =	Br^- moved ______
$Fe(CN)_6^{\equiv}$ Ag^+	D → AD band A → AD band	D_1 = D_2 =	$Fe(CN)_6^{\equiv}/Ag^+$ =	$Fe(CN)_6^{\equiv}$ moved ______

MW_1	MW_2	MW_2/MW_1	$R_1/R_2 = \sqrt{MW_2/MW_1}$	Observed Ratio R_1/R_2
Cl^- = 35	Ag^+ = 108			Cl^-/Ag^+ =
Br^- = 80	Ag^+ = 108			Br^-/Ag^+ =
$Fe(CN)_6^{\equiv}$ = 212	Ag^+ = 108			$Fe(CN)_6^{\equiv}/Ag^+$ =

TABLE 5–2. RESULTS FROM THE OSMOSIS EXPERIMENT

was solution B hypo or hypertonic B was

Which solution was hypertonic (A or B)? hypotonic

Which solution gained water (A or B)? B

Which solution had the higher osmotic potential (A or B)? B

Which substance inside the membrane reacted with the iodine? starch

Which substance(s) moved out through the membrane? glucose

What test that you learned in Topic 1 could you use to test for the presence of glucose outside? Benedicts solution

If you swallowed some of solution A, which substance(s) could be absorbed directly by the cells of your gut, assuming that they had the same permeability as the dialysis tubing? glucose & ~~starch~~

Which substance(s) would have to be further digested? starch

If your dialysis bag was flaccid or limp at the beginning of the experiment, at the end was it (very flaccid, flaccid, turgid, very turgid)? turgid

Which of the following statements best describes the situation at equilibrium if you let the system stand for a long time? 1

1. No molecules move across the membrane.
2. All molecules cross the membrane equally often in either direction.
3. Molecules to which the membrane is permeable cross equally often in either direction.
4. Only water molecules cross the membrane equally often in either direction.

TABLE 5–3. RESULTS FROM THE ACTIVE TRANSPORT EXPERIMENT

	FLASK A	FLASK B
Initial color of suspension		
Final color of suspension		
Color of cells		
Color of solution		
Is dye inside cells?	XXXXX	
Did Na_2CO_3 enter cells?	XXXXX	
Is the membrane permeable?	XXXXX	
Color of acidified suspension		XXXXX
Color of cells		XXXXX
Did neutral red enter cells?		XXXXX
Did acetic acid enter cells?		XXXXX

What term describes a membrane that is permeable to some substances but not to others? ______________________________

Are dead cells permeable or impermeable to acetic acid? ______________________________

TOPIC 6

Cellular Respiration

OBJECTIVES

When you have completed this topic, you should be able to:

1. Distinguish anaerobic from aerobic respiration.
2. Define fermentation and give its starting materials and end products.
3. Measure CO_2 production by fermenting yeast.
4. List the starting materials and end products of each of the following processes: glycolysis, citric acid cycle, electron transport system.
5. Describe an experiment to measure the respiration rate of a living plant or animal.
6. State the importance of the electron transport system and the role of O_2 in it.
7. Describe an experiment that will measure the reaction of oxygen with an enzyme of the electron transport system.
8. Describe an experiment that will show the effect of an inhibitor on the electron transport system.

TEXT REFERENCES

Chapters 6, 7.
Energy, ATP, oxidation reduction reactions, coenzymes, glycolysis, citric acid cycle, electron transport system, fermentation.

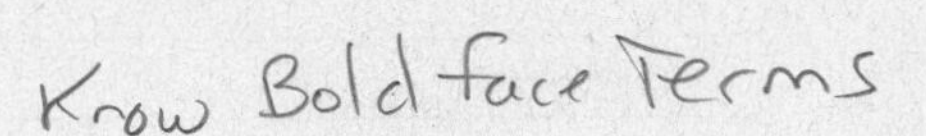

INTRODUCTION

Heterotrophs obtain energy by oxidizing organic molecules and coupling the oxidation reactions to the synthesis of ATP. The ATP is then used to carry out metabolic reactions necessary to maintain the organism's physical integrity, and to support all its other activities.

Some organisms are able to exist in the absence of molecular oxygen, and may even be harmed by the presence of oxygen. They still carry out oxidation reactions but do not use molecular oxygen itself.

Define **oxidation:**
loss of electrons
____

Anaerobic respiration in the absence of oxygen is called **fermentation.** It begins with an energy-rich substance such as glucose, utilizes the enzymes of **glycolysis,** and finally produces ethanol and CO_2, or a mixture of organic acids and other compounds. There is a net gain of only two molecules of ATP for each molecule of glucose oxidized in fermentation.

Most organisms use molecular oxygen in the process of **cellular respiration.** Pyruvic acid, produced by the reactions of glycolysis, enters another set of reactions, the citric acid cycle (tricarboxylic acid cycle) in which the carbon skeleton is completely oxidized to CO_2. The pyruvic acid also loses a molecule of CO_2 as it enters this cycle. The hydrogen carriers, NAD^+ and FAD, are reduced by the reactions of glycolysis and the citric acid cycle. NADH and $FADH_2$ then transfer their hydrogen atoms to a third set of enzymes called the **electron transport chain.** The flow of electrons through this system is coupled to the synthesis of ATP. The very last enzyme in the chain, **cytochrome oxidase,** reacts directly with molecular oxygen to produce water. This step is the point where the oxygen that organisms take in is used and where **metabolic water** is produced. The purpose of the oxygen is to return the cytochrome oxidase to an oxidized form so that it can pick up more electrons from its neighboring electron carrier. Aerobic respiration of a molecule of glucose usually gives a net yield of 36 molecules of ATP.

FERMENTATION IN YEAST

Yeasts are eukaryotic fungi that are commercially very important. They are necessary for the production of beer, wine, bread, and industrial chemicals. In this experiment you will study the production of CO_2 during the fermentation of various carbohydrates by yeast cells.

What molecule is split in the reaction that produces the CO_2?

$C_6H_{12}O_6$+12

When carbohydrate is fermented by yeast, only one-third of the carbons are released in the form of CO_2. The rest of the carbon is released as molecules of H_2O.

Why is it advantageous for an organism to be able to carry on fermentation?

In case their is no sunlight

Study the experiment shown in Figure 6–1 for measuring fermentation under anaerobic conditions. Work with a partner.

□ Obtain a test tube rack, four clean, dry test tubes, four graduated pipets, four Pasteur pipets, and pieces of Parafilm or clay.
□ Label the tubes #1–#4.
□ Fill the tubes as follows:

Tube #1: 2 mL of sucrose + 2 mL of yeast suspension.
Tube #2: 2 mL of galactose + 2 mL of yeast suspension.
Tube #3: 2 mL of molasses + 2 mL of yeast suspension.
Tube #4: 2 mL of water + 2 mL of yeast suspension.

□ Mix the solution in tube #1, then draw it up in a graduated pipet, and place your finger over the top of the pipet.
□ Seal the opposite end of the pipet with a piece of Parafilm or clay.
□ Using the Pasteur pipet, completely fill the graduated pipet to overflowing with solution from tube #1.
□ Place the pipet *upside down* in tube #1, and set the tube back in the rack.
□ Repeat for the other test tubes.

As gas forms during the fermentation, it will rise in the pipet and collect at the plugged end. Note the exact level of the gas in each pipet every minute for about 20 min, and record your readings in Table 6–1 at the end of this topic.

Did the type of carbohydrate being fermented affect the rate of gas formation? yes

Which carbohydrate supported the fastest rate of fermentation? molasses

Your instructor may ask you to vary this basic experiment by using different concentrations of carbohydrate or yeast, or by adding inhibitors.

RESPIRATION IN PLANTS AND ANIMALS: CO_2 PRODUCTION

In aerobic respiration the carbon skeleton of glucose is completely broken down to CO_2. The CO_2 must be quickly removed via the transport and gas exchange systems, so that it does not build up and make the body fluids so acidic that the organism cannot function.

In which reactions is CO_2 released during respiration?

aerobic

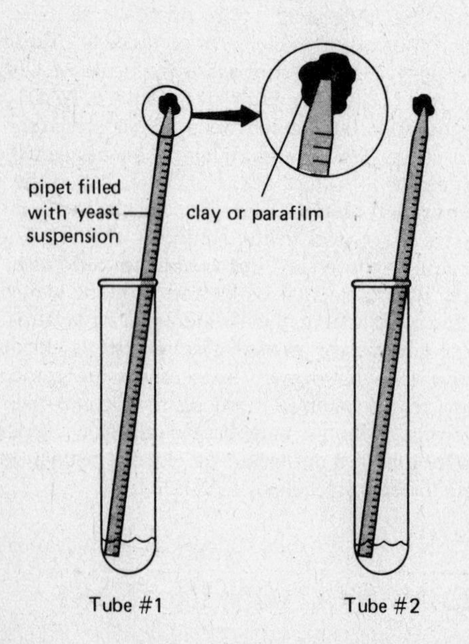

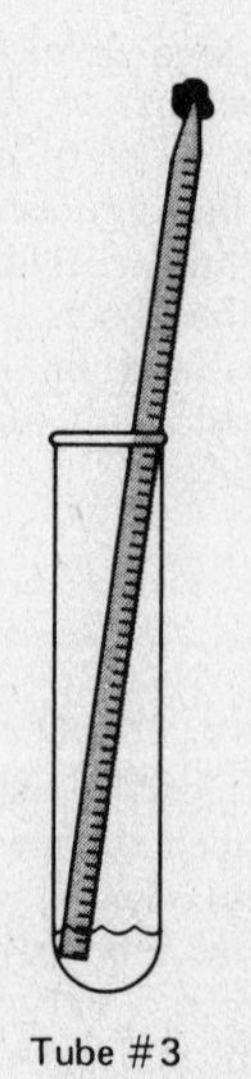

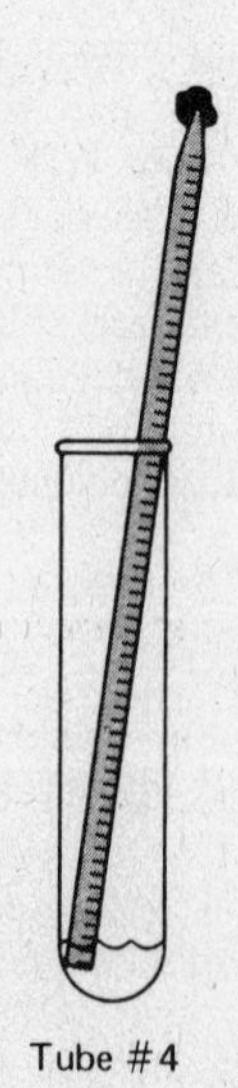

Figure 6–1. Fermentation experiment. Fermentation by yeast within the pipet produces CO_2. The pipet must be completely filled with yeast suspension at the beginning of the experiment.

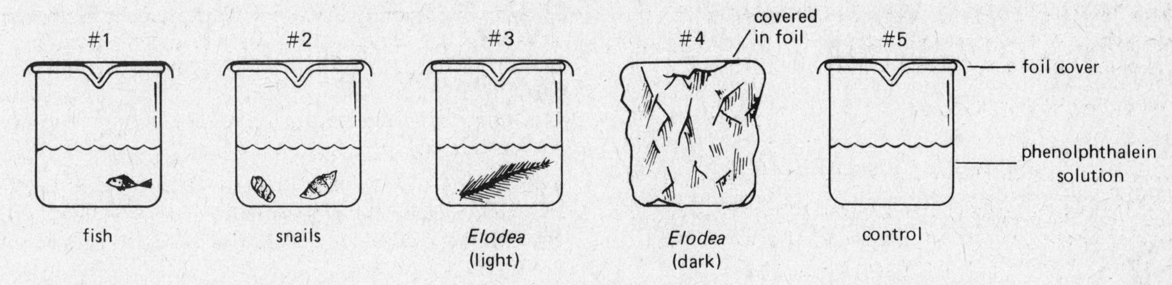

Figure 6–2. Respiration experiment. These aquatic organisms produce CO_2. Each 100 mL beaker contains 25 mL of phenolphthalein solution and its top is covered with aluminum foil. Beaker #4 is completely covered with foil to exclude light.

Why does CO_2 make the tissues acidic?

CO_2 in water gives off carbonic acid H_2CO_3

In this experiment you will measure the rate of CO_2 production in several organisms. This rate is an indicator of the respiration rate, or metabolic rate. A diagram of the experiment is shown in Figure 6–2.

Record your results in Table 6–2.

□ Obtain five 100-mL beakers, and label them #1–#5.
□ Set up the beakers as follows:

	Phenolphthalein*	Organism
Beaker #1	25 mL	1 fish
Beaker #2	25 mL	2 snails
Beaker #3	25 mL	5 cm of *Elodea*
Beaker #4	25 mL	5 cm of *Elodea*
Beaker #5	25 mL	None

□ Cap beakers #1, #2, #3, and #5 with aluminum foil, and note the time.
□ Completely cover beaker #4 with aluminum foil so that the *Elodea* is kept dark, and do not open this beaker until the end of the experiment.
□ Note the changes in the color of the phenolphthalein solutions in the beakers (except #4) every 10 min for 30 min.
□ At the end of 30 min, return the living organisms to the proper containers on the instructor's desk.
□ Add HCl (0.0025 M) drop by drop to beaker #1 until the solution just barely turns colorless. Thoroughly mix the solution after each drop is added, and record the total number of drops needed or the volume of acid used from the buret.
□ Repeat for beakers #2, #3, #4, and #5.
□ If you counted drops, convert the number of drops to millilitres by dividing by 20.

As CO_2 is produced, the pH of the solution will slowly decrease. This change can be measured using the pH indicator, phenolphthalein. This dye is prepared in a solution at pH 10.0, in which it is red. The addition of a certain number of drops of HCl will lower the pH sufficiently so that it becomes colorless (below pH 9.2). If CO_2 has been produced, the H_2CO_3 in the solution will reduce the number of drops of acid needed to turn the solution colorless. The more CO_2 produced, the less HCl needed.

Before calculating the CO_2 production of your organisms, you will need to know their approximate volumes.

□ Obtain a 25-mL graduated cylinder, and fill it with water exactly to the 20-mL mark using a Pasteur pipet.
□ Place one fish in the water so that it is completely below the surface. (If they are very small, use several fish, and then divide the volume by the number of fish.)
□ Note exactly how far above the 20-mL mark the water rose. The difference in water level with and without the organism in the water is the approximate volume.
□ Repeat for the other organisms.
□ If an empty snail shell is available, measure the volume of this nonliving part of the snail, and subtract it from the total volume to determine the volume of the living organism more accurately.

Use the following equation to calculate the CO_2 production in μM/mL/hr:

$$\frac{[\text{mL HCl (control)} - \text{mL HCl (expt.)}] \times \text{2.5 } \mu\text{M HCl/mL}}{\text{volume of organism (mL)} \times \text{time (hr)}}$$

5mM

Which organism had the highest metabolic rate?

Fish Snail

Which one had the lowest rate? (Light!) Elodea

Explain your results with *Elodea* in the dark versus in the light:

Dark uses up stored sugars giving off CO_2 while (light) uses CO_2

Which *Elodea* experiment was a better measure of the metabolic rate? Dark Elodea

Why?

*Your instructor may ask you to use only water for the incubation period. In that case use 25 mL of water, incubate the organisms for 30 min, then add phenolphthalein as directed by your instructor. Omit color observations during the incubation.

THE ELECTRON TRANSPORT CHAIN: CYTOCHROME OXIDASE

Most of the ATP in a cell is made in the electron transport system by **oxidative phosphorylation.** The last enzyme, **cytochrome oxidase,** donates its electrons to catalyze the conversion of molecular oxygen to water:

$$2e^- + 2H^+ + \frac{1}{2}O_2 \xrightarrow{\text{cytochrome oxidase}} H_2O$$

The water produced in this way is called **metabolic water** because it is formed by the organism's metabolism, and not taken in from its environment.

If oxygen is removed from the electron transport system, all of the component enzymes will remain in their reduced forms, and electron flow will cease. No ATP can be synthesized by oxidative phosphorylation, and the organism must rely on anaerobic respiration (glycolysis) only. Many organisms cannot survive without a functioning electron transport system. Therefore substances, such as potassium cyanide or sodium azide, that block these enzymes of the chain are lethal poisons.

Where in the cell are enzymes of the electron transport sytem located?

In this experiment you will study cytochrome oxidase from beans. The enzyme will be in the form of a crude extract from bean seedlings. Record your results in Table 6–3.

□ Obtain four test tubes and label them #1–#4.
□ Add the following reagents to the tubes:

	Bean Extract (Cytochrome Oxidase)	Sucrose	Sodium Azide
Tube #1	—	4 mL	—
Tube #2	2 mL	2 mL	—
Tube #3	2 mL	1.8 mL	5 drops (.02 mL)
Tube #4	2 mL boiled*	2 mL	—

*Use boiled enzyme if it is available; otherwise, add 2 mL bean extract to the tube, and then hold it in a boiling water bath for 10 min before adding the other reagents.

□ Add 5 drops (0.2 mL) of Nadi reagent A to each tube. Mix.
□ Add 5 drops (0.2 mL) of Nadi reagent B to each tube. Mix.
□ Observe the tubes every 10 min for 30 min, and note the color changes.

The two components of the Nadi reagent will reduce the cytochrome c present in the bean extract. If the cytochrome oxidase is active, it will oxidize cytochrome c, and pass the electrons to oxygen. More of the reagent will then be used to keep the cytochrome c reduced. When the reagent becomes oxidized, it turns blue. The electrons are flowing from the dye to oxygen. (See Figure 6–3.)

Sodium azide is a potent inhibitor of cytochrome oxidase, and will block the reaction of this enzyme with molecular oxygen.

Did inhibition of cytochrome oxidase block the reaction of cytochrome c with the Nadi reagent?

If the electron transport system were blocked at an early step, but oxygen and active cytochrome oxidase were present, would the reaction of the Nadi reagent continue in the absence of electron flow from the other enzymes? Explain your answer.

When you are finished with your work, return all animals and materials, and wash your glassware.

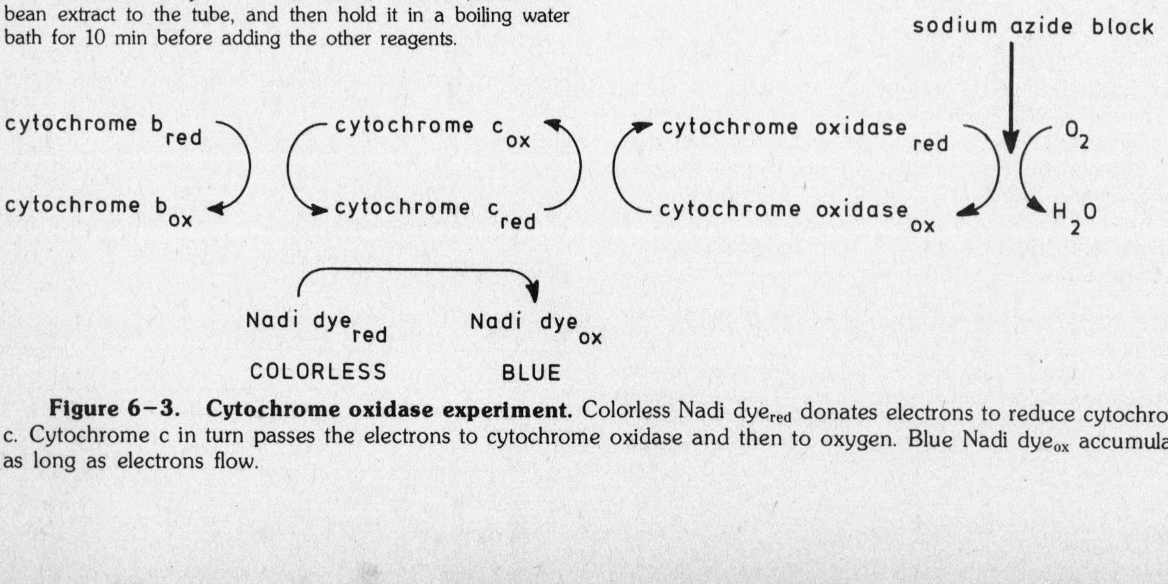

sodium azide block
cytochrome b_{red}
cytochrome b_{ox}
cytochrome c_{ox}
cytochrome c_{red}
cytochrome $oxidase_{red}$
cytochrome $oxidase_{ox}$
O_2
H_2O
Nadi dye_{red}
COLORLESS
Nadi dye_{ox}
BLUE

Figure 6–3. Cytochrome oxidase experiment. Colorless Nadi dye_{red} donates electrons to reduce cytochrome c. Cytochrome c in turn passes the electrons to cytochrome oxidase and then to oxygen. Blue Nadi dye_{ox} accumulates as long as electrons flow.

EXPLORING FURTHER

Lehninger, A. L. *Biochemistry*. 2nd ed. New York: Worth, 1975.

Scientific American Articles

Amerine, Maynard A. "Wine." August 1964 (#190).
Dickerson, Richard E. "Cytochrome c and the evolution of energy metabolism." March 1980 (#1464).
Hinkle, Peter C., and Richard E. McCarty. "How cells make ATP." March 1978 (#1383).

TOPIC 7

Photosynthesis

OBJECTIVES

When you have completed this topic, you should be able to:

1. Identify the pigments involved in photosynthesis by paper chromatography.
2. Give the function of chlorophyll and the accessory pigments.
3. Determine the wavelengths most useful for photosynthesis by determining the absorption spectrum of a pigment extract.
4. Distinguish between the electron transport and carbon fixation reactions of photosynthesis.
5. Demonstrate the effect of changing the level of light on electron transport as measured by O_2 production.
6. Demonstrate the effect of changing the level of light on electron transport by measuring reduction of a dye in the Hill reaction.
7. Explain why stimulating carbon fixation affects electron transport as measured by O_2 production.
8. Describe an experiment designed to measure the extent of carbon fixation as a function of light.

TEXT REFERENCES

Chapter 8.
Chloroplasts, electromagnetic radiation, photosynthetic pigments, photochemical reactions, Photosystems I and II, electron transport, chemiosmosis, photophosphorylation, carbon fixation.

Know boldface Terms

INTRODUCTION

The overall reaction of photosynthesis can be summarized as two closely linked processes. Light-dependent electron transport produces ATP and NADPH, which feed into the assembly line of the Calvin cycle where enzymes carry out **carbon fixation** by converting CO_2 to glucose. A simplified diagram of the process is shown in Figure 7–1.

PHOTOSYNTHETIC PIGMENTS: CHROMATOGRAPHY

If you have never carried out chromatography, turn back to Topic 1 and read the description of chromatography. You need to understand the following terms: **adsorption, matrix, chromatogram, origin, solvent front,** and ***R_f***.

In this experiment the matrix will be the cellulose of Whatman #1 filter paper, and the solvent is petroleum ether:acetone (9:1). Although substances do not have to be colored to be separated by chromatography, it is a little more dramatic when you can actually see the different substances as they pull apart to form distinct spots. You will begin by applying a green mixture of **PIGMENT** containing **chlorophylls** and **accessory pigments** to the paper, and will obtain a separation of the mixture into several distinct components at the end of the chromatography.

Preparing the Chromatogram

A correctly prepared chromatogram is shown in Figure 7–2.

☐ Obtain a rectangle of filter paper (about 10 × 15 cm), and lay it on a clean surface, *touching the paper as little as possible* and only at the edges.

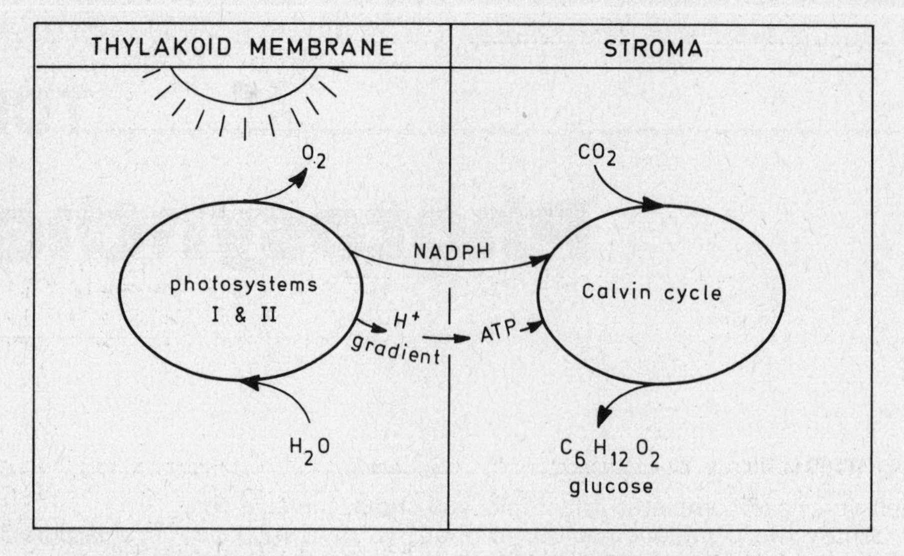

Figure 7–1. Photosynthesis. ATP synthesis depends on a hydrogen ion gradient setup as a result of electron transport. The Calvin cycle is tightly coupled to electron flow and quickly stops if light is not available.

□ Use a pencil to mark a straight line 1.5 cm from and parallel to the long edge of the paper; this is the origin.
□ Make three pencil dots on the origin that are at least 3 cm apart.
□ Place a support (such as a pencil) under the top edge of the paper so that the origin doesn't touch the surface.
□ Use a capillary pipet to spot **PIGMENT** onto the dots by touching the pipet to the paper as briefly as possible so that your spots will be quite small.

DO NOT SPOT CHLOROPLASTS OR CHLOROPHYLL EXTRACT ONTO THE PAPER. USE ONLY PIGMENT FOR THIS EXPERIMENT. SMALL SPOTS ARE BETTER THAN BIG SPOTS.

□ Use forceps to hold the paper with three wet spots in front of the blower until the spots are dry, or blow on them yourself. Touch the paper with your fingers as little as possible.
□ Again make a small spot on each dot with the capillary pipet, and dry.
□ Repeat until you have made a minimum of five applications on each dot.
□ Dry the chromatogram thoroughly before starting your chromatography.

Running the Chromatogram

□ Form your dry chromatogram into a cylinder, as shown in Figure 7–3, and staple it at the top and bottom so that the edges are close together but *do not touch or overlap.*
□ Obtain a chromatography jar and cover. Add enough solvent to the jar to just cover the bottom (about 20 mL). Replace the cover on the jar, and close the solvent stock bottle.
□ When the solvent has had a chance to evaporate and saturate the air in the chromatography jar, pick up your chromatogram with forceps, and set it gently in the center of the jar. It should stand up straight and *not touch the sides of the jar.*
□ Cover the jar and leave it undisturbed during the chromatography.
□ Check the jar every few minutes, and note the progress of the solvent as it wets the paper and moves toward the top.

DO NOT LET THE SOLVENT REACH THE TOP OF THE PAPER.

□ After about 20 min, the solvent will be close to the top of the paper, and you must immediately stop the chromatography. Use the forceps to remove the chromatogram. Replace the cover on the jar.
□ *Immediately* mark the **solvent front** (edge of the part wet by the solvent) with a pencil. You will not be able to calculate any R_f values if you fail to do this before the rapidly drying solvent dries.
□ Dry your chromatogram with the blower, or hang it up with clothespins for a few minutes before analyzing it.

Analyzing the Chromatogram

The number of spots that you observe will depend on the exact condition of the extract that you used. These are the major pigments in the extract:

Chlorophyll *a*: a blue-green pigment
Chlorophyll *b*: a yellow-green pigment
Carotenes: yellow-orange accessory pigments
Xanthophylls: yellow accessory pigments

□ Outline each visible spot with a pencil, and note its color on the chromatogram.
□ Make a dot with your pencil in the center of each spot. If the spot is irregular or band-shaped, use your judgment as to where the approximate center might be if it were a round spot.

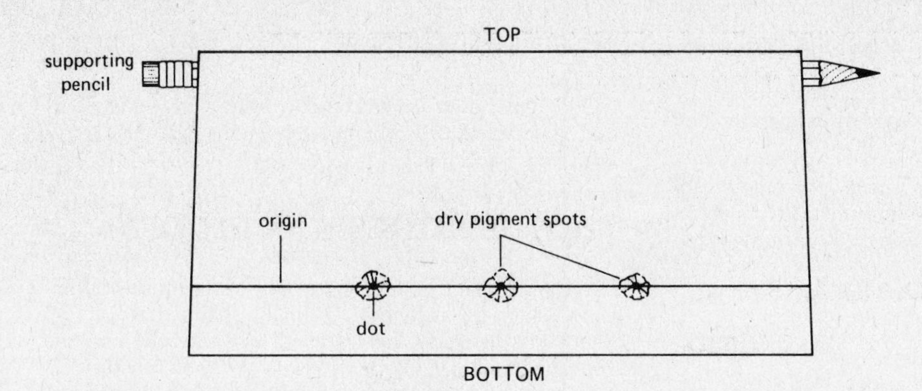

Figure . Chromatography otosynthetic pigments aper chromatogram aft tting three samples.

□ Take your chromatogram into a dark room, or put it into a darkbox, and view it under long-wavelength ultraviolet light.

NEVER LOOK DIRECTLY AT ULTRAVIOLET LIGHT. IT CAN DAMAGE YOUR EYES.

□ Note the bright pink **fluorescence** of chlorophyll.
□ Use the difference in fluorescence to help distinguish the chlorophylls and any other fluorescent compounds present. Outline the spots with pencil.
□ Note whether there are any dark areas that contain compounds that absorb UV light, but do not show fluorescence. Outline them with pencil.
□ Finally, carefully measure the distance from the origin to the solvent front, and record it in Table 7–1 at the end of this topic.
□ Measure the distance from the origin to the dot in the center of each spot observed in visible or UV light.
□ Calculate the R_f for each spot, and describe its appearance under visible and under UV light.
□ If you can, identify the major pigments, and record your R_f values for them in the table.
□ Tape the chromatogram into the space provided.

These pigments are located in the thylakoid membranes of the chloroplasts where they normally would absorb light, become chemica cited, and pass electrons to the electron carrier hotosystems I and II. When they are free, howev they are on the chromatography paper, there a molecules to accept the electrons, and the UV l xcited molecules release their energy as photons light during fluorescence. Only the molecules of n compounds are able to give up the energy as Those of other compounds lose it in the form of

PHOTOSYN ETIC PIGMENTS: ABSORPTI SPECTRUM

Only the ligh is absorbed by chloroplasts can be used in pho thesis. As you have already seen, the pigments of een plant look green to the eye because they perm en light to pass through, but absorb the red and light. The particular wavelengths of light that are **orbed** by a certain substance form a pattern ca **absorption spectrum.** The spectrum is detern by illuminating a solution of the substance with wavelength of light in turn, and measuring the abs n in each case. Using visible light, you will deter the visible spectrum for a mixture of pigments ex d from chloroplasts. The spectrum can then be

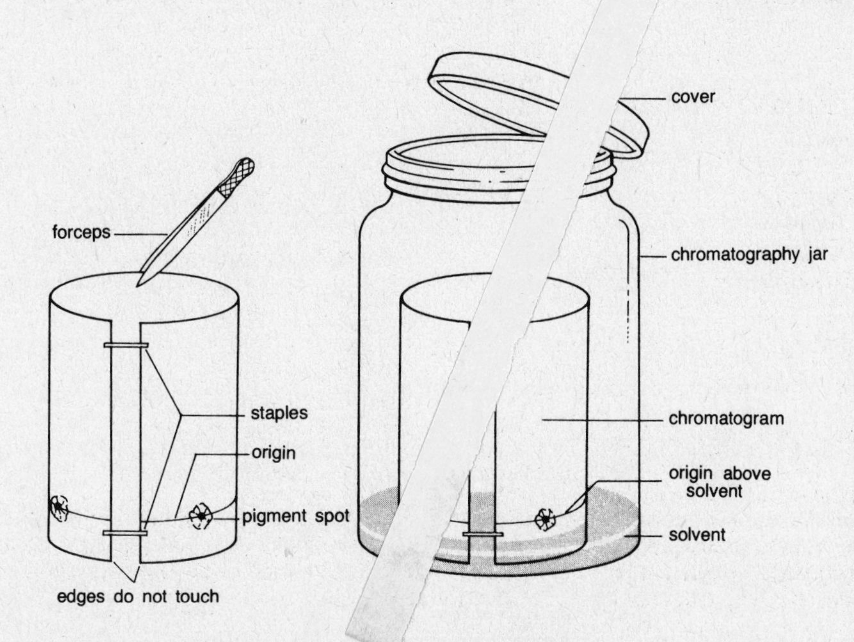

Figure 7–3. Chromatography of photosynthetic pigments. A paper chromatogram ready to run has been stapled so that its edges do not touch. The dry chromatogram should be put into the jar so that it does not touch the sides and the origin should be above the level of the solvent. Keep the jar covered.

plotted as a graph of absorbance versus the wavelength or color of light used.

In this experiment you will have to *be careful not to spill the solvent* on or into the spectrophotometer because it will dissolve the plastic.

□ Turn on the Spectronic 20 spectrophotometer to warm up for a few minutes while you turn back to Topic 2, and review the procedure for using this instrument (p. 10).
□ Fill a spectrophotometer cuvette halfway with 80% acetone to use as your blank.
□ Fill a second tube halfway with diluted **CHLOROPHYLL EXTRACT.**
□ Set the wavelength at 400 nm, and adjust the needle to ∞ absorbance with the left knob.
□ Insert your blank, and adjust the needle to zero absorbance with the right-hand knob.
□ Insert the tube containing chlorophyll, and take your first reading. It should be in the vicinity of 0.8. If it is not, ask your instructor for help. Record the absorbance in Table 7–2.

The zero and ∞ absorbance will change each time you adjust the wavelength, so you will have to reset the spectrophotometer with and without the blank for each of your readings.

□ Change the wavelength to 405, reset the spectrophotometer, and take your second reading; enter the absorbance in Table 7–2.
□ Continue your readings every 5 nm until the readings have become quite low. Then switch to every 10 or 20 nm.
□ When the absorbance starts to rise, switch back to taking readings every 5 nm, and continue until you reach 700 nm.

The purpose of taking the spectrum is to get the shape as accurately as possible, especially around the regions of greatest absorbance (low 400s and high 600s). When the absorbance rises and then falls again, the peak is called the **absorption maximum,** and the wavelength at which it occurs is used as a characteristic for identifying a compound. If you are not sure where at least two maxima are in your spectrum, go back to the regions of high absorbance and take some more readings.

□ Plot your data on the grid provided in Table 7–2, and label each absorbance maximum with its corresponding wavelength.

PHOTOSYNTHESIS IN *ELODEA*

In an intact plant both electron transport and carbon fixation will go on as long as the plant is illuminated, and it is possible to study the effect of light intensity and the concentration of CO_2 on the rate of photosynthesis. A simple way to measure photosynthetic rate is to observe the rate at which oxygen is produced as a result of splitting water. An actively photosynthesizing plant will produce an abundance of oxygen bubbles.

□ Obtain a healthy green sprig of *Elodea,* and measure back about 10 cm from the tip.
□ Wipe a sharp razor blade with alcohol, let it dry, and make a diagonal cut across the stem of the plant as gently as you can. Save the tip.
□ Put the tip into a test tube upside down so that the cut end of the stem is about 3 cm from the top of the tube.
□ Immerse the whole tube in the aquarium, or fill it with tap water until it is completely full and the cut stem is well under the surface of the water.
□ Place the tube 50 cm from the floodlamp, making sure that there is a container of water between the lamp and *Elodea* to act as a heat sink.

Your setup should look like that shown in Figure 7–4.

□ Wait a few minutes for the plant to become adjusted. Bubbles should appear at the cut end of the stem. If there are hardly any bubbles after 15 min, ask your instructor for help.
□ When there are two or more bubbles per minute, begin counting, and record the bubbles produced in a 3-min interval in Table 7–3.
□ Repeat for another 3-min interval. Calculate the average bubbles per minute for the light intensity at 50 cm.

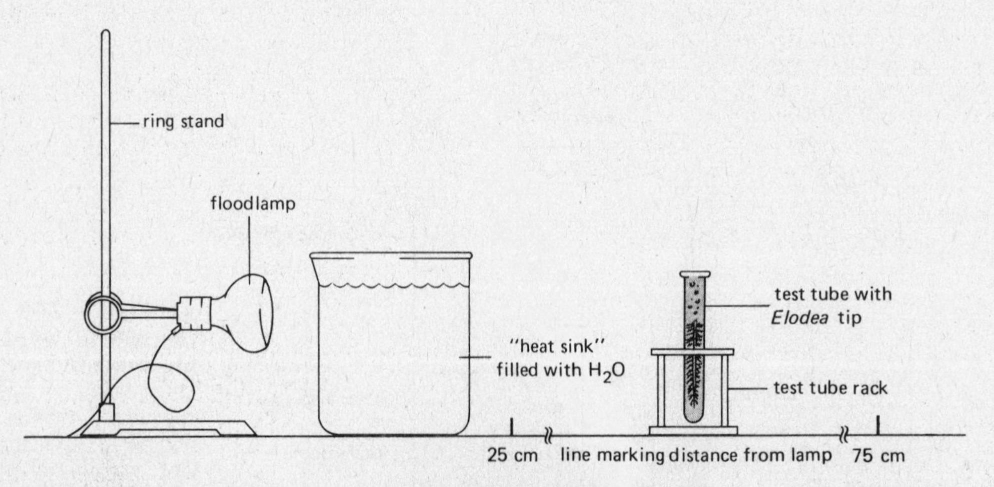

Figure 7–4. Experimental setup to measure oxygen production. *Elodea* can be exposed to different light intensities by moving it towards or away from the light source. The rate of oxygen evolution at different light intensities can be measured by counting the rate of bubble production. The "heat sink" must be used so that the water in the tube doesn't heat up and kill the plant.

□ Move the tube to 25 cm, and allow it to adjust. Repeat your measurements and record the results.
□ Now move the plant to 75 cm. Allow it to readjust, and repeat your measurements.

Did you see a different rate of oxygen evolution when the light intensity was increased or decreased?

increased

□ Replace the water in the tube with 0.5% $NaHCO_3$ (sodium bicarbonate, which will act as a source of CO_2 for the plant).
□ Place the plant back at 25 cm, and measure the rate of bubble release for two 3-min intervals.

Does the initial reaction of CO_2 in carbon fixation require light? NO

Is oxygen produced as a result of carbon fixation or of electron transport?

carbon fixation

Can electron transport continue to produce NADPH if there is no CO_2? Why or why not?

Can the Calvin cycle enzymes continue to fix CO_2 when the light is turned off? Why or why not?

Yes

THE HILL REACTION

In 1937 Robin Hill showed that chloroplasts continue to carry out electron flow and oxygen production in the absence of CO_2 as long as they are provided with an electron acceptor.

What is the fate of the electrons when CO_2 is present?

Hill used a dye that changes color when it becomes reduced, so that the rate of photosynthesis could be measured by observing the rate at which the dye changed color. In this experiment you will use a blue dye nicknamed DCIP (2,6-dichlorophenolindolphenol), which is blue when oxidized, and becomes colorless after it has been reduced:

$$DCIP_{ox} + 2e^- + 2H^+ \rightarrow DCIP \cdot H_{2red}$$

The experimental tube will contain "living" chloroplasts, buffer, and the blue dye, and will be held in the light.

What change will you look for in the experimental tube?

What should the control tube contain?

Should you hold it in the light?

The chloroplasts for this experiment are diluted in sucrose buffer so there are 100 μg of chlorophyll per millilitre, and they should be kept cold until you are ready to use them. Be sure to use the suspension labeled **CHLOROPLASTS.**

□ Set the wavelength of the spectrophotometer at 620 nm.
□ Obtain three spectrophotometer tubes and label them #1, #2, and #3.
□ Add these reagents in the order given (left to right) in the table below.
□ Cover the tubes with Parafilm or stoppers and mix the ingredients by inverting the tube several times.
□ Immediately cover the tubes with aluminum foil, and keep them covered except when they are being measured.
□ Adjust the spectrophotometer to ∞ absorbance, insert tube #1, and adjust the absorbance to 0.
□ Insert tube #2, and record its absorbance in Table 7–4. Cover the tube and wait 3 min.
□ Again read the absorbance of tube #2.

Did any change in the dye occur in the dark?

□ Choose one light intensity (1000, 1500, or 2000 foot-candles), or take the one assigned to you, and hold your tube at the appropriate position for exactly 30 sec. Cover the tube.
□ Read the absorbance; cover the tube again.
□ Hold the tube in the light for another 30-sec period, and again read the absorbance.
□ Continue until you have 12 readings.
□ Since the experiment is meaningless without the control, read the absorbance of tube #3 (no chloroplasts).
□ Hold the tube for 3 min at the same light intensity that you used in the experiment.
□ Read the absorbance again.
□ Graph your results in Figure 7–5 as percent DCIP reduced as a function of seconds exposed to the light. Label the curve with the light intensity that you used.

Join forces with students who used the other light intensities available and pool your results for comparison. Work together to answer the questions on the reverse of your graph (p. 60).

	Chloroplasts	Cold Buffer	Distilled Water	DCIP
Tube #1: blank	0.5 mL	3.0 mL	1.5 mL	—
Tube #2: reaction	0.5 mL	3.0 mL	0.5 mL	1.0 mL
Tube #3: control*	—	3.0 mL	1.0 mL	1.0 mL

*A tube identical to tube #2 but held in the dark would be an equally good control but would use up the chloroplasts.

CARBON FIXATION

Although carbon fixation can occur in the absence of light, it will quickly slow down and stop because the ATP and NADPH will soon be used up. A leaf or part of the plant that is not illuminated will obtain food from the other parts of the plant, but will not be able to carry on photosynthesis itself. Roots, tree trunks, and the other nongreen parts of plants must routinely obtain carbohydrate from the green photosynthetic parts. In this experiment you will study the effect of depriving part of a green leaf of light. The extent of carbon fixation will be measured as the amount of the storage carbohydrate, starch, that is formed.

□ Fold a piece of black paper in half, and cut out two copies of a small square, rectangle, or some other interesting shape.
□ Stick a label on one of the pieces, and write your initials and the date on it.
□ Use a paper clip to fasten the black shapes to the top and bottom surfaces of a leaf of a plant such as *Coleus* so that they are matched up exactly.
□ Place the plant in sunlight until the next laboratory period.

WAIT ONE WEEK.

□ Remove the leaf that was partially covered and take off the black paper.
□ Drop the leaf into a beaker of boiling 95% alcohol in the hood, and boil it until it turns white.
□ Remove the leaf, and place it in a finger bowl.
□ Flood the leaf with iodine solution, and let it absorb the iodine for a few minutes.
□ Rinse the leaf with tap water, and observe which parts stain darkly.

What happens when the leaf is boiled in alcohol?

____

____

What part of the leaf stained dark? ______________

What carbohydrate causes the dark staining?

____

Why does carbon fixation occur only in the light?

____

____

If you tested the leaf for glucose, would you expect to find any in the unstained part of the leaf? __________

When you finish your work, discard any debris, and clean your glassware thoroughly.

EXPLORING FURTHER

Baker, Jeffrey, J. W., and Garland E. Allen. *Matter, Energy and Life. An Introduction to Chemical Concepts.* 4th ed. Reading, Mass.: Addison-Wesley, 1981.

Galston, A. W., P. J. Davies, and R. L. Satter. *The Life of the Green Plant.* 3rd ed. Englewood Cliffs, N.J.: Prentice-Hall, 1980.

Lehninger, A. L. *Biochemistry.* 2nd ed. New York: Worth, 1975.

Scientific American Articles

Björkman, Olle, and Joseph Barry. "High-efficiency photosynthesis." October 1973 (#1281).

Govindjee, and R. Govindjee. "The primary events of photosynthesis." December 1974 (#1310).

Miller, Kenneth R. "The photosynthetic membrane." October 1979 (#1448).

Stein, William H., and Stanford Moore. "Chromatography." March 1951 (#81).

Stoeckenius, Walther. "The purple membrane of salt-loving bacteria." June 1976 (#1340).

Woodwell, George M. "The carbon dioxide question." January 1978 (#1376).

Student Name ______________________ **Date** ____________

TABLE 7–3. DATA ON OXYGEN EVOLUTION BY ***ELODEA***

TAP WATER

Distance	Bubbles/3 min	Avg. Bubbles/min
25cm	3	1
25 cm	3	1
50 cm	1	.33
50 cm	2	.7
75 cm	1	.33

$NaHCO_3$

Distance	Bubbles/3 min	Avg. Bubbles/min
25	15	5
25	15	5
50	6	2

SUMMARY

Treatment	Distance	Avg. Bubbles/min
None	75 cm	.33
None	50 cm	1
$NaHCO_3$	25 cm	5
None	25 cm	1

What effect did decreasing light (increasing distance) have on the rate of photosynthesis?

The rate of photosynthesis went down

What effect did increasing CO_2 ($NaHCO_3$) have on the rate of photosynthesis?

The rate went up

TABLE 7–4. DATA FROM THE HILL REACTION

1. The average initial absorbance of tube #2 was _______________.
2. Calculate the change from this value for each reading by subtracting the **new** absorbance from the **initial** absorbance.
3. Calculate the % DCIP reduced = (change × 100)/initial.

TUBE #2 REACTION (extra space provided)

Seconds of Light	(1) Absorbance	(2) Change in Absorbance	(3) % DCIP Reduced
0 (initial)		0	0
0 (3 min dark)			
30			
60			

TUBE #3 CONTROL

Seconds of Light	(1) Absorbance	(2) Change in Absorbance	(3) % DCIP Reduced
0		0	0
180			

*For more accurate results, this value should be determined for each point, and subtracted.

Student Name ____________________ **Date** ____________

HILL REACTION AT A LIGHT INTENSITY OF __________ FOOT-CANDLES

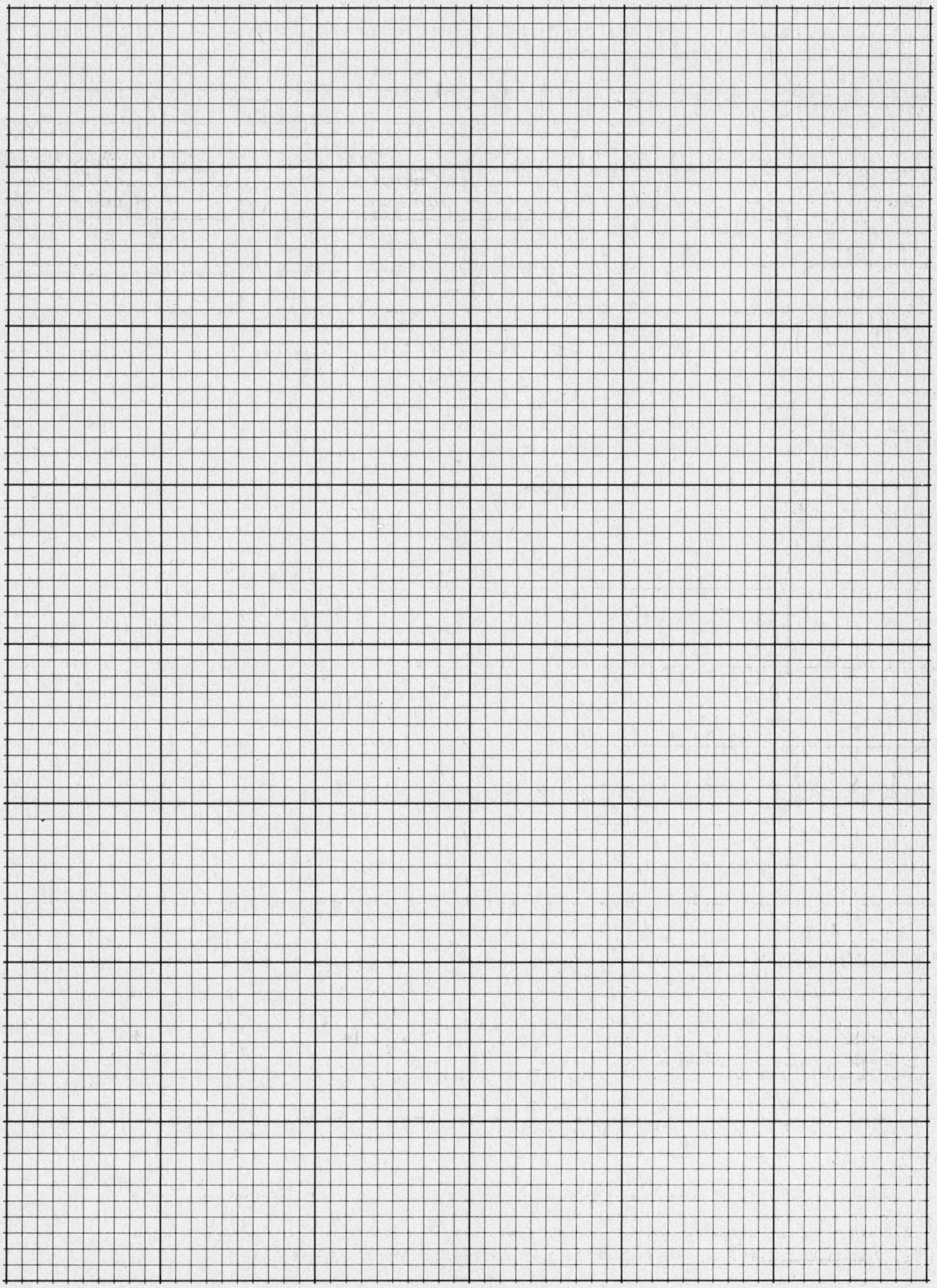

Figure 7–5. Hill reaction. Plot the percent DCIP reduced as a function of seconds exposed to light. Be sure to label both axes of your graph.

Compare your curve with those of other class members who used different light intensities. Your instructor may ask you to draw additional curves on your own graph.

Did the rate of dye reduction depend on the intensity of the light? ________________

Did the lowest light intensity used still show more dye reduction than seen in the control? ________________

If the light intensity could be increased even more without heating up the chloroplasts, would the rate of photosynthesis keep increasing indefinitely? Explain your answer.

____

____

TOPIC 8

Cell Reproduction

OBJECTIVES

When you have completed this topic, you should be able to:

1. Describe the four stages of the cell cycle.
2. Draw the structure of a chromosome before replication and after replication.
3. Distinguish between sister chromatids and homologous chromosomes.
4. Describe what you look for in the microscope during interphase and in each stage of mitosis.
5. Identify in the microscope which stage of mitosis an animal or plant cell has reached.
6. Describe the main differences between mitosis in a plant cell and in an animal cell.
7. State when and in which tissues meiosis occurs.
8. Use models to show how meiosis differs from mitosis in terms of chromosome movement.
9. Explain in terms of DNA synthesis, synapsis, crossing over, and cytokinesis how meiosis differs from mitosis.
10. Identify the haploid and diploid stages of meiosis in a diploid organism such as yourself.

TEXT REFERENCES

Chapter 11; Chapter 36 (E,F).
Chromosomes, haploid and diploid cells, cell cycle, mitosis, meiosis, crossing over, spermatogenesis, oogenesis, fertilization.

Know bold face terms

INTRODUCTION

Cells come into existence through the division of their parent cells, and most cells divide in turn to produce daughter cells. Usually this occurs during **mitosis,** when the genetic material is duplicated, and one copy is passed on to each daughter cell. Mitosis is generally followed by **cytokinesis,** or cytoplasmic division, in which the rest of the cell divides in half to form two new cells. Sometimes cytokinesis does not occur, and a cell with many nuclei may form. In animals such a cell is called a **syncytium,** and in plants it is termed a **coenocyte.**

Mitosis is part of the cell cycle. Trace one turn of the cycle in Figure 8–1. During G_1 (first gap) the decision to begin the process of cell division is made. DNA replication occurs in the S (synthesis) period. At this time each chromosome doubles so that it consists of two **sister chromatids** joined by the **centromere.*** Although the chromosomes are uncoiled and therefore invisible in the light microscope at this time, chemical techniques can be used to show that the chromosomes are replicating. The replicated chromosomes that appear at the beginning of mitosis are double-stranded; one strand enters each of the daughter cells. The two daughter cells produced by mitosis are exactly equivalent to the parent cell, and can be thought of as carbon copies.

In animals and plants that reproduce sexually, two sex cells or **gametes** fuse during fertilization to produce the **zygote.** The zygote will always have twice as much genetic material as the gamete. Therefore, at some stage before the next fertilization, the genetic material must be reduced by half so that the amount of nuclear material stays the same from one generation to the next. This reduction in nuclear material occurs through the process of **meiosis.** Only germ cells carry out meiosis. The gametes resulting from this process contain only one-half of the genetic material of their parent cells.

In this exercise you will study both mitosis and meiosis. It is essential that you thoroughly understand both processes and the differences between them before undertaking the topic genetics.

*The centromere is sometimes called the kinetochore. The two terms are equivalent, but only the former will be used here for consistency.

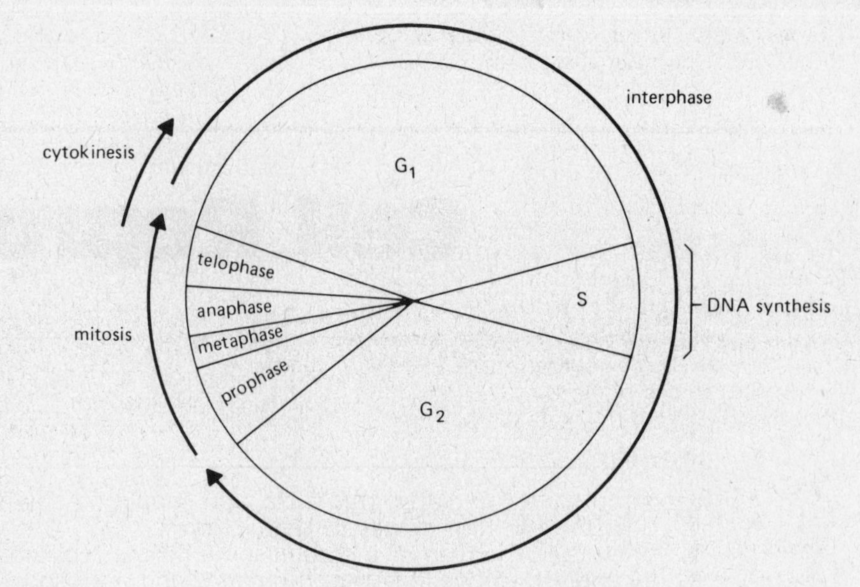

Figure 8–1. The cell cycle. The decision to divide occurs in G_1, the chromosomes are replicated in S, the preparations for cell division are made in G_2, the nucleus divides in mitosis, and the cell cytoplasm divides in cytokinesis.

MITOSIS IN AN ANIMAL CELL

An animal cell about to undergo mitosis has already replicated its DNA during the S period of interphase.

How many strands does each chromosome have? 2 2 strands

What are the strands called? chromatids

The **centrioles,** pairs of organelles, found only in animal cells, have also replicated. The daughter centrioles move to the opposite ends or **poles** of the cell, and their position determines how the chromosomes will move during mitosis. The chromosomes always move toward the centrioles, and the cell always divides in a plane perpendicular to a line joining them.

Centrioles have an interesting structure closely related to that of cilia and flagella. They are little cylinders made of nine triplets of microtubules, and each centriole is made of two such cylinders at right angles. They determine the orientation of the **mitotic spindle,** which is also composed of microtubules and extends from one centriole to the other. During mitosis, the chromosomes move along fibers in the mitotic spindle, and gather at the poles of the cell. The position of the centrioles determines the orientation of the **astral rays,** microtubules extending out from the centrioles in all directions. The function of these astral rays is not known.

The centrioles, mitotic spindle, and astral rays are collectively called the **mitotic apparatus,** and take up much of the interior of the cell when it is in mitosis. Mitosis requires a great deal of energy, so there are many mitochondria associated with the mitotic apparatus to provide ATP for synthesis and chromosome movement.

□ View a prepared slide of animal cells undergoing mitosis, such as the cells in whitefish embryos, first under low, and then under high power in your microscope.
□ Locate cells that are undergoing mitosis, some of which have large, prominent mitotic spindles and condensed chromosomes.
□ Using the following descriptions and Figure 8–2 as your guide, locate cells which represent interphase, each of the stages of mitosis and cytokinesis.

Interphase

During interphase, the genetic material is in the form of greatly extended fibers of DNA that form a tangled mass called **chromatin.** The fibers are too small to be seen in the microscope so the nucleus looks homogeneous. The nucleus is bounded by the double **nuclear membrane,** and contains **nucleoli,** which are centers for the synthesis of RNA. They may not be visible in every cell.

How many nucleoli do you see in the interphase cells? one

Prophase

As the cell enters prophase, the chromatin starts to condense into discrete chromosomes, the nucleoli disappear, and the nuclear membrane breaks down. By the end of prophase, the centrioles have migrated to opposite poles, the chromosomes are moving to the center of the cell, and the mitotic spindle starts to appear.

Metaphase

The spindle forms, and the chromosomes take up positions on the **metaphase plate,** an imaginary plane in the middle of the cell halfway between the two poles. They are located on the outside of the spindle, so they would be arranged in a circle if you viewed them from one of the poles. Each double-stranded chromosome is attached to a **spindle fiber** at its centromere, and its flexible arms are free to move about.

Anaphase

Metaphase ends and anaphase begins when the centromere of each chromosome divides. Each sister chromatid now has its own centromere, and is called a chromosome, so that there are twice as many chromosomes as there were during metaphase.

If a certain cell has three pairs of chromosomes (N = 3), how many chromosomes does the cell have during metaphase? N=6

How many chromosomes does it have during anaphase? 3

Spindle fibers, attached to the centromeres, cause chromosomes to move, with the arms of the chromosomes trailing passively behind. A chromosome without a centromere will not move during anaphase, and is randomly left behind in one of the two daughter cells.

Telophase

The chromosomes gather at the poles in telophase, and begin to form the daughter nuclei. Since each original chromosome in the cell contributed a chromosome identical to itself to each new nucleus, the two daughter nuclei will be identical to the nucleus of the parent cell. The nuclear membrane begins to reform, the nucleoli reappear, and the chromosomes begin to unwind and disappear from view. At the end of telophase, the nuclei look like interphase nuclei, and in fact have entered interphase of the next cell cycle.

Cytokinesis

Cytokinesis, or division of the cytoplasm, usually accompanies telophase. In animal cells the membrane constricts in a ring around the middle of the cell, and eventually pinches the cell in two. In embryos this ring is called the **cleavage furrow.** The cytoplasm and all its constituents are passively divided so that each cell gets about half the materials and organelles in the cell.

Why couldn't the chromosomes be passively divided when the cell pinches in two?

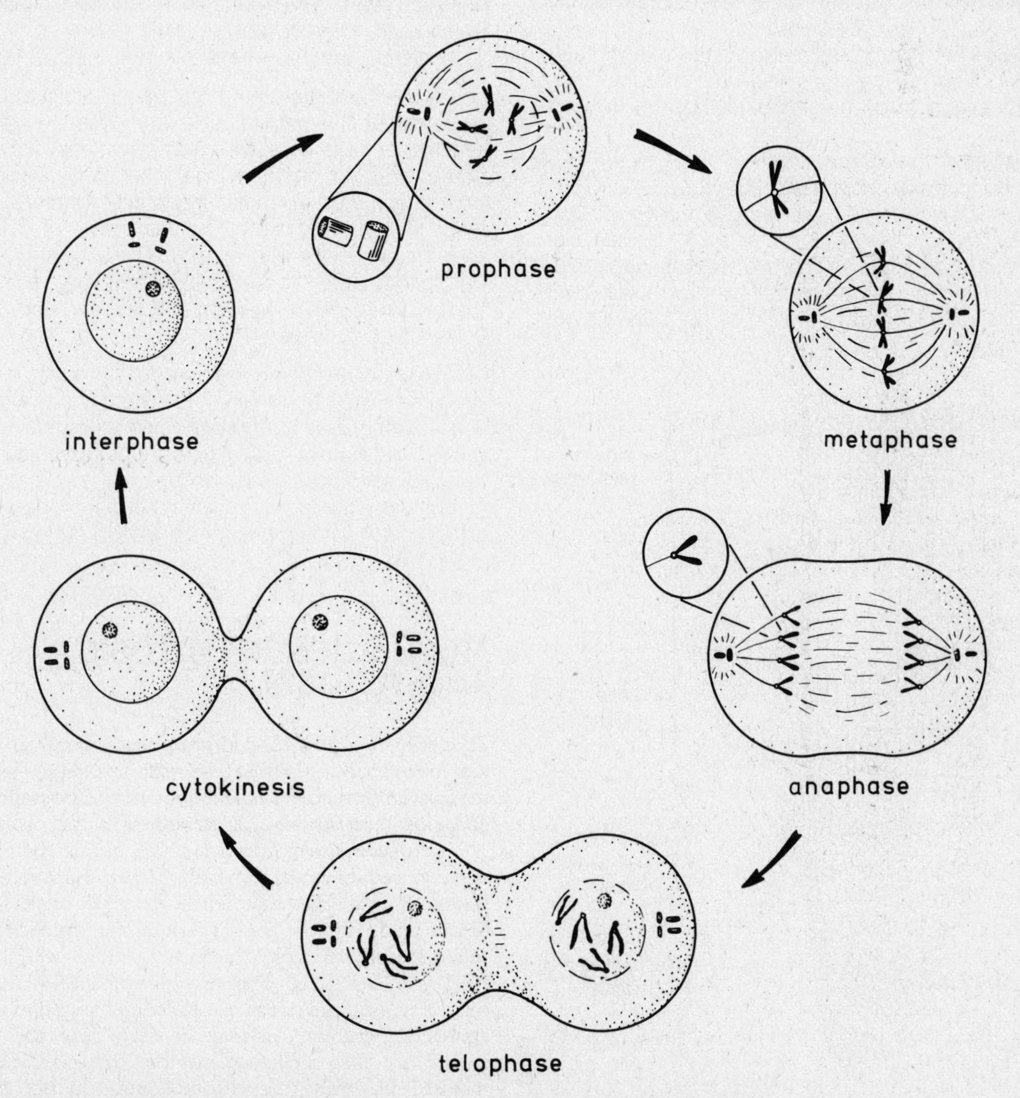

Figure 8–2. Mitosis. The stages of mitosis in an animal cell. Since mitosis is a continuous process, each stage blends into the next. Note the centrioles and the astral rays not found in plant cells.

MITOSIS IN A PLANT CELL

□ View a prepared slide of plant cells undergoing mitosis, such as the root tip cells of an onion, first under low and then under high power. In the onion root tip, look at the cells near, but not at, the very tip of the root in the region marked "X" in Figure 8–3. Avoid the cells in the very center.
□ Locate cells in interphase, each stage of mitosis, and in cytokinesis. If the section does not happen to pass through the nucleus of a cell, that cell will appear to have no nucleus, and should be skipped.

Note the following differences between mitosis in the plant cells compared with mitosis in the animal cells that you have seen:

1. **Interphase.** The nuclei look similar to the animal cell nuclei, but may have a different number of nucleoli.

How many nucleoli do the plant cells have?

____

2. **Prophase.** There are no centrioles in plant cells, and you will not see any astral rays. Otherwise the events of prophase are similar.
3. **Metaphase.** Events are similar to those in the animal cell. You will see the spindle but no astral rays.
4. **Anaphase.** Events are similar to those in the animal cell.
5. **Telophase.** Events are similar to those in the animal cell. In plants, telophase cells can be easily distinguished because you will see two nuclei within a cell that is about the same size as the other cells that have only one nucleus. The two new daughter cells together are the same size as the parent cell, and look like mirror images. Growth of the cells will occur later on.

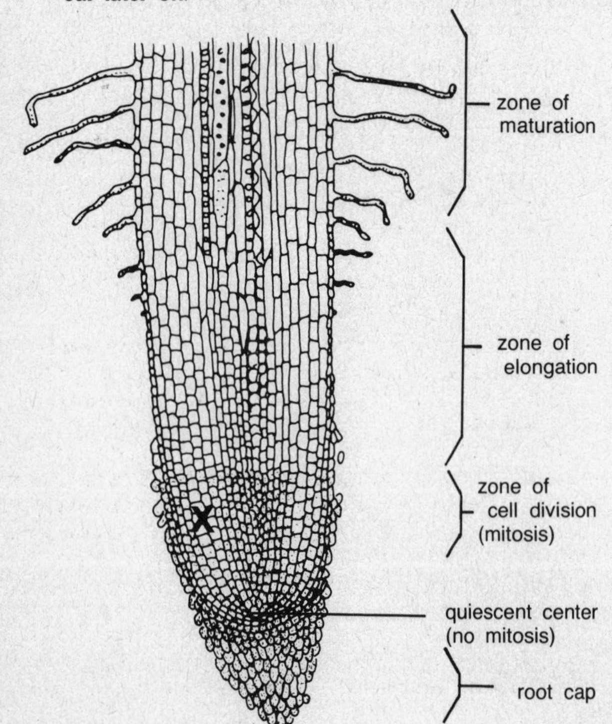

Figure 8–3. Onion root tip. Look at the position marked X to find dividing cells.

6. **Cytokinesis.** This process is completely different in plant cells. The parent cell does not pinch in or change shape at all. Instead, membrane vesicles form in the metaphase plate, and coalesce to form a cell plate, or **middle lamella,** dividing the two daughter cells. Cellulose is laid down on either side of this plate to eventually form cell walls between the two new cells. Each cell then begins to elongate in its long dimension until it is as big as the parent cell, and can enter another cycle of mitosis.

Working with your partner, determine the stage of mitosis in 100 cells, and record your results in Table 8–1 at the end of this topic.

□ Move your slide until you have the field filled with an area where there are many cells undergoing mitosis (you should be viewing under high power).
□ Place the pointer on a vertical row of cells, or make a mental note as to which row you are going to count.
□ Indicate the stage of each cell in turn to your partner as you go down the row. Do not leave out any cells unless there is no nucleus visible at all.
□ As each cell is staged, your partner should make a check next to that stage in the table.
□ When you have counted about 50 cells, change places, and let your partner stage the rest.
□ Continue until you have counted exactly 100 cells.

It is known that onion root tip cells take about 16 hr to complete the mitotic cycle. By determining the percentage of cells in each stage of mitosis and in interphase, you can calculate the amount of time spent in each stage. For example, if 10 cells out of 100 were found to be in prophase, the percentage of cells in prophase is $10/100 \times 100 = 10\%$. This shows that any one cell spends 10% of the time in prophase, so it spends 0.10×16 hr or 1.6 hr (1 hr and 36 min, since 0.1 hr = 6 min) in that stage.

□ Calculate the percentage of cells in each stage, and enter the results in Table 8–1. (There is no calculation if you counted exactly 100 cells; just enter the number of cells.) You may be asked to use the combined totals of the class for your calculation.
□ Calculate the hours and minutes spent in each stage, if the entire cell cycle takes 16 hr (0.1 hr = 6 min).

MECHANICS OF MITOSIS AND MEIOSIS

To clarify the difference in chromosome movement during mitosis and meiosis, you will use pipe cleaners of various lengths to represent the sister chromatids of the chromosomes, and beads to represent the centromeres.

In **diploid** organisms such as human beings, there are two sets of chromosomes (2N) in the adult. Meiosis produces cells that have only a single set of chromosomes (N), which are consequently **haploid.** Thus meiosis converts diploid cells to haploid cells. In Figure 8–4 you see a cell with two chromosomes undergoing mitosis compared with a cell containing two chromosomes undergoing meiosis. In each case the chromosomes are first replicated so that they are double. In meiosis, however, the chromosomes do not act independently but join together in a **homologous pair** during meiosis I, the first meiotic division. In a cell with

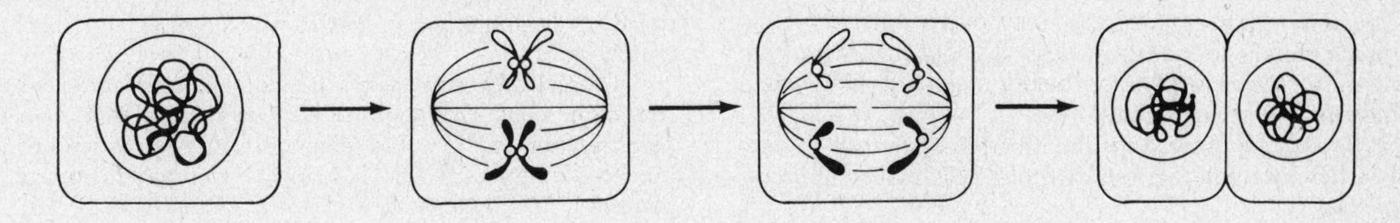

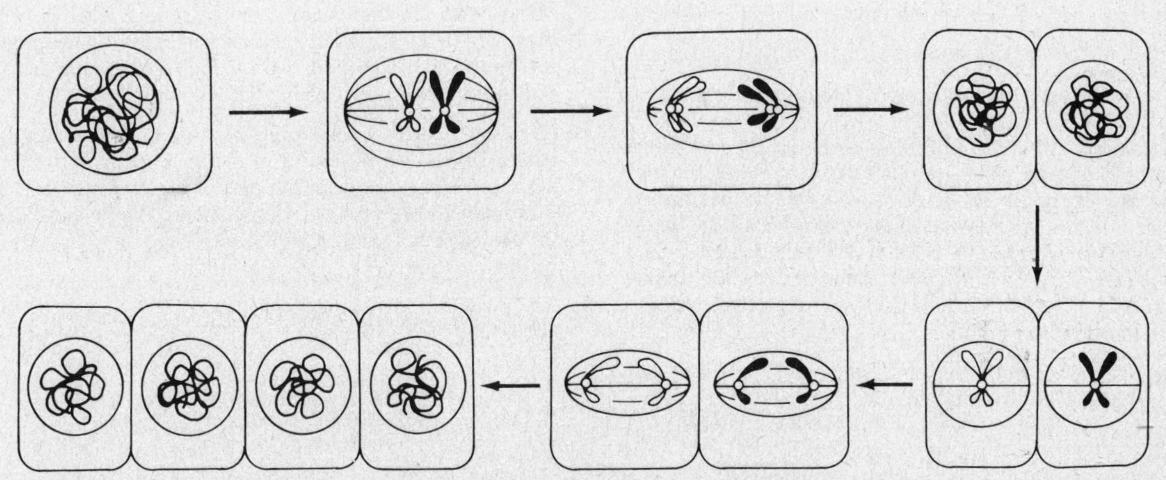

Figure 8–4. Mitosis and meiosis compared. A cell with 2 chromosomes (2N = 2) undergoes mitosis and meiosis very differently. Mitosis produces two daughter cells that are identical to the diploid parent, whereas meiosis produces four haploid cells that are genetically different from the parent cell.

many chromosomes, all of them would join in homologous pairs. During the first meiotic division, the two members of the pair separate, but the chromosomes do not split in half as they do in mitosis, and they remain double-stranded. Thus there is no need for DNA synthesis before meiosis II. The two sister chromatids of each chromosome separate in meiosis II, so that each daughter cell receives one single-stranded chromosome. The four daughter cells produced when a cell undergoes meiosis are usually not identical, in contrast to the carbon copy daughters produced during mitosis. These four daughter cells are haploid (N) rather than diploid (2N), and contain only one set of the genetic material.

□ Insert two identical pipe cleaners into a bead to construct a chromosome that has already replicated.
□ Make a second, **homologous chromosome** from pipe cleaners the same length as those in the first chromosome. The second chromosome does not need to be the same color as the first, but it must be the same length, and its sister chromatids must be identical to each other (they are carbon copies).
□ Turn to Figure 8–7 at the end of this topic, which shows parent and daughter cells in metaphase, and place each chromosome on a *separate* spindle fiber, as it would appear in mitosis.
□ Take another bead for each chromosome to represent its second centromere.
□ Separate the sister chromatids and give each one a centromere so that it is a new full-fledged chromosome.
□ Separate the two daughter chromosomes, placing one in each daughter cell, and repeat the process for the second homologous chromosome.

Are the chromosomes in the daughter cells genetically identical to those that you started with except that they have not yet replicated? ______________

If not, ask your instructor for help!

During mitosis, each chromosome acts independently, and there is never any pairing of homologous chromosomes. During meiosis, there is **pairing of homologous chromosomes** and the daughter cells formed are *not identical to the parent cell.*

□ Form a homologous pair of replicated chromosomes that are the same length, but different colors. The sister chromatids of each chromosome must be identical in length *and* color.
□ Place the chromosomes in the center of the spindle side by side on the same spindle fiber. The two chromosomes are in **synapsis,** which is a close side-by-side position, and their four strands form a **tetrad** or four-stranded structure. Synapsis is the key event in meiosis and will result in **genetic variation** in the daughter cells.
□ Move one whole chromosome to each of the daughter cells. You do not need more beads because the chromatids *do not separate* during the first meiotic division.

How many chromosomes does each daughter cell have?

____

□ Separate the sister chromatids of one chromosome, and give each one a bead to represent its centromere. The new chromosomes can now move apart into the daughter cells formed in meiosis II. Note that there was *no chromosome replication* before meiosis II. Otherwise, meiosis II is exactly like mitosis.

□ Repeat the division for the other daughter cell so that you have formed the four daughter cells that result from meiosis II.

How many chromosomes are there in each cell resulting from meiosis? ______________

Are these cells haploid or diploid? ______________

What process will restore the diploid number of chromosomes in the next generation? ______________

□ Starting with four replicated chromosomes, carry out a mitotic division.

□ Starting with four replicated chromosomes, in which the homologous pairs are different colors, and in which one pair is longer so that you can tell the pairs apart, carry out the two meiotic divisions. Make a note of the chromosomes found in each of the four daughter cells.

□ Arrange the chromosomes differently on the spindle in meiosis I so that the products after meiosis II will now be different than before.

How many different types of daughter cells can you create using both arrangements?

____

□ In Figure 8–7 draw in the chromosomes of a cell undergoing meiosis. Start with four chromosomes (2N = 4), two long and two short, and draw one member of each homologous pair with dotted lines for chromatids, and the other with solid lines.

MEIOSIS IN AN ANIMAL CELL

In animals, meiosis occurs only in the gonads.

Where does meiosis occur in a woman? ______________

What is this process called?

____

Where does meiosis occur in a man? ______________

What is this process called?

____

In males, the sperm are formed much as you illustrated in the section on mechanics. Each cell undergoing **spermatogenesis** will produce four sperm cells. Note that each sperm will have a genetically identical twin. In **oogenesis,** however, a single, large egg cell is produced which carries the energy stores for the development of the zygote after fertilization. Nuclear division occurs in the same way, but cytoplasmic division is very unequal. A nucleus with a small amount of membrane and cytoplasm is discarded after meiosis I as the **first polar body.** A **second polar body** is discarded after meiosis II. If the first polar body undergoes division, there may be two or three polar bodies stuck onto the outside of the **ovum,** or egg cell. Thus, in the female meiosis gives rise to a single gamete, the ovum, and polar bodies that have no function and degenerate. Figure 8–5 illustrates the processes of oogenesis and spermatogenesis.

The roundworm *Ascaris* has large, easily removed ovaries in which the stages of meiosis can be seen. This parasitic organism is easily obtained from the intestine of horses. (Another species of *Ascaris* is a common human parasite in warm climates, and it is a serious human pest that sometimes causes intestinal blockage and lung infection.)

In *Ascaris,* as in many organisms, meiosis is not completed until after fertilization. Then meiosis I occurs, followed by meiosis II, fusion of the egg and sperm nuclei, and the first division of the zygote. Cells in progressively later stages of meiosis and early development are found in section of the oviduct and uterus farther and farther from the ovary, (see Figure 8–6).

□ Study meiosis in the prepared slide, noting that the earlier stages will be found in the sections at the top of the slide, which are from the part of the oviduct closer to the ovary. A set of slides rather than a single slide may be provided for your study of meiosis.

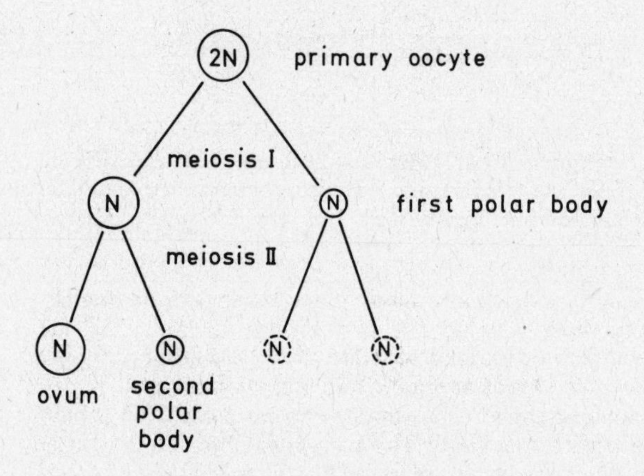

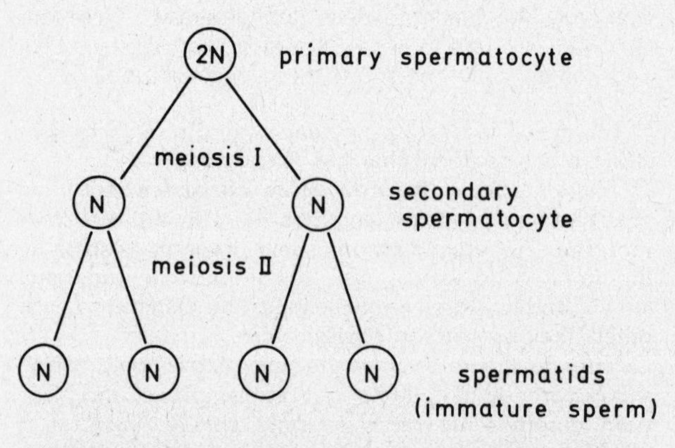

Figure 8–5. Oogenesis and spermatogenesis compared. Oogenesis produces one ovum or egg cell but spermatogenesis produces four sperm cells for each cell entering meiosis.

Sperm Cell

These triangular-shaped cells are small in relation to the egg cells of the female. The top of the sperm cell must enter the egg cell before meiosis can begin.

□ Locate a free sperm cell.

Primary Oocyte (or Future Egg Cell)

This is the cell that will undergo meiosis after the sperm cell penetrates. It is a diploid cell, and has no shell around it.

□ Locate a primary oocyte.
□ Locate a primary oocyte into which a sperm has penetrated. The sperm head becomes modified into a sphere once it enters the oocyte.

Meiosis I

In *Ascaris* there are four chromosomes (2N = 4).

How many pairs of chromosomes will there be in meiosis I? ___2___

Each pair of chromosomes forms a **tetrad** as the homologous chromosomes undergo **synapsis.** The tetrads move to one edge of the cell where the spindle forms during prophase I and metaphase I. This is the only time during the entire life cycle of a sexually reproducing organism that the members of a homologous pair of chromosomes come together. While the chromosomes are in synapsis, and the four sister chromatids are close together in the tetrad, sister chromatids of the two chromosomes may exchange parts of their arms in the process of **crossing over.** There is no gain or loss of genetic material, but the genes on one chromatid are exchanged one-for-one with those on the other chromatid. The shuffling of genes during crossing over results in more variety among the gametes than would otherwise occur. When crossing over is going on, there will be crosslike structures visible in the microscope, called **chiasmata.** Special techniques are required in order to see them.

□ Locate a cell in metaphase I in which there are prominent tetrads. The sperm nucleus may or may not be visible in the section of cell that you are viewing, but there cannot be a polar body on the cell surface.

The pairs of chromosomes separate during anaphase I, and the two daughter nuclei form during telophase.

Cytokinesis I

During this process, a piece of membrane enclosing one daughter nucleus pinches off to form the first polar body. The large cell is now called the **secondary oocyte.**

□ Locate a secondary oocyte that has a polar body attached to its surface.

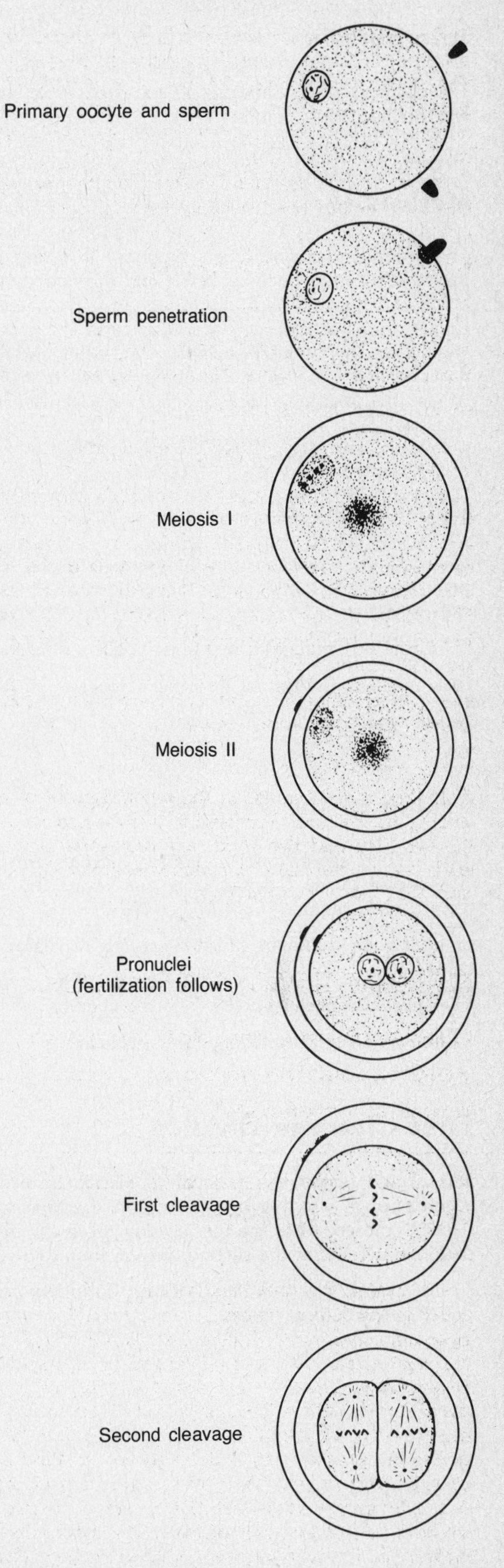

Figure 8–6. Meiosis in *Ascaris*. Fertilization precedes meiosis in this species.

Interkinesis

The cell prepares for meiosis II, and there is *no replication* of the chromosomes.

Meiosis II

The secondary oocyte enters prophase II during which the chromosomes become visible and the nuclear membrane, if it has reformed, disappears. In metaphase II the spindle forms at the cell surface, and the chromosomes line up at the metaphase plate. The sister chromatids separate and become chromosomes as the centromeres divide in anaphase II. Daughter nuclei reform in telophase II.

Cytokinesis II

One daughter nucleus is pinched off to form the second polar body, and the large cell remaining is called the egg cell, or ovum.

□ Locate an ovum with two polar bodies on its surface.

How many chromosomes does the nucleus of an *Ascaris* ovum have? ______________

Fusion of the Pronuclei

The haploid egg nucleus and sperm nucleus are called **pronuclei.** They become large and clear, and move toward each other.

□ Locate an ovum in which there are two large pronuclei.

The pronuclei finally fuse to form the zygote.

□ Locate a zygote with one large nucleus.

First Cleavage Division

The zygote begins the process of mitotic division, or **cleavage,** in which it begins to form the multicellular *Ascaris* embryo. After the first division there will be two daughter cells, after the second division four, and so on.

□ Locate a zygote that has divided to form two cells of equal size with large nuclei.

How many chromosomes does each of these cells have?

____

SPERM CELLS

Your instructor will indicate the source of the living sperm that you will study. It may be from a bull, frog, or another source.

If frog sperm is used, you may be asked to remove the testes from the frog. Instructions for pithing the frog are given in Topic 23, and a diagram of the internal organs can be found in Topic 19.

□ Make a wet mount of sperm suspension or a bit of tissue from the frog testis teased apart in Ringer's solution, and examine it under low, then high power in the microscope.

Are the sperm cells motile or immotile?

____

What is the shape of the sperm cells?

____

Why does a sperm cell contain many mitochondria?

____

____

____

Sperm are produced by maturation of spermatids within the seminiferous tubules of the testis. In spermatogenesis, haploid spermatids are produced from diploid spermatocytes (see Figure 8–5).

When you are finished with your work, return all prepared slides, clean up your work area, and store your microscope.

EXPLORING FURTHER

Giese, A. C. *Cell Physiology.* 5th ed. Philadelphia: W. B. Saunders, 1979.

Sloboda, Roger D. "The role of microtubules in cell structure and cell division." *American Scientist* 68:290, 1980.

Scientific American Articles

Lazarides, Elias, and Jean Paul Revel. "The molecular basis of cell movement." May 1979 (#1427).

Mazia, Daniel. "The cell cycle." January 1974 (#1288).

Student Name ______________________________ **Date** ________________

TABLE 8–1. DATA ON MITOTIC STAGES IN PLANT CELLS

Make a check in the appropriate box for each cell staged. Then total the checks.

Number of Cells in Stage	Total	% in Stage	Hr and Min in Stage
Interphase:			
Prophase:			
Metaphase:			
Anaphase:			
Telophase:			

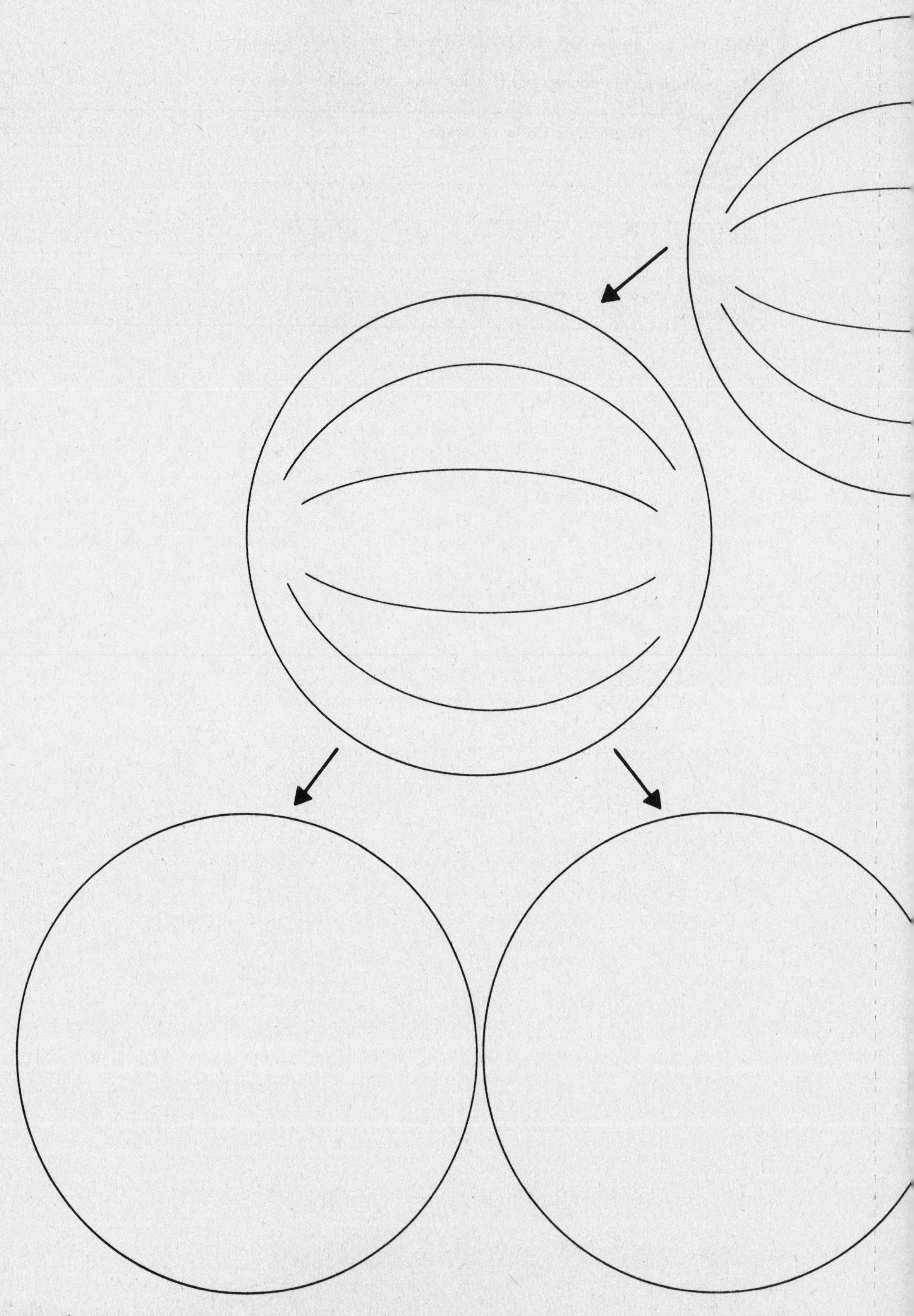

Figure 8–7. Diagram of cell reproduction.

Student Name ______________________ **Date** ______________

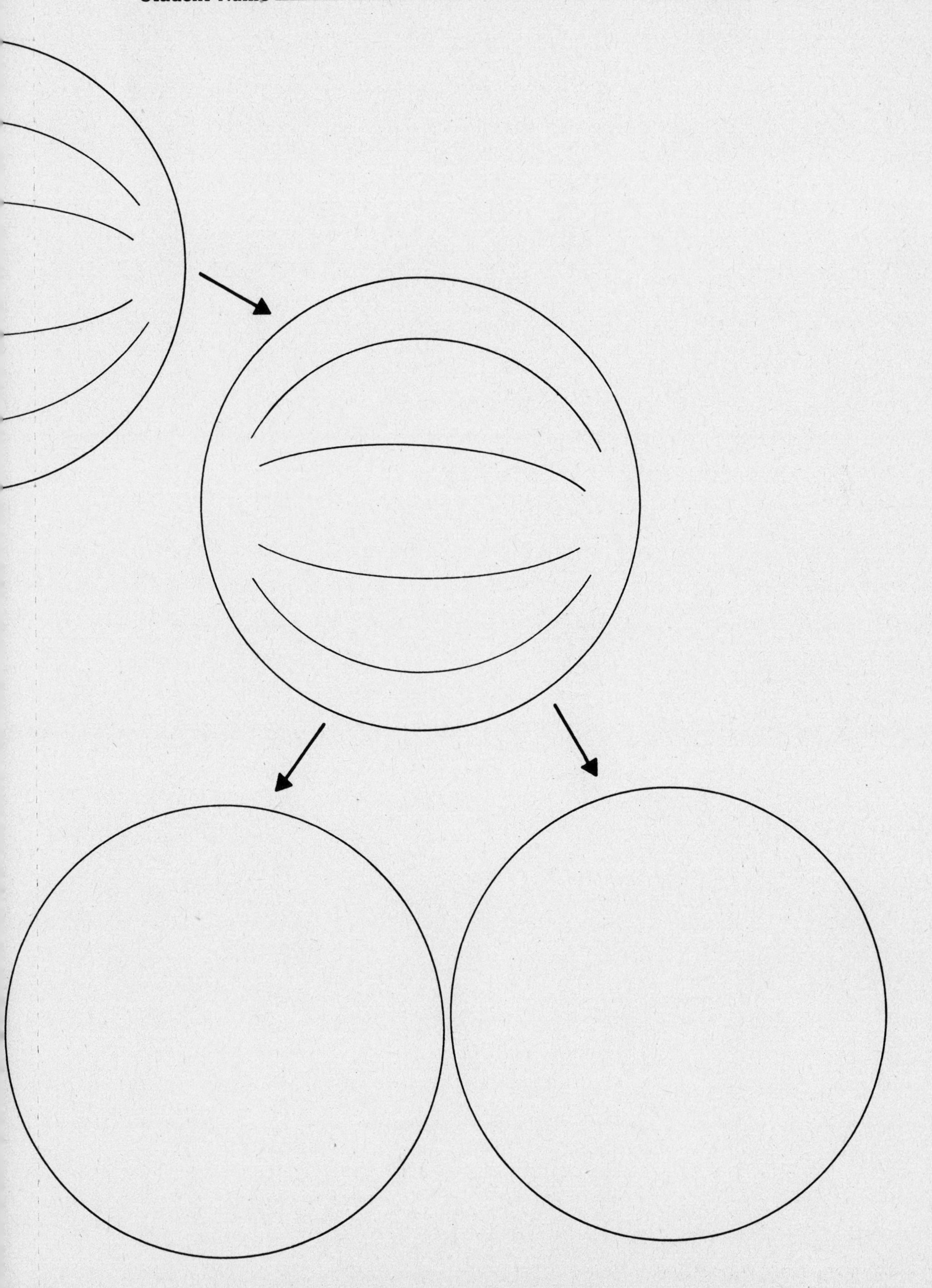

TOPIC 9

Mendelian Genetics and the Analysis of Data

OBJECTIVES

When you have completed this topic, you should be able to:

1. Explain why *Drosophila* is a good experimental animal for genetic experiments.
2. Describe the morphology of or recognize each stage of the *Drosophila* life history, including the adult male and female.
3. Define the following terms: parental (P_1) generation, first filial (F_1) generation, second filial (F_2) generation, dominant, recessive, allele, homozygous, heterozygous, monohybrid cross, dihybrid cross, segregation, and independent assortment.
4. Use a Punnett square to describe a monohybrid cross and a dihybrid cross in which the two genes show independent assortment; give the genotypic and phenotypic ratios expected in each case.
5. Define linkage, and describe a dihybrid cross in which the two genes are completely linked; give the expected genotypic and phenotypic ratios for the offspring.
6. Define sex linkage; use a Punnett square to describe a cross of a homozygous female with a mutant male, and give the expected phenotype and genotype of the offspring.
7. Give the reciprocal cross and explain why the reciprocal cross is necessary when the mutation might be either dominant or recessive.
8. Given the results from a cross of homozygous wild type and mutant flies, determine the mode of transmission of the genes involved.
9. Analyze the results using the chi-square test to see how closely they fit Mendel's laws.

TEXT REFERENCES

Chapter 12 (G); Chapter 13; Chapter 14 (A,F,G); Chapter 28 (D).
Insects, metamorphosis, Mendelian genetics, dominance, law of segregation, law of independent assortment, mutation, sex determination, sex linkage.

INTRODUCTION

In this laboratory you will have the opportunity to work with the fruit fly, *Drosophila melanogaster*. This tiny fly is one of the most important organisms used by geneticists in studying the mechanism of inheritance. The advantages of using *Drosophila* as an experimental animal are:

1. It has a short life cycle (10 days at 25°C).
2. A single healthy female can produce several hundred offspring.
3. It is easily cultured in the laboratory in small bottles on simple culture media.
4. It is small.
5. There are many different easily recognized variations of *Drosophila*.
6. It has a small number of chromosomes ($2N = 8$). In addition, the cells of the salivary glands of the larva possess enlarged chromosomes which have been important in mapping the location of certain genes.

In this exercise you will study the life history and morphology of *Drosophila melanogaster*, and analyze the results of a genetic cross in which two different types of *Drosophila* were mated.

LIFE HISTORY

The fruit fly is an insect of the order Diptera (two-winged flies). The Diptera all undergo complete metamorphosis, which means that the immature forms bear no resemblance to the adult. The flies and other insects that undergo complete metamorphosis show four different stages in their life histories (see Figure 9–1):

1. The **egg** is laid by the adult female (♀) on the food surface, and develops into the wormlike larva.
2. The white wormlike **larva,** or maggot, is characterized by its voracious appetite and rapid growth. During the course of its growth the larva will **molt,** or shed its skin, two times. The periods between molts are called **instars.** The third instar ends as the outer skin darkens and hardens, forming the pupal case.
3. During the **pupal** stage, the larval tissue breaks down, and the new winged adult is formed.
4. The **adult** fly that emerges from the pupal case is light-colored and elongated with unexpanded wings. The wings must expand, dry in the air, and stiffen before they can be used for flight.

The adult stage is the reproductive stage, and the adults will mate about 6 hr after emergence. The female (♀) receives about 4000 sperm from the male (♂), and stores these in special sacs. The sperm are released gradually as the eggs are produced. Up to a thousand eggs may be produced by a healthy female.

□ Examine a bottle containing a *Drosophila* culture, and find an example of the larva, pupa, and adult.

MORPHOLOGY

Handling the Flies

Because the fruit fly is small, it will be necessary to immobilize the flies so they can be observed with the dissecting microscope. Ether will be used to anesthetize them. Ether is highly explosive if ignited.

LIGHTING MATCHES OR LIGHTERS AND SMOKING IN THIS LABORATORY ARE FORBIDDEN.

Obtain an anesthetizer, re-etherizer (glass Petri dish), ether, camel hair brush, thumper (rubber pad), dissecting needle, and a 3 × 5 in. card or filter paper.

The procedure for etherizing the flies is as follows (see Figure 9–2):

1–3. Add ether to the anesthetizer bottle by putting a few drops of ether on the cotton or foam until it is moist, but not wet. Recap the ether bottle immediately.

4. Take the vial of wild type (++) flies provided by the instructor, and shake the flies down into the bottom by tapping it lightly on the thumper. Note that + is the symbol used for a normal or wild type trait, so ++ means a fly with all normal traits.

5–6. Now take the cotton plug out of the vial, and *quickly* place the empty anesthetizer over the top.

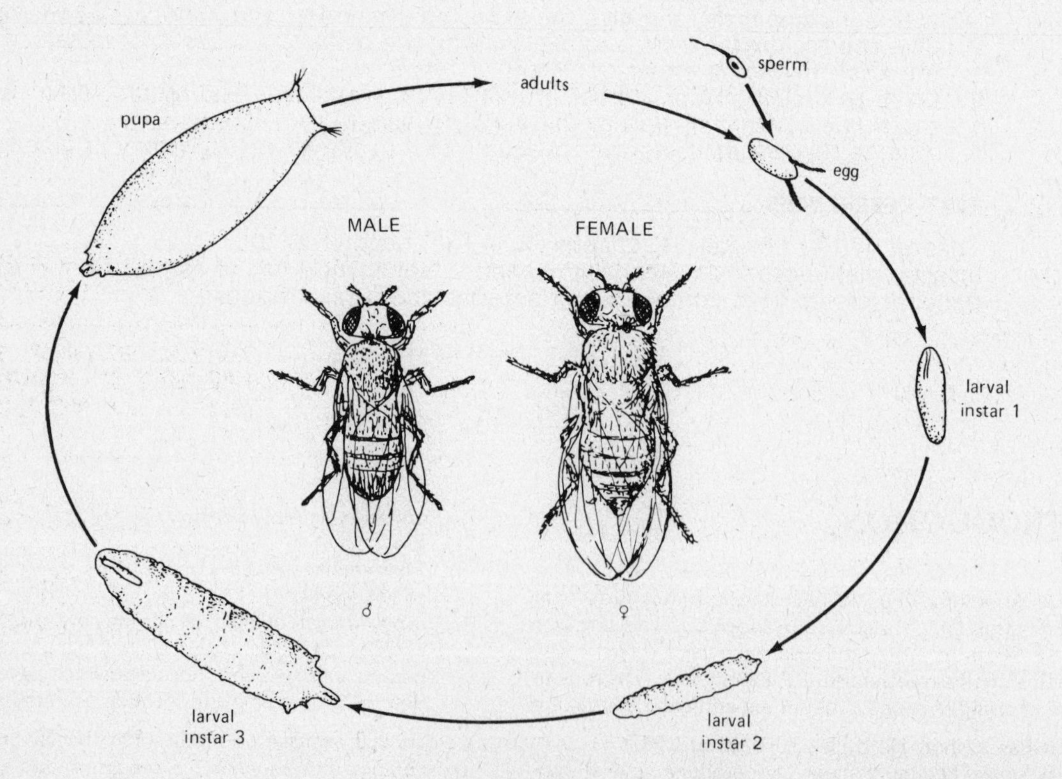

Figure 9–1. Life history of *Drosophila*. At room temperature the cycle from when the egg is laid to hatching (eclosion) of the adult from the pupa takes about 14 days. (Redrawn from William T. Keeton, Michael W. Dabney and Robert E. Zollinhofer, *Laboratory Guide for Biological Science,* illustrated by Paula D. Santo Bensadoun and Monica Howland. With the permission of W. W. Norton and Company, Inc., © 1968 by W. W. Norton and Company, Inc.).

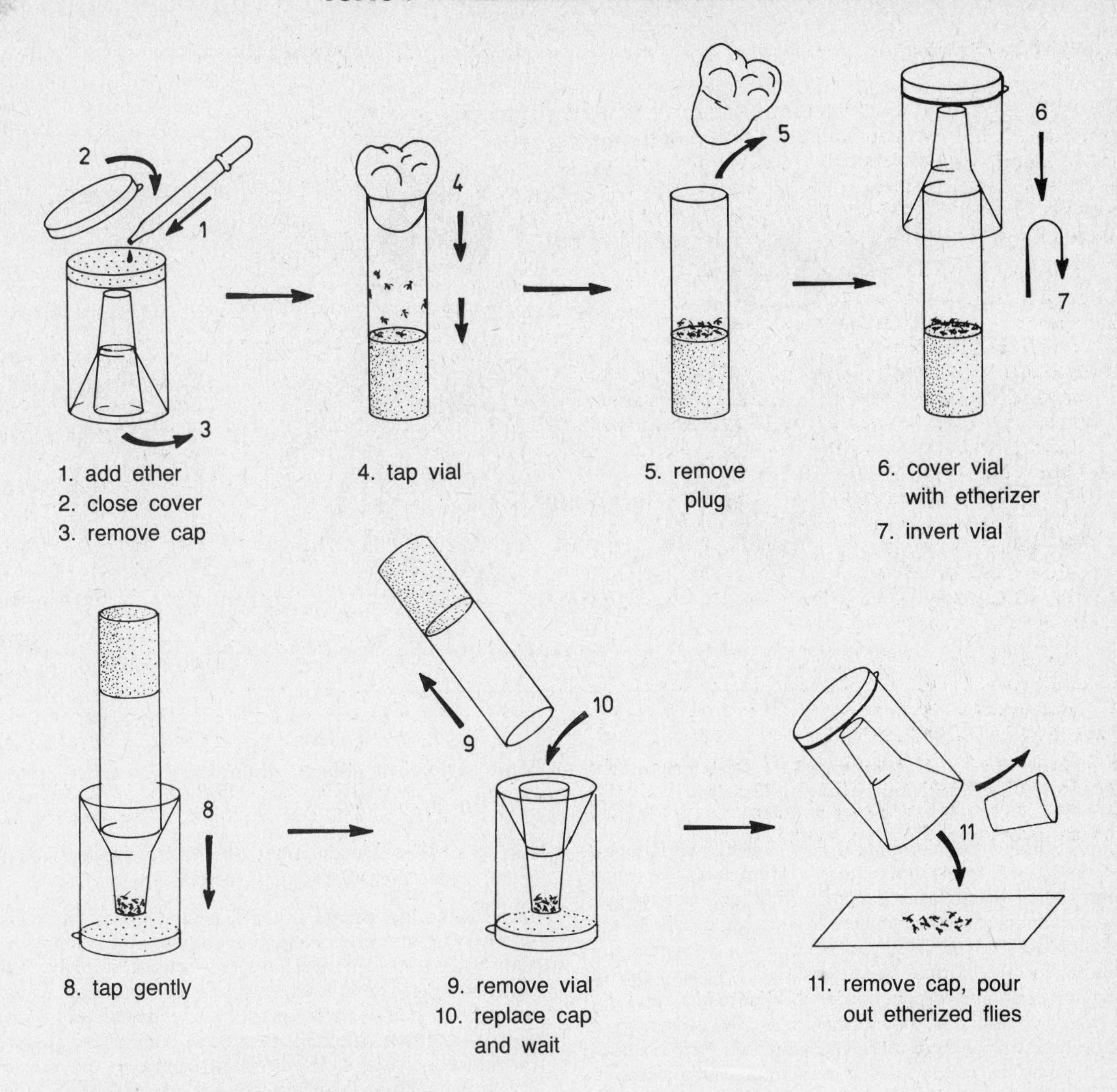

Figure 9–2. Etherizing *Drosophila*.

7–9. Reverse the position of the two bottles, and shake the flies into the anesthetizer by tapping *lightly* on the thumper as shown in Figure 9–2. Pounding too heavily will cause the food to loosen and fall into the anesthetizer. After the flies have been shaken into the anesthetizer, quickly remove the culture bottle, replug the vial, and cap the anesthetizer.

10–11. Watch the flies closely because they will slow down and stop moving in about 30 sec. After they have stopped moving, count to five slowly, and then pour the flies out onto the 3 × 5 in. card. It is important to follow this procedure exactly because overetherizing will kill the flies. You can tell if you have overetherized them by the position of the wings: dead flies have their wings extended at right angles to their bodies.

12. Prepare the re-etherizer by adding a few drops of ether to the pad in the Petri dish and setting the dish on the table with the pad side down.

13. The flies will remain immobilized for 5 to 10 min. When the flies begin to move, they must be re-etherized by covering them with the re-etherizing plate. Wait for the animals to stop moving, and resume your observations.

DO NOT ALLOW REVIVING FLIES TO ESCAPE AND INFEST THE LABORATORY. DISPOSE OF ALL UNWANTED FLIES IN THE MORGUE PROVIDED (A PAN OF MOTOR OIL).

Now put the card containing the flies under the dissecting microscope. It is helpful to draw a line halfway across the card, and label the halves male (♂) and female (♀). After you learn to sex the flies, you can brush them onto the appropriate half of the card.

Use Figure 9–3 as your guide in identifying your male and female flies.

The Head

The members of the class Insecta have three divisions of the body: head, thorax, and abdomen.

□ Examine the head. The eyes of *Drosophila* are of two

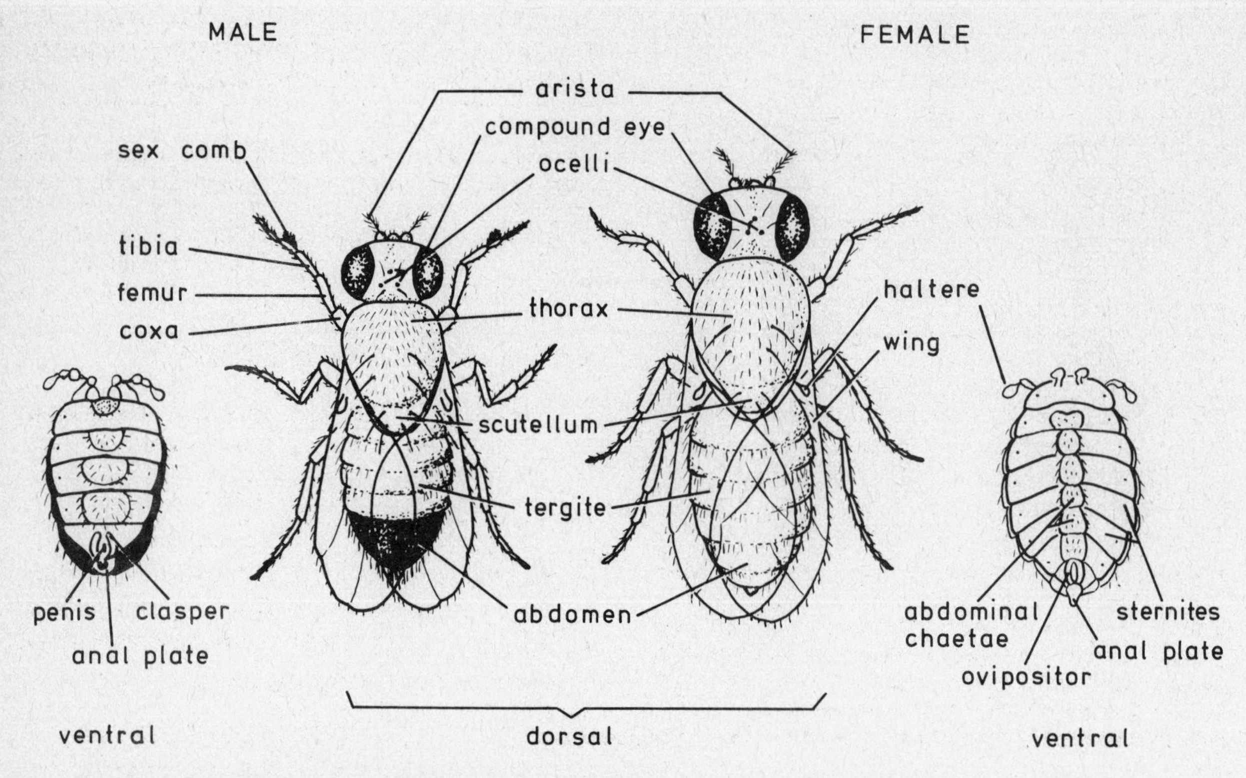

Figure 9–3. Male and female *Drosophila*. The sex comb of the male and characteristics of the abdomen can be used to distinguish the male and female.

types, simple and compound. The **compound eyes** are large and made up of many different facets called ommatidia, which form a honeycomb type of arrangement. Pay particular attention to the color: this is the normal dull red or wild type color (remember that the symbol for any normal characterisic is +). Two types of pigment, red and brown, occur in *Drosophila*, and together give the eye the normal color. Mutations affect the amount of each of these pigments and their distribution; many different possible eye color mutants result. The three simple eyes are called **ocelli,** and form a triangle on the top of the head between the two compound eyes. These apparently act to increase the brain's sensitivity to light, but they do not function in the formation of a visual image. Notice the **antennae.** Each antenna has a fleshy base and a featherlike **arista.** Examine the head from a side view, and look at the mouth parts. In the flies they have been modified to form a tubular **proboscis,** through which food is sucked up. If you press on the side of the head gently, the fly will usually protrude its proboscis.

The Thorax

The thorax consists of three fused segments, each bearing a pair of legs. Examine the legs. From the body outward each leg consists of the **coxa,** the very small **trochanter,** the **femur** (the largest part), the **tibia,** and the five-segmented **tarsi** (singular, **tarsus**). Using the highest magnification on your microscope, look closely at the first tarsal segment of the foreleg. In the male there is a clump of enlarged black bristles. This so-called **sex comb** is found only in the male, and is an important characteristic used in differentiating the sex of the flies. Keep looking until you locate the sex comb. (*Hint:* males are smaller than females.)

□ Examine the **wings** of the flies, and notice the veins, which give support to the filamentous wings. Many insects have two pairs of wings that function in flight. The flies, however, have only one pair of functional wings; the second pair is represented by the **halteres,** a pair of small bulbous structures located in back of the base of the functional wings. If you press lightly on the side of the thorax with a dissecting needle and remove the needle, the wings will become erect, and you can easily locate these rudimentary wings. The halteres probably function as stabilizers in flight.

□ Look down on the dorsal (or top) surface of the fly's thorax, and identify the shield-shaped posterior part, or **scutellum,** with its four large scutellar bristles.

There are many genetically defined variations of the thorax. Some important variations are differences in color of the thorax, shortness or absence of scutellar bristles, and altered wing structure.

The Abdomen

Abdomens of male and female flies show several differences that are important in distinguishing between the sexes. Sexing the flies may seem difficult at first, but with practice you will find it quite easy. Some *Drosophila* research workers can supposedly sex flies on the wing! While you are not expected to gain such proficiency, you will have to learn to sex the flies accurately. Here are the differences to look for (in addition to the male sex comb, which you have already seen):

1. **Claspers.** The male has dark-colored claspers on

the underside of the abdomen which are not present in the female.

2. **Size.** The male is smaller than the female.
3. **Shape.** The male has a blunt and rounded posterior, whereas that of the female is longer and pointed.
4. **Color.** The entire rear portion of the male abdomen is black, whereas in the female the alternating light and dark bands continue to the tip.

□ First examine the abdomen of a female more closely. On the dorsal (or top) surface the segmented plates, or **tergites,** will be seen as seven bands. The last segment of the abdomen is modified by the presence of the **anus** and the **ovipositor,** which forms the pointed posterior mentioned in (3) above. On the ventral surface find the small **sternites,** the midventral plates (one per segment). The tergites and sternites form a protective armor for the abdomen. Note that the tip of the female's abdomen is pointed and very light in color underneath.
□ Now examine the abdomen of a male fly. Notice that the male has six tergites, compared with the female's seven. The last segments are modified to form the genitalia. Brush the male fly onto his back to observe the genitalia. Just in front of the anus, find the **claspers,** hooklike appendages on the anal plate that aid in copulation. If you press lightly on the abdomen with a needle, you may be able to see the **penis** protruding between the claspers.

THE DARK, BLUNT ABDOMEN, WITH ITS DARK-COLORED CLASPERS UNDERNEATH IN THE MALE, IS THE BEST TRAIT TO USE IN DISTINGUISHING MALE FROM FEMALE FLIES.

□ After you have finished examining the wild type (normal ++) flies, obtain a vial of mutant flies from your instructor. It is helpful to examine the mutant and wild type side by side so the differences can be more clearly seen.

When you are ready to return the flies to the vial (if they are to be saved), do the following:

□ Bend the 3 × 5 in. card slightly, and brush the flies gently onto the glass of the *horizontal* vial as shown in Figure 9–4. Use a creased paper funnel if possible (Figure 9–4).

LEAVE THE VIAL ON ITS SIDE UNTIL THE FLIES RECOVER BEFORE MOVING THE VIAL.

This precaution is important because the surface of the food source is moist, and the flies may stick to the food and suffocate.

If the flies are dead, or if they are not to be saved, dispose of them in the morgue (a pan of motor oil).

THE GENETICS EXPERIMENT

The purpose of this experiment is to determine whether the mutant alleles are dominant or recessive, and how they are inherited.

You will be asked to study either cross A or cross B. These are **reciprocal crosses,** in which the mutations in the mutant parent are exactly the same, but the mutant parent is male in one case and female in the other.

Cross A: virgin wild type (++) females × mutant males
Cross B: virgin mutant females × wild type males (++)

Why must the female flies be virgin?

____

____

Remember that all parent flies used in this experiment are homozygous. The wild type parents are homozygous for normal alleles, and the mutant stock is homozygous mutant for two genes. Your instructor will give you information concerning the mutant traits you will be studying.

The procedure for setting crosses is as follows:

1. Remove all flies from stock bottles of wild type and mutant flies. Discard these flies because they have already mated.
2. Within the next 8 hr, collect and etherize the flies that have emerged. The females collected within 8 hr of emerging will be physiologically incapable of mating, and will be virgin.
3. Sex the flies, and separate the males from the females. Store them in separate vials if they are not to be used immediately.
4. Place three virgin females and three males of the

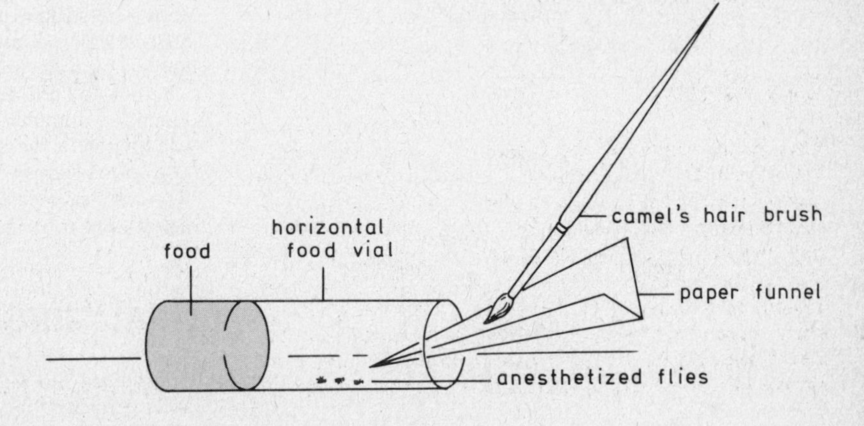

Figure 9–4. Returning flies to the vial. Sleeping flies must never come in contact with the sticky food, so keep the vial horizontal.

type called for in the cross into a horizontal food vial, and label the vial with the cross, the date, and your initials.

5. Wait until all of the flies have recovered from the ether, and then incubate the vial at 24–26°C for two weeks.

If you are going to set the cross yourself, your instructor will provide you with stock bottles of the parental flies and instructions as to when you will collect the virgins and set the cross. Cooperate with students doing the other cross so that you do not waste the flies that you do not need.

If the cross has been done for you, your instructor will give you a bottle or vial containing the adult flies resulting from a cross made two weeks ago. These flies represent the F_1 generation from the homozygous wild type and mutant parents.

Your first task will be to examine these offspring (the F_1), and decide whether the mutant traits are dominant or recessive to wild type. When the wild type and mutant stocks are mated, the offspring will receive one copy of each gene from each parent. If the offspring look like the mutant parent, the mutation is **dominant;** if the offspring look like the wild type parent, the mutation is **recessive:**

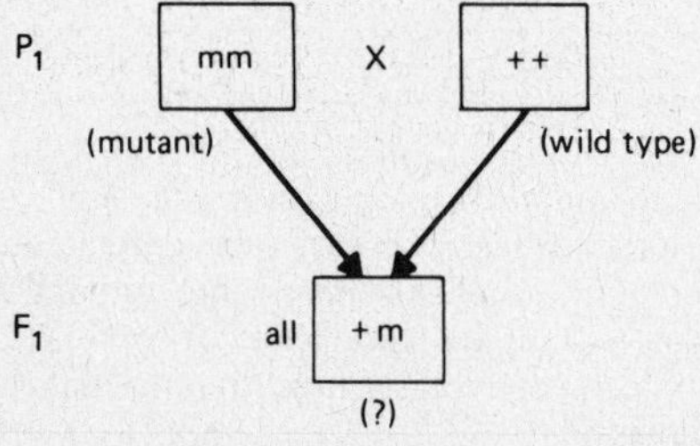

You will consider each mutation in turn, and decide whether it is dominant or recessive. The second part of the experiment is to determine the mode of inheritance of these genes. In *Drosophila* there are three pairs of **autosomes** and a pair of **sex chromosomes** (2N = 8). If the mutation is located on one of the autosomal chromosomes, it will show autosomal inheritance, and it will not matter which parent passed the mutation to the F_1 offspring. Cross A and cross B will give identical results:

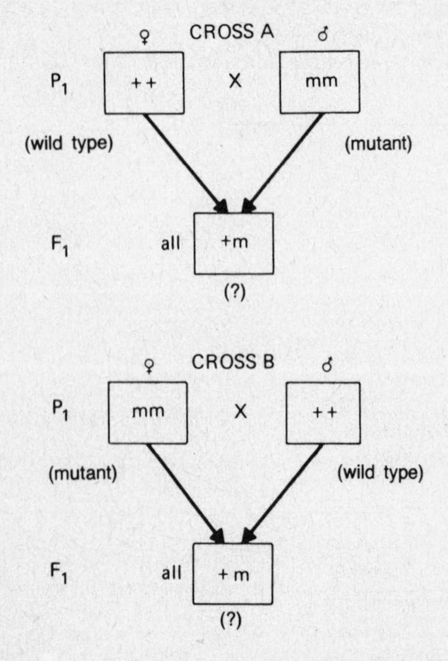

If, on the other hand, the mutation is **sex-linked** and located on a sex chromosome, it *does* matter which parent carried the mutation, and cross A and cross B will give different results:

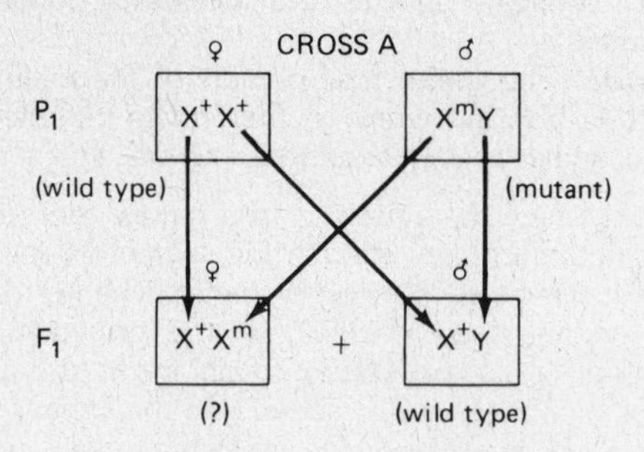

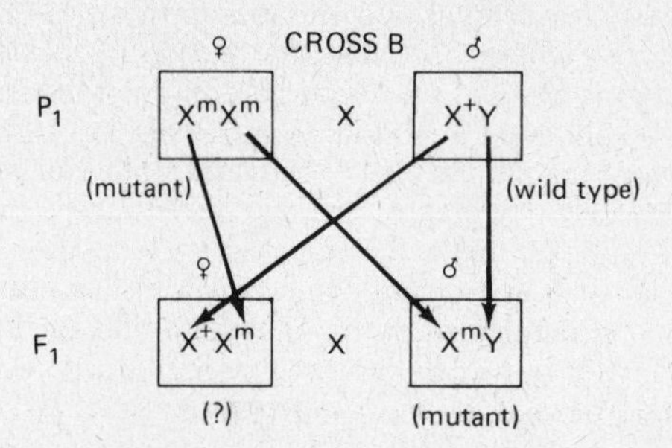

If the mutation is dominant, F_1 females, but not F_1 males, from cross A will show the mutation, but if it is recessive, F_1 males, but not F_1 females, from cross B will show it. In each case the reciprocal cross will not show any difference between the sexes.

What are the possible arrangements of the two mutant genes on the chromosomes?

1. Both of the mutant genes are on the same autosome, so they are inherited together, and are said to be **linked.** Any fly that inherits one mutant allele will also inherit the second one.
2. Both of the mutant genes are on the X chromosome. These traits are sex-linked, and must show the pattern of sex-linked inheritance.
3. The mutant genes are on different autosomes. They are inherited independently of each other, and neither shows the sex-linked pattern of inheritance.
4. One of the mutant genes is on the X chromosome, and the second is on an autosome. The genes will be inherited independently of one another, and will show different patterns of inheritance.

You are to examine the F_1 and determine from the data the dominance relationships and which of the above possibilities is the correct one. Remember that, as a class, you will analyze reciprocal crosses. The information from both crosses will be necessary to determine whether or not sex linkage is involved. Cross A will identify a dominant mutation as sex-linked, while cross B will identify a recessive mutation that is sex-linked. Since you don't know beforehand whether the mutation is dominant or recessive, it is necessary to do both reciprocal crosses.

ANALYZING THE CROSSES

Anesthetize the flies, and examine them under the dissecting microscope. Brush the flies onto the center line on the card, and sex them one by one, pushing them to the proper side of the card as you determine the sex. Are all the males similar? Are all the females alike? Do the males resemble the females? Do the offspring resemble the mutant or wild type parents for each mutant trait?

Record your observations of the F_1 sample under cross A or cross B (whichever applies) in the analysis chart at the end of this topic. Put a check in the appropriate column for each fly, and total the classes. **Check the results with your instructor before you discard the flies.** When you are satisfied with your results, follow instructions for disposing of the flies, and put your work place and equipment back in order. Before starting your analysis, consult with the students who did the reciprocal cross, and record these data also.

REPORT AND STATISTICAL ANALYSIS

Your instructor may ask you to prepare a formal report on your results. Follow the instructions in Appendix A for preparing this report.

The data that you obtained may be tabulated for the class as a whole, and statistically analyzed to see how well they fit the results expected from the most likely pattern of transmission. Instructions for using the chi-square test are given in Appendix B. In your report, show that you understand this test by applying it to the data for at least one trait in the cross that you studied.

EXPLORING FURTHER

Buchsbaum, R. *Animals without Backbones.* 2nd rev ed. Chicago: University of Chicago Press, 1975.

Srb, A. M., R. D. Owen, and R. S. Edgar. *General Genetics.* San Francisco: W. H. Freeman, 1965.

Scientific American Articles

Beadle, George W. "The ancestry of corn." January 1980 (#1458).

Bennet-Clark, H. C., and A. W. Ewing. "The love song of the fruit fly." July 1970 (#1183).

Benzer, S. "Genetic dissection of behavior." December 1973 (#1285).

Crow, James F. "Genes that violate Mendel's rules." February 1979 (#1418).

Hadorn. "Fractionating the fruit fly." April 1962 (#1166).

Kornberg, Roger D., and Aaron Klug. "The nucleosome." February 1981.

Student Name ______________________________ **Date** ____________________

ANALYSIS CHART

Record the results from your cross and from the reciprocal cross in the following tables; then analyze the results by completing the rest of the form.

CROSS A

Data from a cross of wild type females × ____________________ males.

Phenotype	Females	Males	Totals
Totals			

CROSS B

Data from the reciprocal cross which was ____________________ females × wild type males.

Phenotype	Females	Males	Totals
Totals			

Now, keeping in mind the information about the parental generation and your F_1 data from both crosses, draw your conclusions concerning the dominance and mode of transmission of the mutant genes you are studying. To guide you in your analysis, ask yourself the following questions.

TRAIT 1

Trait 1 = ______________________________

Is the mutant dominant or recessive to wild type? ______________________________

Did the mutant show a sex difference in either cross? ______________________________

If so, in which cross? ______________________________

This trait is inherited as

(a) autosomal dominant ________

(b) autosomal recessive ________

(c) sex-linked dominant ________

(d) sex-linked recessive ________

TRAIT 2

Trait 2 = ______________________________

Is the mutant dominant or recessive to wild type? ______________________________

Did the mutant show a sex difference in either cross? ______________________________

If so, in which cross? ______________________________

This trait is inherited as

(a) autosomal dominant ________

(b) autosomal recessive ________

(c) sex-linked dominant ________

(d) sex-linked recessive ________

ARRANGEMENT ON CHROMOSOMES

Your data may allow you to tell whether these two mutations are linked, that is, carried on the same chromosome.

Do both mutations show sex-linked inheritance? ______________________________

If so, they must both be on the X chromosome, and are therefore linked to each other.

Does one but not the other show sex-linked inheritance? ______________________________

If so, they must be unlinked because one is on the X chromosome, and the other is on an autosome.

To decide whether the two mutant traits are always inherited together in an autosomal pattern, you must do another cross.

What kind of cross should this be?

____

If so, they are located on the same autosomal chromosome.

Which conclusion(s) can you draw from your data?

(a) the mutations are linked and are on the X chromosome ________

(b) the mutations are unlinked because one is on the X chromosome ________

(c) the mutations are linked and are on an autosomal chromosome ________

(d) the mutations are unlinked and neither one is sex-linked ________

(Conclusions (c) and (d) cannot be distinguished on the basis of your F_1 results alone.)

Student Name ______________________________ **Date** ______________

CROSS DIAGRAMS

When you have decided on the mode of inheritance for each of the two mutant genes, assign symbols for each gene, and diagram your cross (A or B) below. Show your work in the Punnett square.

What types of progeny would be expected?

____

____

Why do you have to do the reciprocal cross?

____

____

If you knew ahead of time that the mutation was recessive, which cross (A or B) would you do to decide whether the mutation was sex-linked or not? ______________________________

DIAGRAM OF CROSS (A or B) __________

Trait #1: ______________________

Type of inheritance for Trait #1: ______________________

Symbol for mutant allele: __________

Symbol for normal allele: __________

P_1

	♀		♂
Genotype		×	
Phenotype			

F_1

	♀		♂
Genotype		+	
Phenotype			

Trait #2: ______________________

Type of inheritance for Trait #2: ______________________

Symbol for mutant allele: __________

Symbol for normal allele: __________

P_1

	♀		♂
Genotype		×	
Phenotype			

F_1

	♀		♂
Genotype		+	
Phenotype			

TOPIC 10

Human Genetics

OBJECTIVES

When you have completed this topic, you should be able to:

1. Determine your own possible genotypes for several physical traits.
2. Define antigen, antibody, and antiserum.
3. Give the possible genotypes of persons with the following blood types: A, B, AB, O, Rh^+, Rh^-.
4. Give the blood types that would be compatible in a transfusion with each of the blood types in Objective 3.
5. Describe a Barr body and explain why it is found only in normal female somatic cells.
6. Explain why sex-linked recessive traits in mammals occur much more frequently in males than in females.
7. Define karyotype, centromere (kinetochore), long arm, and short arm.
8. Use diagrams to explain how nondisjunction in the male or in the female can lead to the following karyotypes: XO, XXY, XXX, XYY.
9. Analyze an unknown karyotype and determine whether it represents a normal male, a normal female, or an abnormal genotype.

TEXT REFERENCES

Chapter 14; Chapter 33 (D).
Sickle-cell anemia, metabolic defects, multiple alleles, human blood groups, human sex chromosomes, nondisjunction, karyotype.

INTRODUCTION

Humans may be of much more interest to the average person than fruit flies, but it is not easy to study variation in human genes by conventional methods. There are problems in finding traits that are controlled by a single gene because it is unethical to deliberately mutagenize human populations. There are also problems in finding the appropriate matings, reciprocal crosses, and so forth. Few are eager to donate blood, tissues, or information about their families and themselves in order to push back the frontiers of human genetics. Much of the early work in this field depended, therefore, on the painstaking analysis of **pedigrees** in which the pattern of appearance of a certain trait in several generations of a family showed how the allele for that trait was inherited. Family photographs, diaries, and the recollections of older family members were used to identify the phenotypes of as many family members as possible. If enough information existed, it was possible to determine the pattern of inheritance of the allele by eliminating those patterns not consistent with the pedigree.

A human pedigree is illustrated in Figure 10-1. Males are indicated by the squares and females by the circles. Open symbols are used for individuals with a normal phenotype and filled symbols for individuals who show the trait phenotypically. A T connecting the symbols for a man and woman represents their marriage and offspring. Use what you have learned about Mendelian genetics to determine whether this trait is inherited as a dominant or recessive and whether or not it is sex-linked. One approach is to check each possibility in turn, one at a time, until you either reach an inconsistent result or show that your choice is consistent with the entire pedigree. Repeat for each of the other possibilities. Record your conclusion on the answer sheet at the end of this topic. If you have not already done so, be sure to do problem 5 in the Self-Quiz at the end of Chapter 14 in your text.

It is now known that in addition to many physical characterisics, single genes control many biochemical traits such as blood type and the presence of certain enzymes. More than 150 human diseases are known to be caused by mutant alleles for important proteins and en-

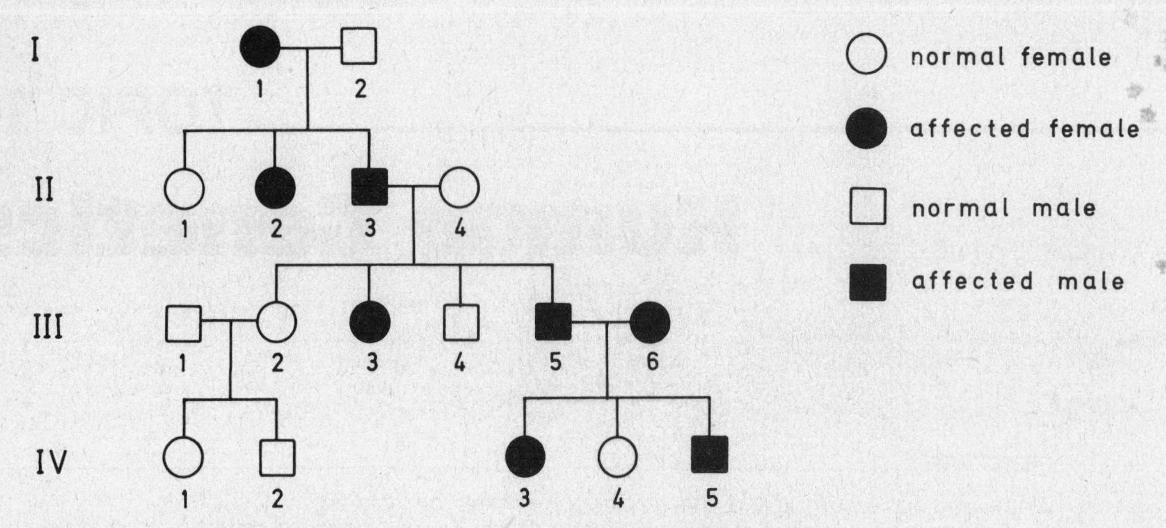

Figure 10–1. A human pedigree. Open symbols represent individuals with a normal phenotype and solid symbols represent affected individuals.

zymes. Some familiar examples are sickle-cell anemia, hemophilia, and phenylketonuria (PKU).

Now that the genetic basis of these diseases is understood, it is sometimes possible to design tests to determine whether prospective parents carry mutant alleles that can cause a genetic disease in their child. If there is a risk of genetic disease, the parents may ask for genetic counseling to maximize their chances of having a normal child. In some cases **amniocentesis** may be done during pregnancy. In this procedure some of the amniotic fluid is withdrawn to obtain a sample of fetal cells. The chromosomes from the amniotic sac surrounding the fetus can be studied by **karyotype analysis** to see whether there is a normal complement of chromosomes and whether the fetus is male or female. If the disease is known to be caused by a sex-linked recessive allele, males will have a much greater risk of having the disease than females. If the child is male, a single mutant allele on the X chromosome will cause the child to be affected. If the child is female, however, a single mutant allele will be recessive, and the child's phenotype will be normal. A female must have the recessive mutation on both X chromosomes in order to have the disease, and that is not very likely to happen. The cells can also be grown in culture and tested for **biochemical genetic defects.** If there is a serious untreatable or fatal genetic defect, the parents may decide to terminate that pregnancy. Some genetic diseases can be cured or controlled by medical treatment, but many cannot.

PHYSICAL GENETIC TRAITS

Many human physical traits are genetically controlled. The genes on the chromosomes of males seem to determine that the child will become male, and other genes determine eye color, hair color, height, skin color, and many other traits that distinguish individuals from one another. Several characteristics are known to be controlled by single gene differences. Some of these are listed below and illustrated in Figure 10–2.

1. **Pigmented iris:** The *P* allele for a pigmented iris (green, hazel, brown, or black eyes) is dominant over the *p* allele for lack of pigment (gray or blue eyes).
2. **Tongue rolling:** The *R* allele for the ability to roll the tongue into a U shape is dominant over the *r* allele for lack of this ability.
3. **Bent little finger:** The *B* allele for a bent little finger is dominant over the *b* allele for a straight little finger.
4. **Widow's peak:** The *W* allele for a widow's peak is dominant over the *w* allele for a straight hairline.
5. **Thumb crossing:** The *C* allele for crossing the left thumb over the right thumb when you interlace your fingers is dominant over the *c* allele for crossing the right thumb over the left.
6. **Attached ear lobes:** The *a* allele for attached ear lobes is recessive to the *A* allele for unattached ear lobes.
7. **Hitchhiker's thumb:** The *h* allele for the ability to bend the last joint of the thumb back at an angle of 60° or more is recessive to the *H* allele for the lack of ability to bend the thumb back this far.

□ Determine your own phenotype for each of the above characteristics and record your phenotypes in Table 10–2 at the end of this topic.
□ Fill in your possible genotypes for each of the traits.

If you have the trait determined by the recessive allele, there is only one possible genotype. If your trait is determined by the dominant allele, you should list both possible genotypes, unless one of your parents is homozygous recessive. In that case you are heterozygous.

HUMAN BLOOD GROUPS

The first chemical trait in humans to be understood was the determination of the ABO antigens on red blood cells. Any substance that causes the immune system to produce **antibodies** is an **antigen** or **anti**body-**gen**erating substance. Many different types of substances can be antigens. When a foreign antigen enters the body, antibodies corresponding specifically to the antigen will be formed. Antibodies are proteins that are specific for one particular antigen and bind molecules of

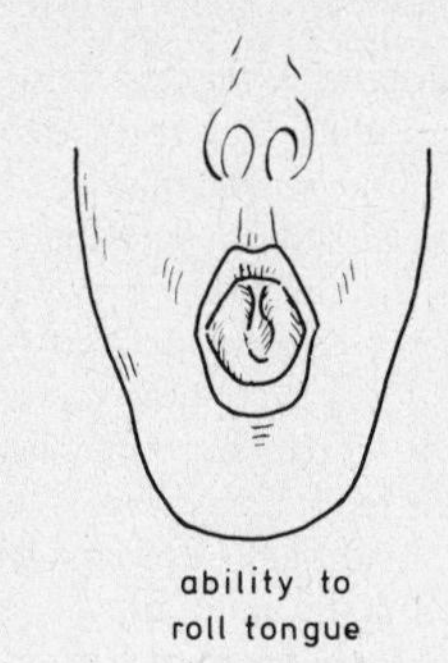

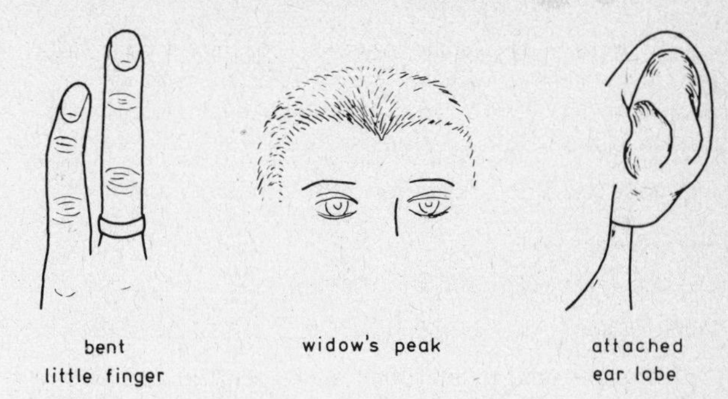

Figure 10–2. Human genetic traits. These traits are each controlled by a single gene.

the antigen together into clumps. If the antigens are attached to red blood cells, as are the ABO antigens, the red blood cells will clump in the presence of the antibody:

A antigen will clump with anti-A antibodies.
B antigen will clump with anti-B antibodies.
AB antigen will clump with both anti-A and anti-B antibodies.
O antigen clumps with neither anti-A nor anti-B antibodies.

Another antigen, which is also present on red blood cells, is called the Rh antigen. Blood cells with this antigen are Rh^+, and those without it are Rh^-.

Rh^+ cells will clump with anti-Rh antibodies.
Rh^- cells will not clump with anti-Rh antibodies.

There are many other antigens on red blood cells that are also genetically determined.

Because it is easy to obtain a small sample of blood from a large number of people, the genetic basis of blood groups was soon well understood. The type of antigens of the ABO group that are present are determined by **multiple alleles,** any one of which may occur at a particular locus, or position, on a chromosome. Instead of just two possible alleles at this locus, there are three—namely, I^A, I^B, and i. The I^A allele determines the A antigen, the I^B allele determines the B antigen, the i allele determines neither antigen. Any individual has two alleles for ABO blood groups and could have any of the following genotypes.

Genotype	Blood Type (Phenotype)	Antigen Present	Antibodies in the Blood
I^AI^A	A	A	Anti-B
I^Ai	A	A	Anti-B
I^BI^B	B	B	Anti-A
I^Bi	B	B	Anti-A
I^AI^B	AB	A and B	None
ii	O	None	Anti-A, Anti-B

Because a normal person never makes antibodies against his own antigens, the antibodies present in the blood never react with the antigens present on the red blood cells. A person with A antigen on his red blood cells will have anti-B antibodies but never anti-A antibodies. Otherwise the red blood cells would clump together.

Blood groups are medically important because only compatible blood can be used during a blood transfusion. The transfusion will be incompatible if the red blood cells of the donor can clump by reacting with the antibodies of the recipient. The safest type of transfusion is between individuals of the same blood type. In this case, the red blood cell antigens and the antibodies are identical, and there is no possibility of a clumping reaction. If the donor's red blood cells contain any antigen, however, the blood can be given only to individuals who lack the corresponding antibodies. Type A blood, for example, can be given only to persons with type A blood or type AB blood. If the recipient's blood were of type B or type O, there would be anti-A antibodies in it that would clump the donor red blood cells. The antibodies of the donor are greatly diluted during a transfusion and can be ignored. The mechanism of clumping during an incompatible transfusion is shown in Figure 33–16 in your text.

Use the following procedure to determine your ABO blood type. The anti-A serum that you will use contains anti-A antibodies and will clump or agglutinate red blood cells that have A antigen. The anti-B antiserum contains anti-B antibodies and will clump cells that have B antigen.

DO NOT PRICK YOUR FINGER IF YOU HAVE A BLOOD CLOTTING DISORDER OR HAVE BEEN EXPOSED TO HEPATITIS. TO AVOID THE RISK OF SERUM HEPATITIS, DO NOT TOUCH ANY MATERIALS THAT MIGHT BE CONTAMINATED WITH ANOTHER PERSON'S BLOOD. NEVER USE A BLOOD LANCET ALREADY USED BY ANOTHER PERSON.

If you are unable or unwilling to volunteer to have your finger pricked, work with a partner who doesn't mind doing it. Work quickly so your blood doesn't clot during the test.

□ Obtain a clean slide. Mark it with two circles labeled A and B.
□ Squeeze one of your fingers to increase the circulation, and wipe it with sterile cotton moistened with 70% ethanol.
□ Without touching the "sterile" finger, open a sterile lancet packet from the correct end, and prick your finger or have someone prick it for you.
□ Squeeze out a drop of blood onto each of the circles on your side.
□ Obtain a toothpick and the anti-A antiserum. Add one drop of antiserum to the blood drop in the circle marked A without touching the dropper to the blood, and stir the drop with the toothpick.
□ Obtain another toothpick and the anti-B antiserum. Add one drop of antiserum to the circle marked B, and stir the second drop with the toothpick.

If there is a positive reaction, the blood cells will form large clumps within 1-2 min, and the drop will appear speckled. If the reaction is negative, the drop will remain cloudy, without clumps.

Did your red blood cells react with the anti-A antiserum?

____

Did they react with anti-B antiserum? ______________

What is your ABO blood type? ______________

Another antigen that may be present on red blood cells is the Rh antigen. There are several dominant alleles (*Rh*) that determine the presence of this antigen and a recessive allele (*rh*) that determines the absence of the antigen. Persons normally do not have anti-Rh antibodies whether they have the antigen or not. These are the possible genotypes and blood groups:

Genotype	Blood Group (Phenotype)	Antigen Present	Antibodies in the Blood
Rh/*Rh*	Rh^+	Rh	None
Rh/*rh*	Rh^+	Rh	None
rh/*rh*	Rh^-	None	None

If an Rh^- person comes in contact with Rh^+ blood cells, the immune system will begin producing anti-Rh antibodies. This might happen after a blood transfusion or after a pregnancy in which an Rh^- mother carries an Rh^+ child. If the person then comes in contact with Rh^+ blood cells a second time, the antibodies will cause the cells to clump. This might happen during a second pregnancy of an Rh^- woman carrying a second Rh^+ child. The antibodies of the mother might enter the circulation of the baby and cause a disease called **erythroblastosis fetalis.** The baby must have an immediate transfusion of antibody-free blood when it is born to stop the clumping reaction.

Determine your Rh blood type using anti-Rh antiserum, which clumps red blood cells containing Rh antigen.

□ Obtain a clean slide.
□ Prick your finger as you did before, or use the same finger, and squeeze a drop of blood onto the slide.
□ Obtain the anti-Rh antiserum and a toothpick.
□ Add 1 drop of the serum to the blood without touching it, and stir the drop with a toothpick.
□ Note whether there is a clumping reaction.

Did your blood clump with anti-Rh antiserum?

____

What is your Rh blood type? ______________

When you are finished with these tests, dispose of all lancets, slides, and debris in the container marked **BLOOD TEST WASTES.**

SICKLE-CELL ANEMIA

The presence of normal beta chains in hemoglobin A is determined by the *A* allele. The mutant allele at this locus that causes sickle-cell anemia is called *S*. The mutant hemoglobin S molecules have a greater tendency to aggregate when the concentration of oxygen is low. The aggregates that form distort the shape of the red blood cells and cause them to block capillaries in the joints and internal organs (see Figure 10–3). The lack of circulation that results causes a painful **sickle-cell crisis** which can be dangerous and even fatal. The disease of sickle-cell anemia, which occurs mainly in people of African descent, can be controlled to a certain extent by medical treatment, but there is no cure at the present time.

The allele that causes this disease shows incomplete dominance with the normal allele so that a person who is heterozygous has both hemoglobin A and hemoglobin S in his red blood cells. The heterozygous person is a **carrier** for the disease, but usually does not have any symptoms. The normal hemoglobin can apparently compensate for the abnormal hemoglobin S except under conditions of extreme stress. A person who is homozygous for the *S* allele is **affected** and has **sickle-cell anemia.** Until recent advances in medical treatment became available, most homozygous children would die early in childhood.

If two persons heterozygous for the *S* allele marry, what is the chance of their child having sickle-cell anemia?

____

□ Examine the prepared slide of normal blood cells and cells from a person with sickle-cell anemia.
□ Draw a few of the abnormal cells in the space provided at the end of this topic.

HUMAN SEX CHROMOSOMES

In humans, sex and other traits are determined by the sex chromosomes. Males have one X chromosome and a tiny Y chromosome which contains very few genes.

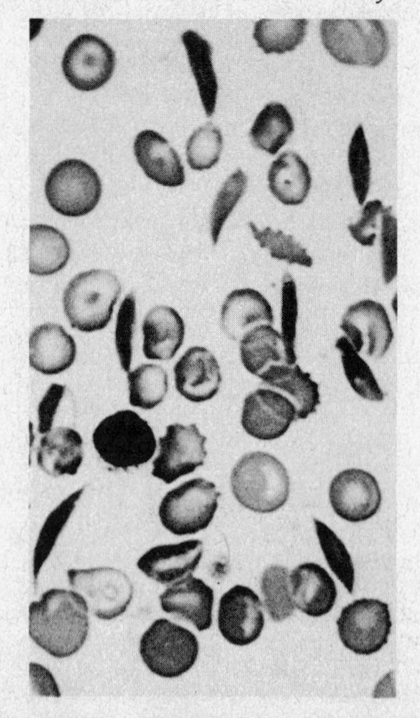

Figure 10–3. Sickled red blood cells. Cells containing hemoglobin S will assume bizarre shapes when they are deprived of oxygen. As a result, capillaries will become blocked and cause painful symptoms in the patient. (Photo courtesy of Carolina Biological Supply Company.)

Females have two X chromosomes. However, early in the development of a female embryo, one of the X chromosomes becomes inactive in each cell, so that only one of the two X chromosomes is actually active. The inactive chromosome becomes condensed, but is still replicated and passed on to all the daughter cells. The condensed X chromosome can be seen in certain cells as a **Barr body.** In epithelium, for example, many of the cells will show the Barr body as a darkly staining mass just inside the nuclear membrane. In other cells, such as certain white blood cells, the condensed chromosome forms a **drumstick** which protrudes from the nucleus.

Preparations of epithelial cells from the oral cavity are easy to obtain and can be used to screen for chromosomal abnormalities in the sex chromosomes. If a person is a normal female there will be one Barr body in many of the cells. There should be no Barr bodies in a normal male. If there is an extra X chromosome, it will be condensed because there is always only a single active X chromosome in human cells. An abnormal male with an extra X chromosome will have one Barr body, and an abnormal female with an extra X chromosome will have two Barr bodies.

How many Barr bodies would you expect to find in a person with four X chromosomes (XXXX)?

____

Determine whether your epithelial cells contain Barr bodies using this procedure. About 35% of the cells should have visible Barr bodies if you are female.

□ Obtain a clean slide and a tongue depressor or clean toothpick.
□ Add a drop of methylene blue dye to the slide.
□ Scrape the inside of your cheek with the tongue depressor or toothpick, and stir the cells into the dye.
□ Add a cover slip and examine the cells under high power in the microscope.

Do your cells have Barr bodies? ______________

□ Exchange slides with someone of the opposite sex, and note whether Barr bodies are present.

Draw a cell with a visible Barr body and one with no Barr body in the space provided at the end of this topic.

This type of test is used to determine whether athletes competing in certain important sporting events such as the Olympic games are genetically of the appropriate sex for the events they are entering. Persons are usually not allowed to compete in women's events unless they are genetically female.

HUMAN KARYOTYPES

A **karyotype** is the display of a person's chromosomes. Because each cell is supposed to have the same set of chromosomes, in theory, a karyotype could be made from any cell in metaphase. Karyotypes from adults can be made easily from cheek cells or white blood cells. The karyotype can also be determined during fetal development by studying cells removed from the amniotic fluid during amniocentesis. In that case, care must be taken not to confuse cells of the mother with cells from the fetus.

After the cells have been stained and squashed, the individual chromosomes are apparent and can be photographed. The chromosomes are cut out, matched in pairs, and arranged in a chart according to size and shape. The karyotype of a normal male is shown in Figure 10–4.

The pairs of chromosomes are first sorted into groups of about the same shape and size. Note that each chromosome consists of two sister chromatids joined in the region of the **centromere.** The position of the centromere divides the chromosome into two **long arms** and two **short arms.** Sometimes the arms are almost equal in length, and sometimes the long arms are much longer than the short arms. The pairs are then arranged within the groups in order of decreasing size. The X chromosome is a large chromosome, about as large as pairs 6-7, with arms of about equal length. It is placed over the X in the chart. The Y chromosome is the smallest one, even smaller than those in group G, and it is placed over the Y. A normal female would have two X chromosomes and no Y chromosome. Note that the pairs within a group are very similar and cannot be told apart easily. The exact order of the pairs within the group doesn't matter much in karyotype analysis, but the chromosomes must be placed in the correct group according to their size and the relative lengths of their arms.

Karyotype analysis is often used to determine whether a particular disease or syndrome is due to an abnormal complement of chromosomes. In the course of studies carried out for this purpose, or **medical genetics,** it has become apparent that there is a lot of variation among apparently normal people in the exact arrangement of their chromosomes. This discovery has opened up the field of **somatic cell genetics,** in which the complexity of the interactions of normal human genes is being investigated.

In 1959 Lejeune* discovered that children with Down's syndrome (Mongolism) have three copies of chromosome #21. Since then many human genetic disorders have been traced to an abnormal chromosome complement. It has also been found that up to 40% of spontaneously aborted fetuses have abnormal chromosomes, with the XO karyotype (Turner's syndrome) being particularly common.** The physical and mental defects associated with these karyotypes are shown in Table 10–1.

An abnormal number of chromosomes in an individual is often the result of an error in meiosis during which an egg or sperm cell with an incorrect number of chromosomes is produced. When the homologous chromosomes pair during meiosis I, it is possible that the members of the pair fail to separate properly during metaphase I, and so both members of the pair eventually end up in the same gamete. The failure of the chromosomes to separate is called **nondisjunction,** and it will result in gametes with extra or missing chromosomes. When the gametes are fertilized, the offspring will have extra or missing chromosomes also. Suppose, for example, that the X chromosomes stayed together during meiosis I in a female. The resulting egg might have two X chromosomes. After fertilization of such an egg, the zygote would have either three X chromosomes or two X chromosomes and one Y chromosome, depending on the type of sperm that fertilized the egg.

---

*Lejeune, Turpin, and Gautier, *Ann. de Genet.* 1, 1959.

**C. Stern, *Principles of Human Genetics*, 3rd ed. San Francisco: W. H. Freeman, 1973, p. 131.

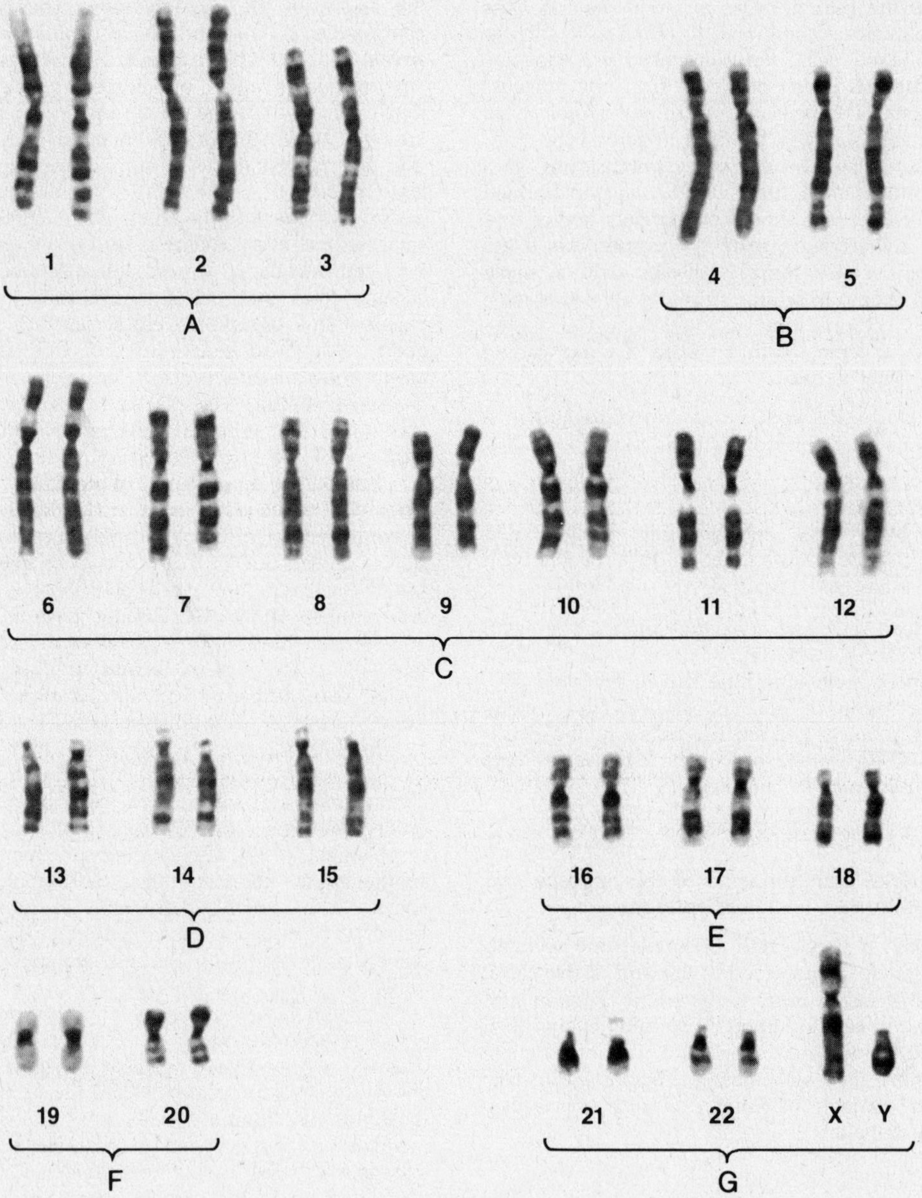

Figure 10–4. Human karyotype. These are the chromosomes found in each cell of a normal male. (Courtesy of Jorge J. Yunis, M.D., Medical Genetics Division, University of Minnesota Medical School.)

TABLE 10–1. CHARACTERISTICS ASSOCIATED WITH DEFECTIVE KARYOTYPES

Syndrome	Karyotype	Characteristics
Klinefelter's male	XXY	Unusual body proportions and sterility; subnormal mental ability
Turner's female	XO	Short stature; webbing of the neck; may have low mental ability and sterility
Super female	XXX	May have low mental ability; fertile
"Cri du chat"	defective #5	Catlike cry; severe physical and mental abnormalities; nonlethal
Patau's syndrome	extra #13	Physical abnormalities; lethal soon after birth
Edward's syndrome	extra #18	Unusual features of the head and fingers; often dies in infancy
Down's syndrome	extra #21	Characterisic facial features; low mental ability; stocky build; sometimes heart defects

Alternatively, the egg might have no X chromosomes after nondisjunction.

What types of zygotes could such an egg form?

____

□ Obtain from your instructor a copy of a chromosome squash of a human cell.
□ Record the number of this squash in Table 10–4.

The karyotype of the cell corresponds to one of the following patterns:

TABLE 10–2. HUMAN KARYOTYPES

Phenotype	Karyotype	Number	✓
Normal male	XY	46	✓
Normal female	XX	46	✓
Klinefelter's male	XXY	47	✓
Turner's female	XO	45	✓
Super female	XXX	47	✓
"Cri du chat" male	Partial deletion of short arm of #5	46	✓
Patau's female	3 copies of #13		
Edward's female	3 copies of #18		
Down's male	3 copies of #21	47	✓

Normal males and females have 46 chromosomes in each cell. There are 22 pairs of autosomes and either a pair of X chromosomes or one X and one Y. Determine the number of chromosomes that you would find in cells of persons with the chromosomal defects listed, and enter the numbers in Table 10–2.

□ Count the number of chromosomes in your chromosome squash and record the number in Table 10–4.
□ Check off the karyotypes that the squash might represent in Table 10–2.
□ Cut out the chromosomes and arrange them in order of decreasing size. Use a ruler if possible.
□ Match up pairs of similarly shaped chromosomes and place them in the correct groups of Table 10–4.
□ Rearrange the pairs of autosomes and the sex chromosomes until you have a reasonably good version of one of the karyotypes that you have checked off in Table 10–2. The order of chromosomes within each group is not important, but the chromosomes within a group must have the same general shape with respect to the position of the centromere and must be of approximately the same size. A chromosome in group B, for example, must be larger than any in group C, smaller than any in group A, and its centromere located towards the top rather than in the middle.

NOTE THAT THE SEX OF EACH KARYOTYPE LISTED IS SPECIFIED.

□ Determine the sex of your karyotype and check that you have arrived at one of the possible phenotypes.
□ Tape the chromosomes in place on the chart.
□ Ask your instructor to check your work, or hand in your chart before you leave.

EXPLORING FURTHER

deGrouchy, J., and C. Turleau. *Clinical Atlas of Human Chromosomes.* New York: John Wiley and Sons, 1977.

McKusick, Victor A. *Mendelian Inheritance in Man: Catalogs of Autosomal Dominant, Autosomal Recessive and X-linked Phenotypes.* 5th ed. Baltimore: Johns Hopkins, 1975.

Patau, K. "The identification of individual chromosomes, especially in man." *Am. J. Human Genetics* 12: 250, 1960.

Roth, E. F., Jr., *et al.* "Sickling rates of human AS red cells infected in vitro with *Plasmodium falciparum* malaria." *Science* 202: 650–652, 1978.

Saxen, L., and J. Rapola. *Congenital Defects.* New York: Holt, Rinehart and Winston, 1975.

Stern, C. *Principles of Human Genetics.* 3rd ed. San Francisco: W. H. Freeman, 1973.

Sutton, H. E. *An Introduction to Human Genetics.* 2nd ed. New York: Holt, Rinehart and Winston, 1975.
Winchester, A. M. *Heredity, Evolution and Humankind.* St. Paul, Minn.: West, 1976.

Scientific American Articles

Anderson, W. French, and Elaine G. Diakumakos. "Genetic engineering in mammalian cells." July 1981.
Brady, R. O. "Hereditary fat metabolism diseases." August 1973.
Cerami, Anthony, and Charles M. Peterson. "Cyanate and sickle-cell disease." April 1975 (#1319).
Clarke, C. A. "The prevention of rhesus babies." November 1968 (#1126).
Croce, Carlo M., and Hilary Koprowski. "The genetics of human cancer." February 1978 (#1381).
Friedman, Milton J., and William Trager. "The biochemistry of resistance to malaria." March 1981.
Fuchs, Fritz. "Genetic amniocentesis." June 1980 (#1471).
Glass, H. Bentley. "The genetics of the Dunkers." August 1953 (#1062).
Macalpine, Ida, and Richard Hunter. "Porphyria and King George III." July 1969 (#1149).
McKusick, Victor A. "The mapping of human chromosomes." April 1971 (#1220).
Ruddle, Frank H., and Raju S. Kucherlapati. "Hybrid cells and human genes." July 1974 (#1300).

Student Name ______________________ **Date** ______________

TABLE 10–3. HUMAN PHYSICAL GENETIC TRAITS

Determine your phenotypes and genotypes for the traits listed in the following table.

Trait	Dominant Allele	Your Phenotype	Your Genotype(s)
Pigmented iris	*P* = pigment present	Pigmented	PP or Pp
Tongue roller	*R* = ability to roll	roll	RR Rr
Bent little finger	*B* = bent little finger	bent	BB or Bb
Widow's peak	*W* = widow's peak present	present	WW Ww
Thumb crossing	*C* = left over right	right over left	cc
Attached ear lobes	*A* = unattached ear lobes	attached	aa
Hitchhiker's thumb	*H* = < 60° angle of the thumb	NO	hh
PTC	T = taster	Taster	TT or Tt

In the spaces below, draw red blood cells in sickle-cell anemia, a cell with a Barr body, and a cell without a Barr body.

red blood cells in sickle-cell anemia

- sickle cell

cell with Barr body

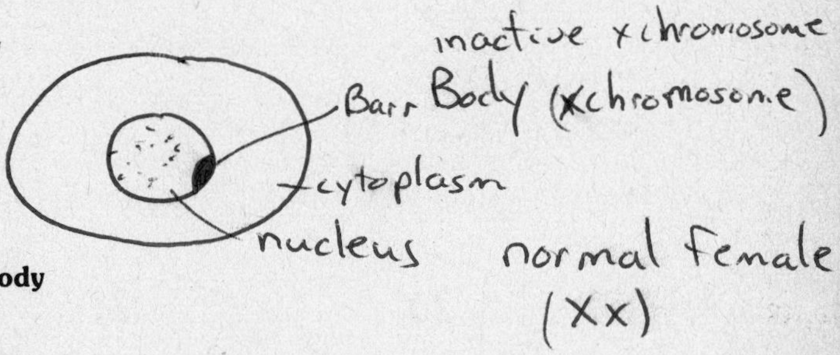

inactive x chromosome
Barr Body (x chromosome)
cytoplasm
nucleus
normal female (XX)

cell without Barr body

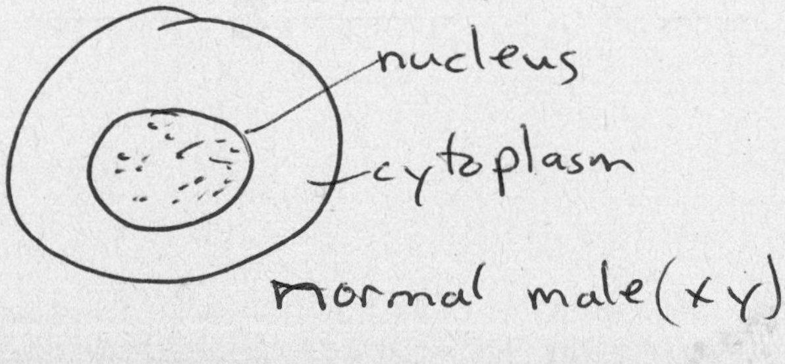

nucleus
cytoplasm
normal male (XY)

TABLE 10–4. KARYOTYPE ANALYSIS CHART

Chromosome squash # ____________________

Karyotype of the squash: ____________________

Corresponding phenotype: ____________________

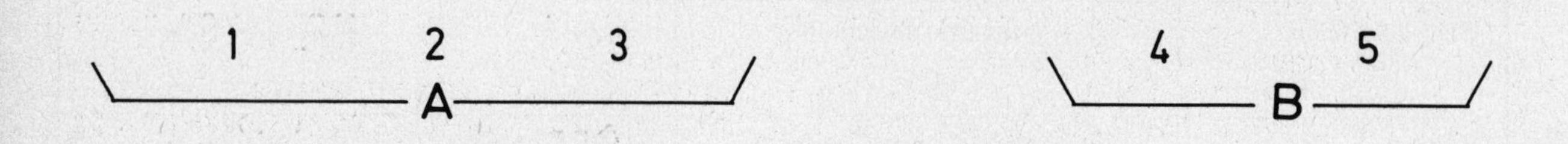

6 7 8 9 10 11 12

C

13 14 15

D

16 17 18

E

19 20

F

21 22

G

X Y

TOPIC 11

Population Genetics

OBJECTIVES

When you have completed this topic, you should be able to:

1. State the Hardy-Weinberg Law and the conditions under which it applies.
2. Determine the homozygous frequency of a recessive allele in a sample assumed to be in Hardy-Weinberg equilibrium and calculate the approximate numbers of heterozygous and homozygous dominant members expected in the sample.
3. Define evolution.
4. Describe a model experiment with two alleles in which the selection against the alleles is different and predict the change in frequency of the less advantageous allele.
5. Define genetic drift.
6. Describe a model experiment to test whether a certain population is small enough to show genetic drift.

TEXT REFERENCES

Chapter 15 (A,B).
Hardy-Weinberg Law, natural selection, gene flow, genetic drift.

INTRODUCTION

Hardy and Weinberg discovered that sexual reproduction alone will do nothing to change the frequencies of alleles in a population provided that certain conditions are met:

1. There must be no natural selection against either allele (or no difference in natural selection).
2. Mating between members of the population must be random. The population must be large enough in size so that each generation represents a random sample.
3. There must be no net mutation of one allele to the other.
4. There must be no gene flow due to individuals entering or leaving the population through emigration and immigration.

If these conditions are perfectly met, the frequency of each allele will remain exactly the same from generation to generation. Furthermore, the frequencies of the dominant and recessive phenotypes can be determined if the frequency of either phenotype or of either allele is known.

Suppose that the frequency of the *A* allele is X, and there are two alleles for this gene. The frequency of the second allele, *a*, has to be equal to $1 - X$ because the sum of the two alleles must be 100%, or a frequency of 1. The chance of getting the genotype *AA* is the chance of getting one *A* allele times the chance of getting a second or X^2. To avoid confusion, the frequency of the dominant allele is called p, so the frequency of the *AA* genotype is p^2.

The chance of getting the *aa* genotype is the chance of getting one *a* allele times the chance of getting the second. The frequency of the recessive allele is called q, so the chance of the *aa* genotype is q^2. A heterozygote can be formed by getting a recessive, then a dominant allele ($q \times p$); or by getting a dominant, then a recessive allele ($p \times q$). Therefore the chance of ending up with the *Aa* genotype is $qp + pq$ or $2pq$. In summary, the frequencies of genotypes in each and every generation of a population meeting Hardy-Weinberg conditions can be predicted, once p and q are known, as follows:

Frequency:	p^2	+	$2pq$	+	q^2
Genotype:	*AA*		*Aa*		*aa*

Note that the frequency terms represent the terms of the binomial expansion $(p + q)^2$. If more than two alleles are involved, more unknowns can be added, and each term will represent the frequency of a corresponding genotype.

The frequency of phenotypes can then be stated once the frequencies of the genotypes are known:

Dominant phenotype ($A-$) = $AA + Aa = p^2 + 2pq$
Recessive phenotype (aa) = $aa = q^2$

Because the frequencies of the dominant and recessive phenotypes must add up to 1, the second frequency is known as soon as one or the other is calculated.

TASTE TEST

In actual populations it is usually not possible to tell the genotypes of all the individuals. If the *A* allele for a certain trait is dominant over the *a* allele, the *AA* and *Aa* genotypes will have the same phenotype. The **phenotype** is what can be easily observed or measured. If the frequency of the recessive phenotype (*aa*) is measured ($aa = q^2$), q^2 is known. If the frequency of the dominant phenotype is known ($AA + Aa = p^2 + 2pq$), q^2 can be calculated because the dominant and recessive phenotypes must add up to 1:

$$q^2 = 1 - (p^2 + 2pq)$$
$$aa = 1 - (AA + Aa)$$

This means that the recessive phenotype frequency is equal to 1 minus the dominant phenotype frequency.

In this experiment you will determine your phenotype for a certain trait and use the results from the entire class to calculate the approximate number of individuals in the class who are homozygous dominant or heterozygous for the trait, assuming the class is a population meeting the Hardy-Weinberg conditions.

Phenylthiocarbamide (PTC) is a chemical that tastes bitter to some people (tasters) but not to others (nontasters).

□ Place a piece of PTC-impregnated paper on your tongue and moisten it.

Do you detect a bitter taste other than the taste of the paper itself? ______________

□ Chew the paper a bit and roll it around on your tongue.
□ Classify yourself as a taster or nontaster, and record the results of this test in Table 11–2 at the end of this topic.
□ Also record the class results and the total number of students in the class.

The ability to taste PTC is genetically determined by the dominant alleles T^e (early tasters) and T^l (late tasters). (Late tasters can detect PTC only after chewing the paper slightly.) These dominant alleles will be lumped together as *T*. Those people homozygous for the recessive allele *t* are nontasters. Assume that your class is a reasonably accurate sample of an ideal randomly mating population, and use the Hardy-Weinberg Law to calculate the frequencies of *t* and *T* in the class. Once you know *t* and *T*, calculate the percentage of *TT* and *Tt* genotypes in the class.

How many individuals are expected to be homozygous dominant (*TT*)? ______________

How many are heterozygous (*Tt*)? ______________

□ Record your results in Table 11–2.

Sodium benzoate (SB) is another chemical that only some people can taste. Its taste may seem sour, sweet, or bitter. Repeat the test with a piece of paper impregnated with sodium benzoate.

Are you a taster for sodium benzoate?

____

Which type of taste did you detect, if you are a taster?

____

□ Record your observation and the class results in Table 11–2.

What is the frequency of the *t* allele, assuming *t* is a recessive allele for nontaster? ______________

Is this frequency different from that of *t* for PTC tasting?

____

Variation in the ability to taste PTC and sodium benzoate has no obvious survival value today, and yet there is ethnic variation in allele frequencies between different groups. For instance, 63% of Arabs are tasters for PTC, whereas 98% of Native Americans can taste it. At least one condition for the Hardy-Weinberg Law was not met with respect to this gene in the world population.

How do you think this ethnic variation came about?

____

____

____

____

How could this variation in allele frequency be associated with survival potential for the ability to detect a bitter taste?

____

____

____

____

____

Assuming there was no selective advantage for this trait, what other factor might account for the ethnic differences?

____

NATURAL SELECTION

Evolution is defined as a change in gene (allele) frequency. As you have seen, if a population meets the Hardy-Weinberg conditions, it is in genetic equilibrium, so there will be no change in gene frequency and therefore no evolution. The usefulness of the concept of genetic equilibrium is that we can measure changes in gene frequency that do occur and try to discover which of the stipulated conditions has been altered in order to give that change. Is the observed change in gene frequency over a period of time or in different populations due to mutation, small population size, selective mating, migrating individuals, or natural selection? For most genes,

natural selection is almost always by far the most important factor in bringing about changes in gene frequency. This experiment will illustrate how rapidly natural selection can work.

In the English peppered moth, *Biston betularia,* the normal moth is a mottled color, but there is a dominant mutation that gives the moth a black color. Dark individuals with the mutation are called the **melanic** form, or *Biston betularia carbonaria.* The dark moths were much more conspicuous when they rested during the day on the light-colored background of woodland trees, and they were more likely to be eaten by the birds. In other words, natural selection against the dark form was greater than that against the light form. As a result of the widespread pollution that followed in the aftermath of the Industrial Revolution, however, the trees became darkened due to the loss of light-colored lichens. Lichens are extremely sensitive to air pollution, and cannot survive under polluted conditions. Eventually the dark forms were no longer at a selective disadvantage. In fact, their dark coloration was now better protection against predation than the normal coloration. As a result, the frequency of the dark allele and the percentage of dark moths increased.

The elegant experiments of H. B. D. Kettlewell in the 1950s showed that dark moths really do have an advantage over normal moths in actual field studies. When normal moths were released in a polluted environment, 25% of them were recovered, but when dark moths were released under the same conditions, 52% of them were recovered. He went on to show that more white moths than black moths were actually eaten by birds.

First Round

In this experiment we will exaggerate the difference in selection against the two alleles:

Selection against the normal phenotype (*bb*) = 90%
Selection against the dark phenotype (*BB* + *Bb*) = 50%

□ Obtain a small plastic bag with 50 white beans and 50 dark beans. You may also need a coin to flip.

These 100 beans represent your **initial gene pool.** This is the gene pool that would be present if you mixed 25 homozygous dark (*BB*) and 25 homozygous light (*bb*) moths together and allowed them to mate randomly.

□ Shake up the beans to simulate random mixing of the gametes during the first season of mating.
□ Reach into the bag, and (without looking!) take out two beans: this is the first individual.
□ Set these beans aside, and repeat 49 times until you have used up all the beans, arranging the *BB*, *Bb*, and *bb* genotypes in groups as you draw.
□ Select against the phenotypes by removing 50% of the *BB* genotype and 50% of the *Bb* genotype (birds can't tell the difference). Then remove 90% of the *bb* genotype. Round off the number of individuals to be removed to the nearest integer. For example, if you happen to calculate that 4.5 individuals should be removed, flip a coin to decide whether to remove 4 or 5 individuals.
□ Count up the number of light and dark beans left, and calculate the percentages of each.
□ Enter the results in Table 11–3 under generation 1.

If you are not confident about how to proceed, follow the example shown in Table 11–1.

The numbers corresponding to the boxed numbers in the example are the ones you should enter in Table 11–3 under generation 1.

Second Round

Begin the second round by replenishing your beans so that you have 100 beans with the same percentages of light and dark alleles that you had after the first round.

□ Shake up the beans.
□ Select 50 individuals.
□ Remove 50% of the *BB* and *Bb* genotypes and 90% of the *bb* genotype.
□ Calculate the percentages of light and dark beans left after selection, and enter the results in the table under second generation.

TABLE 11–1. CHANGE IN GENE POOL FOR FIRST ROUND

Intial gene pool:	[50*b*] [50*B*]	*b* frequency = [50% (0.50)]
First draw:	9 = *bb*, 32 = *Bb*, 9 = *BB*	
Selection against "moths":	0.9 × 9 = (8) of the *bb*'s must be removed	
	0.5 × 32 = (16) of the *Bb*'s must be removed	
	0.5 × 9 = (5) of the *BB*'s must be removed	
Remaining "moths":	1 = *bb*, 16 = *Bb*, 4 = *BB*	
Gene pool after first round:	[18*b*] [24*B*]	
First round *b* frequency:	18*b*/(18*b* + 24*B*) × 100	*b* frequency = [43% (0.43)]
Add in:	+ 25*b* + 33*B*	
Second round gene pool:	43*b* 57*B*	*b* frequency = 43% (0.43)
You are now ready for the second draw.		

Third Round

Begin the third round by replenishing your beans so that you have the new percentages, and repeat the procedure as described in the second round until you have five generations.

When you have finished five rounds, you will be very tired of drawing beans, but will have enough data to make a meaningful graph of the decline in the frequency of the *b* allele over five generations. Use the grid in Table 11–3 to plot the *frequency of b* against the *number of generations.*

GENETIC DRIFT

In very small populations the selection of alleles making up the individual genotypes in the next generation will not be completely random, and the frequencies of the alleles will fluctuate even without mutation, migration, or selection. This type of evolution is due to **genetic drift** and it becomes more important as population size decreases.

It is possible that the frequency of one allele may drop so low that it disappears from the next generation altogether. The situation where only one of the two alleles is left is termed **gene fixation.**

Measure genetic drift in the following experiment without natural selection.

First Round

□ Put 50 light beans and 50 dark beans in the bag, and shake it up to simulate mating.
□ Draw 10 pairs of beans from the bag to represent 10 individuals.
□ Count the light and dark beans and calculate the percentages. Enter the results in Table 11–4 under generation 1.

Second Round

□ Begin by replenishing your beans so that you have 100 beans in the percentages that you had at the end of the first round.
□ Mix the beans.
□ Draw 10 pairs of beans from the bag to represent 10 individuals.
□ Count the light and dark beans, and calculate the percentages. Enter the results in Table 11–4 under generation 2.

Third Round

□ Replenish your beans, and repeat as in the second round until you have completed five rounds. Enter the results in the table.

Plot the frequency of *b* as a function of generation on the grid below Table 11–4. This is the kind of genetic drift that you might expect to see in a real population that is too small to avoid sampling error due to small population size

Did either your light or dark alleles disappear?

____

What factors could reverse gene fixation in a real population?

____

____

____

Return your bean bag, and dispose of any taste paper debris at the end of this lab.

EXPLORING FURTHER

Mayr, Ernst. "Darwin and natural selection." *Am. Scientist* 65: 321, May 1977.
Stern, C. *Principles of Human Genetics.* San Francisco: W. H. Freeman, 1973.
Wilson, E. O., and W. H. Bossert. *Primer of Population Biology.* Sunderland, Mass.: Sinauer, 1971.

Scientific American Articles

September 1978: This issue is devoted to Evolution.

Allison, A. C. "Sickle cells and evolution." August 1956 (#1065).
Ayala, F. J. "The mechanisms of evolution." September 1978 (#1407).
Bishop, J. A., and L. M. Cook. "Moths, melanism and clean air." January 1975 (#1314).
Brower, L. P. "Ecological chemistry." February 1969 (#1133).
Eckhardt, Robert B. "Population genetics and human origins." January 1972 (#676).
Kimura, Motoo. "The neutral theory of molecular evolution." November 1979 (#1451).
Lewontin, Richard C. "Adaptation." September 1978 (#1408).
Mayr, Ernst. "Evolution." September 1978 (#1400).
Wills, C. "Genetic load." March 1971 (#1172).

Student Name ______________________________ **Date** ______________

TABLE 11–2. DATA ON PTC AND SB TASTING

Data from the PTC Test

Individual results: Taster ____________ Nontaster ____________

Class results:

Number of tasters ______ = ______ % tasters (*TT* + *Tt*)

Number of nontasters ______ = ______ % nontasters (*tt*)

Total ______

q^2 = ______ ; q = ______ ; frequency of *t* allele = ______

1 − q = p = frequency of *T* allele = ______

p^2 = genotype of *TT* = ______ % of the class

2pq = genotype of *Tt* = ______ % of the class

If class member is known to be a taster, what is the chance that he or she is homozygous for the *T* allele?

% *TT* × 100/(% *TT* + % *Tt*) = ____________ %

Data from the SB Test

Individual results: Taster ____________ Nontaster ____________

Class results:

Number of tasters ______ = ______ % tasters (*TT* + *Tt*)

Number of nontasters ______ = ______ % nontasters (*tt*)

Total ______

q^2 = ______ ; q = ______ ; frequency of *t* allele = ______

1 − q = p = frequency of *T* allele = ______

p^2 = genotype of *TT* = ______ % of the class

2pq = genotype of *Tt* = ______ % of the class

If class member is known to be a taster, what is the chance that he or she is a carrier for the *t* allele?

% *Tt* × 100/(% *TT* + % *Tt*) = ____________ %

TABLE 11–3. FLUCTUATION IN FREQUENCY OF *b* ALLELE DUE TO INCREASED SELECTION AGAINST IT

	B			*b*		
Rounds	**#**	**%**	***Frequency***	**#**	**%**	***Frequency***
Initial	50	50	.5	50	50	.5
Generation 1						
Generation 2						
Generation 3						
Generation 4						
Generation 5						

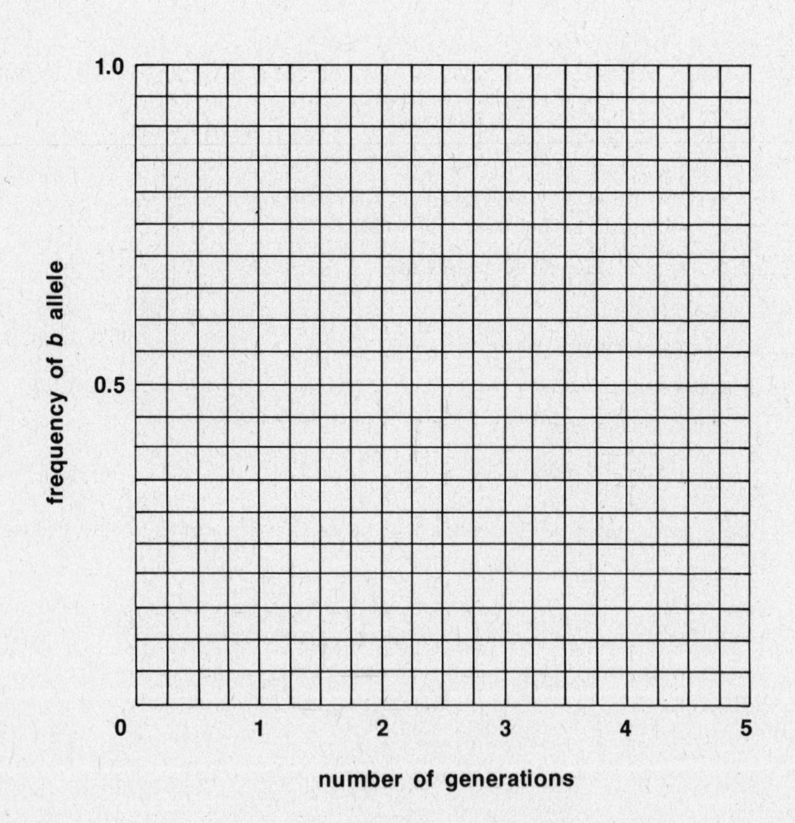

Student Name ______________________________ **Date** ____________________

TABLE 11–4. FLUCTUATION IN FREQUENCY OF *b* ALLELE DUE TO GENETIC DRIFT

	B			*b*		
Rounds	**#**	**%**	***Frequency***	**#**	**%**	***Frequency***
Initial	50	50	.5	50	50	.5
Generation 1						
Generation 2						
Generation 3						
Generation 4						
Generation 5						

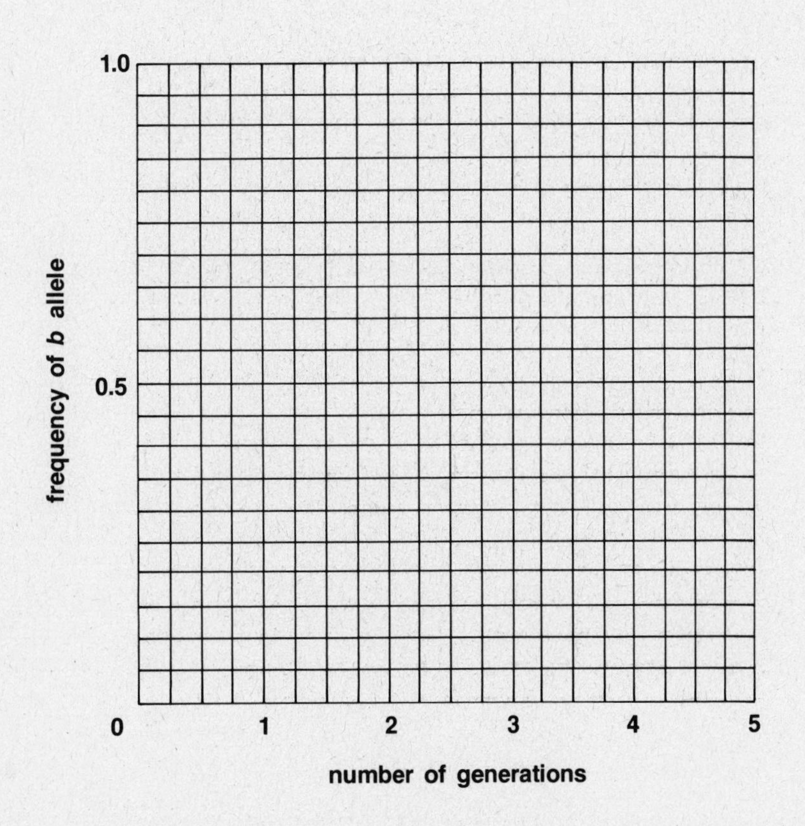

TOPIC 12

Classification of Organisms

OBJECTIVES

When you have completed this topic, you should be able to:

1. Define taxonomy and taxon.
2. List the major taxa in the correct hierarchy from most general to least general.
3. Correctly classify an unknown organism using a simple dichotomous key.
4. Define numerical taxonomy.
5. Arrange a set of imaginary organisms into a taxonomic array to show their possible evolution.

TEXT REFERENCES

Chapter 20 (A–D).
Species, binomial nomenclature, taxonomy, Five Kingdoms.

INTRODUCTION

In order to understand some of the relationships between living organisms, biologists use the process of classification called **taxonomy.** Taxonomy is simply the ordering of a set of organisms according to some system. Animals are classified according to the following scheme:*

Kingdom
Phylum
Class
Order
Family
Genus
Species

Can you classify yourself (*Homo sapiens*) according to this scheme?

*An easy way to remember this scheme is by a mnemonic device such as "Kindly Professors Cannot Often Fail Good Students."

Kingdom ____________
Phylum ____________
Class ____________
Order ____________
Family ____________
Genus ____________
Species ____________

Plants are classified in the same way except that Division is used instead of Phylum.

The members of each taxonomic group, or **taxon,** all share certain characteristics. Within the taxon, the members may be divided into smaller groups whose members share certain characteristics that are not found in the members of the other groups.

The presence or absence of certain traits will allow you to determine whether an organism belongs to a certain taxon, which more specific taxon it belongs to within the first taxon, which taxon within the second taxon, and so on. These traits, however, are not always the ones that are obvious in an intact animal or plant. Assignment to a certain taxon may depend on the careful study of an organism's biochemistry, embryology, or parts of its life cycle that are difficult to observe.

In order to identify the taxon to which a certain organism belongs in the field or without extensive study, we use a shorthand guide to readily discernible features called a **dichotomous key** (dichotomous means "two forks"). Such a key divides a certain group of organisms into two groups on the basis of a certain trait, then divides each of those groups into two groups, and so on, until the desired level of identification is reached.

Because the purpose of a key is to identify something quickly, each key will be limited in scope. If it starts out with a broad group of organisms, say, invertebrates or algae, it will not be able to identify the species of each organism but perhaps only the family or order. If the key is more limited in content, limited, for example, to spring wildflowers of a certain geographical area, it will be able to include genus- and species-level identification. The person using a key can use his prior knowledge or the constraints of geographical region or season to choose a key that will give as precise an identification as possible.

Dichotomous keys can be used in series to identify organisms of special interest when the first key is very general. A person studying seashore life might use a key to marine invertebrates in conjunction with a key to bivalves to identify a certain clamlike organism.

Consider the dichotomous scheme shown in Figure 12–1 for grouping all organisms. Each step in keying out an organism is simply a yes or no answer to successive questions in a key, part of which would look like this:

1a. Is the organism motile? (2a)
1b. Is the organism stationary? (3a)
2a. Is the organism aquatic? (4a)
2b. Is the organism terrestrial? (5a)
3a. Is the organism aquatic? (6a)
3b. Is the organism terrestrial? (7a)
4a. Does the organism have ray fins? (class Osteichthyes, or bony fish)
4b. Does the organism lack ray fins? (9a)
5a. Does the organism have four legs? (10a)
5b. Does the organism have zero, two, six, eight, or many legs? (11a)
6a. Is the organism green? (12a)
6b. Is the organism not green? (13a)
7a. Does the organism have a woody stem? (14a)
7b. Does the organism have a herbaceous stem? (15a)

To use the key, you would begin with question 1a. If the answer to this question were yes, you would proceed to question 2a; if the answer were no, you would go to question 1b. You would then continue from 1b to 3a to 6a, and so on. Three consecutive yes answers would identify an organism as a fish in the class Osteichthyes, but the identification of most organisms, including other fish, would require several more questions. A more detailed key would be necessary to identify the fish more specifically.

In today's laboratory exercise you will use simple dichotomous keys to identify members of a class that are similar in their general features but easily distinguished when you know what to look for.

CONIFER KEY (SUBDIVISION SPERMOPSIDA, CLASS CONIFERAE)

Conifers are plants that usually have evergreen leaves and produce gametes in male and female cones.

□ Select one of the specimens of a conifer on demonstration in the lab, and key it out using the key in Table 12–1. This key includes some of the conifers that will be found in your region of the country. Your instructor may give you additional information about species that may be especially common in your area. To use the key, begin with question 1a. If this item applies, go to question 2a; if not, go to 1b. Continue from 1b to 13a, from 13a to 14a, and so forth.

Specimen number: ______________

Identification: ______________

□ If living conifers are available in the field for class study, select a specimen and key it out in your spare time.

Characteristics of field specimen:

____
____
____
____
____
____
____
____

Identification: ______________________________

Your instructor may ask you to key out other preserved or live specimens.

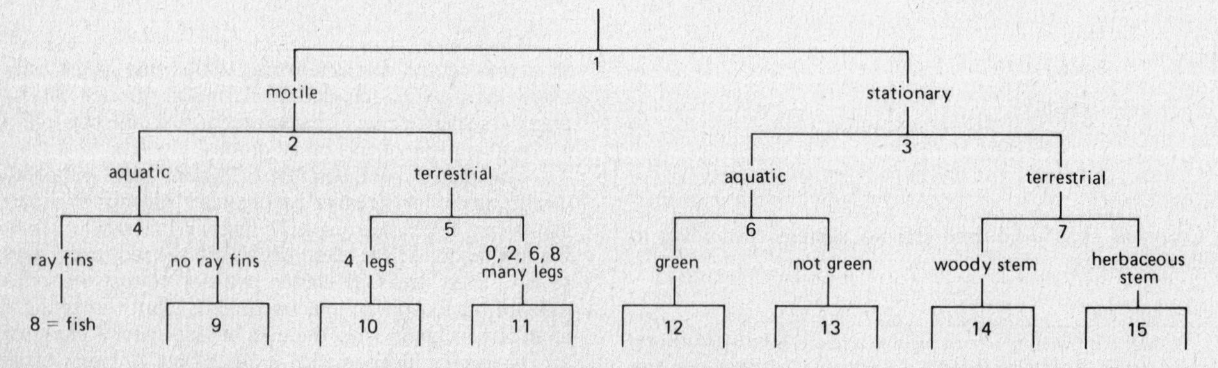

Figure 12–1. A simple dichotomous key.

TABLE 12–1. A KEY TO THE CONIFERS

1a.	Leaves are in clusters.	(2a)
1b.	Leaves are borne singly along the stem.	(13a)
2a.	More than 5 leaves in a cluster.	(3a)
2b.	Leaves occur in clusters of 2–5; persistent for several seasons.	(4a)
3a.	Leaves are deciduous (fall off in winter), soft, flattened.	(Larch or Tamarack)
3b.	Leaves are persistent, stiff, 4-sided (ornamental in warmer areas).	(True cedar)
4a.	Leaves mostly 5 in a cluster.	(White pine)
4b.	Leaves 2–3 in a cluster.	(5a)
5a.	Leaves mostly 3 in a cluster.	(6a)
5b.	Leaves mostly 2 in a cluster.	(8a)
6a.	Twisted needles, less than 5 in. long.	(Pitch pine)
6b.	Straight needles, more than 5 in. long.	(7a)
7a.	Cones very thorny (South only).	(Loblolly pine)
7b.	Cones not thorny (South only).	(Longleaf pine)
8a.	Limbs orangeish in color.	(Scotch pine)
8b.	Limbs not conspicuously orange.	(9a)
9a.	Needles more than 3 in. long.	(10a)
9b.	Needles less than 3 in. long.	(11a)
10a.	Bark grayish in color.	(Austrian pine)
10b.	Bark reddish-brown or brown (North).	(Red pine)
	Bark reddish-brown or brown (South)	(Shortleaf pine)
11a.	Cones have very long thorns.	(Mountain pine)
11b.	Cones have tiny thorns or no thorns.	(12a)
12a.	Twigs not whitened (North).	(Jack pine)
12b.	Twigs whitened (South).	(Scrub pine)
13a.	Leaves are scalelike or triangular and opposite or whorled.	(14a)
13b.	Leaves needlelike or flattened.	(17a)
14a.	Branchlets flattened; all leaves scalelike, opposite.	(White cedar, or Arbor vitae)
14b.	Branchlets not flattened.	(15a)
15a.	Leaves needlelike to triangular in tight spirals (frost-free).	(Monkey puzzle tree)
15b.	Leaves needlelike in whorls or pairs.	(16a)
16a.	Leaves 3-sided in whorls of 3 on 3-sided twigs.	(Dwarf juniper)
16b.	Leaves 3-sided in pairs in 4 rows on 4-sided twigs.	(Red cedar)
17a.	Leaves soft on green branchlets (like compound).	(18a)
17b.	Leaves rigid on brownish branchlets (solitary).	(20a)
18a.	Leaves dark green, persistent.	(Redwood)
18b.	Leaves light yellow-green, deciduous.	(19a)
19a.	Leaves to 1 in., subopposite.	(Dawn redwood)
19b.	Leaves less than 1 in., alternate (South).	(Bald cypress)
20a.	Leaves 4-sided with persistent woody petiole.	(21a)
20b.	Leaves flattened in cross section.	(25a)
21a.	Young twigs are hairy.	(22a)
21b.	Young twigs are not hairy.	(23a)

22a. Leaves are dark or yellow-green. (Red spruce)
22b. Leaves may be blue-green, with white powder. (Black spruce)

23a. Leaves are sharp and at right angles to the stem. (Blue spruce)
23b. Leaves point forward. (24a)

24a. Leaves dark green and branchlets droop. (Norway spruce)
24b. Leaves blue-green and branchlets do not droop. (White spruce)

25a. Leaves do not have stomata underneath; seed in red berry (shrub). (26a)
25b. Leaves have two lines of stomata (tree). (27a)

26a. Leaves ¾ in. or less with inconspicuous midrib. (American yew)
26b. Leaves 1 in. or longer with prominent midrib. (Japanese yew)

27a. Leaves round-tipped, whitened beneath. (28a)
27b. Leaves pointed, green beneath. (29a)

28a. Leaves are attached by slender stalks; twigs rough. (Eastern hemlock)
28b. Leaves are not attached by stalks; twigs smooth. (Balsam fir)

29a. Leaves are dark green and straight. (Douglas fir)
29b. Leaves are light bluish-green and strongly curved. (Fir)

NUMERICAL TAXONOMY

The real problem with classification comes in deciding which characteristics should be considered in setting up a given category, or taxonomic group. All the organisms within the category will have some characteristics in common, but none of the organisms would be expected to have every one of the characteristics associated with the category.

The introduction of computers has made it possible to consider large numbers of characteristics in determining the degree of relatedness of different organisms. Classification is based not on the relative importance of certain characteristics, but rather on the number of characteristics held in common. This type of classification is called **numerical taxonomy,** or **quantitative taxonomy.** It has been widely used in the classification of organisms, diseases, plant communities, politicians, psychological types, socioeconomic neighborhoods, and so forth.

An experimental system to test the procedure of numerical taxonomy was devised by Joseph H. Camin. It involves a set of imaginary animals, christened "Caminalcules," which was generated "according to rules known so far only to him, but believed to be consistent with what is generally known of evolutionary principles."*

Study the set of 29 Caminalcules illustrated in Figure 12–2, and try to arrange the set into some meaningful pattern of groups with anatomical features in common. For example, individuals with eyes might be grouped separately from those without eyes. Then animals with two eyes might have given rise to animals with a single eye or vice versa.

Your instructor will either provide you with a set of cutouts corresponding to Figure 12–2 or will ask you to cut up Figure 12–2 yourself so that you can try out different arrangements.

Arrange all of the individuals into a branching tree to show how they might have arisen through evolution from a common ancestor. The goal is to construct the tree so that it requires the *fewest number of postulated evolutionary steps.* A change from organism #7 to #15 (or vice versa) requires a single step, whereas a change from #7 to #8 would require two steps.

When you are satisfied with the arrangement that you have worked out, record the results in Table 12–2 in the form of an evolutionary tree. Use numbers to identify the individual Caminalcules, and work with a pencil so that you can change your mind if necessary. Consult you instructor to see how your tree compares with others produced by the class. Your instructor also has a computer-generated result which you can check when you have finished your own scheme.

Return all materials when you are finished with your work for this laboratory.

*From "Numerical taxonomy" by Robert R. Sokal.

EXPLORING FURTHER

Blackwelder, R. E. *Taxonomy: A Text and Reference Book.* New York: John Wiley and Sons, 1967.

Hosie, R. C. *Native Trees of Canada.* 8th ed. Don Mills, Ontario, Canada: Fitzhenry and Whiteside Ltd. in cooperation with the Canadian Forestry Service and the Canadian Government Publishing Center, Supply and Services, 1979.

Petrides, George A. *Field Guide to Trees and Shrubs.* New York: Houghton-Mifflin, 1973.

Savory, T. *Animal Taxonomy.* London: Herremann Educational, 1970.

Whittaker, R. H. "New concepts of kingdoms of organisms." *Science* 163: 150, 1969.

Whittaker, R. H., and L. Margulis. "Protist classification and the kingdoms of organisms." *Biosystems* 10: 3–18, 1978.

Scientific American Articles

Sokal, Robert R. "Numerical taxonomy." December 1966 (#1059).

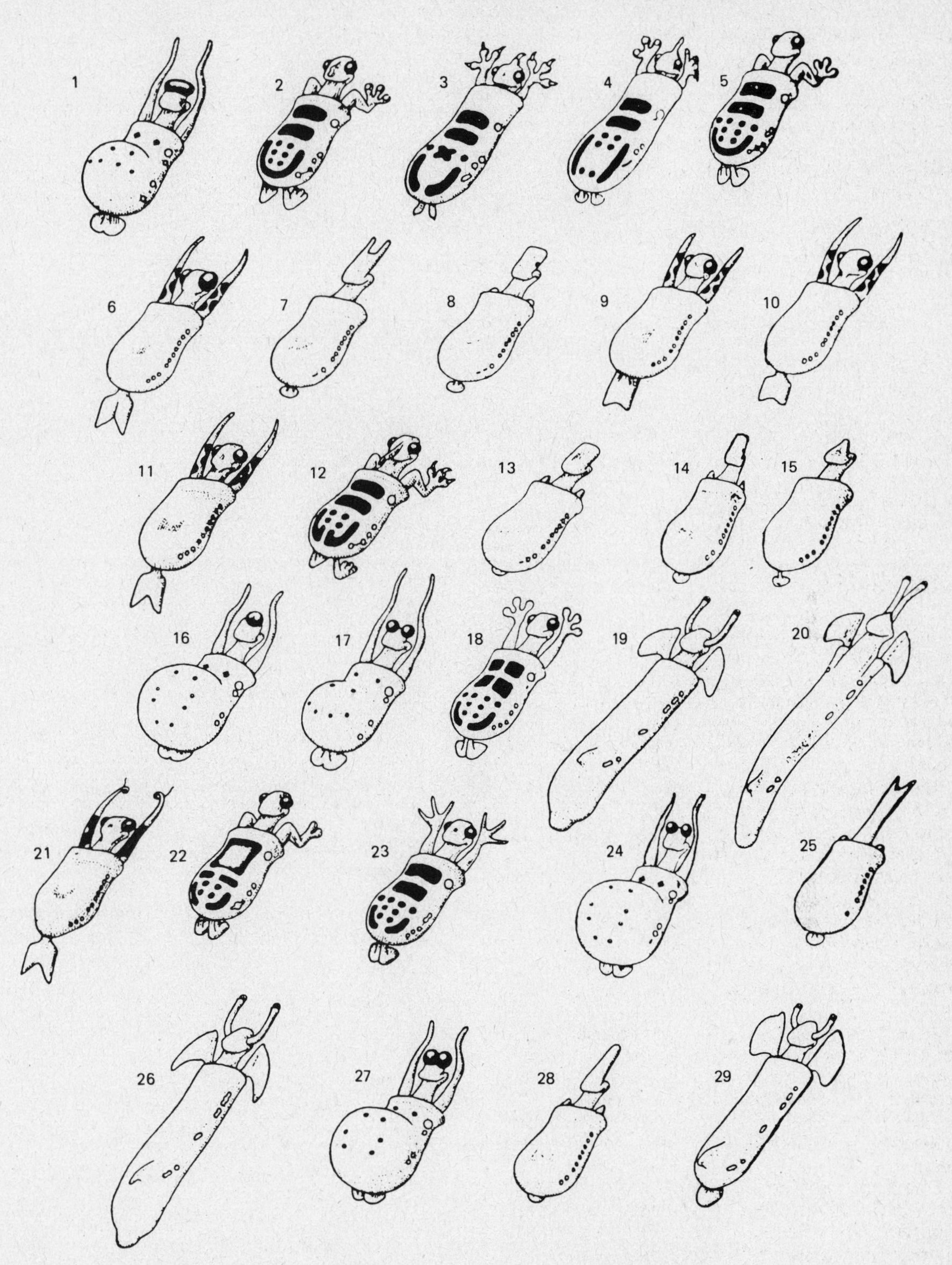

Figure 12–2. Caminalcules. Imaginary organisms ready to be classified. (From "Numerical taxonomy" by Robert R. Sokal.)

Student Name ______________________________ **Date** ______________

TABLE 12–2. CLASSIFICATION OF CAMINALCULES

Use the identifying numbers to record your classification scheme in the form of a branching diagram.

Introduction to Living Organisms (Topics 13-19)

TEXT REFERENCES

Chapter 20 (D–E).
Five Kingdoms: Monera, Protista, Fungi, Plantae, Animalia; fossil record.

As more information about the evolutionary history and biochemistry of organisms has become available, the old division of organisms into plants, animals, and protists was no longer sufficient. Although the boundaries between the kingdoms listed below are somewhat unclear, the following scheme* is used in Topics 13–19:

Kingdom Monera: All prokaryotic organisms (bacteria, blue-greens).

Kingdom Protista: Eukaryotic unicellular heterotrophs and most unicellular autotrophs (protozoa, diatoms, dinoflagellates, euglenoids).

Kingdom Fungi: Eukaryotic heterotrophs that absorb food, mostly multicellular (yeasts, molds, mushrooms).

Kingdom Animalia: Eukaryotic multicellular heterotrophs that ingest food (sponges to vertebrates).

Kingdom Plantae: Eukaryotic autotrophs, usually multicellular, with cell walls of cellulose (algae, mosses, ferns and fern allies, seed plants).

The characteristics of the kingdoms given here are to provide a general idea of the type of organism that you might find in each one, and are by no means definitive. (A yeast, for example, is a unicellular heterotroph, but it is placed in the kingdom Fungi because it is much more closely related to fungi than to the protozoans or unicellular algae.)

In the discussion of the five kingdoms in Topics 13–19 you will become familiar with the diversity of living organisms. The structures and life histories are given in considerably more detail than you may be required to learn. This will allow you some flexibility to concentrate more on those organisms that you find most interesting and that the instructor is able to provide. The point of these laboratories is not to have you learn a stupendous number of biological "facts," but rather to introduce you to the different ways that organisms cope with their common biological needs: protection, nutrition, excretion, reproduction, and dispersal.

*See R. H. Whittaker, "New concepts of kingdoms of organisms," *Science* 163: 150–160, 1969.

TOPIC 13

Introduction to Monera and Protista

OBJECTIVES

When you have completed this topic, you should be able to:

1. Describe the characteristics of prokaryotic organisms.
2. List the types of organisms that are prokaryotic.
3. Demonstrate the occurrence of bacteria in the environment by culturing samples from locations where they might be found.
4. Using bioluminescent bacteria as an oxygen indicator, explain the effect of bacteria on the level of oxygen in a closed system.
5. Describe the features of blue-greens that distinguish them from bacteria and discuss their role in nature.
6. List the features of eukaryotic cells that distinguish them from prokaryotic cells.
7. Describe the most common eukaryotes inhabiting the termite gut in terms of their appearance and locomotion and the phylum to which each belongs.
8. Describe the animal-like and plantlike characterisics of *Euglena,* a photosynthetic protist that also has features of a flagellate.
9. Distinguish algae that belong to the diatoms from those that belong to the dinoflagellates and state the importance of these organisms in nature.

TEXT REFERENCES

Chapters 22, 23.
Prokaryotes, cyanobacteria, eubacteria, pathogenic bacteria, dinoflagellates, euglenoids, diatoms, heterotrophic protists.

INTRODUCTION

In this exercise you will study the variety of single-celled organisms that exists, and learn something about their importance to larger organisms. Prokaryotes are either true bacteria (Eubacteria) or blue-green bacteria (Cyanophyta). They belong to the kingdom Monera, which includes all the prokaryotes. Eukaryotes arose from prokaryotes about 1.45 billion years ago and gave rise to the immense variety of higher organisms that we have today, but the prokaryotes have persisted and are vital in maintaining the stability of the biosphere at the present time.

Single-celled eukaryotes are a heterogeneous group of organisms included in the kingdom Protista. The heterotrophic protists are the protozoa that are divided into phyla primarily on the basis of their means of locomotion. The autotrophs are divided into phyla on the basis of the composition of their cell walls and the types of chlorophyll that they possess. Unfortunately, these divisions are somewhat unclear because some heterotrophs are obviously closely related to certain autotrophs. Sometimes an autotroph can lose its chloroplasts and continue to exist perfectly well as a colorless carbon copy of its former self.

BACTERIA IN THE ENVIRONMENT

Bacteria are so small (often only about 1 μ in size) that their internal structure is not visible in the light microscope.

Because of their small size, identification of bacteria depends on their shape, their arrangement in groups, the appearance of their colonies on culture plates, and the metabolic reactions that they can carry out.

Shape

The form of a bacterium may be a sphere (**coccus**), a rod (**bacillus**), or a spiral (**spirillum**). Any of these forms may have one or more **flagella** and thus be motile.

□ Review the basic bacterial forms by examining a prepared stained slide under oil immersion.

Colonies: Airborne Bacteria

Any organism of sufficient size is covered with a collection of bacteria called its "natural flora." Most of these bacteria are harmless (**commensal**) or even helpful (**mutualistic**), but some of them are **pathogens** which have the potential of causing disease. Even otherwise harmless bacteria can cause infection and disease if they gain access to parts of our body that are normally protected. The harmless bacteria of our natural flora help protect our bodies by maintaining conditions that are unfavorable for the growth of pathogens.

Many bacteria on our bodies, in the air, and in our immediate environment can be grown in the laboratory or **cultured** by providing them with a suitable medium that simulates the conditions to which they are accustomed. The medium can be either a liquid broth or a solid gel that is made by adding agar to a liquid medium and pouring the sterilized liquid into a flat Petri plate where it hardens and can be kept sterile. When sterile medium is **inoculated** by adding a sample of bacteria, the bacteria will multiply rapidly due to the lack of competition and will soon exhaust the nutrients. A liquid broth will become cloudy due to the presence of billions of cells, and an agar Petri plate will become covered with dots of **colonies,** or clones, consisting of 10^9 or so cells all descended from a single cell. The size, shape, color, and appearance of the colony help identify the type of bacterium producing it. **Indicators** may be added to the agar to show whether a certain reaction has been carried out. Any bacterium can be identified if you can find conditions under which the bacterium will grow in the laboratory and have the patience to carry out a sufficient number of tests.

IN THE FOLLOWING EXPERIMENT, FOLLOW THE DIRECTIONS GIVEN BY YOUR INSTRUCTOR FOR SEALING THE CULTURE PLATES AND FOR DISPOSING OF THEM PROPERLY. NEVER OPEN PLATES THAT HAVE BEEN INCUBATED BECAUSE THEY MAY CONTAIN PATHOGENIC BACTERIA.

Using nutrient agar plates, test for the presence of airborne bacteria.

□ Label the plate AIR, and add your initials and the date.
□ Place the plate near a source of moving air such as an open window, a hallway with traffic, or an air conditioner, and open it fully for 5 min.
□ Close the plate and place it upside down in the incubator at 37°C for 24–48 hr.
□ Label a second plate COUGH, and add your initials and the date.
□ Open the plate and cough over it with your mouth open three times.
□ Close the plate and place it upside down in the incubator at 37°C for 24–48 hr.
□ Label a third plate SUPPRESSED COUGH, and add your initials and the date.
□ Cough over the plate three times but either keep your mouth closed or use a handkerchief. Close the plate.
□ At the end of the experiment record the number of colonies on each plate, and enter the results in Table 13–1 at the end of this topic.
□ Dispose of the plates properly.

Are you or the air in general a better (or worse!) source of bacteria? ______________________________

____

Did suppressing your cough cut down on the number of bacteria that you spread? ______________________________

____

Metabolism: Oral Bacteria

BRING A TOOTHBRUSH TO THE LABORATORY FOR THIS EXERCISE.

The natural flora of your mouth is important in determining the susceptibility of your teeth to decay. *Lactobacillus* is a type of bacterium that contributes to conditions favoring decay by producing acid which lowers the pH at the tooth surface. Even though this bacterium does not actually cause the decay, there is a strong statistical correlation between the activity of *Lactobacillus* and susceptibility to decay. It is possible to test for the presence of this bacterium by using Snyder's medium, which contains an indicator (bromcresol green) that turns yellow when acid is produced.

□ If you have not already looked at some of your own bacteria in the microscope, turn to Topic 4 and do so now, following the directions for obtaining and staining a sample.
□ Obtain 2 sterile test tubes, 2 sterile pipets, 3 tubes of Snyder's medium, and some sugarless gum.
□ Place the tube in a boiling water bath for 10 min to melt the agar and then keep it warm (45°C) until it is inoculated.
□ Chew the sugarless gum without swallowing, and drool into the test tube until you have at least 1 mL of saliva liquid, not counting the bubbles.
□ Use a sterile pipet to transfer 0.2 mL of saliva to the tube of agar.
□ Cover it and mix the saliva with the medium by tapping the tube while holding it firmly with the other hand.
□ Allow the tube to cool at room temperature until the agar has gelled.

□ If you have brought your toothbrush, brush your teeth thoroughly, repeat the saliva collection in a second tube, and inoculate a second tube of medium using a fresh sterile pipet.
□ Incubate the tubes at 37°C and also incubate a third uninoculated tube as a control.
□ After 24 hr, 48 hr, and again after 72 hr, record the color of each tube in Table 13–2.
□ Use the interpretation chart to determine your susceptibility to tooth decay.

BIOLUMINESCENT BACTERIA

Some marine bacteria can produce light under favorable conditions in the presence of oxygen. These bacteria are quite plentiful in sea water and are also symbiotic in the light organs of certain fish. The fish provide the bacteria with nutrients and oxygenation and then utilize the light produced by the bacteria for their own purposes—for example, communicating, attracting prey, and defending against predators.

The reaction requires free **oxygen,** the enzyme **luciferase,** and reduced **flavin mononucleotide** (FMN_{red}):

$$\text{aldehyde} + FMN_{red} + O_2 \xrightarrow{\text{luciferase}} \text{acid} + FMN_{ox} + \text{light}$$

The bacteria must be actively growing to produce enough enzyme and reduced FMN to carry out the reaction in which aldehyde and FMN are oxidized by oxygen.

□ Fill a test tube two-thirds full with an active culture of luminescent bacteria such as *Photobacterium fischeri*.
□ Cork the tube, and take it into a dark closet or darkroom.
□ Shake the tube gently to mix it well.

What color of light is produced? ______________

□ Hold the tube very still and watch the light fade away.

What part of the tube becomes dark most quickly?

____

____

Which component of the reaction is being used up so that the cells become dark? ______________________

____

□ Test your answer by gently swirling the tube again.

Why does mixing the tube allow the cells to light up?

____

____

____

____

Do the cells at the air interface ever become dark?

____

These bacteria are very sensitive indicators of the amount of oxygen in the environment. There is still oxygen in the part of the tube that is dark, but not enough to support bioluminescence.

How is oxygenation of bodies of water such as ponds and streams maintained so that fish and other aerobic organisms can survive?

____

____

____

____

CYANOBACTERIA (BLUE-GREEN BACTERIA)

Blue-greens are widespread in the environment and can exist under extremely harsh conditions in which organic nutrients are entirely lacking. They are a problem in the laboratory because they are able to contaminate distilled water supplies that are exposed to light by using the light for energy and obtaining minerals from the traces that remain in the distilled water. Most blue-greens carry out photosynthesis with the production of oxygen and thus help recycle CO_2 into organic compounds. Many blue-greens also fix nitrogen by converting atmospheric N_2 into ammonia (NH_3), which can be used to synthesize amino acids. The ability of these forms to obtain energy, carbon, and nitrogen from the air makes them the most nutritionally independent organisms on earth. In addition, their prokaryotic features make them very hardy and allow them to compete successfully with eukaryotes. When conditions favor the rapid multiplication of blue-greens, they tend to form a mat on the surface of a body of water and cause the suffocation of the aerobic organisms. Their gelatinous capsules and toxins make them poor food for predators. When the oxygen is depleted, anaerobic bacteria flourish and produce foul-smelling gases, such as hydrogen sulfide, that make the water unfit for drinking or recreation.

□ Examine a drop of *Anabaena* culture under high power in your microscope.

Is this organism in the form of single cells, colonies, or filaments? ______________

Are all the cells alike? ______________

The larger cells are called **heterocysts** and are specialized for nitrogen fixation. (They may not be present in your sample.)

What color are the majority of the cells? ______________

What color are the heterocysts? ______________

Do you see chloroplasts? ______________

Blue-greens are not always blue green in color, but they always contain chlorophyll *a*, as do green plants. The accessory pigments may give them a red, purple, or brown color. The pigments are located on membranes, but there are no membrane-bound chloroplasts.

HETEROTROPHIC PROTISTS IN THE TERMITE

There are four groups of heterotrophic protists:

1. **Phylum Sarcodina:** Protists that move by means of pseudopods like *Amoeba.*
2. **Phylum Ciliophora:** Protists that have cilia, like *Paramecium.*
3. **Phylum Zoomastigina:** Protists that have one or more whiplike flagella, each of which is equivalent to a very long cilium. *Trichomonas* is a flagellated protist that is responsible for human vaginal infection.
4. **Phylum Sporozoa:** Parasitic protists that may have evolved from flagellates and sarcodines and that are no longer motile.

Examples of the first three types can be seen in any environment that favors the growth of protozoans, such as the internal organs or intestine of a higher animal. Sporozoans are generally parasites of mammals.

Termites need a healthy population of protists in their hindgut in order to survive on their normal diet of wood, which is very rich in cellulose. Certain flagellates produce the enzyme **cellulase,** which hydrolyzes the wood fragments ingested by the termite. The flagellates in turn release some of their nutrients into the gut of the termite. The termite absorbs and lives on these nutrients. The termite and the flagellate are **mutualistic symbionts.** Young termites must become infected with the flagellates at an early age or they will starve to death, and because termites shed their hindgut and foregut lining and content at each molt, termites that are isolated after molting will not become reinfected and will also starve.

□ Place a termite in a drop of physiological saline (0.85%) on a slide, and press on the abdomen with the forceps to hold it in place.
□ Pull off the head with another forceps; most of the intestine will come out with the head. Separate the free part of the intestine from the head and abdomen.
□ Discard the rest of the termite, and tease apart the intestine with needles or insect pins.
□ Add a cover slip and observe the intestinal flora under low power in your microscope.

Which are the most prevalent: amoebalike organisms, flagellates, or ciliates?

____

□ Add a drop of Protoslo and/or a drop of methylene blue stain and observe under high power.

Some of the common symbionts of termites are shown in Figure 13–1. Try to identify one or more of them in your slide.

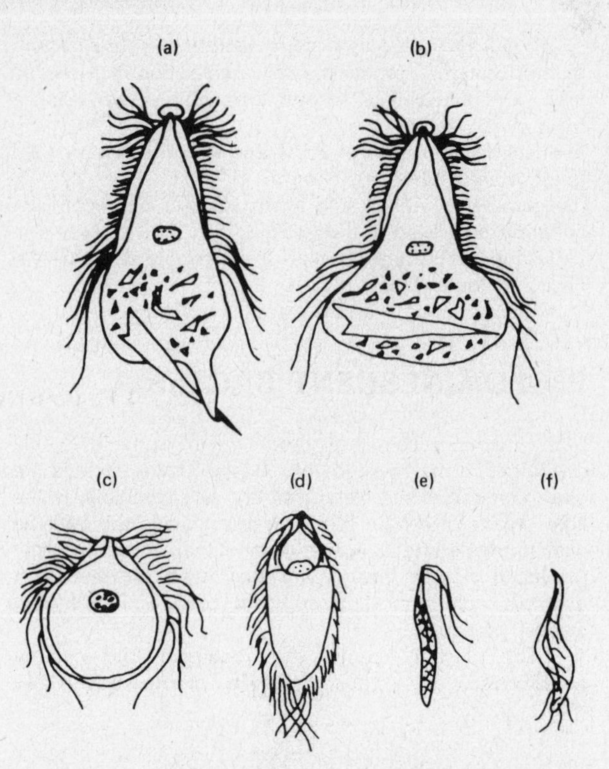

Figure 13–1. Termite gut flagellates. Some common flagellates that you might find inhabiting the termite gut. (a) *Trichonympha collaris* ingesting a particle of wood; (b) *Trichonympha collaris* digesting wood particles; (c) *Trichonympha sphaerica*; (d) *Trichonympha agilis*; (e) *Streblomastix strix*; (f) *Dinenympha gracilis.* (a), (b), and (c) as seen under low power; (d), (e), and (f) under high power. (Drawings (a) and (b) after *Animals without Backbones* by R. Buchsbaum by permission of the University of Chicago Press. Copyright University of Chicago 1948, 1975; (c)–(f) after G. M. Bousch and H. C. Coppel, p. 313 in *Insect Diseases,* Vol. II, ed. by G. E. Cantwell. New York: Marcel Dekker, 1974. Courtesy of Marcel Dekker, Inc.)

PHYLUM EUGLENOPHYTA: *EUGLENA*

You have already examined an example of a unicellular green alga, *Chlamydomonas,* in Topic 4 where you saw how such cells may have formed colonies. Other photosynthetic protists, such as *Euglena,* are not as closely related to green plants; *Euglena* may be related to the simple ancestral flagellates that gave rise to both modern plants and animals. A diagram of *Euglena* is shown in Figure 13–4 at the end of this topic.

Euglena has structures in which starch is stored, a characteristic of green algae. Like *Chlamydomonas,* it has a photosensitive eyespot which allows it to perceive and respond to light. Also like *Chlamydomonas, Euglena* has chlorophylls *a* and *b* contained in chloroplasts, and a flagellum, but it does not have a thick cell wall composed of cellulose. Instead it has a tough elastic pellicle outside the cell membrane. This allows *Euglena* to move by changing its shape and wriggling along the substrate in a unique form of locomotion called **euglenoid movement.** Furthermore, this organism is capable of absorptive nutrition just like a heterotroph. If it is deprived of light, but given a rich medium, it will survive. Its photosynthetic pigments will be lost, and it will not be able to resume autotrophy if the nutrients are removed. This conversion of an autotroph to its sister heterotroph can also be caused artificially by treatment with certain drugs.

Would you expect *Euglena* to show a positive or negative response to light? _______________

□ Cover a culture dish of *Euglena* with black paper or plastic, and make a small hole on one side.
□ Place the dish so that the hole is near a source of strong light such as a lamp or sunlit window.
□ Examine the dish at the end of the period, and observe whether the *Euglena* have migrated toward or away from the source of light.

What kind of response does *Euglena* show?

____

□ Make a wet mount of a *Euglena* culture, add a drop of Protoslo, and observe under high power.
□ Try to locate the **eyespot** near the attachment of the flagellum.

What color is the eyespot? _______________

□ Locate the **reservoir** (gullet) in which the base of the flagellum is attached.
□ Locate the **contractile vacuole** near the reservoir.
□ Identify the green **chloroplasts** which take up much of the space within the cell.
□ Locate the large **nucleus.**
□ Note the location of the **cell membrane,** and locate the **pellicle** which gives the cell support at the outer boundary of the cell.
□ Scan the slide for cells that may have become attached to the substrate. Observe their method of locomotion.

Does *Euglena* have a rigid cell wall? _______________

□ Stain the slide with Lugol's solution, and allow a few minutes for the stain to penetrate.
□ Locate the **paramylum bodies** which stain dark. Paramylum bodies contain paramylum, a storage form of glucose.
□ Study the **flagellum,** which should now be clearly visible.

How is the base of the flagellum attached?

____

□ Label these structures in Figure 13–4.

PHYLUM PYRROPHYTA: DINOFLAGELLATES

This phylum contains the **dinoflagellates,** which are primitive autotrophs and most of which contain chlorophylls *a* and *c* and carry on photosynthesis. Their accessory photosynthetic pigments often give dinoflagellates a red color. The chromosomes contain DNA but not protein, and the division of the nucleus during reproduction of the cell is a bizarre intermediate between prokaryotic and eukaryotic nuclear division. Most have two flagella, and some have distinctively formed cellulose armor covering the cell membrane. Figure 13–2 illustrates some dinoflagellates.

One species of *Gonyaulax* produces a toxin that kills fish and mussels that ingest it. Other species are bioluminescent and cause the "sparks" in sea water at night during the hot summer months. *Noctiluca,* as its name (= night light) implies, also causes flashes of bioluminescence at night. Each cell produces repeated bursts of light in response to the agitation of wave action or the movements of a nighttime swimmer.

□ Examine a prepared slide of some dinoflagellates in your microscope.

How big are the organisms?
_______________ micrometres

PHYLUM CHRYSOPHYTA: DIATOMS

Most members of this phylum are **diatoms,** which make up the bulk of the phytoplankton in the sea and in fresh water. Some diatoms occur in soil as well. These tiny autotrophs contain chlorophylls *a* and *c,* and their accessory photosynthetic pigment, fucoxanthin, gives them a greenish-brown or golden brown color. Some forms have a gliding motion, but most are nonmotile. Other related organisms may have flagella.

The intricate skeletons of these organisms are impregnated with silica and often take the form of two overlapping **valves** that fit together like a round, flat box and its cover. Because the skeletons are highly

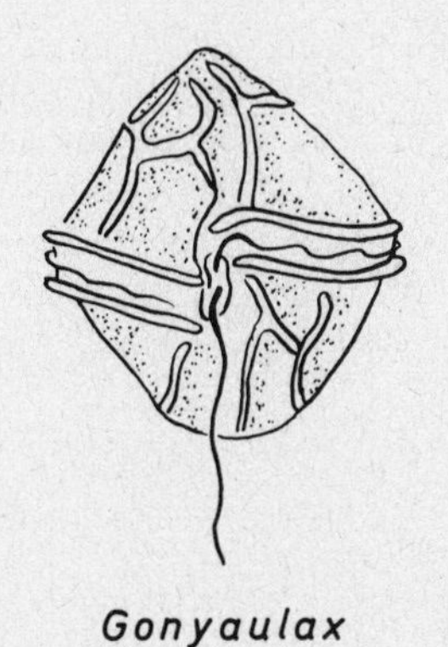

Gonyaulax

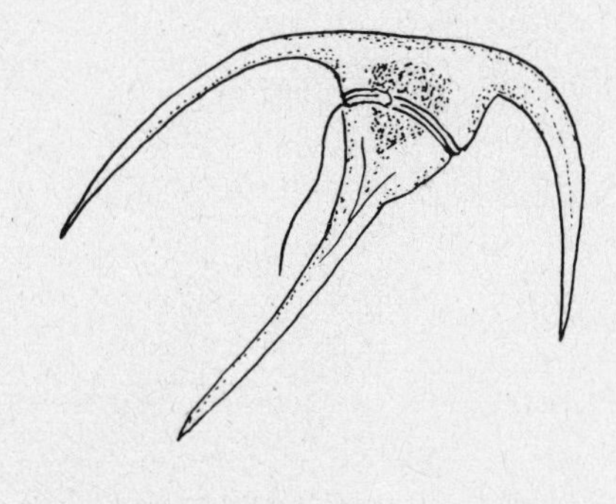

Ceratium

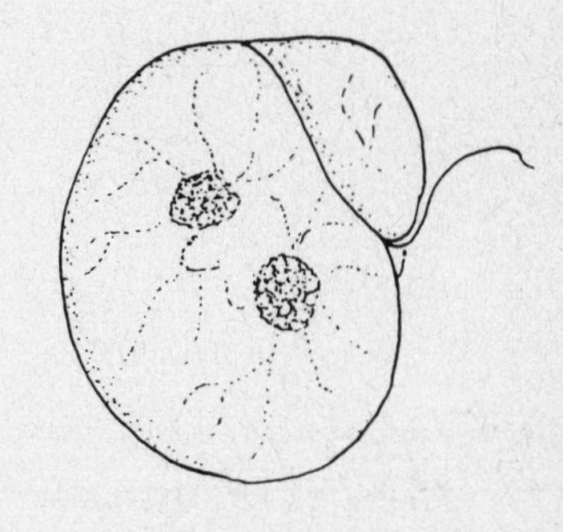

Noctiluca

Figure 13–2. Dinoflagellates. Dinoflagellates are unicellular autotrophs and belong to Phylum Pyrrophyta.

resistant to decomposition, they accumulate in huge quantities and produce beds of **diatomaceous earth.** This material is an excellent abrasive and is widely used in industrial products such as fine polishes and toothpaste. A few of these delicate microorganisms are shown in Figure 13–3.

□ Examine a wet mount or prepared slide of diatoms.
□ If you are observing living cells, identify a cell that is pointed at either end.

Does the cell show any movement? ______________

□ Make a wet mount of diatomaceous earth and examine it for round, pillbox-shaped diatoms. Such diatoms are marine and therefore found only in the ocean.

Was your sample of diatomaceous earth of marine origin? ______________

When you are finished with your work, return the prepared slides, clean up your work place, and store your microscope properly.

EXPLORING FURTHER

Barnes, R. D. *Invertebrate Zoology.* 4th ed. Philadelphia: W. B. Saunders, 1980.
Buchsbaum, R. *Animals without Backbones.* 2nd rev ed. Chicago: University of Chicago Press, 1975.
Kudo, R. P. *Protozoology.* Springfield, Ill.: Charles C Thomas, 1954.
Jahn, T., and F. Jahn. *How to Know the Protozoa.* Dubuque, Iowa.: W. C. Brown, 1949.
Mandel, I. D. "Dental caries." *American Scientist* 67:680, 1979.
Stanier, R. Y., E. A. Adelberg and J. L. Ingraham. *The Microbial World.* 4th ed. Englewood Cliffs, N.J.: Prentice-Hall, 1976.

Scientific American Articles

Abraham, E. P. "The beta-lactam antibiotics." June 1981.
Adler, J. "The sensing of chemicals by bacteria." April 1976 (#1337).
Barghoon, Elso S. "The oldest fossils." May 1971 (#895).
Berg, H. "How bacteria swim." August 1975 (no offprint).
Bonner, John Tyler. "How slime molds communicate." August 1963 (#164).
Braude, A. I. "Bacterial endotoxins." March 1964 (no offprint).
Brill, Winston J. "Biological nitrogen fixation." March 1977 (#922).
Cairns, John. "The bacterial chromosome." January 1966 (#1030).
Costerton, J. W., G. G. Geesey and K.-J. Cheng. "How bacteria stick." January 1978 (#1379).
Devoret, Raymond. "Bacterial tests for potential carcinogens." August 1979 (#1433).
Fraser, David W. and Joseph E. McDade. "Legionellosis." October 1979 (#1447).
Gilbert, Walter and Lydia Villa-Komaroff. "Useful proteins from recombinant bacteria." April 1980 (#1466).
Jannasch, Holger W. and Carl O. Wirsen. "Microbial life in the deep sea." June 1977 (#926).
McCosker, John E. "Flashlight fishes." March 1977 (#693).
Schopf, J. William. "The evolution of the earliest cells." September 1978 (#1402).
Walsby, A. E. "The gas vacuoles of blue-green algae." August 1977 (#1367).
Woese, Carl R. "Archaebacteria." June 1981.

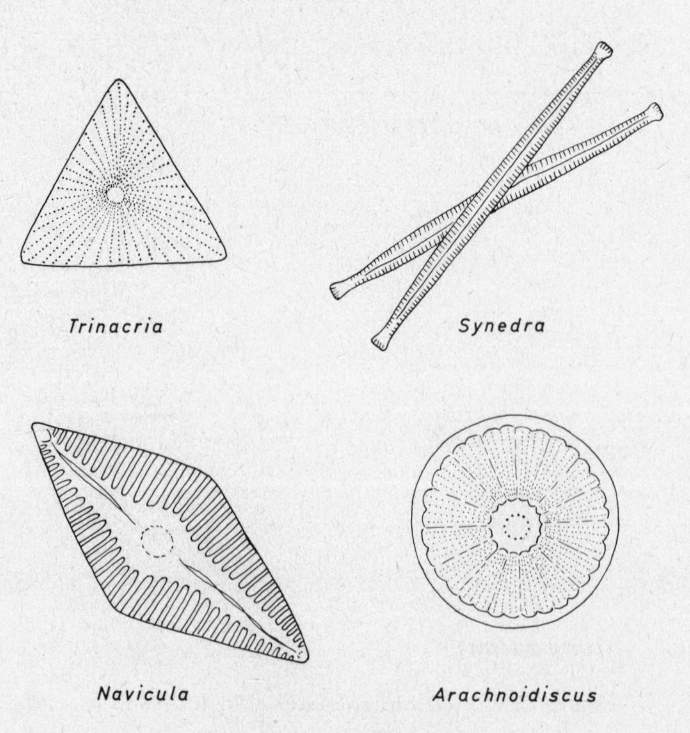

Figure 13–3. Diatoms. One of the chief groups represented in phytoplankton is the diatoms. They belong to Phylum Chrysophyta.

Student Name ______________________ **Date** ______________

TABLE 13–1. CULTURES OF BACTERIAL SAMPLES

Nutrient Agar	Source of Inoculum	Amount of Growth or Number of Colonies
Plate 1	AIR	
Plate 2	COUGH	
Plate 3	SUPPRESSED COUGH	

TABLE 13–2. CULTURES OF ORAL BACTERIA

A. DATA

Tube	Source of Inoculum	Color after 24 hr*	Color after 48 hr*	Color after 72 hr*
1	Before brushing			
2	After brushing			
3	Sterile control			

*The color is considered "green" if there is no change or only slight change and green is dominant. The color is considered "yellow" if green is no longer dominant.

B. INTERPRETATION

Color after 24 hr	48 hr	72 hr	Susceptibility to Caries
Yellow			Marked
Green	Yellow		Moderate
Green	Green	Yellow	Slight
Green	Green	Green	Very low

How likely is it that your oral flora will make your teeth susceptible to decay? ______________

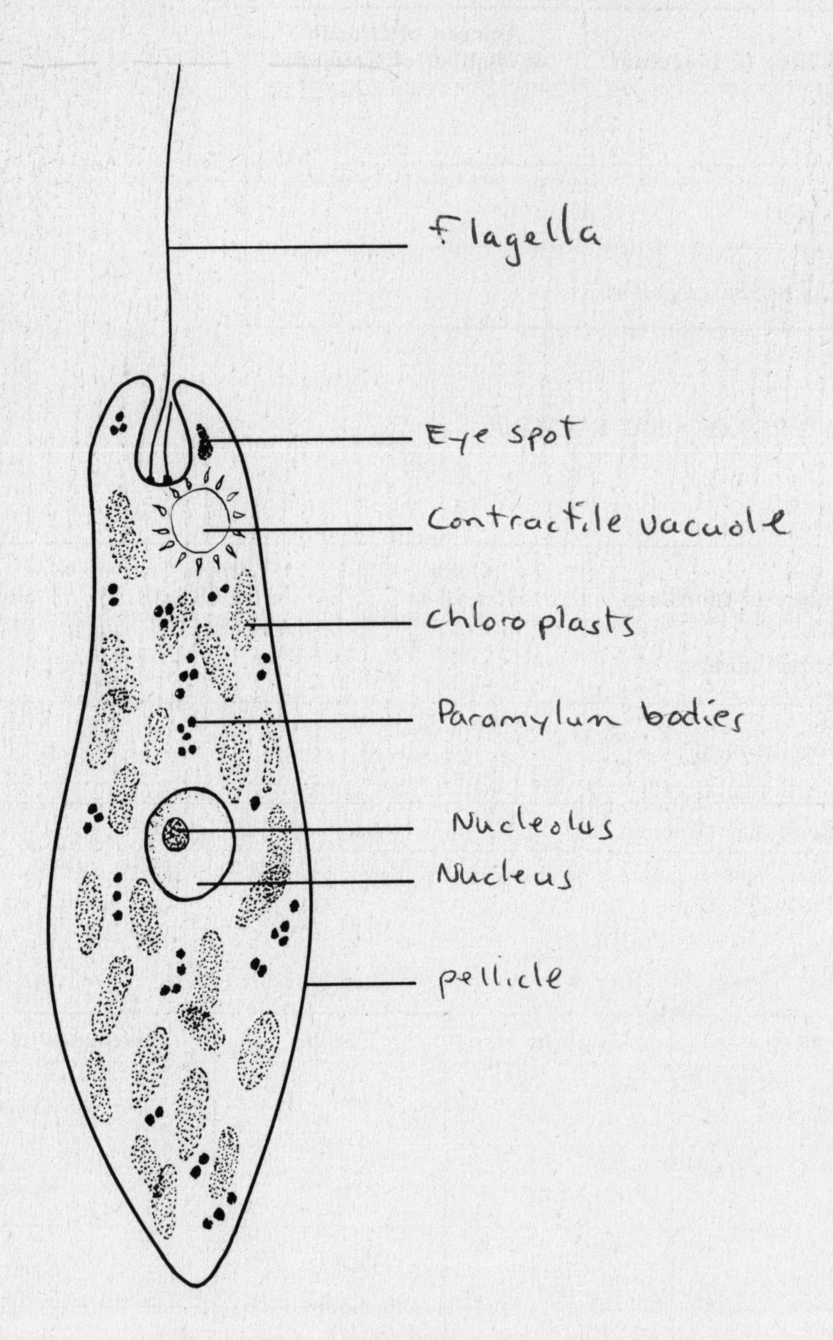

Figure 13–4. ***Euglena.*** Label the indicated structures.

TOPIC 14

Introduction to Fungi and the Lower Plants

OBJECTIVES

When you have completed this topic, you should be able to:

1. List the plantlike characteristics of fungi.
2. List the characteristics that set fungi apart from the members of the plant kingdom.
3. Name three classes of true fungi (Eumycophyta) and an example from each class.
4. Describe the two main types of nutrition found in the fungi.
5. List the three divisions of algae, and give the characteristics of each division.
6. Define alternation of generations in plants, and give an example of an alga in which the sporophyte is dominant, another in which the gametophyte is dominant, and one in which neither the sporophyte nor the gametophyte is dominant.
7. Name the processes giving rise to plant spores, plant gametes, and plant zygotes.
8. Define: dikaryon, mycelium, coenocyte, hyphae, fruiting body, gill, basidium, ascus, zygospore, conidiospore, sporangium.
9. Define: holdfast, stipe, blade, air bladder, zoospore.
10. State some commercially important uses of fungi and algae.

TEXT REFERENCES

Chapter 24; Chapter 25.
Alternation of generations, zygomycetes, ascomycetes, basidiomycetes, slime molds, lichens, red algae, brown algae, green algae.

INTRODUCTION

Algae and fungi are grouped together because they both have relatively simple bodies. Their cells show little differentiation, and the egg cells are not protected by a layer of sterile jacket cells. The algae are all photosynthetic and the fungi are heterotrophs.

Most members of these groups are multicellular, but unicellular stages are involved in their life histories. A generalized life history for algae includes both stages (Figure 14–1). The multicellular stage may be haploid or diploid (N or 2N). Sometimes a multicellular haploid stage is followed by a multicellular diploid stage, and then by multicellular haploid, and so on. In this case, the alga is said to show **alternation of generations:** the haploid and diploid stages alternate with each other.

In plants and fungi, unlike in animals, the cells produced by meiosis are **spores.** If the haploid organism that grows from the germinating spore produces gametes, the gametes arise by differentiation, not by meiosis. (The cells are *already* haploid.) Fusion of haploid cells or nuclei produces the diploid stage of the next generation. In algae and higher plants, the diploid plant that produces spores is called the **sporophyte** ("spore plant"), and the haploid plant that produces gametes is called the **gametophyte** ("gamete plant").

THE FUNGI

Fungi are heterotrophic organisms that have many plantlike features. They are nonmotile and have rigid cell walls that may contain cellulose. Because they can neither carry out photosynthesis nor ingest food, they must obtain nutrients by **absorption** through their cell membranes. If they feed on living cytoplasm, they are **parasites,** and if they feed on dead organisms or detritus, they are **saprobes.** Such fungi, the decomposers, are vital in recycling the dead bodies of organisms so that the nutrients in them can be used by living plants and animals.

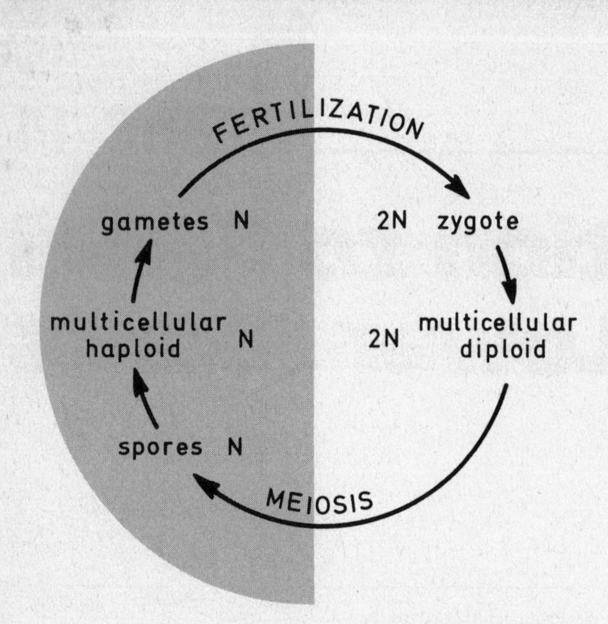

Figure 14–1. Generalized life history. Algal life histories may include one or both multicellular stages.

The true fungi are classified as the **Division Eumycophyta.** You will study examples of true fungi from the zygomycetes, ascomycetes, and basidiomycetes.

Rhizopus (Class Zygomycetes)

The black bread mold *Rhizopus* forms a mat, or **mycelium,** of threadlike **hyphae.** The nuclei are haploid, and the fungus lives in a terrestrial environment.

□ Obtain a Petri dish containing an asexually reproducing mycelium of *Rhizopus.*
□ Examine the mold under the dissecting microscope without opening the dish.
□ Note the tiny black spheres of the **sporangia** that grow at the ends of some erect hyphae. The sporangia release spores.
□ Make a wet mount of mycelium with sporangia, and examine it in the compound microscope.
□ Note the cluster of rootlike hyphae at the base of the hypha bearing the sporangium. These hyphae anchor the hypha and are called **rhizoids.**
□ Note the horizontally growing hyphae called **stolons.**

A fungus without cross walls is said to form a **coenocyte,** in which all of the nuclei are in a common cytoplasm. If there are cross walls, the fungus is **septate.**

Is the mycelium of *Rhizopus* coenocytic or septate?

____

The haploid spores released by the sporangia are formed by mitosis, and their dispersal is a form of asexual reproduction. The spores will give rise to new haploid mycelia.

□ Obtain another culture of *Rhizopus* in which two mating types (+ and −) of mycelia have fused and undergone sexual reproduction.
□ Note the dark line in the mycelium where the + and − mating types fused. The dark color is due to the formation of **zygospores.**
□ Make a wet mount of a bit of mycelium from the black region, or view a prepared slide, and look for cross bridges between the hyphae.
□ Examine such a cross bridge for the presence of a dark sphere, the **zygospore.**

When the two types of mycelia come into contact, the bridge is formed and the cell wall breaks down so that the cytoplasms of the two sides are in contact. A nucleus from each side moves into the bridge, and the haploid nuclei fuse to form a diploid nucleus. The zygote nucleus divides several times, and then undergoes meiosis to form a sporangium containing haploid spores. These spores, unlike the ones produced asexually, will be genetically different from either parent. Sexual reproduction in *Rhizopus* is shown in Figure 14–2.

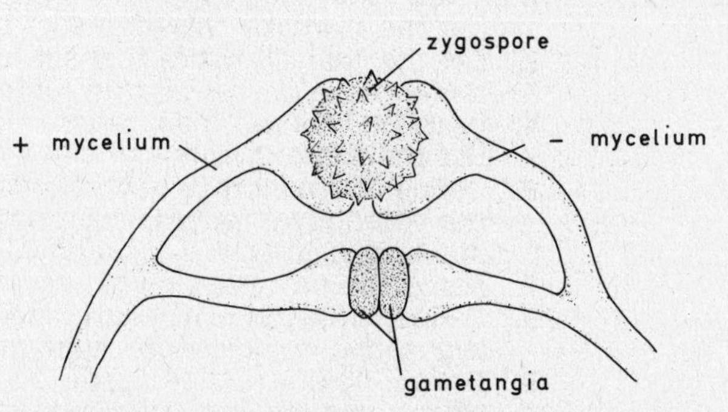

Figure 14–2. Sexual reproduction in *Rhizopus.* The diploid zygospore will under meiosis to produce a new haploid mycelium.

Yeast (Class Ascomycetes)

This fungus is a unicellular ascomycete. It has an unusual method of reproducing asexually by budding off. It also reproduces sexually by forming a saclike **ascus** containing spores produced by meiosis. The life history of a yeast is shown in Figure 14–3.

□ Make a wet mount of a drop of actively growing yeast culture, and observe the cells in the microscope under high power.
□ Note the cells that are producing smaller buds or short chains of buds.

Yeasts are important in the baking and brewing industries and are used to produce important industrial chemicals.

What by-product of yeast fermentation is important in baking? ________________

What by-product is important in brewing?

____

Which of these by-products is produced in both aerobic and anaerobic respiration? ________________

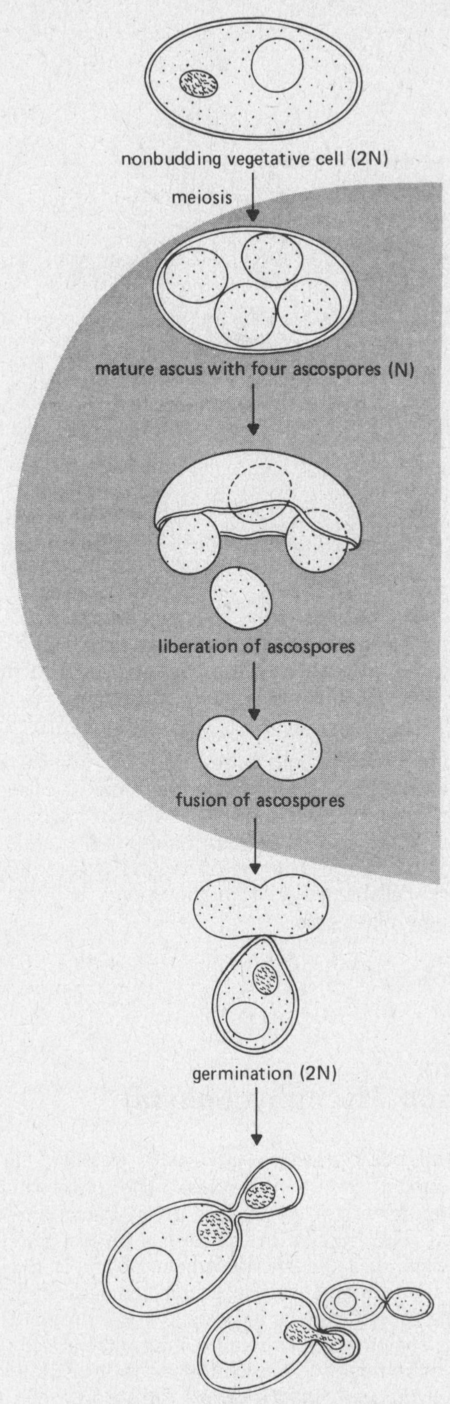

Figure 14–3. The life history of yeast, an ascomycete. (Redrawn from Harry W. Seeley, Jr., and Paul J. Van Demark, *Microbes in Action: A Laboratory Manual of Microbiology,* 2nd ed. San Francisco: W. H. Freeman and Company, © 1972.)

Cup Fungi (Class Ascomycetes)

Cup fungi such as *Peziza* (and also the mushrooms and their relatives) are the fungi with which we are the most familiar because they are the most conspicuous. The large part of the fungus that we see is formed of masses of hyphae fused together in the **fruiting body.** Each fruiting body is supported by an underground mycelium that is much larger than itself. (See text Figure 24–10.) Part of the fruiting body is called a **dikaryon** (dikaryon means "two nuclei") because each of its cells has two haploid nuclei. The cells in this part are different from diploid cells and are designated N + N rather than 2N. During sexual reproduction, some of the cells of the fruiting body undergo nuclear fusion to produce true diploid (2N) nuclei. Meiosis, which follows, produces haploid (N) **ascospores.** The large size of the fruiting body helps disperse the spores. The spores germinate to produce mycelia, and different mating types (+ and −) of mycelia interact to form a new dikaryon and fruiting body.

□ Obtain and examine a preserved specimen of a cup fungus such as *Peziza.* The cup shape of the fruiting body allows rain to collect and splash around inside where the ascospores are produced and aids in splashing out the ascospores.
□ Examine a cross section of a cup fungus under high power.
□ Locate an ascus.

How many spores does an individual ascus contain?

____

Cup fungi can also reproduce asexually by forming chains of conidiospores at the tips of the hyphae. These tiny, airborne spores are the result of mitosis rather than meiosis. They pervade the entire biosphere, and germinate at every opportunity.

Ascomycetes are biologically interesting because some of them have become obligate symbionts with algae to form **lichens.** (Lichens will be considered later in this exercise.) Ascomycetes such as morels and truffles are gourmet delicacies, and the mold *Penicillium* is the source of the antibiotic penicillin. Other ascomycetes such as smuts and rusts are serious agricultural pests.

Mushroom (Class Basidiomycetes)

The common edible mushroom (*Agaricus campestris*) is the fruiting body of a basidiomycete. As in the cup fungus, it is formed by the fusion of masses of hyphae. In sexual reproduction, underground mycelia of two mating types fuse and form a mycelium with two nuclei per cell. This mycelium is an N + N dikaryon. As in the ascomycetes, the mycelia are septate.

□ Obtain a common mushroom, and make a cross section of it with a razor blade.
□ Note the **stalk** and the **cap** that bears the **gills** on its undersurface.
□ Examine the gills with the dissecting microscope.

The reproductive cells, the **basidia,** are located on the surface of the gills. Each basidium undergoes nuclear fusion to form a fused nucleus which immediately enters meiosis. The haploid nuclei form **basidiospores,** which are released from the surface of the gills. Each spore will germinate to form a new mycelium. Sexual reproduction in basidiomycetes is described in Figure 14–4.

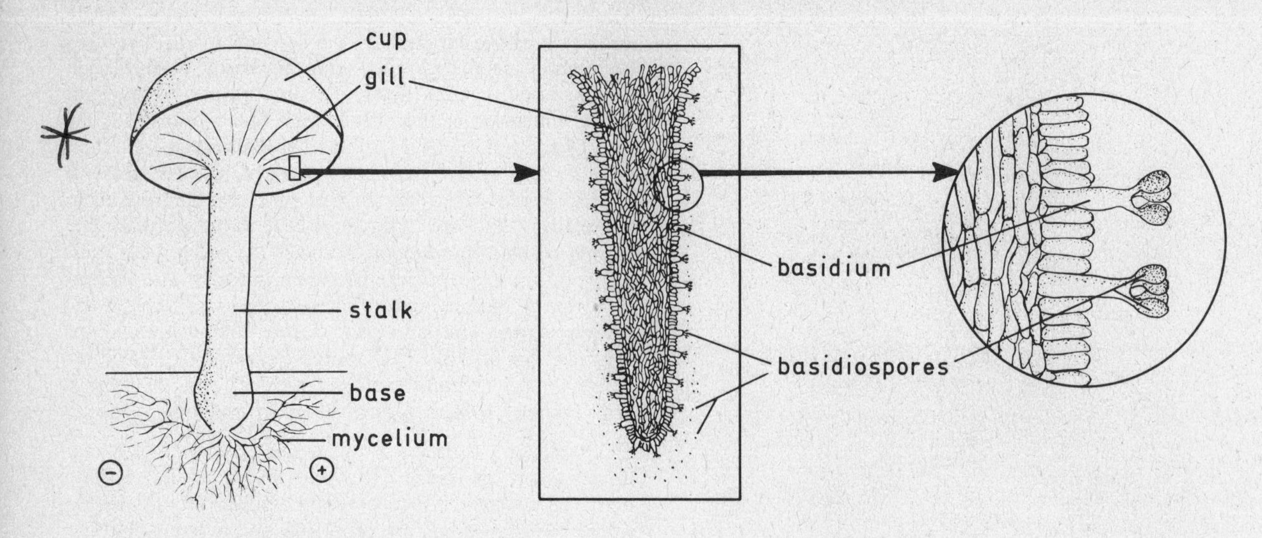

Figure 14–4. Basidiomycete reproduction. The mushroom is the fruiting body formed when + and − mycelia meet and fuse.

□ Make a wet mount of a small piece of gill, or examine a prepared slide through the cap of a mature mushroom such as *Coprinus.*
□ Note the small, dark basidiospores on the edges of the gills. Each basidium produces four basidiospores.

Which cells in the life history of a basidiomycete can undergo meiosis? ______________________

What is the difference between a diploid organism and a dikaryon?

____

____

□ Examine the plasmodium under low power (100×).
□ Note the "veins" of streaming cytoplasm with many nuclei.

If conditions become unfavorable, the plasmodium will dry up and turn into a resting stage known as a **sclerotium.** Pieces of dried sclerotium can be used to start new cultures of a plasmodium. *Physarum* can also reproduce by sporangia that form on its surface and contain haploid spores. Two types of haploid cells (amoebae and swarm cells) emerge from the spores and fuse to produce new plasmodia.

Slime Molds (Division Myxomycophyta)

Slime molds are grouped separately as the myxomycetes because they have more animal-like features than the other fungi. The **cellular slime molds (Class Acrasiomycetes)** have amoeboid cells that can aggregate to form fruiting bodies. Each fruiting body produces spores that disperse and germinate into more amoebae.

□ Examine the demonstration culture of a cellular slime mold such as *Dictyostelium* in the dissecting microscope.
□ Locate a sluglike **grex** (pseudoplasmodium), an aggregate of many amoebae that creeps over the agar leaving a slimy trail.
□ Locate a **fruiting body** (sorocarp) which is supported by a thin stalk. When mature, it will burst to release the spores.

Reproduction in *Dictyostelium* is shown in Figure 24–21 in your text.

Acellular slime molds (Class Myxomycetes) have the form of a gigantic coenocyte called a **plasmodium** in which hundreds of nuclei share a common cytoplasm.

□ Obtain a Petri dish containing a culture of an acellular slime mold such as *Physarum,* and place it on the stage of your compound microscope.

Lichens (Division Mycophycophyta)

These symbiotic organisms are usually classified by the type of fungus that they contain. The photosynthetic algal component may be a blue-green bacterium or a green alga. The hyphae of the fungus protect and support the alga, and the photosynthetic ability of the alga provides nourishment, so the lichen can flourish where neither the fungus nor the alga could live by itself. These organisms are pioneers and can colonize harsh and unfavorable environments. They contribute to the breakdown of rocks in the early stages of soil formation.

There are three basic types of lichens. **Foliose** lichens have a leaflike cell body and are the type most commonly seen. **Crustose** lichens live as a very thin coating that is tightly bound to its substrate. **Fruticose** lichens have a fruiting body that is a holdover from the fruiting body of the ascomycete fungus that is one of the symbiotic partners. The internal structure of lichens is shown in Figure 14–5.

□ Examine the specimens of crustose, foliose, and fruticose lichens on demonstration.
□ Note that the color of lichens is highly variable.
□ Examine a prepared slide showing a lichen in cross section. Alternatively, with your instructor's permission, tease apart a small bit of lichen, make a wet mount, and examine it in the compound microscope.

□ Note the hyphae and the algal cells.

Lichens reproduce by **fragmentation,** in which a piece of the organism simply breaks off and is passively moved to a new location, where it grows into another lichen. Lichens are important in colonizing harsh environments in which other organisms cannot survive, and they are widespread in the tundra region where they are an important source of food for herbivores.

THE ALGAE

Red Algae (Division Rhodophyta)

These phtosynthetic plants may be more closely related to the blue-greens than are the other plant divisions. The plant body is usually filamentous or leaflike, and it is attached to the surface by a **holdfast.** Red algae contain only chlorophyll *a* and the accessory pigments phycocyanin and phycoerythrin (a red pigment). There is no flagellated form in the life history, and these algae are usually marine.

□ Examine the specimen of red alga on demonstration.

What is the body form of the demonstration alga?

____

The life history of a red alga includes a dominant haploid stage; sometimes a diploid zygote develops into a multicellular diploid stage as well.

Commercially important substances such as carrageenan and agar are obtained from species of red algae, and these algae may be important in building up and maintaining the integrity of coral reefs. Red algae can live at greater depths than any other algae.

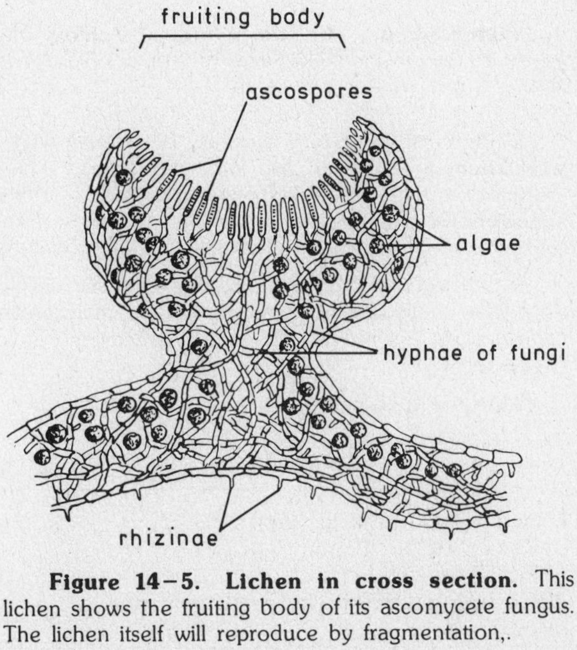

Figure 14–5. Lichen in cross section. This lichen shows the fruiting body of its ascomycete fungus. The lichen itself will reproduce by fragmentation,.

Brown Algae (Division Phaeophyta)

These algae contain chlorophylls *a* and *c* and the carotenoid fucoxanthin, and they may be related to the unicellular diatoms. All of the forms are multicellular, and they are the largest and most complex of the algae.

□ Examine the specimen of a brown alga such as *Fucus* on demonstration.
□ Locate the **holdfast,** the **stipe** (stalk), and the **blade** ("leaf").

Does your specimen contain **air bladders** filled with gas? _______________

□ Make a cut across one of the blades to see its internal structure.

Are all parts of the blade photosynthetic?

____

In algae of substantial size, the air bladders float the photosynthetic blade near the surface where it can obtain more light. The outer surface of the blade is covered by a gelatinous coating to prevent it from drying out when it becomes exposed to air.

□ Examine the tips of the blades for swollen sacs called **receptacles;** eggs and sperm are produced within the receptacles.

The life history of *Fucus* is similar to the life cycle of animals. The plant itself is diploid, and cells within the receptacles undergo meiosis to produce eggs and flagellated sperm, thereby skipping the multicellular haploid stage. The gametes fuse to form zygotes, and then each zygote grows into a new diploid alga.

The giant brown algae, the kelps, are the source of alginate, a thickening agent important in making ice cream, toothpaste, and many other products. Brown algae have also been used for fertilizer and livestock feed.

Green Algae (Division Chlorophyta)

Green algae contain unicellular and multicellular organisms and are thought to be more closely related to higher plants than the other algae. They contain chlorophylls *a* and *b* and carotenoids. The plant body may have a holdfast, but its structure is usually simpler than that of a brown alga.

Some of the unicellular and multicellular forms are very closely related. For example, the individual cells of *Gonium* or *Pandorina* are almost identical to those of the unicellular alga *Chlamydomonas.* These organisms may have arisen from unicellular ancestors whose cells failed to separate after mitosis. *Gonium* in turn may have given rise to more elaborate colonies such as that of *Volvox,* in which there are hundreds of cells and some division of labor.

□ Make a wet mount of *Chlamydomonas,* and view it under high power in your compound microscope.
□ Note the **flagella,** the **chloroplast,** and the spot of red pigment.

How is *Chlamydomonas* different from *Euglena*? (*Euglena* is described in Topic 13 and illustrated in Figure 13–4.)

____

____

During sexual reproduction, two haploid individuals fuse to form a diploid zygote. The zygote immediately undergoes meiosis to produce spores without going through a multicellular diploid stage. The spores develop into new vegetative organisms. The zygote is therefore the only diploid stage in the life history.

□ Make a wet mount of *Gonium* and examine it under low power.

How many cells are there, on the average, in a colony?

____

Do the cells of *Gonium* appear any different from those of *Chlamydomonas*? ______________

The gametes produced by *Gonium* in sexual reproduction are flagellated but are different in size.

Volvox is a colony of hundreds of haploid algal cells.

□ Place a drop of *Volvox* culture on a slide without adding a cover slip, and view it under your dissecting microscope.

What structures propel the colony about?

____

□ Add a cover slip and view the immobilized colony under high power in your compound microscope.
□ Note the spacing of the cells. They are connected by thin threads of cytoplasm that allow the swimming movements of the colony to be coordinated.

Refer to the life history of *Volvox*, which is illustrated in Figure 14–6.

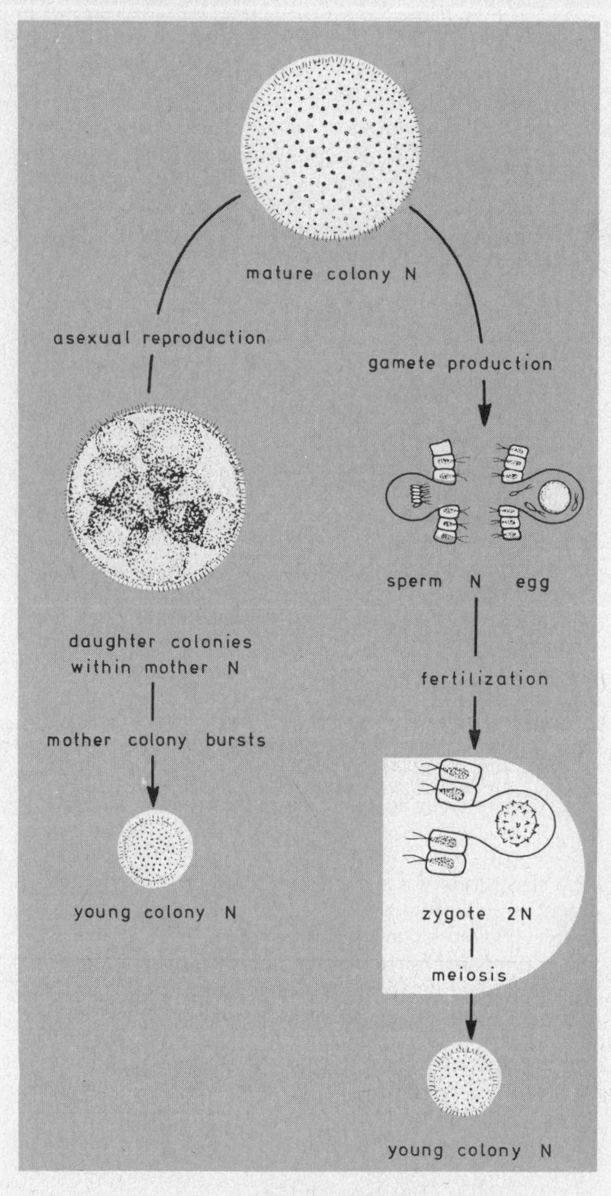

Figure 14–6. The life history of *Volvox*. The zygote is the only diploid stage of this colonial green alga.

Volvox produces egg and sperm cells, and the flagellated sperm must swim to the nonmotile egg. The diploid zygote undergoes meiosis to form a new haploid colony. In asexual reproduction, vegetative cells are released inside the mother colony where they develop into new daughter colonies. The daughters are released when the old colony breaks open.

A filamentous alga, *Ulothrix*, has a multicellular haploid stage, although its life history is otherwise similar to that of *Chlamydomonas*.

□ Make a wet mount of living *Ulothrix*, or examine a prepared slide of this organism with your compound microscope.

What is the shape of the individual cells?

____

Are all of the cells alike? ______________

□ Look for the holdfast at one end of the filament. (It may be broken off.)

What is the shape of the single chloroplast?

____

Do you see storage organs (pyrenoids)?

____

Ulothrix can reproduce asexually by forming flagellated **zoospores.** It can also form tiny gametes which are smaller than the zoospores and fuse to form **zygospores.** The zygospores undergo meiosis to form zoospores, each of which will grow into a new filament.

Another green alga, *Ulva*, has a multicellular diploid stage that is identical in appearance to its multicellular haploid stage. Its history follows the pattern shown in Figure 14–1.

□ Examine a speciman of *Ulva*, or sea lettuce.

What is the form of the body?

____

Is there a stalk, as in the brown algae?

____

Which stage do the motile spores grow into?

____

Which stage develops from the zygote?

____

Which stage is present in *Ulva* but is absent in a brown alga such as *Fucus*?

____

Although plant life histories may at first seem very confusing, they must all conform to the general pattern given in Figure 14–1. Meiosis always takes place in a diploid cell (or dipoloid nucleus, as in fungi) and must produce haploid cells, which are almost always spores. It may be helpful to learn an example of each type of life cycle:

1. Neither multicellular haploid nor multicellular diploid stage present:

   ____

2. Multicellular haploid stage present:

   ____

3. Both multicellular haploid and multicellular diploid stages present:

   ____

4. Multicellular diploid stage present:

   ____

Which type of life cycle corresponds to that of humans and other animals? _______________

Return all demonstration materials and store your microscopes properly before leaving the laboratory.

EXPLORING FURTHER

Alexander, M. *Microbial Ecology*. New York: John Wiley and Sons, 1971.

Alexopoulos, C. J., and H. C. Bold. *Algae and Fungi*. New York: Macmillan, 1967.

Bold, H. C. and M. J. Wynne *Introduction to the Algae: Structure and Reproduction*. Englewood Cliffs, N.J.: Prentice-Hall, 1978.

Dawson, E. Y. *Marine Botany: An Introduction*. New York: Holt, Rinehart and Winston, 1966.

Prescott, G. W. *How to Know the Freshwater Algae*. 3rd ed. Dubuque, Iowa: W. C. Brown, 1978.

Smith, Alexander. *The Mushroom Hunter's Field Guide*. Ann Arbor: University of Michigan Press, 1980.

Scientific American Articles

Bonner, J. T. "How slime molds communicate." August 1963 (#164).

Gibor, A. "Acetabularia: A useful giant cell." November 1966 (#1057).

Lamb, I. M. "Lichens." October 1959 (#111).

Litten, W. "The most poisonous mushrooms." March 1975.

Niederhauser, J. S., and W. C. Cobb. "The late blight of potatoes." May 1959 (#109).

Preston, R. D. "Plants without cellulose." June 1968 (#1110).

Strobel, Gary A., and Gerald N. Lanier. "Dutch elm disease." August 1981.

TOPIC 15

Introduction to the Higher Plants: Embryophytes

OBJECTIVES

When you have completed this topic, you should be able to:

1. Define embryophyte and list the characteristics of these plants that link them to the green algae.
2. Describe the problems faced by plants that moved onto the land.
3. Describe adaptations of land plants to each problem listed in Objective 2.
4. Give the adaptations of bryophytes that allow them to live on land.
5. List the problems of terrestrial life that bryophytes have not solved.
6. Describe the life history that applies to club mosses, horsetails, and ferns.
7. Compare horsetails and ferns with respect to the following features: size of presently living species, stem, leaves, roots, vegetative propagation, type of gametophyte, and type of sporangium.
8. Given a plant specimen or description, place it correctly in one of these groups: Bryophyta, Lycopsida, Sphenopsida, Filicineae, Gymnospermae, or Angiospermae.
9. Name the embryophyte groups in which the gametophyte is the dominant generation.
10. Describe the gametophyte of a fern and a conifer.
11. Give reasons why gymnosperms persisted, even though angiosperms became dominant.

TEXT REFERENCES

Chapter 26.
Adaptation to life on land, mosses, club mosses, horsetails, ferns, gymnosperms, angiosperms.

INTRODUCTION

Although some algae and fungi can eke out a living on land, their relatively undifferentiated bodies never allowed them to develop adaptations for a truly terrestrial life. Most members of the plant groups that you will be studying in this exercise live exclusively on land. They have been able to cope in some fashion with the resulting problems and have taken advantage of the benefits. Land plants are thought to have arisen from the green algae. Both groups contain chlorophylls *a* and *b*, and they usually store their food reserves in the form of starch. An alga that developed some kind of waterproofing and rootlike structures to draw water up from the soil would have been able to live out of water on the shore. Such a plant may have given rise to the more highly adapted land plants.

All of these plants show alternation of generations. Sometimes there is a large photosynthetic gametophyte and a tiny sporophyte. In other cases the sporophyte is a large photosynthetic plant that propagates the species through dispersal of spores or seeds as well. The reproduction of many land plants still involves a flagellated sperm that must swim to the egg through a film of water. These plants are restricted to a moist environment for fertilization to occur. All of the land plants, however, do have an important adaptation for reproduction on land: the zygote is held within the female reproductive organ where it is surrounded by a layer of sterile jacket cells. These cells protect the embryo while it starts to develop into a new sporophyte within the female reproductive organ. The plants that have this feature during reproduction are called **embryophytes.**

MOSSES (DIVISION BRYOPHYTA, CLASS MUSCI)

Bryophytes (mosses and liverworts) are different from the other land plants in that they lack vascular tissue. Because true roots, stems, and leaves must have vascular tissue, the organs of a moss are described as "rootlike," "stemlike," and "leaflike." As a result, mosses are limited in several important ways. They cannot grow large in size because they have no strong supporting tissue to keep their bodies erect. They must live in damp places because their rootlike rhizoids cannot penetrate deep into the soil to extract the moisture they need. Finally, their photosynthetic parts above ground and the nonphotosynthetic parts below ground must be close together because there is no vascular tissue for the transport of nutrients and minerals. The gametophyte is the dominant stage in the moss life history.

□ Obtain a single moss plant (gametophyte), and examine it under the dissecting microscope.
□ Note the "**stem**," "**leaves**," and **rhizoids.**

Is any part of the moss plant nonphotosynthetic? If so, which part?

____

____

What are the functions of the rhizoids?

____

____

____

□ Make a wet mount of a "leaf."

Are there guard cells or stomata? ____________

□ Use a razor blade to make a very thin section of the "leaf" or obtain a prepared slide and view it in the compound microscope.

Do you see xylem, phloem, or large air spaces in the "leaf"? If so, which ones?

____

Is there a cuticle? ____________

During reproduction the flagellated sperm produced by the male moss plant (gametophyte) swims to the female gametophyte and fertilizes the egg. The zygote grows into a sporophyte and remains attached at the tip of the female plant. (See Figure 15–4 at the end of this topic.) Usually the sporophyte is nonphotosynthetic and obtains its nutrition from the female gametophyte. The spores are formed in a capsule on the end of a long stalk and are released from as high up as the plant can possibly reach (3–4 cm).

Why should the spores be released as high as possible?

____

____

____

□ Obtain a moss plant that has a sporophyte growing on it, if living material is available. If not, obtain a prepared slide of a moss sporophyte.
□ Note the long **stalk** and the **capsule.**

Is either part of the sporophyte photosynthetic?

____

□ Mount the capsule in a drop of water on a slide, and remove the tip of it with teasing needles. Observe in the dissecting microscope.
□ Try to look into the capsule. You will see a circle of toothlike projections that help disperse the spores.
□ Tease apart the capsule to locate the spores.

Are the spores haploid or diploid? ____________

Is the sporophyte haploid or diploid?

____

What process gave rise to the spores?

____

Label the parts of the female moss plant shown in Figure 15–4.

The spore germinates into a threadlike **protonema,** which eventually forms the buds of the new gametophyte.

Mosses cope with the dryness of land by living in a damp environment and retaining as much water as possible. The mat of many moss plants crowded close together acts like a sponge to hold water and cut down evaporation. Some mosses even have special cells that are adapted to absorb water when it is available. Sphagnum moss, for example, has large, reinforced cells with pores in their cell walls for taking up water.

□ Obtain a 100-mL beaker and a weighed 1- to 5-g sample of dry sphagnum moss (note exact weight). Your instructor may ask you to weigh your own sample.
□ Fill the beaker about two-thirds full with water, and make a mark at the surface of the water with a wax pencil.
□ Immerse the moss in the beaker, and allow it to soak up as much water as possible.
□ Remove the moss with a forceps.
□ Fill a 100-mL graduated cylinder with water to exactly the 100-mL mark, and add water from it to your beaker until the water level again reaches the pencil mark.
□ Note how much water was needed to refill the beaker. This amount is equivalent to the amount removed by the moss.
□ Calculate how many times its weight of water the moss can absorb:

$$\text{absorption} = \frac{\text{mL (g) of water removed}}{\text{dry weight of moss (g)}} = \_\_\_\_\_\_$$

Sphagnum moss is widely used by commercial greenhouses and home gardeners to hold moisure around the roots of plants, especially during shipping.

CLUB MOSSES (DIVISION TRACHEOPHYTA, SUBDIVISION LYCOPSIDA)

The remaining plants to be studied in this laboratory all belong to **Division Tracheophyta** and have true **vascular tissue.** Club mosses and horsetails are fern allies because all these plants have a similar life cycle. They have a large sporophyte with true vascular tissue, and the spores grow into tiny gametophytes that are independent of the parent sporophyte. There is a flagellated sperm and the zygote grows into a new sporophyte. Although present-day ferns and their allies do not have woody tissue, in former times they did and reached the proportion of trees. Some species of tree ferns have survived to the present.

□ Examine a living or preserved club moss such as *Lycopodium,* and note the leaves, stems, and roots.

What is the form of the leaf?

____

What is the value of this adaptation?

____

□ Cut through the tip of one of the branchlets.

Is the interior hollow or solid? ______________

The leaf remains green through the winter, so it is **evergreen,** as are some ferns, most gymnosperms, and some angiosperms. Possession of evergreen leaves is therefore not a good characteristic for classifying plants.

The club moss reproduces vegetatively by its underground stem, the **rhizome,** which sends up shoots to form new photosynthetic stems.

□ Examine the rhizome of your specimen.

Does it have leaves? ______________

During sexual reproduction cones called **strobili** form at the tips of the branches, and the spores are produced within them in **sporangia.** The sporangium always forms on the surface of a leaf called a **sporophyll.** (See Figure 15–4.)

□ Examine the strobilus, if your specimen has one. It consists of a cluster of sporophylls.

After the spore germinates, growth of the gametophytes and fertilization go on underground. Sexual reproduction in club mosses is so slow that most propagation is by the spreading out of the rhizomes. Label the parts of the club moss shown in Figure 15–4.

Although ancient club mosses included very large trees, the present species are all small. It is illegal to collect them in most areas because they are so few in number.

HORSETAILS (SUBDIVISION SPHENOPSIDA)

Fossil horsetails grew as large as trees, but there is only a single genus of small plants that survives today.

□ Examine a living or preserved specimen of a horsetail such as *Equisetum hyemale* or *Equisetum arvense.* (In *E. arvense* the green shoots that look like pine needles are the branches. Look for the tiny leaves on these branches.)

Describe the **leaves:**

____

____

Where do they occur?

____

Is the **stem** photosynthetic? ______________

□ Note the ribs along the sides of the hollow stem, and the joints or **nodes.**

The stem continues underground as a rhizome. The stalks die in the fall, but the **rhizome** survives the winter and sends new shoots up in the spring. The horsetail spreads vegetatively as the rhizome grows and sends up more shoots. True roots grow from the rhizome.

If your specimen is *E. arvense,* sexual reproduction involves separate nonphotosynthetic **fertile shoots,** which bear conelike **strobili** at the tips and are different in structure from the green **sterile shoots.** (See Figure 15–4.) The spores are formed within the sporangia, which are within the strobili. If your specimen is *E. hyemale,* the strobili are found at the ends of the photosynthetic shoots. Winglike **elaters** on the spores aid in their dispersal by the wind after they are shed from the sporangium.

□ Examine a strobilus. The sporangia contained within it form the spores.

The spores germinate into nonphotosynthetic gametophytes which grow underground. After fertilization, the zygote grows into another sporophyte. Label the parts of the horsetail shown in Figure 15–4. If you studied *E. hyemale,* label the parts described above on the left-hand shoot only.

Although this group is not very important today, during the time of the first settlers and the pioneers, horsetails were called scouring rushes and were used as an abrasive for cleaning pots and for polishing. The stems of the horsetails contain grains of silica (the material of which sand is made), which are good for scouring.

FERNS (SUBDIVISION PTEROPSIDA, CLASS FILICINEAE)

Although you would never name a child "Horsetail" or "Liverwort," you might choose the name "Fern" in honor of these beautiful woodland plants. The life history of a fern is shown in Figure 15–1. It is similar to those of the fern allies (club mosses and horsetails). The fern gametophyte, however, is an independent, photosynthetic plant that produces both eggs and sperm.

□ Obtain a fern plant (sporophyte) and study its structure.

□ Note the deeply lobed leaves, called **fronds.**

Do the fronds have many veins? ______________

Ferns, gymnosperms, and angiosperms are grouped as **Subdivision Pteropsida** because of their large, many-veined leaves.

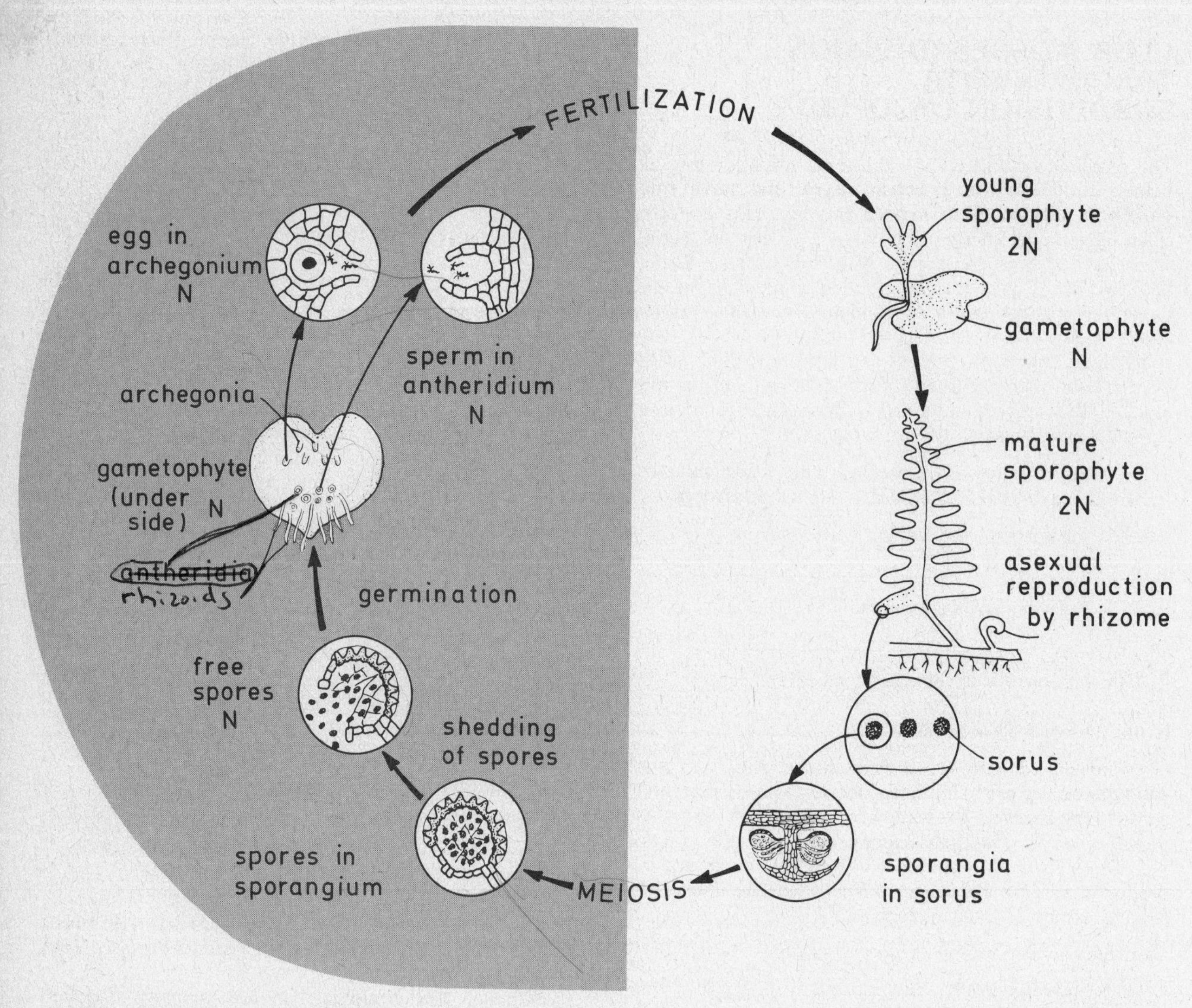

Figure 15–1. Life history of a fern. The sporophyte is dominant and the small gametophyte (valentine plant) is independent and photosynthetic.

□ Note the **petiole** (stalk) of the frond. It attaches the frond to the underground stem.
□ Follow the petiole underground to where it joins the horizontally growing stem, the **rhizome.**

Fronds of most ferns do not survive the winter, but the rhizome does. It sends up new fronds in the spring in the curled up form known as **fiddleheads.** Each season the rhizome grows longer and sends up new fronds to spread the fern by vegetative propagation.

□ Examine a prepared cross section of a fern stem and note the arrangement of the vascular tissue.

Certain leaves called sporophylls bear the sporangia in which the spores are produced. Sometimes there are separate, fertile fronds, but more often the ordinary fronds take on the reproductive function.

□ Examine the undersides of your fronds for the presence of **sori,** which are little brown or colored dots. A sorus is a cluster of sporangia.
□ Examine a prepared slide showing the sorus of a fern in cross section.
□ Locate a **sporangium,** which is a sphere-shaped capsule suspended from a stalk.
□ Note the heavy, segmented ring around most of the sphere. This is the **annulus,** which is sensitive to moisture and helps release the spores by snapping the capsule shut suddenly.
□ Note the **spores** within the capsule of the sporangium.
□ If your frond has sori, scrape the surface of one into a drop of water, make a wet mount, and examine the sporangia in the microscope.
□ Locate the spores that are visible through the transparent walls, and find the annulus, which has thick cell walls.
□ Find the place in the annulus that is not thick-walled. This is where the capsule will break open to release the spores.

After the sporangium is mature, the cells of the annulus dry out, and the annulus straightens out to form a rod, thereby breaking open the capsule. At a certain point the annulus becomes so strained that it snaps back and flings the spores out of the capsule. The spores are thus actively spread out away from competition with the parent plant.

Each spore grows into a small plant called a **pro-**

thallus, which is heart-shaped. (It is sometimes called a "valentine plant.") It bears both an **archegonium** (female) and **antheridium** (male) and produces both eggs and sperm. The prothallus is anchored onto the substrate by **rhizoids.** The sex organs and rhizoids are all located on the undersurface. The sperm must swim over to the eggs which are fertilized and retained within the archegonium.

□ Examine the living prothallus of a fern if living material is available; otherwise look at a prepared slide.
□ Locate the **archegonia,** which are vase-shaped. They may not be present if you are looking at a living prothallus.
□ Locate the globe-shaped **antheridia.**

Why is it advantageous to have the sex organs located beneath the plant?

____

____

The young sporophyte grows right out of the archegonium. The leaves grow upward from the apical notch of the prothallus and the rhizome, and the roots grow downward.

□ Examine living or preserved young fern sporophytes.
□ Note that the prothallus persists even after the sporophyte is well established.

Ferns have survived in spite of competition with the seed plants, and they are plentiful and widespread today. Most ferns are small, but there are some tree ferns still in existence.

GYMNOSPERMS (SUBDIVISION SPERMOPSIDA, CLASS CONIFERAE)

Angiosperms and many gymnosperms have mastered the problems posed by living on land by doing away with the flagellated sperm. Instead the male gametophyte produces large numbers of **pollen grains** which are not susceptible to drying out and which can be spread about by the wind. In addition, the young sporophyte embryo does not have to develop into a mature sporophyte at once but can remain **dormant** in a seed until the conditions become favorable because it is enclosed with its food supply inside a tough seed coat. The production of pollen and seeds allowed these plants to reproduce in the absence of free water and was an important factor in their evolutionary success.

Gymnosperms and angiosperms have **woody tissue** which supports the stem of the plant so that it can reach great size. A large, taller plant has an advantage in obtaining sunlight for photosynthesis and in dispersing its seeds. These plants also have **hormones** that control the growth and development of the plant organs. They were able to become highly adapted to the environment and to change in response to the seasons by producing seeds and becoming dormant at the right time because of the control of these processes by hormones.

Gymnosperms include the cycads, which look like palm trees but bear cones, the ginkgo tree, *Ephedra,* and the familiar conifers.

□ Obtain a branch of a conifer, or observe one in a field demonstration.
□ Note the needlelike **leaves** and the **bark** on the branches and trunk.
□ Note that the leaves are **evergreen** and do not drop off in the fall as do the leaves of **deciduous** trees. (A few conifers are deciduous, the larch (tamarack), for example.)
□ Examine a prepared slide of a cross section of a leaf (needle). Compare what you see with Figure 15–2.

What is the advantage of evergreen leaves?

____

Why are evergreen leaves usually needlelike or covered with a thick protective cuticle?

____

____

Seed bearing in conifers is restricted to the **sporophylls** that make up the **cone.** Gymnosperm seeds are located on the surface of the sporophyll rather than within a fruit as are angiosperm seeds. Male and female cones are separate and produce small, nonphotosynthetic gametophytes that are nourished by the sporophyte. Thus the gametophyte can live wherever the sporophyte lives. Fertilization and early development occur within the female gametophyte.

□ Locate some developing cones which are green and soft. The larger ones are the female **seed cones.** The brownish male **pollen cones** are smaller and usually occur in clusters; they are present for only a short time.
□ Examine a male pollen cone. Each sporophyll or scale bears a pollen sac, also called the **microsporangium** because meiosis takes place there.

Pollen is formed within the pollen sac, and each grain can develop into a tiny, parasitic male gametophtye. During the formation of a pollen grain, a cell undergoes meiosis to form four microspores. Each microspore grows into a pollen grain containing several cells by means of mitosis. One cell nucleus becomes the

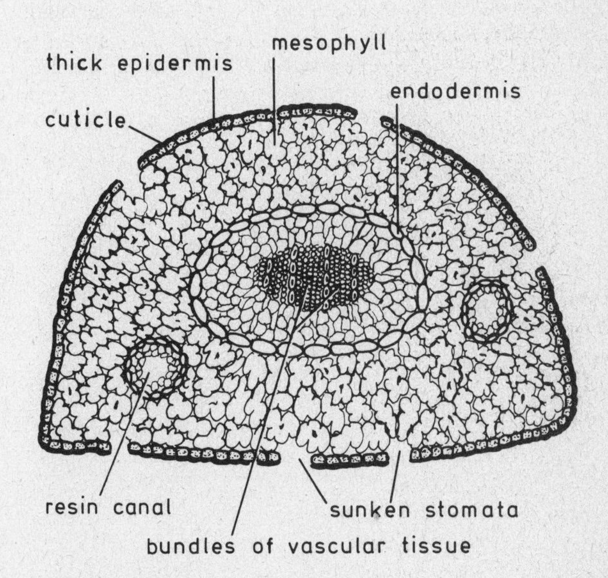

Figure 15–2. Pine leaf in cross section. This evergreen leaf has a reduced surface area and a thick cuticle to conserve moisture.

reproductive nucleus (sperm nucleus), and one gives rise to the pollen tube (tube nucleus), which allows the sperm nucleus to reach the egg. When the mature pollen grain is released from the pollen cone, a few of the grains will reach the scales of a female cone and land on a mature ovule (immature seed). This is the process of **pollination.** After pollination, the pollen tube germinates, grows through the tissue of the female gametophyte, and allows the sperm nucleus to fertilize the egg.

□ Examine a prepared slide of pollen from a conifer.
□ Note the winglike projections of the mature grains that help them to travel further in the wind.

What are the advantages of a pollen grain over a flagellated sperm cell?

____

____

□ Examine an older **seed cone;** remove one scale.
□ Look for two **ovules** containing **megasporangia** on its upper surface. If the ovules have already matured into seeds, they may have been shed, leaving two hollow impressions where they once grew.

Within each megasporangium, meiosis followed by mitosis gives rise to the cells that make up the parasitic female gametophyte including several egg cells. When pollen lands on the sticky surface of the ovule, the pollen tube grows through an opening in the ovule called the **micropyle** and on through the tissues of the female gametophyte until it reaches an egg cell. The sperm nucleus enters the egg in the process of **fertilization,** and the new embryo starts to develop within the megasporangium. A diagram of fertilization is shown in Figure 15–3. The seed of the pine can be understood as an extended pine family in which the young sporophyte is contained within the tissue of its parental female gametophyte, and the gametophyte in turn is contained within the reproductive organs of the grandparental sporophyte. Although several eggs may be fertilized within the female gametophyte, one embryo will win out and be released in the mature seed. The seed contains stored food as well as the embryo, has a tough seed coat, and is carried by the wind.

The structure of angiosperms will be studied in more detail in a later exercise (Topic 30). Reproduction in angiosperms is similar to that in gymnosperms. **Pollen** from the stamen lands on the pistil, and a **pollen tube** grows down to the ovule at its base. The **female gametophyte** within the ovule consists of only a few cells, including a haploid **egg cell** and a special diploid cell. This cell is also fertilized and forms the **endosperm,** a source of nutrients that is incorporated into the seed. **Double fertilization** of both the egg cell and the endosperm cell is unique to angiosperms. (See Figure 30–7.) The outer layer of the ovule forms the **seed coat** to protect the embryo and endosperm. Finally, the base of the pistil containing the ovules—namely, the ovary—develops into a fruit. The fruit may be fleshy, as in a peach, or hard, as in a bean pod. The fruit serves to protect or disperse the seeds.

For your own review, complete Table 15-1 at the end of this topic by checking each feature that applies to the groups you have studied. List the characteristics of all embryophytes in the space below the table.

Before leaving the laboratory, be sure to return all demonstration materials and store your microscopes properly.

EXPLORING FURTHER

Banks, H. *Evolution and Plants of the Past.* Belmont, Calif.: Wadsworth, 1970.

Bold, H. C. et al. *Morphology of Plants.* 4th ed. New York: Harper and Row, 1980.

Jensen, W. A., and F. B. Salisbury. *Botany: An Ecological Approach.* Belmont, Calif.: Wadsworth, 1972.

Weier, T. E., C. R. Stocking and M. G. Barbour. *Botany: An Introduction to Plant Biology.* 5th ed. New York: John Wiley and Sons, 1974.

Scientific American Articles

Altschul, Siri von Reis. "Exploring the herbarium." May 1977 (#1359).

Echlin, Patrick. "Pollen." April 1968 (#1105).

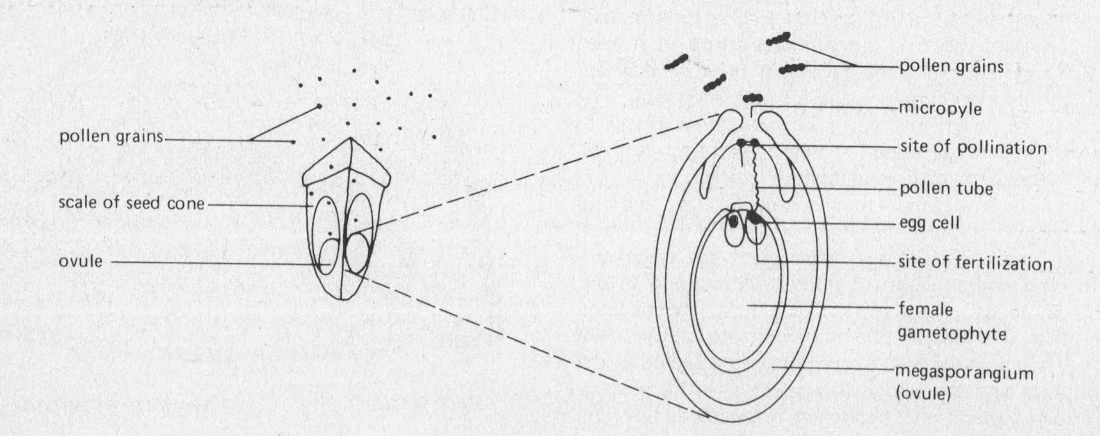

Figure 15–3. Fertilization in a pine cone. Pollination occurs when the pollen grains reach the surface of the ovule and the pollen tube begins to grow. Fertilization occurs when the sperm nucleus from the pollen tube unites with the egg nucleus within the female gametophyte. These two events may be separated by a long period of time.

Student Name ______________________ **Date** ______________

TABLE 15–1. CHARACTERISTICS OF THE EMBRYOPHYTE GROUPS

Characteristics	Musci (Moss)	Lycopsida (Club Moss)	Sphenopsida (Horsetail)	Filicineae (Fern)	Gymnospermae (Conifer)	Angiosperm Flowering Plants
Cuticle	No	✓ Yes	Yes	Yes	Yes ✓	✓
True roots, stems, leaves	No	Yes	Yes	Yes	Yes	Yes
Flagellated sperm	Yes	Yes	Yes	Yes	No Pollen	No
Rhizoids in gametophyte	Yes	No	No	Yes	No	No
Vascular tissue	No	Yes	Yes	Yes	Yes	Yes
Rhizomes	No		No yes	yes		No
Separate male and female gametophyte	Yes	Yes	Yes	No	Yes	Yes
~~Parasitic~~ or nonphotosynthetic sporophyte	Yes		Yes			
Photosynthetic gametophyte and sporophyte				Yes		
~~Parasitic~~ or nonphotosynthetic gametophyte		Yes	Yes		Yes	Yes
Airborne spores	Yes	Yes	Yes	✓		
Pollen					Yes	Yes
Embryo protected by seed coat					Yes	Yes
Seed enclosed within fruit						Yes
Woody vascular tissue					Yes	Yes

Characteristics common to all embryophytes:

During reproduction the sporophyte is nurtured there (by female)

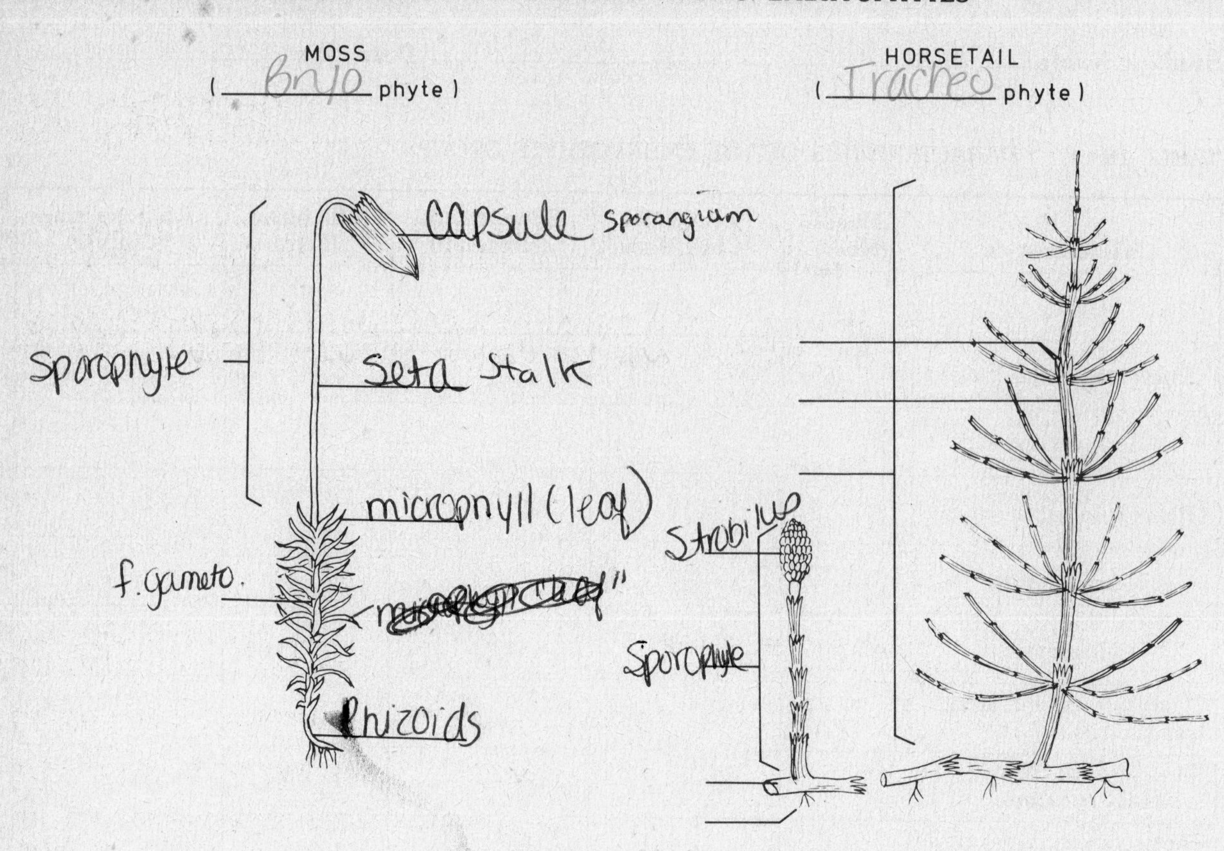

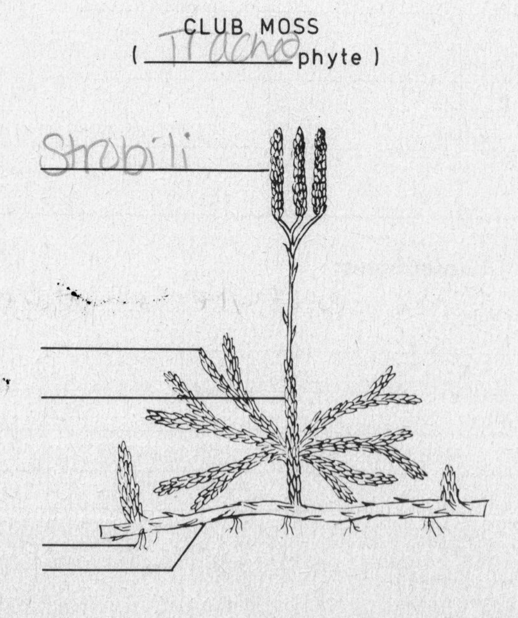

Figure 15–4. The dominant stage of a moss, a club moss, and a horsetail. Label the dominant stage as a sporophyte or gametophyte under its name, and label the parts of each plant marked with label lines.

TOPIC 16

Introduction to the Lower Invertebrates

OBJECTIVES

When you have completed this topic, you should be able to:

1. Describe the advantages of multicellularity, unicellularity, larger size, radial symmetry, bilateral symmetry, the mobile stage of life history, the sessile stage of life history, the sense organs, cephalization, and the nervous system.
2. Define coelom, coelomate, and acoelomate, and state the advantage that a coelom gives to an organism possessing it.
3. Give the distinguishing features of members of the phyla Cnidaria, Platyhelminthes, Nematoda, and Rotifera.
4. Describe the following structures and give the phylum or phyla in which each of them is found: tentacles, nematocysts, gastrovascular cavity, flame cells, coelom, polyp, medusa, and gonads.

TEXT REFERENCES

Chapters 27, 28 (A).
Larva, cnidarians, alternation of generations, body symmetry, coelom, flatworms, nematodes, rotifers.

INTRODUCTION

As competition between unicellular organisms became intense, larger and more complex organisms had a selective advantage, and multicellular organisms came into existence. **Multicellularity** allowed differentiation of cells that performed different specific functions. Different types of cells became organized into **tissues** and the tissues into **organs,** specialized structures adapted to carry out a certain function or process.

In organisms with more sophisticated organs and organ systems, the body plan changed to accommodate the new complexity. In nematodes, rotifers, and more complex organisms, there is a space between the gut and body wall called the **coelom.** This space allows the gut and body to move independently and provides a protective space in which the internal organs can be contained. When the coelom is not fully lined by mesoderm, as in nematodes and rotifers, it is a **pseudocoelom.** In annelids, molluscs, and the other organisms that we will be studying in subsequent exercises, the coelom is a true coelom because it is fully lined by mesodermal tissue.

Compare the body plans shown in Figure 16–1 and be sure you can relate each type to the phyla in which it is found.

PHYLUM CNIDARIA: *HYDRA*

Cnidarians have a saclike body consisting of two cell layers surrounding a cavity called the **gastrovascular cavity.** In the **polyp** form the organism is more or less sessile, and the opening to the cavity is directed upwards. In the motile **medusa** form the opening is underneath. Both forms usually have fingerlike extensions of the body called tentacles, and specialized cells called **nematocysts** are present. There is much cellular specialization in the form of reproductive, sensory, secretory, and primitive nerve cells. Movement is accomplished by the contraction of muscle fibers.

Hydra is an example of a polyp that never produces a medusa stage. Its reproductive cells form gametes. The fertilized zygote will eventually develop into a small *Hydra*. *Hydra* can also reproduce by budding off a small

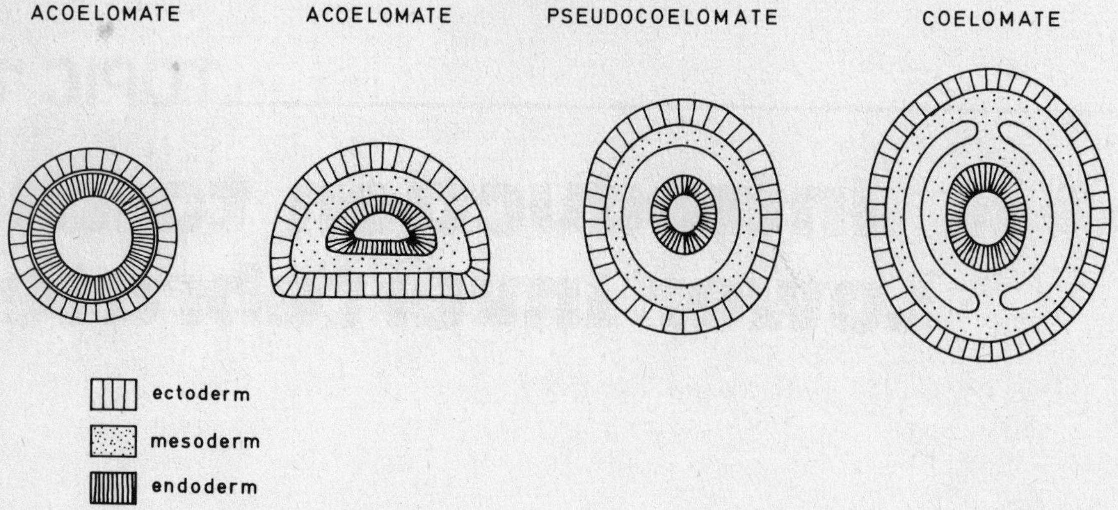

ACOELOMATE ACOELOMATE PSEUDOCOELOMATE COELOMATE

ectoderm
mesoderm
endoderm

Figure 16–1. Body plans. Acoelomates with two cell layers are the cnidarians. Acoelomates with three cell layers are the flatworms. Pseudocoelomates such as the nematodes and rotifers have a body cavity, but it is between the inner and middle layers of cells. In true coelomates, the body cavity is formed within the middle cell layer and the organs are protected and supported by this layer.

Hydra, complete with tentacles, from the side of its body. The internal structure of *Hydra* is shown in Figure 16–2.

□ Examine a living *Hydra.* Locate the **basal disc** by which it is attached, the cylindrical **body,** and the crown of **tentacles.**

What kind of symmetry does *Hydra* show?

____

□ If brine shrimp are available, add a drop of *washed* brine shrimp to your *Hydra* and observe its eating behavior. (Salt water will quickly kill the *Hydra* by upsetting its water balance.)

The tentacles are covered with nematocysts which sting and entangle the prey so that it can be brought to the mouth and ingested.

Is the shape of the *Hydra* different after ingestion?

____

Is digestion of the food intracellular or extracellular?

____

What happens to the wastes after the edible parts of the brine shrimp have been absorbed?

____

____

____

____

____

____

____

□ Examine a prepared slide of nematocysts in the microscope.

The nematocyst consists of a tiny thread coiled up inside a capsule. When it is triggered by the touch of prey, the thread turns inside out and then entangles or paralyzes the prey organism. Firing of the nematocysts is under nervous control. Toxins in the nematocysts of cnidarians such as fire coral and the Portuguese man-of-war can be harmful or even fatal to humans.

Other cnidarians, such as the jellyfish *Gonionemus,* have a medusa stage as well as a tiny polyp stage in their life histories. Organisms that have both a polyp and medusa stage show **alternation of generations:** the medusa gives rise to the polyp, the polyp then gives rise to the medusa, and so forth.

PHYLUM PLATYHELMINTHES

Flatworms have three body layers, **ectoderm, mesoderm,** and **endoderm,** but there is no body cavity, so they are **acoelomate.** The gastrovascular cavity has a single opening. Flatworms have true organs and organ systems. Motile flatworms show cephalization and bilateral symmetry.

Class Turbellaria

Look at Figure 16–3, which is the outline of a planarian, and note the shape of the anterior (front) and posterior (rear) ends of this animal.

□ Examine a living planarian, such as *Dugesia,* or look at a prepared slide in the microscope.
□ Locate the anterior end, with its earlike **auricles** and pigmented **eyespots.**

Cilia on the underside of the worm are used in its locomotion. The **pharynx, mouth,** and **genital pore** are also located on the underside.

□ Turn the worm upside down to look for these structures.
□ Feed your planarian with a bit of meat. Try to observe the extension of its pharynx as it feeds.

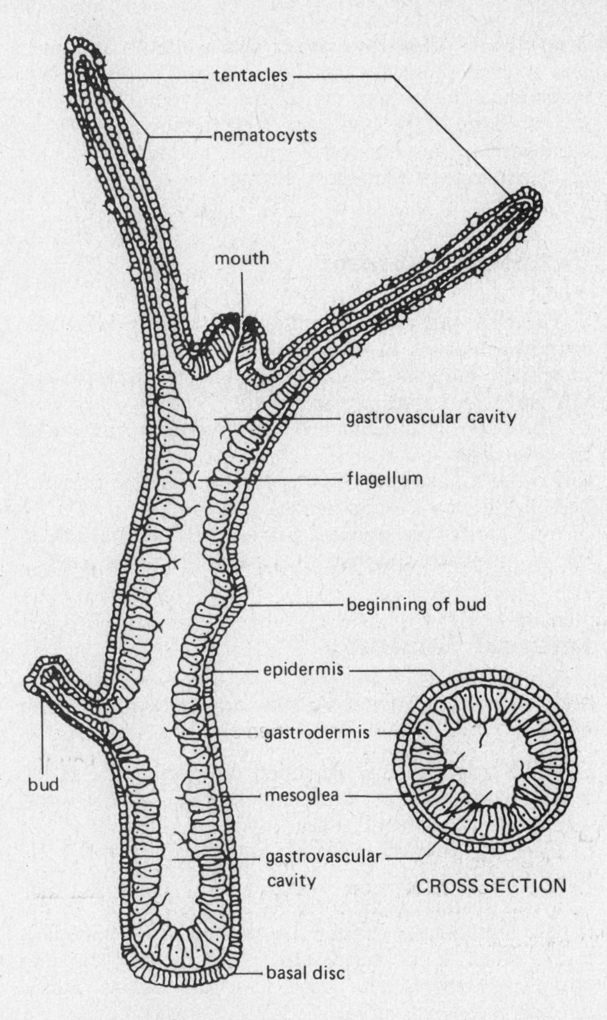

Figure 16–2. ***Hydra.*** Prey is captured by the stinging or entangling nematocysts and stuffed into the mouth by the tentacles. Movement occurs by contraction of muscle fibers in the mesoglea. (Redrawn from Wodsedalek, J. E. and Charles F. Lytle, *General Zoology Laboratory Guide,* 8th Ed. © 1971, 1977, 1981 Wm. C. Brown Company Publishers, Dubuque, Iowa. Reprinted by permission.)

If you are looking at a prepared slide, how many branches does the gastrovascular cavity have?

____

The **excretory system** consists of highly branched interconnecting tubes which are lined with ciliated **flame cells** that move the contents along to the excretory pores. Its major function is probably water balance.

The nervous system consists of nerve cells grouped in nerve cords along each side of the worm and concentrated in the ganglia (two masses near the eyespots), sometimes called the "brain."

An individual worm produces both eggs and sperm but does not fertilize itself. Two individuals mate by transferring sperm to each other during copulation. Fertilization is internal, and eggs are laid in cocoons or jelly. The most common method of reproduction is asexual reproduction by transverse fission in which the animal splits in half behind the pharynx, and each half regenerates the missing parts.

Draw in and label the eyespots, pharynx, mouth, gastrovascular cavity, nerve cords, and "brain" in Figure 16–3.

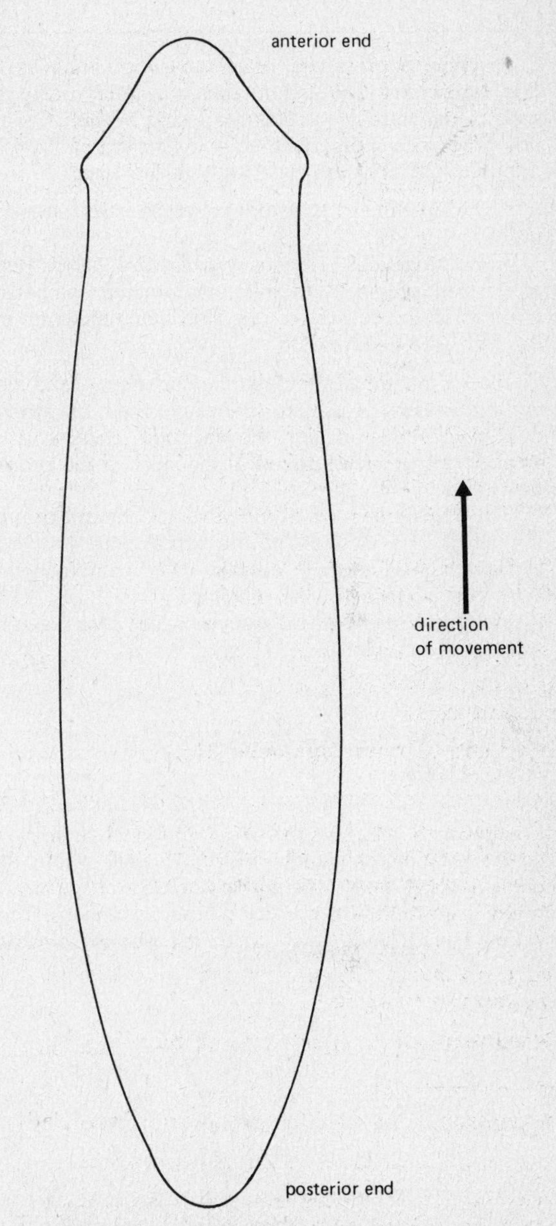

Figure 16–3. A planarian. Draw in and label the parts of the digestive system, nervous system, and sense organs of the flatworm.

Class Cestoda

The members of this class are highly specialized parasitic tapeworms. The adult tapeworm lives in the intestine of a vertebrate and is firmly attached by its **scolex** which contains suckers and hooks. The rest of the ribbonlike body is a series of segments called **proglottids.** Each proglottid contains male and female reproductive organs, nerves, and excretory canals. There is no digestive system.

How does the tapeworm obtain food?

____

____

____

□ Examine a preserved whole tapeworm such as the dog tapeworm, *Dipylidium caninum,* and locate the scolex. Immature proglottids are found at this end, and ripe proglottids stuffed with eggs are broken off from the opposite end and leave the body in the feces.

The structure of a mature proglottid is shown in Figure 16–4.

The proglottids of a tapeworm fertilize themselves or an adjacent proglottid. In an infected animal, competition will lead to the success of one individual tapeworm and the rest will be eliminated.

□ View a mature proglottid in the microscope, and identify the **ovaries** which are centrally located; the **uterus** between the ovaries; and the **vaginas** which lead out through the **genital pores** at the sides of the proglottid.
□ The **testes** fill most of the proglottid. Sperm pass out through the vas deferens to the genital pore.
□ Find the **excretory canals** which run along the sides and posterior of the proglottid.
□ Find the **longitudinal nerves** which pass down the sides of the proglottid.

PHYLUM NEMATODA: *ASCARIS*

Roundworms may be the most abundant animals on earth. They occur by the millions in soil where they damage plants and crops and may reach a large size, as in the giant roundworm which parasitizes humans. These worms have a body cavity called the **pseudocoelom,** a **complete digestive tract** with a mouth and anus, and a thick protective cuticle. The mesoderm contains longitudinal muscles, so the worms can move by a whiplike thrashing of the body, but they cannot creep like an earthworm. There are dorsal and ventral nerve cords but no respiratory or circulatory system.

External Anatomy

□ Examine preserved specimens of male and female roundworms such as *Ascaris.*
□ Identify the male, which has a distinctly curved posterior end with copulatory spicules.
□ The opposite anterior end has a **mouth** surrounded by three **lips.**
□ Locate the **anus,** which is a slit across the posterior end of the worm on its ventral side.
□ Find the female **genital pore** on the ventral side of the worm about one-third of the way from the head.

Internal Anatomy

The internal organs of *Ascaris* are shown in Figure 16–5.

□ Make a longitudinal slit along the dorsal side of the female worm, and pin back the sides. Cover the worm with some water to minimize drying.
□ Find the **intestine** which runs through the body from mouth to anus.
□ In the middle of the intestine, locate the coils of the female reproductive system. The **ovaries** are threadlike and connect to the **oviducts,** which in turn lead into the larger **uteri.** The uteri join just before exiting through the genital pore.

These structures can be seen more easily in cross section. (See Figure 16–5.)

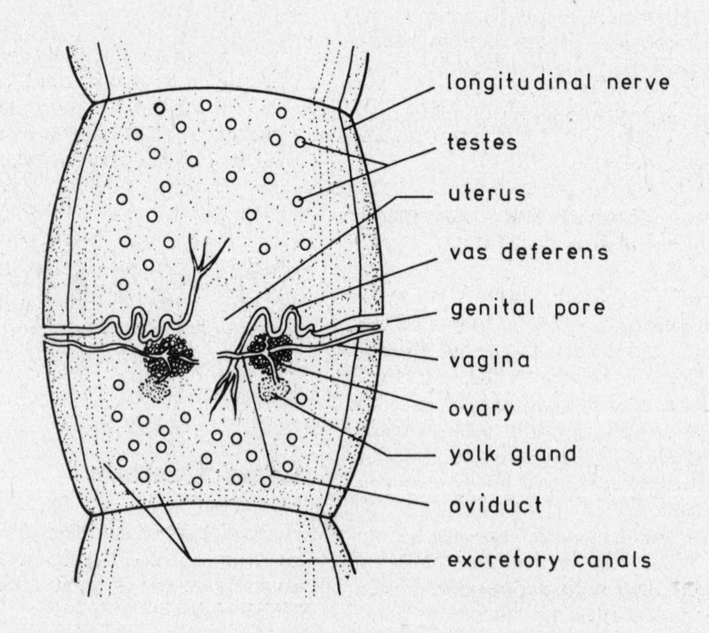

Figure 16–4. Mature proglottid of the dog tapeworm, *Dipylidium caninum.* Copulation between two proglottids results in internal fertilization of the eggs. Fertilized eggs completely fill the ripe proglottid, which breaks off from the tapeworm and eventually bursts to release the eggs.

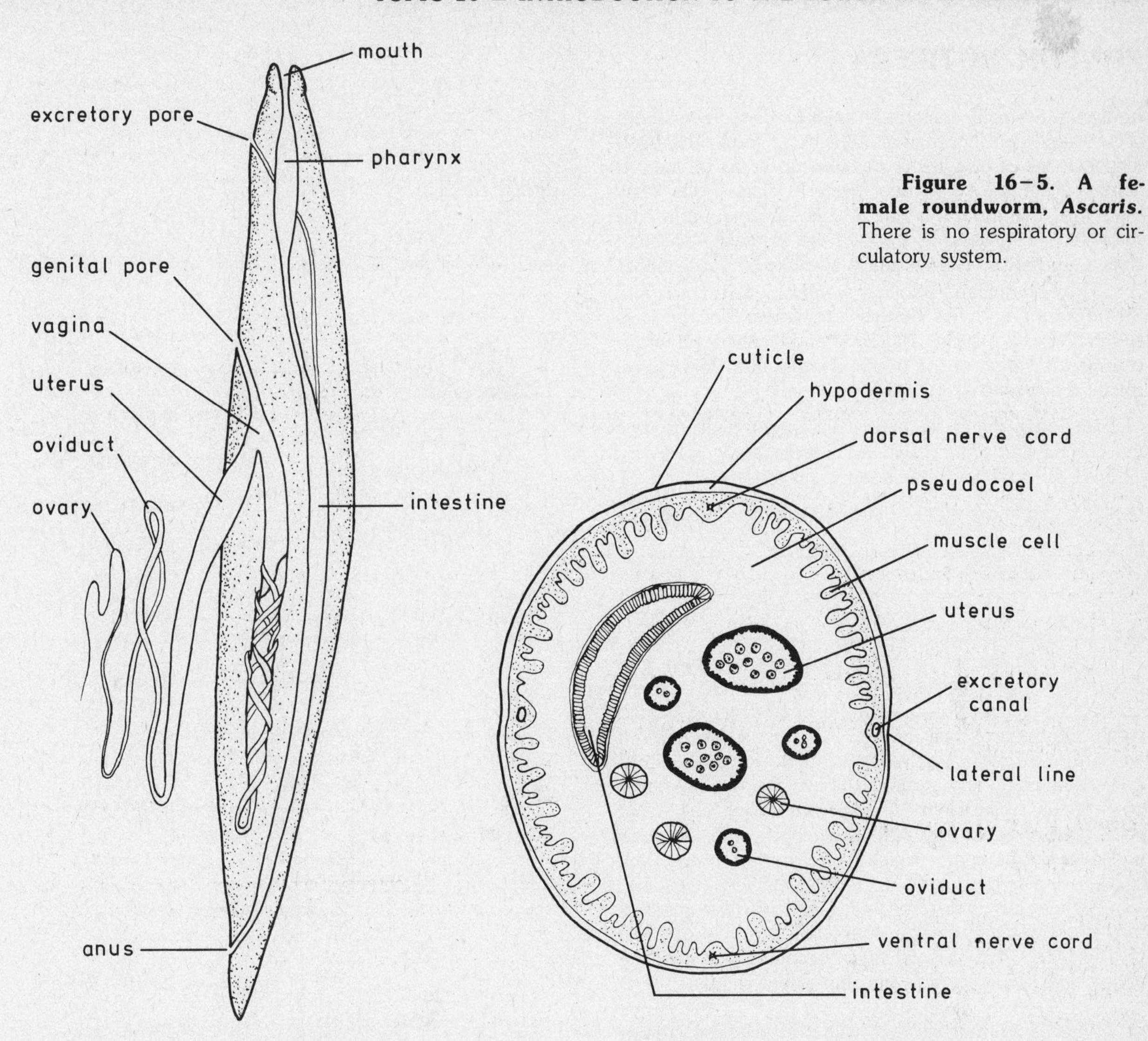

Figure 16–5. A female roundworm, *Ascaris*. There is no respiratory or circulatory system.

□ Examine a cross section of *Ascaris* in the microscope, and find the large, flattened **intestine.**
□ Identify the small, round structures which are sections through the tubes of the **oviduct** and the larger **uterus** which is filled with eggs; the **ovary** is solid.
□ Locate the dorsal and ventral **nerve cords.**
□ Identify the **cuticle** surrounding the worm and the longitudinal **muscle cells** which are the large cells surrounding the body cavity.
□ The space just under the cuticle in either side of the worm is the **lateral line** in which the **excretory canals** are located.

In the male *Ascaris* a threadlike testis leads into ducts that empty into the cloaca at the posterior of the digestive tract. As you already learned in Topic 8, fertilization is internal, and the eggs have passed through the early stages of development before they leave the body of the female.

Anatomy of a Nematode Cyst

A parasite has a great advantage if its eggs or larvae are well protected and can survive adverse conditions until they can reach a new host. In *Trichinella spiralis*, the parasitic roundworm that causes the disease trichinosis, fertilized eggs develop into larvae within the reproductive tract of the female worm and are then deposited within the host's lymphatic system. The larvae are carried throughout the body and those that invade skeletal muscle will form cysts and survive. The cysts have a low metabolic rate and a protective wall. If the host is a pig, the meat might be ingested without being properly cooked, and the cyst walls would be broken down in the digestive tract of the new host. The young worms quickly mature and produce hundreds of offspring. These larvae enter the lymphatic system, burrow into the host's skeletal muscle, and again encyst.

What are two ways in which the disease trichinosis could be controlled?

1. ______________________________

____

2. ______________________________

____

□ Examine a prepared slide that shows the cysts formed in skeletal muscle by young trichina worms.

Do the encysted worms harm their host? ____________

PHYLUM ROTIFERA

Rotifers are small, coelomate animals that move about actively and can ingest small particles of food. The beating of circles of cilia in the **corona** gives the phylum its name (rotifer means "wheel bearer"). These cilia draw food into the mouth. Food is then ground up in the **mastax** and passes on through the stomach, intestine, cloaca, and anus. The corona is also used in locomotion.

The anterior end also has a sensory **antenna,** and the posterior end has **spurs** and **toes.** The body is covered by a cuticle. Muscle bands can contract to change the shape of the body. The structure of a typical rotifer is shown in Figure 16–6.

□ Examine a living rotifer or a prepared slide. Locate the corona and antenna at the anterior end.
□ Find the toes and spurs at the posterior end.

How do you know that a rotifer is not a giant protozoan?

____

____

____

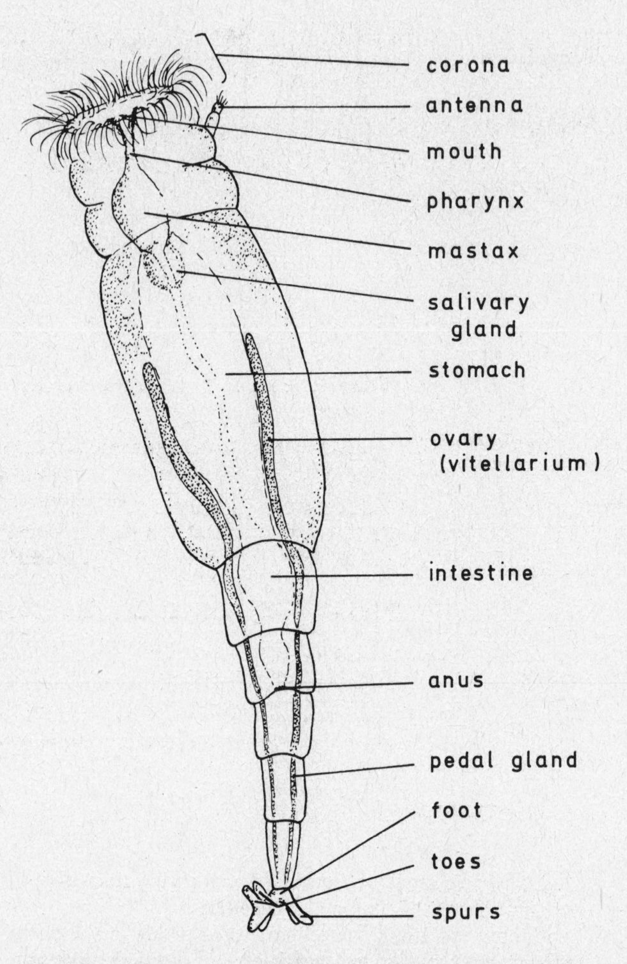

Figure 16–6. A rotifer, *Philodina*. This type of rotifer is always female and reproduces by parthenogenesis.

In certain species of rotifers there are no males, and the unfertilized eggs develop into new female rotifers. The eggs may develop in a brood pouch that is located within the body of their mother. When males are present and are involved in reproduction, they are usually small and degenerate and fertilize the female by the crude process of **hypodermic impregnation:** the male stabs the female in any part of her body with his penis and releases the sperm which have to swim through the body to the eggs in order to fertilize them.

Rotifers can withstand drying and will lose most of their body water to enter a desiccated form that can survive for years. When water again becomes available, the animal will rehydrate and become active shortly.

When you are finished with this exercise, return your materials and store your microscope properly.

EXPLORING FURTHER

Barnes, R. D. *Invertebrate Zoology*. 4th ed. Philadelphia: W. B. Saunders, 1980.
Buchsbaum, R. *Animals without Backbones*. 2nd rev. ed. Chicago: University of Chicago Press, 1975.
Buchsbaum, R., and L. J. Milne. *The Lower Animals: Living Invertebrates of the World*. Garden City, N.Y.: Doubleday, 1962.
Russell-Hunter, W. D. *A Biology of the Lower Invertebrates*. New York: Macmillan, 1968.
Villee, Claude A., W. F. Walker and R. D. Barnes. *General Zoology*. Philadelphia: W. B. Saunders, 1978.

Scientific American Articles

Gierer, A. "Hydra as a model for the development of biological form." December 1974 (#1309).
Goreau, T. F., N. I. Goreau and T. J. Goreau. "Corals and coral reefs." August 1979 (#1434).
Savory, T. H. "Hidden lives." July 1968 (#1112).

TOPIC 17

Introduction to the Higher Invertebrates: Annelids and Molluscs

OBJECTIVES

When you have completed this topic, you should be able to:

1. Define coelom and enumerate the advantages to an organism of having one over not having one.
2. List the major features of an annelid and of a mollusc.
3. Describe the adaptations of the earthworm that allow it to live on land.
4. Describe the advantages and disadvantages of the exoskeleton in a mollusc.

TEXT REFERENCES

Chapter 28 (A–C).
Coelom, annelids, molluscs.

INTRODUCTION

Annelids and molluscs, as well as the animals that we will study in later exercises, have a true **coelom** that arises by the splitting of the middle layer of cells in the body, the mesoderm. (See Figure 16–1.) The mesoderm surrounds all of the organs that develop within the coelom, protecting and supporting them, and it can give rise to layers of muscle so that the gut and other organs can move and function independently from the body wall. The space within the coelom makes room for an independently active circulatory system, internal organs, and large gonads. In some organisms such as the earthworm the fluid in the coelom acts as a hydrostatic skeleton on which the muscles of the body wall can exert pressure during locomotion. In many coelomates, the coelomic fluid helps in circulating nutrients and gas exchange.

PHYLUM ANNELIDA

Annelids have **metamerically segmented** wormlike bodies, and most are divided into segments internally by partitions called **septa.** In polychaetes and oligochaetes the body wall contains circular and longitudinal muscles that squeeze the fluid within each segment to lengthen or shorten the worm during its creeping movements. The **closed circulatory system** contains blood for improved gas exchange. The **complete digestive tract** processes food in one direction from mouth to anus and has specialized parts. The nervous system shows **cephalization** with sense organs and masses of nerve cells called ganglia at the anterior end. The nephridia are true excretory organs which can actively absorb and excrete substances to regulate the composition of the coelomic fluid.

Class Oligochaetae: Earthworm

Oligochaetes (oligochaete = "few bristles") have only four pairs of setae on each segment and lack parapodia. Gas exchange takes place over the entire body surface and is aided by an efficient circulatory system in which the blood is moved by little pumps ("hearts"). Many oligochaetes such as the earthworm have taken up a terrestrial life but have not really adapted to land; they secrete a thick mucus to keep their skin moist and must remain within their damp burrows in the ground unless

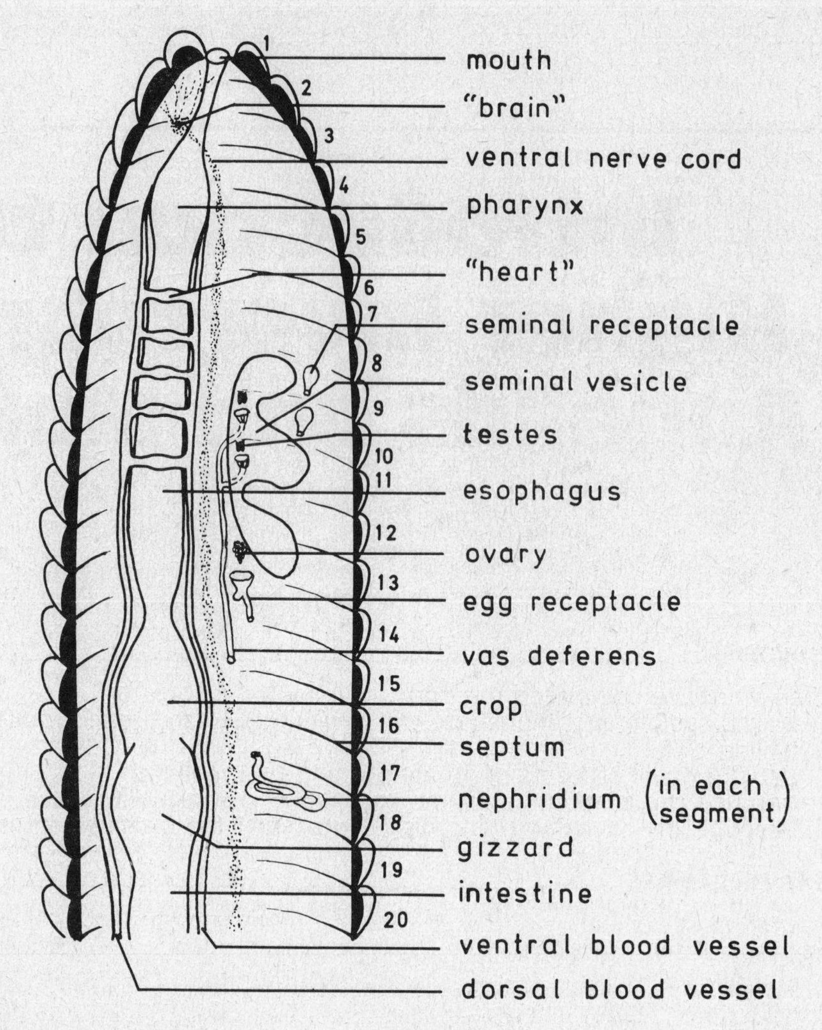

Figure 17–1. Internal structure of the earthworm, *Lumbricus*. The dorsal side is towards the left. Sense organs are lacking, but there is a well-developed nervous sytem, closed circulatory system, complete digestive tract, and excretory organs.

the surface is soaking wet. Earthworms feed on dead vegetation and have a well-developed digestive system to process their diet of decaying vegetation. The head is much reduced to adapt to its burrowing way of life. The individual earthworm is hermaphroditic, producing both eggs and sperm, but does not fertilize itself. Rather, eggs and sperm are shed into a protective cocoon of mucus where external fertilization takes place.

Earthworms, in contrast to polychaetes, do not have a trochophore larva. Instead, a young worm hatches directly from the cocoon.

EXTERNAL ANATOMY AND BEHAVIOR

□ Examine a live or preserved earthworm and find the **clitellum,** an enlarged section near the anterior end that secretes mucus during copulation.
□ Locate the **mouth** under the protruding **prostomium.**
□ Counting back from the prostomium, find the ninth segment, and look for openings of the seminal receptacles between the ninth and tenth and the eleventh and twelfth segments on the worm's ventral surface. Sperm are stored here when worms copulate.
□ The oviducts open on the fourteenth segment to release the eggs.
□ Seminal vesicles open on the fifteenth segment to release sperm.
□ Look at one segment closely, and locate the **setae** on the lower part of the side of the worm.
□ Find the **excretory pores** that occur on the sides of most of the segments. Each pore is the opening of one **nephridium,** or excretory organ.

When two worms mate, they overlap ventral surfaces so that the anterior tip of one worm is a little past the clitellum of the other and vice versa. Sperm are released by both worms but have to migrate over many segments in mucus-covered grooves, so the worms must remain joined for quite a long time. After mating, the mucus secreted by the clitellum passes over the head of the worm and picks up eggs from the oviduct openings and sperm from the sperm receptacles as it moves. The cocoon is deposited in the soil and a young worm hatches from it eventually.

If live earthworms are available, investigate locomotion and other behavioral responses.

□ Place the worm on a moist paper towel in a dissecting pan. Study its locomotion under the dissecting microscope.

Referring to Figure 17–2, which muscles contract to cause elongation of a segment? ____________

Which muscles contract to cause shortening of a segment? ____________

How does the worm respond to the bright light of your microscope?

____

____

□ Touch the front end of the earthworm gently with a probe. What is the response? ____________
□ Hold the probe still in the path of the moving earthworm. How does the worm respond to this foreign object?

____

____

□ Dip your probe in 5% acetic acid. (This is similar to vinegar and not acidic enough to injure the worm.) Move the probe towards the front end of the worm until you get a response.

Describe the response. ____________________

____

How do you think the response is important to the earthworm? ____________________

____

INTERNAL ANATOMY

Use Figures 17–1 and 17–2 as guides to the internal structure of the earthworm.

□ Open the worm from the darker **dorsal side,** and keep it covered with water while you work.
□ Notice the internal segmentation that corresponds to the external segmentation. Identify the internal **septum** between adjacent segments.
□ Locate the **dorsal blood vessel** which carries blood from the anterior to posterior. The circulatory system has hemoglobin and the vessels have one-way valves.
□ Locate the five pairs of pumping blood vessels ("**hearts**") that connect the dorsal and ventral blood vessels.
□ Find the **ventral blood vessel,** the **ventral nerve cords,** and the subneural blood vessel. Blood in these vessels flows from posterior to anterior.
□ Follow the large digestive tube back from the mouth as far as the clitellum, identifying the muscular **pharynx,** the **crop** (about segment 16), the **gizzard** (about segment 19), and the **intestine** (all the rest).
□ Within each segment find the paired **nephridia.** Each of these excretory organs is a coiled tube leading to an **excretory pore.** The beginning of the tube is in the anteriorly adjacent segment. Fluid enters by ciliary action, and the tube excretes waste into the fluid and reabsorbs needed substances as it passes out of the body.
□ Locate the saclike **seminal receptacles** in segments 9 and 10.
□ Sperm made in the tiny testes are stored in the multilobed **seminal vesicles** until copulation. Then they pass out through the **vas deferens.**

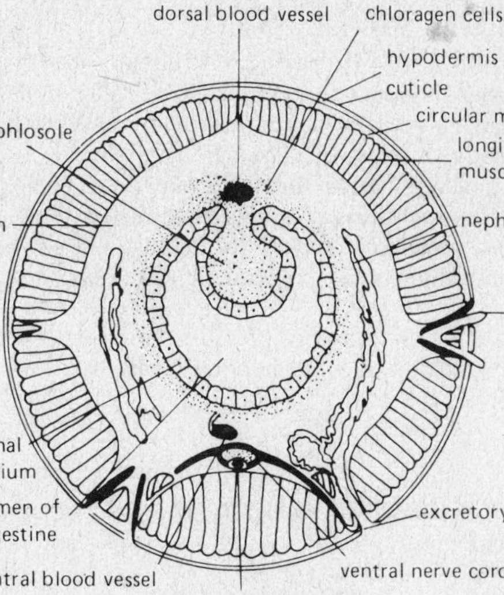

Figure 17–2. Cross section of the earthworm, *Lumbricus*. The surface area of the intestine is doubled by the typhlosole, and the spacious, fluid-filled coelom acts as a hydrostatic skeleton in locomotion. (Redrawn from C. A. Villee, W. F. Walker and R. D. Barnes, *General Zoology*. Philadelphia: W. B. Saunders, 1978.)

□ **Ovaries** are located in segment 13, and the eggs are stored in the adjacent **egg receptacles** before being released through the **oviducts.**

The earthworm does not have eyes or ears, but the epidermis contains light-sensitive cells and sensory bristles that are sensitive to vibration. The paired ventral nerve cords have fast-conducting neurons that allow the worm to retreat rapidly if it senses danger. In the head region, the nerve cords connect to swellings of nerve cells, called **ganglia,** by tracts of fibers that pass on either side of the digestive tract, just in front of the pharynx.

□ Locate the whitish paired ganglia at the anterior end of the ventral nerve cord that represent the earthworm's "brain." They are located dorsal to the digestive tract, and just anterior to the pharynx.
□ Obtain and study a prepared slide of the cross section of an earthworm; compare your slide with Figure 17–2.
□ Locate the intestinal epithelium that forms the gut.
□ Locate the large space it surrounds which is the **lumen** or interior of the intestine; the infolding at its top is the **typhlosole,** which increases the surface area available for absorption.

Notice the large coelom, which is the fluid-filled space between the muscular body wall and intestine, and which contains the body organs. In the earthworm the coelom acts as a hydrostatic skeleton.

□ Locate the blood vessels and the ventral nerve cord.
□ Surrounding the gut are **chloragen cells,** which aid in digestion and may represent primitive liverlike cells.
□ Find the coils of nephridium within the coelom.
□ Surrounding the coelom is a thick layer of longitudinal muscle. It functions to shorten the worm.
□ A thinner layer of circular muscle surrounds the longitudinal muscle. It contracts to lengthen the worm.
□ Locate the **cuticle** on the outside of the worm and the **hypodermis** which secretes it.

PHYLUM MOLLUSCA

Molluscs are thought to have arisen from an annelid ancestor but the present-day forms have tended to lose the segmentation that was the outstanding feature of the ancestral mollusc. These animals have three major body parts: a muscular **foot,** a **visceral mass,** and a hard protective **shell** secreted by the mantle, which lies over the body. There is a complete digestive system and an open circulatory system. The respiratory system usually has gills and the excretory organ has nephridia. Molluscs are true coelomates with a body cavity surrounded by mesoderm.

Class Bivalvia: Clam or Mussel

The body of bivalves, such as clams, mussels, scallops, and oysters, is protected by two shells hinged together that are secreted by the **mantle.** These animals show bilateral symmetry and, if you orient one with the hinge at the top and the "points" at the top left, as shown in Figure 17–3, you will be viewing the left side. Water passes in and out of the gill chamber through siphons which are modified parts of the mantle. The gill chamber is sealed off from the rest of the viscera in which the organs are bathed in an open circulatory system. Pigmented blood is pumped by a two-chambered **heart.** There is a complete digestive tract, a true **nephridium,** and a **foot** for digging or locomotion. Bivalves are filter feeders and there is no radula. They extract edible particles from the water circulating through the gill chamber. Labial palps aid in moving the food trapped in mucus into the mouth.

□ Orient a clam or mussel as shown in Figure 17–3. Focus on the large adductor muscles that close the shell. These are shown in Figure 17–5 and must be cut to open the shell.
□ Insert a tool into the space between the lower edges of the shells, and pry the shells apart enough to insert something to act as a wedge.

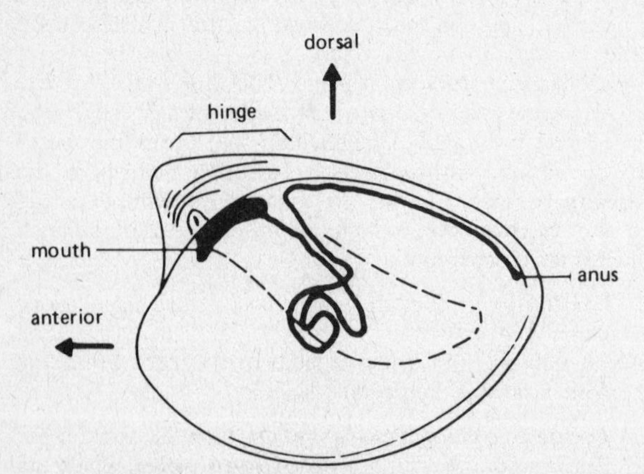

Figure 17–3. A bivalve mollusc. If a clam or mussel is held in this orientation, with the hinge on top and the "points" aiming to the left, the anterior will be to the left and the dorsal part of the animal on top. The approximate location of the digestive tract is shown in black, and the position of the gills is marked by the dashed line.

□ Use a scalpel to cut through the muscles without damaging the internal organs, especially in the region near the hinge. Separate the shells to expose the mantle.

Refer to Figures 17–4 and 17–5 to identify the internal organs.

□ Identify the adductor muscles and the muscular foot. Remove the mantle, which covers the **gills.**
□ Carefully remove the left gills. Slit the bottom of the foot in order to see the organs within it.
□ Trace the digestive tube from the mouth through the **gonad** (yellowish) and **heart** to the anus.
□ Locate the **digestive gland** and the **nephridium** (brown or green).
□ Pass a probe into the **siphons** at the animal's posterior to see where water circulates. Water enters the mantle cavity through the incurrent siphon, circulates through the filaments of the gills, and leaves through the excurrent siphon. The water moves by ciliary action, and particles in the water are the only food of the bivalve.

Because bivalves are filter feeders, any poisonous substances in their food particles will become concentrated in their bodies. During the season that toxic algae are prevalent, it is unwise to risk eating shellfish and the gathering of shellfish for food is usually prohibited by law.

Bivalves are usually either male or female, and they carry out reproduction by shedding a large number of gametes into the water where fertilization takes place. Sometimes the young are brooded in a specialized part of the gill. The embryos develop into motile trochophore larvae.

As in many sessile animals, the bivalves have developed protection in the form of hard shells that can be very tightly clamped together to enclose the soft body. In those forms that are subjected to periodic drying when the tide falls, the shells clamp together tightly to prevent the loss of moisture.

Some bivalves, such as the scallop, can swim after a fashion by closing and opening the valves. The muscle involved in swimming is the part of the scallop that you eat.

When you are finished with your work, dispose of your specimens, return the demonstration materials, and store your microscope properly.

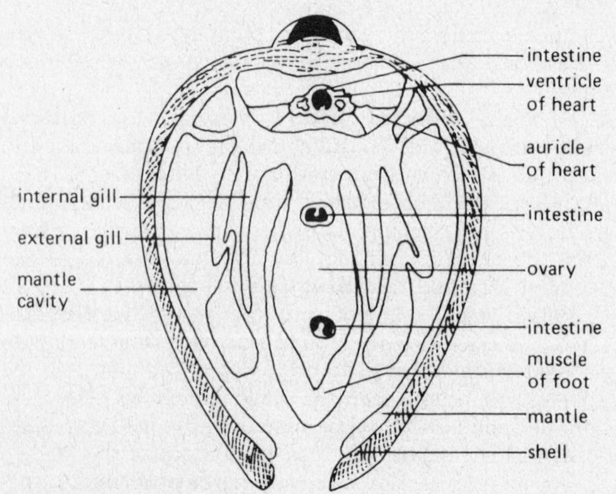

Figure 17–4. Cross section of a bivalve. Water circulates within the mantle cavity, bathing the gills for gas exchange.

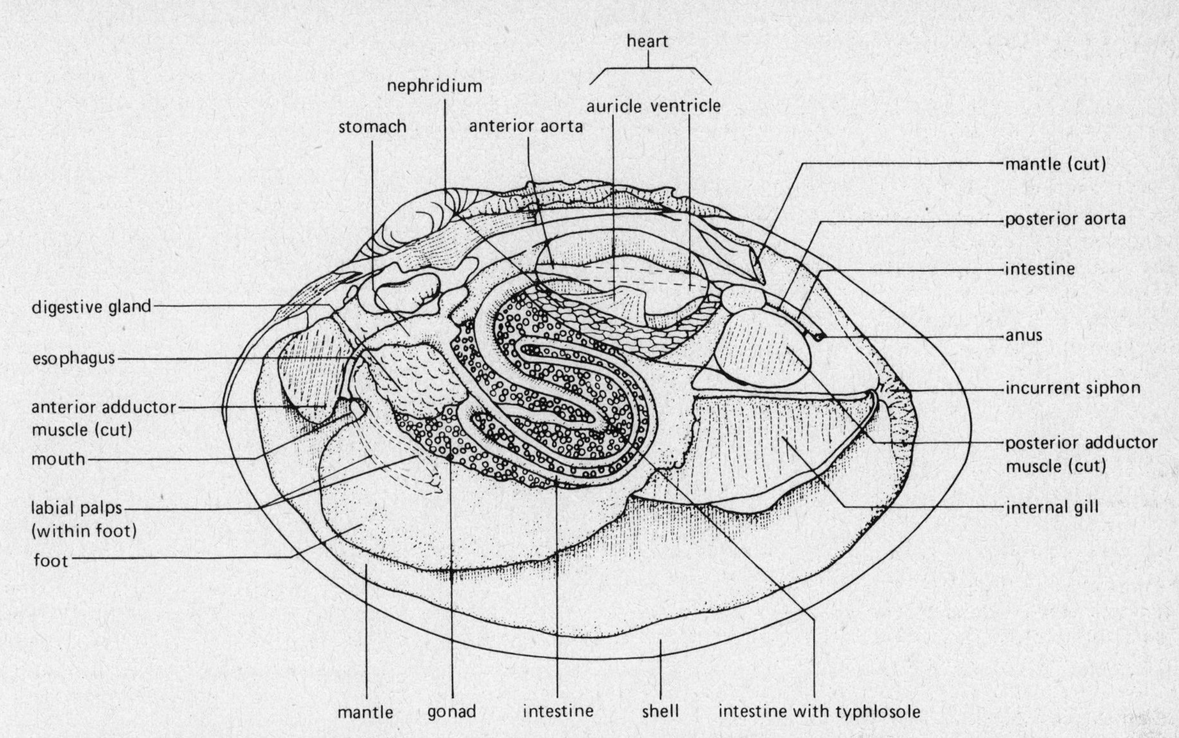

Figure 17–5. Internal structure of a bivalve. The left gill and mantle have been removed to show the internal organs. (Redrawn from Wodsedalek, J. E. and Charles F. Lytle, *General Zoology Laboratory Guide,* 8th Ed. © 1971, 1977, 1981 Wm. C. Brown Company Publishers, Dubuque, Iowa. Reprinted by permission.)

EXPLORING FURTHER

See also Topic 16 references.

Barrington, E. J. *Invertebrate Structure and Function.* Boston: Houghton Mifflin, 1967.

Lane, F. W. *Kingdom of the Octopus.* New York: Sheridan House, 1965.

Morton, J. E. *Molluscs.* 4th ed. London: Hutchinson, 1967.

Scientific American Articles

Boycott, Brian B. "Learning in the octopus." March 1965 (#1006).

Feder, Howard M. "Escape responses in marine invertebrates." July 1972 (#1254).

Nichols, John G., and David Van Essen. "The nervous system of the leech." January 1974 (#1287).

Ward, Peter, Lewis Greenwald and Olive E. Greenwald. "The buoyancy of the chambered nautilus." October 1980.

Willows, A. O. D. "Giant brain cells in mollusks." February 1971 (#1212).

TOPIC 18

Introduction to the Arthropods

OBJECTIVES

When you have completed this topic, you should be able to:

1. List the features of the phylum Arthropoda.
2. Describe adaptations of terrestrial arthropods that allow them to live on land.
3. List three factors that contributed to the success of insects.
4. Define metamorphosis and explain the difference between ametabolous, paurometabolous, hemimetabolous, and holometabolous metamorphosis.
5. Define the following arthropod structures and give the class(es) in which they are found: trachea, gill, book lung, spiracle, hemocoel, green gland, swimmerets, spinnerets, Malpighian tubules, antennule, antennae, labium, mandible, simple eye, compound eye, cephalothorax, carapace, chelicerae.

TEXT REFERENCES

Chapter 28 (D).
Coelom, arachnids, crustaceans, insects.

INTRODUCTION

The majority of animal species are arthropods, and the biomass of insects alone outweighs that of human beings. Arthropods and annelids share a common ancestry, but arthropods have gone on to fill every possible habitat on land and in the sea. Their success is mainly due to the development of a hard **exoskeleton** of chitin, a type of polysaccharide. This skeleton is light compared with the heavy shell of molluscs, and it is jointed, allowing flexibility of the body and appendages (arthropod = "jointed foot"). This versatile exoskeleton protects the animal's body and prevents terrestrial arthropods from drying out. It also serves as an attachment point for muscles so that well-developed locomotion is possible.

The presence of the exoskeleton also creates certain problems, however. Because the skeleton can't grow with the animal, it has to be shed or molted periodically, and the temporarily unprotected animal is vulnerable to attack during and immediately after the molting process. The skin cannot be used for detecting stimuli, so antennae and sensory bristles are used instead.

Although arthropods are coelomate, the coelom is lost during development. The body cavity that you will see has a different origin. The body usually shows some signs of segmentation, and there is pronounced cephalization, with the mouth, nervous system, and sense organs concentrated in the region of the head. The jointed appendages may be modified for other purposes in addition to locomotion. The original ancestral appendage was probably branched, as shown in Figure 18–1. Usually one branch or the other was lost, but sometimes both were retained. Modified appendages are used for chewing, biting, stinging, transferring sperm, chemoreception, and touch perception.

The sexes are separate, and arthropods usually copulate and have internal fertilization. Most arthropods care for the eggs in some way, either by depositing them in or on a food source, or carrying them around on their bodies until the young hatch. The larvae are often very different from the adults and must undergo **metamorphosis,** in which they change into the adult form in a short period of time.

All arthropods have a complete digestive tract, a ventral nerve cord, a body cavity filled with blood that serves as an open circulatory system, a heart, and excretory organs.

CLASS ARACHNIDA

Spiders, ticks, mites and scorpions are the major groups of arachnids. They all have **chelicerae,** appendages for piercing and sucking on their prey, and **pedipalps** that are sensory organs and help hold the prey during feeding. Pedipalps are modified for sperm transfer in males and they are the pincers of scorpions. Arachnids have four pairs of **walking legs.** Most arachnids are carnivorous predators or external parasites. The unsavory reputation of this group is well deserved, as far as humans are concerned: venomous spider bites, scorpion stings, Rocky Mountain spotted fever spread by ticks, scabies mites and lice. There is even a mite that lives at the base of human eyelashes and sometimes causes severe irritations of the eyelids.

□ Examine the variety of preserved arachnids on demonstration. What obvious feature is shared by all arachnids? ______________________________

Do arachnids have antennae? ______________

□ Note the division of the body into the **cephalothorax** and the larger **abdomen.**

Are the legs attached to the cephalothorax or to the abdomen? ______________

External Anatomy: The Spider

These predatory arachnids usually have **poison glands,** which open near the tips of the chelicerae. Once the prey is bitten, it will be quickly paralyzed by the poison. The spider will secrete digestive enzymes onto the prey tissue and then suck up the resulting broth. While spider bites are unpleasant, only a few species can inflict serious or fatal bites on humans.

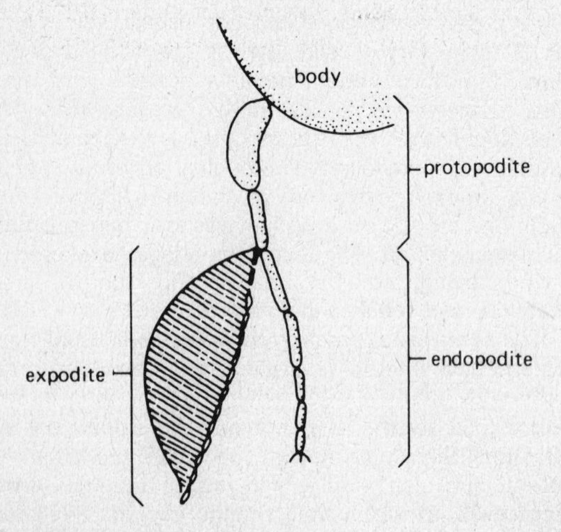

Figure 18–1. The basic arthropod appendage. In most arthropods, one branch or the other of the basic two-branched appendage is lost, but sometimes both branches are retained, especially in the crustaceans.

□ Examine a preserved spider in the dissecting microscope. Identify the cephalothorax with its chelicerae and pedipalps.

□ Examine the **eyes:** there are usually 4–8, and they are connected to the brain.

How many eyes does your specimen have? ______________

The cephalothorax also contains the poison glands and a powerful **sucking stomach** with which the spider ingests its liquid diet.

Spiders spin their webs from a protein that is extruded in liquid form from spinnerets at the tip of the abdomen and polymerizes into silk. The abdomen also contains **book lungs** for gas exchange, gonads, digestive glands and **silk glands.**

□ Look for the spinnerets on your spider specimen.

Manipulation of silk for a variety of purposes probably accounts for the great diversity and success of spiders.

CLASS CRUSTACEA

Crustaceans are mostly marine and are predators or scavengers. They include a wide variety of animals: shrimp, lobsters, hermit crabs, brine shrimp, copepods, barnacles, beach fleas, and wood lice (sow bugs). Along with centipedes, millipedes, and some insects, they have true jaws or **mandibles.** Crustaceans also have two pairs of feeding appendages, the **maxillae,** two pairs of **antennae,** and their appendages are often branched. In crustaceans such as the lobster, the shell is reinforced with calcium carbonate. Pigments in the shell give the animal protective coloration. The loss of one of these heat-sensitive pigments gives cooked lobsters their bright red color by revealing another pigment, which is heat stable.

External Anatomy: The Crayfish

□ Examine a preserved crayfish such as *Procambarus,* and use Figure 18–3 to locate the following structures: **antennae, antennules, rostrum, compound eyes, chelipeds** (pincers), the **carapace,** the covering of the **cephalothorax,** the segmented **abdomen** ("tail"), **walking legs, telson,** and **uropods.**

How many pairs of walking legs does the crayfish have in addition to the chelipeds?

____

□ On the ventral side locate the **swimmerets** on the abdomen, and examine the first pair. They are modified for the transfer of sperm in those mature males that are ready to mate.

Is your crayfish a mature male? ______________

□ Also on the ventral side, locate the **mouth,** with the **mandibles** or true jaws, five pairs of specialized appendages between the mandibles and chelipeds, and the **anus.**

The mandibles tear the food, and the other append-

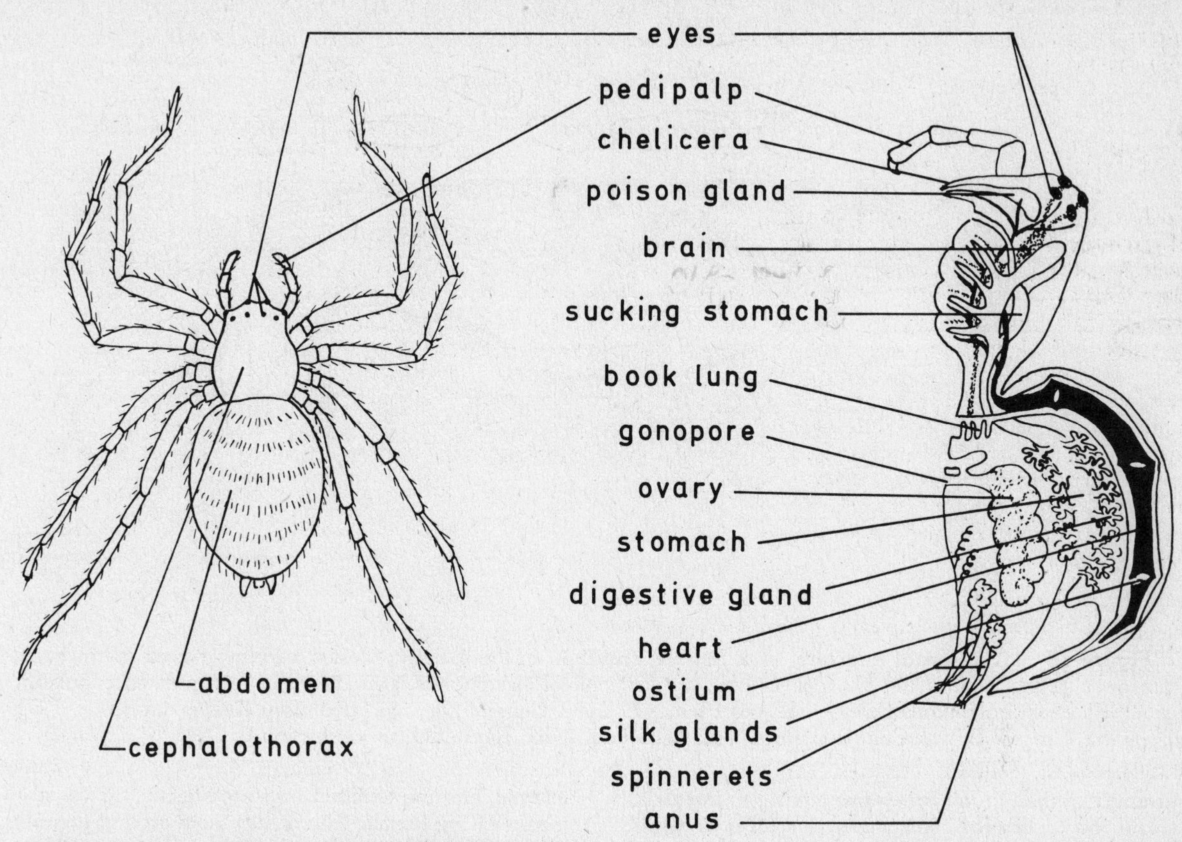

Figure 18–2. Anatomy of a spider. The piercing mouth parts and sucking stomach obtain the liquid diet. Silk is spun from the silk glands and spinnerets in the abdomen.

ages detect and hold the food during feeding. One pair is specialized to move water through the chamber around the gills.

□ Look for the excretory pore at the base of the antenna; it connects to the green gland that you will see later.

Internal Anatomy: The Crayfish

Use Figure 18–4 as your guide to the internal structure of the crayfish.

□ Carefully dissect away the dorsal half of the carapace to expose the internal organs of the cephalothorax.

□ Locate the diamond-shaped **heart** of the open circulatory system. The heart sends blood to the spaces around the organs through several arteries. After passing through the gills, the blood drains back into the heart through the openings in it (ostia).

□ Remove the heart in order to see the **gonads** (ovaries or testes). Sperm exit at the base of the fifth walking legs and eggs leave the oviducts at the base of the third walking legs.

□ Trace the digestive system from the mouth through the **espohagus** to the **cardiac stomach** where food

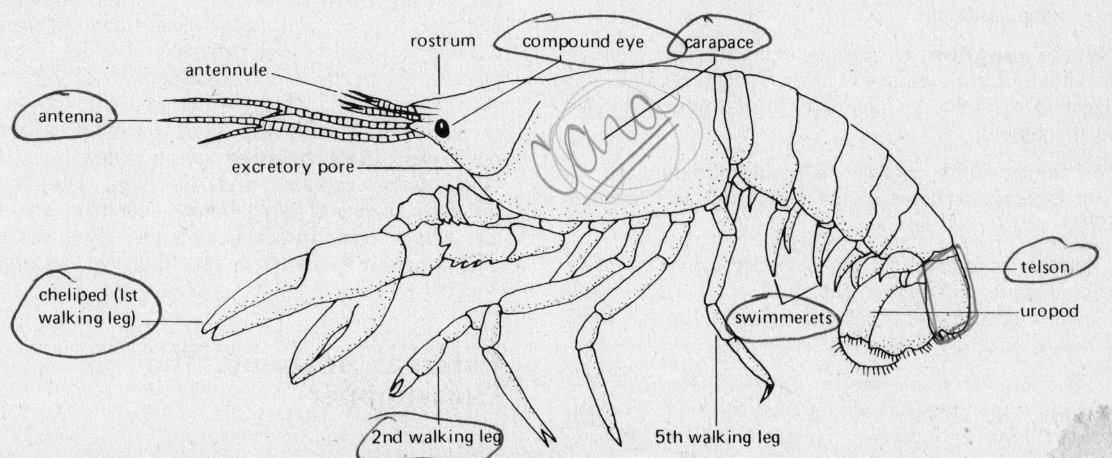

Figure 18–3. External structure of a male crayfish. The sperm ducts open at the base of the fifth pair of walking legs, and the first pair of swimmerets are modified for transferring sperm during copulation.

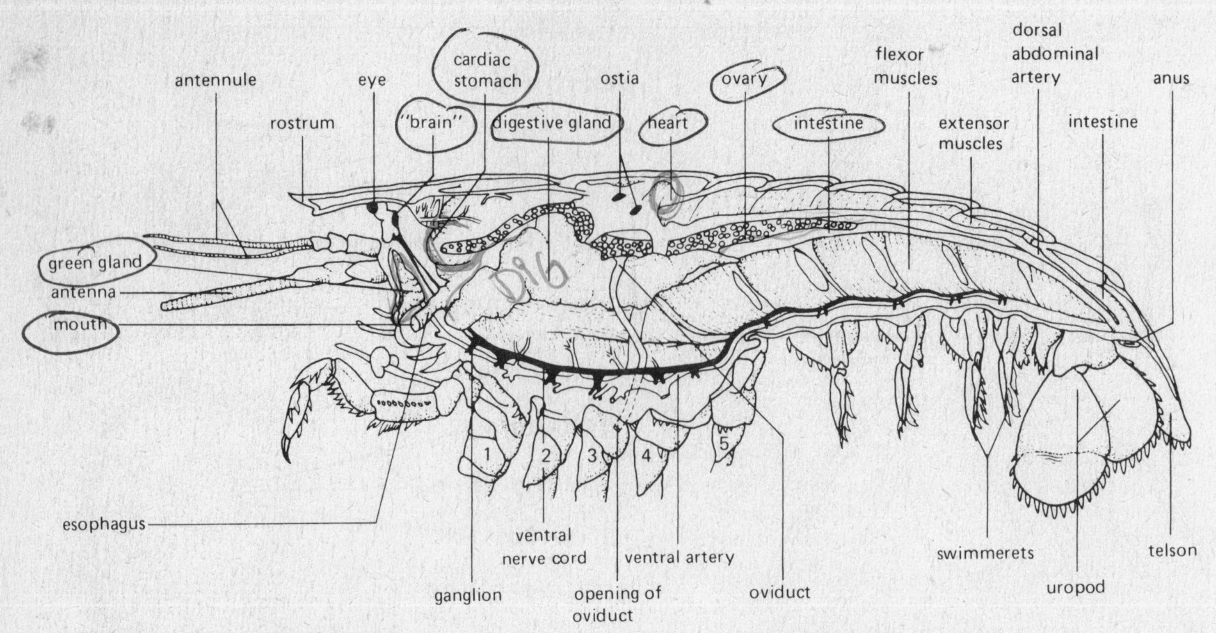

Figure 18–4. Internal anatomy of a female crayfish. In the male, the testes lead into paired sperm ducts, which open at the base of the fifth pair of legs; the first pair of swimmerets is modified to form pointed copulatory organs. (Redrawn from Wodsedalek, J. E. and Charles F. Lytle, *General Zoology Laboratory Guide,* 8th Ed. © 1971, 1977, 1981 Wm. C. Brown Company Publishers, Dubuque, Iowa. Reprinted by permission.)

is ground up, then to the **intestine** leading to the anus.
□ Note the **extensor** (straightens) and **flexor** (curls up) muscles which fill the abdomen ("tail") of the crayfish.
□ Find the **green gland** located just behind each antenna; this excretory organ opens through the excretory pore that you identified before.
□ Note the large liver, or digestive gland, beneath the first part of the intestine.
□ Remove the internal organs to expose the **ventral nerve cord,** which lies along the bottom of the body cavity.
□ Locate a pair of nerves leading from the ventral nerve cord around the esophagus to the large ganglion, or **"brain"** in the head. Short nerves connect eyes, antennae, and antennules to the brain.

The nervous system is the main internal part of the crayfish that still shows segmentation: each segment of the animal has a ganglion that controls most of the functions in that segment.

□ Find a **ganglion,** or swelling, along the ventral nerve cord. It controls the organs in its own segment but is also subject to influence by messages from adjacent ganglia and the brain.

The gills of the crayfish are located outside the body cavity in a space between the body wall and the carapace.

□ Remove a segment of the carapace on one side of the crayfish to expose the gills.
□ Detach one of the walking legs, and note the small gill that is attached to the leg.

The leg and gill are a modified form of the branched appendage shown in Figure 18–1. The leg represents the endopodite, the gill represents the exopodite, and the base of the leg is the protopodite. Branched appendages are common in crustaceans.

Male and female crustaceans copulate, fertilization is internal, and the fertilized eggs are attached to the swimmerets of the female. She carries them around until they hatch into young that resemble the parents. In some marine crustaceans, the eggs hatch into a nauplius larva that bears three pairs of appendages.

CLASS INSECTA

Insects are the most successful invertebrate group and the only group besides vertebrates to inhabit all terrestrial environments. They have come to occupy habitats in which they don't directly compete with vertebrates and have remained small in size. They survive on land by tolerating fluctuations in the salt and water content of the body, and they have developed a unique **tracheal system** for respiration. The insect body consists of three parts: head, thorax, and abdomen. The head has antennae, **compound eyes** (unique to insects), and modified mouth parts. The thorax always bears three pairs of walking legs and one or two pairs of wings, if wings are present. As in the other arthropods, the body cavity is filled with blood, which is pumped through the open circulatory system by the heart. Excretion is performed by **Malpighian tubules** which empty into the gut. Most insects copulate and lay eggs. The young go through a series of molts before becoming sexually mature adults. Some insects have larval stages that are very different from the adult in structure, diet, and habitat.

External Anatomy: The Grasshopper

Grasshoppers are common herbivorous insects found in dry grasslands and open spaces. Use Figure 18–5 to identify the parts of its body.

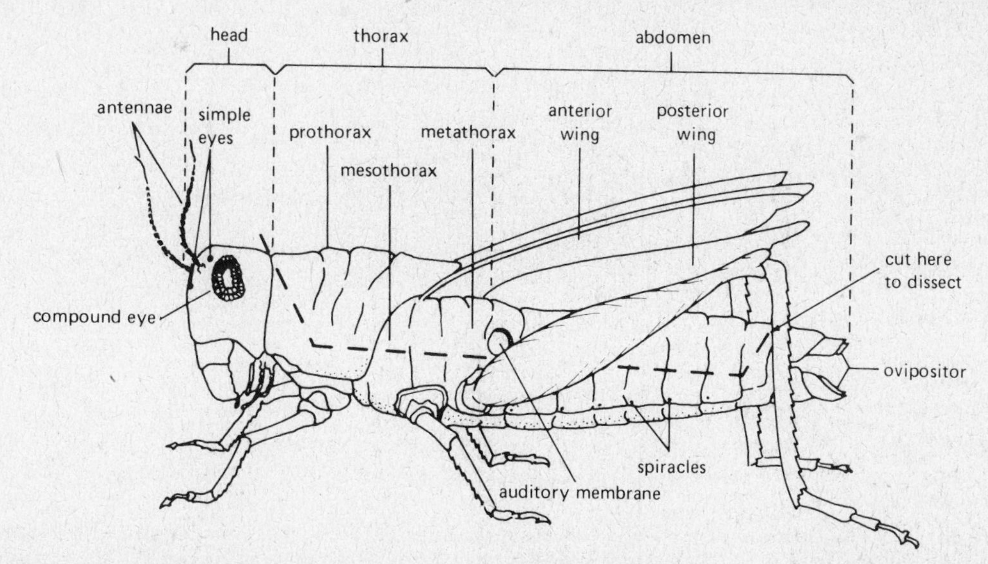

Figure 18–5. External structure of a female grasshopper. The male lacks the ovipositor and has a blunt abdomen. Open the exoskeleton by cutting along the dotted lines. (Redrawn from Peter Abramoff and Robert G. Thompson, *Laboratory Outlines in Biology II.* San Francisco: W. H. Freeman and Company, © 1972.)

□ Examine a preserved grasshopper, and note the large jumping **legs** and the huge compound eyes.

The compound eyes of insects consist of many simple eyes adjacent to each other. The simple eyes cannot form an image but detect light and dark. The combination of the pattern of light and dark on the whole array of simple eyes can be interpreted by the grasshopper's brain as an image. The image on a TV screen is formed by light and dark dots in a similar way. The resolution of the image is not very good compared with that of the vertebrate eye. Insects have color vision.

□ Find the three **simple eyes** (ocelli) on the top of the head. They detect quickly moving objects.

Are the **antennae** jointed? ______________

□ Examine the mouth parts shown in Figures 18–6 and 18–7.
□ Find the **labrum** (upper lip), **labia** (lower lips, usually fused), **mandibles** (true jaws), which surround the mouth, and **maxillae** (second jaws).
□ Just in front of the mouth is the **hypopharynx.** In other insects the hypopharynx may be modified for piercing or sucking.
□ The two pairs of appendages below the mouth (maxillary and labial palps) have sensory receptors.
□ The segmented **thorax** bears the three pairs of **legs** and two pairs of **wings.**
□ Stretch out the wings, and note the difference between the protective anterior wings and the posterior flight wings.
□ The abdomen is blunt in males and has a pointed **ovipositor** for laying eggs at the posterior end in females.

Is your grasshopper a male or female?

____

□ Locate the **spiracles** along the sides of the thorax and abdomen. They connect to the **tracheal system** through which insects carry on gas exchange. The three anterior pairs of spiracles take in air, and the posterior ones let it pass out. The spiracles can be closed, if necessary, to prevent too much water loss under dry conditions.
□ Just above the jumping legs find the **auditory membranes.** These allow the grasshopper to detect sound and communicate with other grasshoppers, especially those of the opposite sex.

Internal Anatomy: The Grasshopper

□ Remove the wings of the grasshopper.
□ Cut along the dotted lines as shown in Figure 18–5, and remove the dorsal part of the exoskeleton.
□ Study the internal part of the exoskeleton that you have cut off, and find the muscles that attach to the wings; these are the **insect flight muscles,** which are rich in mitochondria and perform more efficiently than any vertebrate muscle.

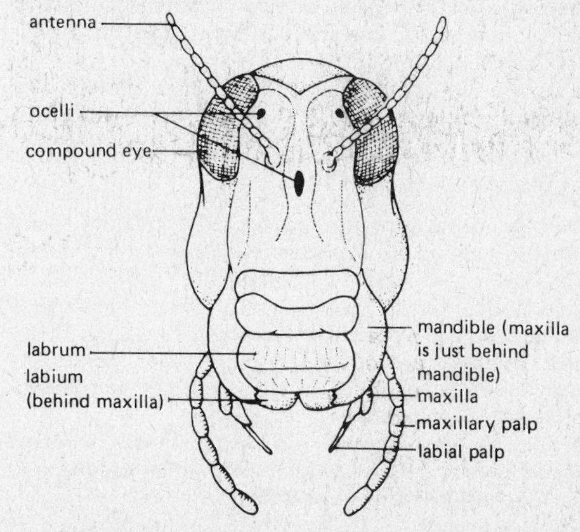

Figure 18–6. Head-on view of a grasshopper. The mouth is located below the labrum, and the hypopharynx (not visible) is between the mandibles and labia.

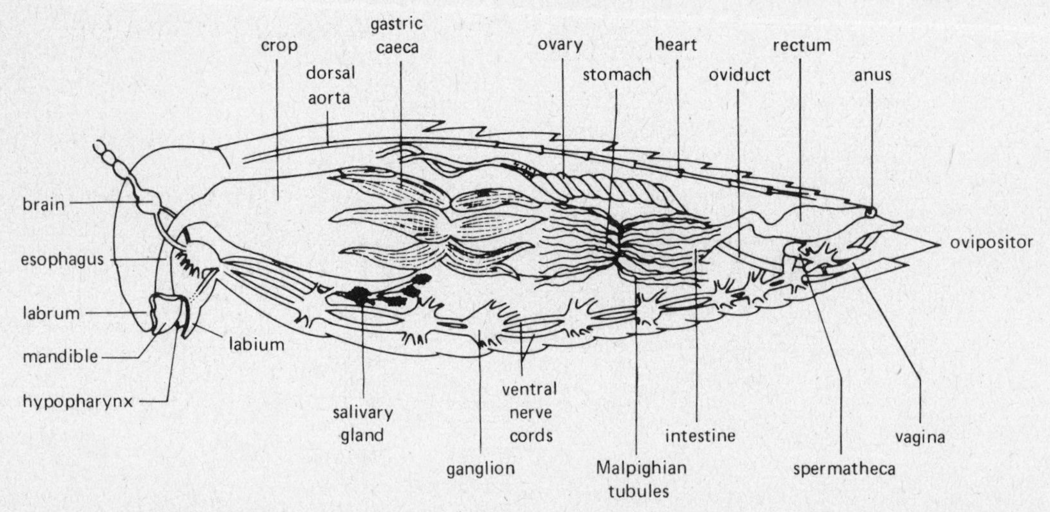

Figure 18–7. The internal organs of a female grasshopper. In the male, the testes lead into seminal vesicles and then to the penis. There is no ovipositor. (Redrawn from Peter Abramoff and Robert G. Thompson, *Laboratory Outlines in Biology II*. San Francisco: W. H. Freeman and Company, © 1972.)

The internal organs are shown in Figure 18–7.

□ Note the muscles along the body wall that move the legs. In an animal with an exoskeleton, muscles at the joints work in pairs to extend (straighten) or flex (bend) the joint.

□ Note the body cavity, or hemocoel, which in life is filled with blood. There is no respiratory pigment because respiration is carried out by the tracheal system.

What is the function of insect blood?

____

____

□ Look for the dorsal aorta and the heart. Blood entering the heart through small openings (ostia) is pumped forward and enters the hemocoel near the brain.

□ Locate the **esophagus** through which food passes from the **mouth** to the large **crop.**

□ Find the outpocketings of the gut, called gastric caeca, at the posterior end of the crop.

□ The large **stomach** leads into the **intestine.** The **Malpighian tubules** excrete nitrogenous wastes, mainly **uric acid,** into the gut at the point where the stomach joins the intestine.

□ Locate the **rectum,** which empties through the **anus;** it reabsorbs water.

The sex organs are located in the posterior part of the abdomen.

□ In the male, locate the **testes,** which produce sperm that are stored in the **seminal vesicles** and eventually deposited in the female by the **penis.**

□ In the female, the sperm are deposited in the **vagina** and stored in the **spermatheca.** Eggs produced in the large **ovaries** are fertilized by the sperm within the body of the female and are deposited with the ovipositor.

□ Remove the digestive organs to see the **ventral nerve cords.** At their anterior end, nerves pass around the esophagus to the **ganglion** of the brain.

Do the ventral nerve cords have ganglia in each segment, as in the crayfish? ____________

Insect Metamorphosis

Metamorphosis differs in the various groups of insects: in some insects the young that hatches from the egg is similar to the adult, but in others it is completely different. There are four types of development:

1. **Ametabolous:** The young are tiny versions of the wingless adult.
2. **Paurometabolous:** The young **nymphs** are similar to the adults, but lack wings and are sexually immature.
3. **Hemimetabolous:** The young **naiads** are aquatic larvae with gills and are unlike the adults.
4. **Holometabolous:** The wormlike **larva** passes through the resting stage of the **pupa** before emerging as a mature adult.

Turn back to the LIFE HISTORY section of Topic 9, and review the life history of *Drosophila,* a fly that belongs to the holometabolous order Diptera.

□ If you have not already done so, examine a culture bottle of *Drosophila* and identify the **larva, pupa,** and **adult.**

□ Note the dark-colored pupae which contain flies about to emerge as adults.

□ Try to find newly hatched adults whose wings are not yet fully extended. They need to dry in the air and stiffen before they can be used for flight.

□ The largest larvae are the ones that are about to pupate, and the smallest ones are newly hatched from the egg. Each stage of larval development is called an **instar,** and the stages are separated by **molting** (*Drosophila* has three instars.).

When you are finished with this laboratory, dispose of your specimens as directed by your instructor, and clean your dissecting tools before putting them away.

EXPLORING FURTHER

Barnes, R. D. *Invertebrate Zoology*. 4th ed. Philadelphia: W. B. Saunders, 1980.

Borror, D. J. and D. M. De Long. *An Introduction to the Study of Insects.* 4th ed. New York: Holt, Rinehart and Winston, 1976.

Buchsbaum, R. *Animals without Backbones.* 2nd rev. ed. Chicago: University of Chicago Press, 1975.

Evans, H. E. *Life on a Little-Known Planet.* New York: E. P. Dutton, 1968.

Russell-Hunter, W. D. *A Biology of the Higher Invertebrates.* New York: Macmillan, 1969.

Wilson, E. O. *Insect Societies.* New York: Academic Press, 1971.

Scientific American Articles

Wicksten, Mary K. "Decorator crabs." February 1980 (#1462).

Horridge, G. Adrian. "The compound eye of insects." July 1977 (#1364).

Hinton, H. E. "Insect eggshells." August 1970 (#1187).

Hölldobler, Berthold K., and E. O. Wilson. "Weaver ants." December 1977 (#1373).

Jones, Jack Colvard. "The feeding behavior of mosquitoes." June 1978 (#1392).

Rothschild, Miriam, Y. Schlein, K. Parker, C. Neville and S. Sternberg, "The flying leap of the flea." November 1973 (#1284).

Wilson, E. O. "Slavery in ants." June 1975 (#1323).

TOPIC 19

Introduction to the Vertebrates

OBJECTIVES

When you have completed this topic, you should be able to:

1. List characteristics that would permit you to distinguish members of the following classes: Chondrichthyes, Osteichthyes, Amphibia, Aves, and Mammalia.
2. Name the class of any vertebrate presented to you.
3. Describe the anatomy of one representative vertebrate in detail, and be able to describe the major features that are similar to and different from those of the other classes of vertebrates.
4. Identify any of the structures in the review list at the end of this topic in any vertebrate in which it occurs.
5. List and explain the problems encountered by water-dwelling vertebrates in adapting to life on land.
6. Explain the adaptations of a representative member of each of the terrestrial classes of vertebrates that allow it to live on land successfully.

TEXT REFERENCES

Chapter 29 (B, E); Chapter 30.
Chordate characteristics, vertebrate characteristics, cartilaginous fish, bony fish, amphibians, reptiles, birds, mammals, transition to land.

INTRODUCTION

These are the classes of Vertebrata that we will be studying in this exercise:

Class Chondrichthyes (cartilaginous fishes: for example, sharks, skates, and rays)
Class Osteichthyes (bony fishes: for example, perch, mackerel)
Class Amphibia (amphibians: for example, frogs, toads, newts, and salamanders)
Class Aves (birds: for example, robin, osprey, penguin, duck, ostrich)
Class Mammalia (mammals: for example, cows, horses, whales, rabbits, humans)

All of the vertebrates have the following characteristics:

1. A vertebral column replaces the notochord. It surrounds and protects the nerve cord and serves as the backbone in locomotion.
2. There is a dorsal, hollow nerve cord.
3. Pharyngeal gill slits are present at some stage in development.
4. A cranium (skull) surrounds and protects the brain and the sense organs in the head region.
5. There are red blood cells (erythrocytes), which contain hemoglobin.
6. There is a pair of nephric kidneys.
7. There is a solid liver.
8. There is a pancreas or pancreatic tissue.
9. There are specific endocrine organs.
10. The eye has a lens and retina.
11. There are olfactory organs, which are usually paired.
12. There are one to three semicircular canals forming part of the inner ear.
13. There is a brain with three primary vesicles dividing it into forebrain, midbrain, and hindbrain.

Before beginning your examination of vertebrate diversity, take a few moments to focus on the vertebra, a small but important structure that has greatly contributed to the evolutionary success of this group.

□ Examine the vertebra on demonstration and identify the parts shown in Figure 19–1.

The dorsal process is one of the set of bumps you feel if you run your hand down a cat's back. The spinal cord runs through the large opening. Lateral displacement of adjacent vertebrae would put pressure on the spinal cord, resulting in severe pain and possibly permanent damage.

The centrum marks the position of the embryonic notochord which has been surrounded and replaced by bone. A spongy intervertebral disc lies between the centra of adjacent vertebrae and allows some flexibility of the backbone. A "slipped" or ruptured disc will change the alignment of the vertebrae and put pressure on the spinal cord.

In this laboratory, you will learn something about vertebrates by observing the demonstrations, dissecting a representative of one of the vertebrate classes, and studying the dissections made by other members of the class. At the end of this exercise there is a list of important structures that you should be able to identify in any vertebrate in which they occur.

Your instructor will provide dissection guides to help you understand the structure of your specimen. If you are going to work with a preserved specimen, rinse off the preservative at the sink before examining the external features. Dissect from the ventral side, first removing the skin by separating it and peeling it back from the abdominal wall. Once the abdomen is open, use your fingers and probe for most of the dissection. You can study much of the internal anatomy by moving organs aside so that you can look underneath. If it is necessary to remove a structure, do it carefully without damaging adjacent organs.

When working with preserved specimens, you should not wear contact lenses. Use eyeglasses or safety glasses, if you have them, and protect your hands with liberal amounts of hand cream before beginning work.

READ THE GENERAL PROCEDURES SECTION IN TOPIC 20 BEFORE BEGINNING WORK ON YOUR SPECIMEN.

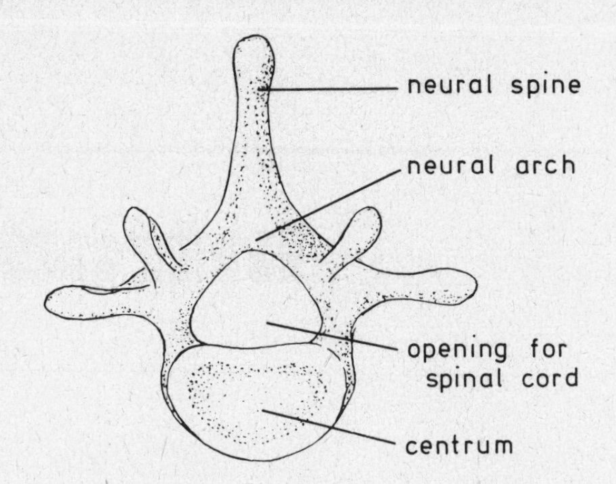

Figure 19–1. A vertebra. In the vertebral column the spinal cord would pass through the large opening and there would be soft discs between adjacent vertebrae to cushion the centra.

CLASS CHONDRICHTHYES

Study the preserved dogfish shark on demonstration. Sharks represent animals that are well adapted to a highly motile, carnivorous way of life:

1. The skeletal elements in the region of the gills have been moved forward to form jaws used for predation.
2. Fins are paired, and the shark's powerful musculature makes it an excellent swimmer and allows it to pursue and catch prey.

Where are the **jaws** located in relation to the head?

____

Feel the skin of the shark. The roughness is caused by **placoid scales,** or dermal denticles, which arise from the dermis of the skin. Modified denticles form the teeth, which can easily be replaced as they become worn or lost. The structure of a shark's jaw and denticles and the internal structure of a single denticle are shown in Figure 19–2.

□ Remove a section of skin on the side of the shark's body to see the segmented musculature underneath.

What other sign of segmentation do you see in the shark's external anatomy?

____

How many pairs of fins does the shark have?

____

If your specimen is a male, the posterior **pelvic fins** will be modified into claspers which transfer sperm during copulation.

Is your shark male or female? __________

Locate the **spiracle** behind the **eye** through which some water may enter the pharynx.

How does the water leave the pharynx?

____

____

The **eye** has a lens and retina with rods and cones so the shark can apparently see color; its vision is not especially keen.

Does the eye have an eyelid? Explain why or why not.

____

____

____

____

____

Along the side of the shark is a light stripe marking the position of the **lateral line organ** which is sensitive to pressure changes in the water. Part of the system in the head region (ampullae of Lorenzini) is modified to sense electrical fields. This sense allows the shark to detect hidden prey. Because strong electrical fields repel sharks, this sense also allows humans to discourage sharks from

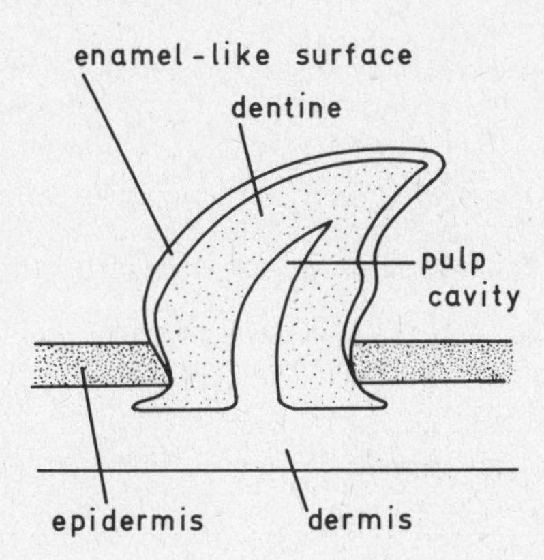

Figure 19–2. The jaw and tooth of a shark. As teeth become worn or are lost, they are replaced by new ones from the rows on the inner sides of the jaws. (Photo: C. Eberhard)

visiting beaches by setting up artificial electric fields in the water.

□ Locate the **olfactory sacs** anterior to the mouth. If you pass a probe into one you will see that it does not connect to the **oral cavity** and is a blind sac. It contains chemoreceptors that allow sharks to sense chemicals in the water.

Within the skull the shark has an inner ear with **semicircular canals.** There are three canals, in the three perpendicular planes. The inner ear detects movement of the head and allows the shark to keep track of its position. All vertebrates have a similar organ of equilibrium.

□ Open the shark's abdomen by cutting around the cloacal area and then towards the head. Cut through the pectoral girdle, and pin back the flaps of the body wall so that the dissection looks like the shark shown in Figure 19–3.

□ Locate the following internal organs (most of these appear in Figure 19–3):

Esophagus. Slit it open to see the papillae inside.

Stomach (J-shaped). Open the stomach to see what the last meal was. Notice the rugae (folds) inside.

Small intestine. Open it to see the **spiral valve** inside.

What is the function of the shark's spiral valve?

____

____

Rectum. Empties into the cloaca.

Rectal gland. Excretes sodium chloride for osmoregulation.

Pancreas. Two lobes.

Liver. This organ is composed of three large oil-filled lobes and is critical in maintaining the buoyancy of open ocean sharks, which must swim or sink.

Gall bladder. A greenish organ located underneath the liver.

Spleen. Triangular shape, alongside stomach.

Kidneys. Long, bandlike organs that extend along almost the entire length of the peritoneal cavity. They drain through the ducts into the cloaca. Most sharks are marine and must conserve water, but the kidneys are not efficient and remove a lot of water as they remove urea from the blood. Sharks osmoregulate by keeping the level of urea in the blood and tissues high, so that water actually tends to enter the body.

Testes or **ovaries.** Located near the anterior ends of the kidneys.

In the female, eggs enter the oviducts and pass down into the uterus, which will be enlarged if the shark is pregnant. In the male shark, sperm from the testes pass through common urogenital ducts to the cloaca. The **claspers** are used to transfer sperm into the female's reproductive tract; therefore, fertilization is internal. In dogfish sharks the young develop in the female's body and then hatch as small sharks, a condition called **ovoviviparous** development. Development lasts up to two years, and many young are born at once.

Cloaca. Common opening of the digestive and urogenital systems. The **urinary papilla** within the cloaca has the opening of the urogenital system.

Heart. Four chambers. Note the **atrium** and **ventricle.** Blood from the body drains into the **sinus venosus** and then into the atrium. The atrium leads into the muscular ventricle, which then forces the blood out through the **conus arteriosus.** The conus arteriosus is also muscular and has internal valves. There is a single circulation because there is only one pathway through the single atrium and ventricle.

Blood from the heart travels to the gills in the **ventral aorta** and **afferent branchial arteries.** It is reoxygenated as it passes through the gill capillaries. It then leaves in the **efferent branchial arteries,** enters the **dorsal aorta,** and is distributed to the body.

Respiration depends on the shark's continuously swimming through the water so that fresh water enters the pharynx from the mouth and spiracles, then passes over the gill capillaries and out through the **gill slits.** The muscular action of the floor of the mouth and in the branchial region helps suck in and expel the water.

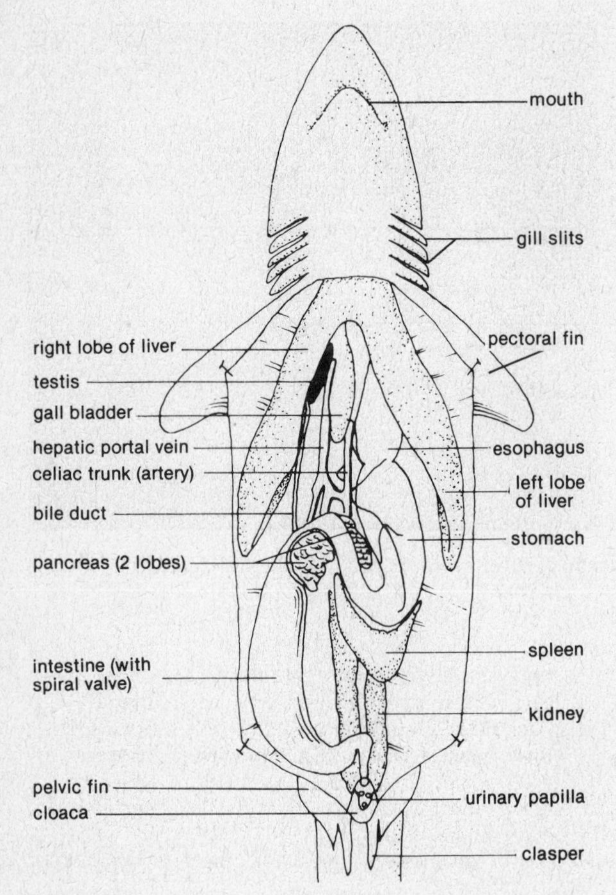

Figure 19–3. Internal anatomy of a male shark.

The brain of the shark is protected by a cartilaginous braincase. Your instructor may ask you to open the braincase and study the anatomy of the shark's brain. The shark's existence is mainly concerned with sensations of smell, sight, and equilibrium with little provision for decision making. This animal is truly an "eating machine" and as such as been highly successful since its ancestors evolved over 300 million years ago during the Devonian period.

CLASS OSTEICHTHYES

Like sharks, many bony fish are fast-swimming carnivores, but they have also evolved into a great variety of other types. Study the preserved perch on demonstration. Handle the specimen carefully because the spines are sharp.

Is the body as flexible as that of the shark?

____

□ Remove a few scales and examine them in the microscope. These are **ctenoid scales** and provide the fish with a tough but flexible covering. All the scales are covered by a layer of very thin skin.
□ Locate the growth rings on the exposed surface of the scale. They can be used to determine the approximate age of the fish.
□ Note the spiny dorsal fins.

How many pairs of fins does the fish have?

____

Most of the propelling force comes from the tail and caudal fin, with the other fins used for balance and direction control.

□ Find the **anus,** which serves only as the opening of the digestive tract, unlike the cloaca of the shark, which is also the urogenital opening. It is in front of the anal fin.
□ The **operculum** covers the gill slits, and water enters through the mouth. The pectoral fins are just behind the opercula.

Where is the **mouth** located in relation to the head?

____

How is this an advantage?

____

____

□ Examine the inside of the mouth.

Do you find teeth or a tongue? ______________

In bony fish that have teeth, the teeth are conical, all alike, and set into the jaw bones.

Is there an eyelid to cover the **eye?**

____

□ Find the pores of the **lateral line organ** along the side of the fish. This organ senses pressure change in the water, as in the shark.
□ The **olfactory sacs** are located anterior to the eyes. They are blind sacs that do not connect with the oral cavity and that contain chemoreceptors used in olfaction.

Within the bony skull is the inner ear with three perpendicular **semicircular canals.** The inner ear is an organ of equilibrium used to detect the fish's position.

Some of the internal organs are shown in Figure 19–4.

□ Open the abdominal cavity of the fish by filleting away the muscles along one side, and locate as many organs as you can.

Liver. A reddish organ in the perch.
Esophagus. Leads to the stomach.
Stomach. Has three pouches called pyloric caeca; empties into the duodenum, the first part of the small intestine.
Intestine. There is no spiral valve, so the intestine is longer than in the shark; exits from the body via the anus.
Spleen. Located next to the stomach.
Swim bladder. This whitish, translucent organ fills the dorsal part of the body cavity. It is derived from the pharynx during development and makes the fish buoyant. As a result, the fish can remain stationary without moving its fins. Gases are secreted into the swim bladder by the red organ, or rete mirabile.
Kidneys. Blackish organs that are diffuse and located dorsal to the swim bladder.
Urinary bladder. Located in the posterior of the body cavity; copious, dilute urine exits through the **urogenital pore** just posterior to the anus.
Testes or **ovaries.** The gonads fill much of the poste-

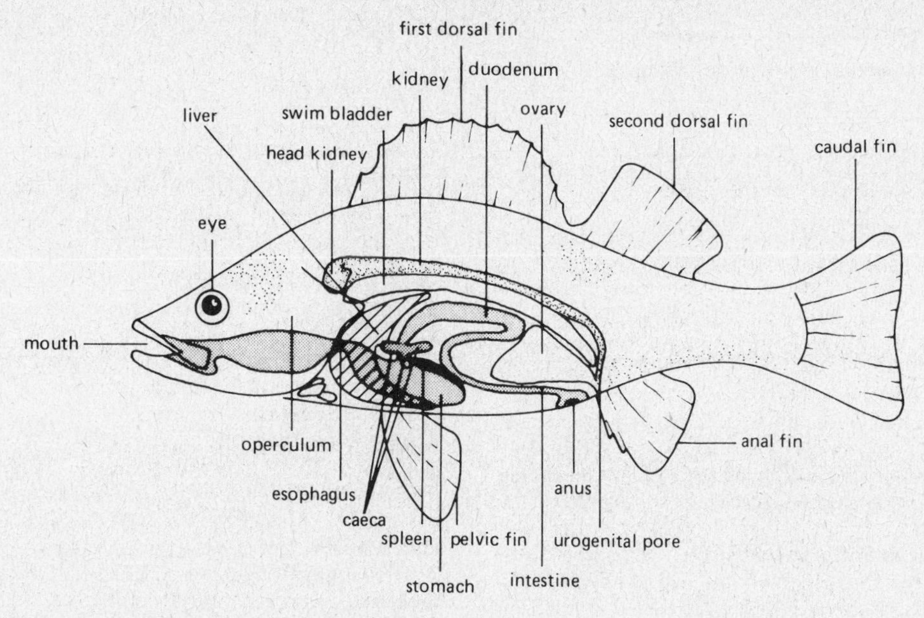

Figure 19–4. Anatomy of a female perch.

rior part of the body cavity. Gametes (eggs and sperm) are shed to the exterior through the urogenital pore.

Most fish reproduce by external fertilization and a great many small eggs are shed. Some fish brood or care for the eggs and young. Many don't, however, and in this case there is often very high mortality among their offspring. This type of reproduction in which the eggs develop externally is **oviparous.**

Heart. This four-chambered organ has an **atrium** and a **ventricle.** Blood enters the heart from the body through the **sinus venosus** and leaves through the muscular **conus arteriosus.** The **ventral aorta** and **afferent branchial arteries** take blood from the heart to the gills where it is reoxygenated in the gill capillaries. It leaves in the **efferent branchial arteries,** enters the **dorsal aorta,** and is sent to the rest of the body. Fish have basically the same pattern of circulation as sharks.

□ Remove the bony **operculum** that covers the gill slits.

What is the function of the operculum?

____

How many **gill slits** does the perch have?

____

□ Notice the **gill rakers** that support and protect the delicate gill filaments.

Water enters the mouth, passes from the pharynx through the gills, aerating the capillaries in the gill filaments so that gas exchange can occur, and finally exits through the operculum. If the fish is not moving, it can pump water through the gills by opening and closing its mouth and using muscles of the pharynx to force water out through the gills.

The brain of the perch is small and deals mainly with smell, sight, and equilibrium. Olfactory nerves leading from the olfactory sacs to the olfactory bulbs are prominent.

□ Examine the teleost skeleton on demonstration.

Notice how the flexible fin rays interact with the fish's skeleton. They make the fish highly maneuverable.

How is the skull different from your own? ___________

____

CLASS AMPHIBIA

Amphibia, Reptilia, Aves, and Mammalia are grouped together as the four classes of **tetrapods,** or **quadrupeds.** The amphibian ancestors of tetrapods evolved from fishlike ancestors that were aquatic, but present-day tetrapods have for the most part taken up a permanent existence on the land and become terrestrial. If you have ever watched a fish out of water suffocate and quickly die, you realize that the evolution from an aquatic to a terrestrial existence must have been gradual. Many adaptations arose during the transition.

Frogs and toads belong to a group of amphibians, the Aneurans, that have no tail and that have a skeleton adapted for jumping. The pectoral and pelvic girdles are fused to the backbone, which is very short.

□ Examine the frog skeleton on demonstration.

Are the vertebrae also fused? ______________

If live frogs are available, study the behavior of the intact animal.

□ Observe its stance and its jumping and swimming motions.

Are the **limbs** attached to the sides or to the ventral side of the trunk? ______________

How many **digits** ("fingers") are there on the forelimbs?

____

On the hindlimbs? _______________

Are the digits longer on the forelimbs or on the hindlimbs? _______________

Do the digits have claws? _______________

What adaptation for swimming do you see?

____

If the frog is a male, there may be an enlarged pad on the side of the inner digit of its forelimb. It helps the male grip the female during mating (amplexus).

Is the skin of the living frog dry or moist? _______________

□ Notice the **cloaca** at the posterior end of the abdomen into which the digestive and urogenital systems open.
□ Study the head region: the **eyes** are very large and adapted to spotting insects, which are the staple of the frog's diet.
□ Find the **nictitating membrane,** which is a transparent eyelid that can close over the eye to keep it clean without blocking the frog's vision.

Are upper and lower eyelids present?

____

□ Locate the external eardrum or **tympanic membrane.** The ear is derived from the lateral line organ, which is still present in aquatic and larval amphibians.

If you are studying a live frog, pith or anesthetize it according to the directions given by your instructor. Pithing instructions are given in Topic 23.

□ Open the mouth of the pithed, anesthetized, or preserved frog, and find the sticky tongue with which it traps insects. It is a true tongue.

Is the frog's tongue attached like yours?

____

□ Pass a hair or some soft probe into a **nostril** (external naris; pl.: nares) to see the connection with the oral cavity. The openings inside are the **internal nares.**
□ Locate the **vomerine teeth,** which are in the roof of the mouth. They hold the prey before it is swallowed by the frog.

Are there other teeth present on the upper jaw?

____

On the lower jaw? _______________

□ Cut the corners of the mouth so that you can open it all the way.
□ Find the large opening into the digestive tract and the small opening to the respiratory tract. The latter is called the **glottis,** and it is located on the floor of the pharynx.
□ Look for the openings at the posterior corners of the mouth that lead into the **vocal sacs** in a male frog.
□ Find the openings alongside the entrance to the esophagus that lead into the **Eustachian tubes** that connect with the middle ear.
□ Open the abdomen of the frog as follows: First cut the skin of the abdominal wall to **one side of the midline** from the pelvic girdle to the throat region. Then cut the abdominal and chest muscles and the pectoral girdle. Continue the cuts to each side at the top and bottom of the incision. Then fold back the flaps so that your frog looks like the dissection shown in Figure 19–5. (This figure shows most of the organs in the frog's abdominal cavity.)
□ Find as many of these organs as you can.

Liver. Large with three lobes.

If your specimen is a female, the abdomen may be filled with masses of dark-colored eggs. Remove them carefully so that you can see the other organs.

Gall bladder. Bile from the liver is stored here and passes through the bile duct to the duodenum.
Esophagus. Leads to the stomach.
Stomach. Pyloric valve separates the stomach from the duodenum of the small intestine.
Small intestine.
Large intestine.
Rectum. Opens into the cloaca.
Fat bodies. Yellow structures with fingerlike projections.
Pancreas. Light-colored lobed organ located near the joining of the stomach and small intestine.
Spleen. Deep red round organ.
Kidney. The kidneys are flattened, reddish organs located in the dorsal part of the abdominal cavity, and extend about half its length. They can form concentrated urine.
Adrenal glands. Yellow bandlike organs on the ventral surface of the kidneys. They are endocrine glands.
Excretory ducts. Carry urine from the kidneys to the cloaca where urine is stored in the bladder. Refer to a more detailed diagram before attempting to locate them.
Bladder. A thin, translucent sac in the ventral part of the posterior abdomen that opens into the cloaca. Water is actively reabsorbed from urine by the bladder.
Testes. Light yellow cylindrical organs located near the anterior ends of the kidneys. Sperm travel through **sperm ducts** into the kidneys, and then through the excretory ducts of the kidneys to the seminal vesicles from which they are released through the cloaca.
Ovaries. Located near the anterior end of the kidneys and attached to the body wall.
Oviducts. Long white coils. The enlarged **ovisac** at the posterior end holds the eggs prior to spawning. During mating, the male mounts on the back of the female, a position called **amplexus,** and eggs and sperm are simultaneously released into the water. External fertilization follows amplexus, and the fertilized eggs develop into tadpoles.

□ If you have a live specimen, observe the beating heart. Locate the **right atrium, left atrium,** and **ventricle** of the three-chambered heart. The large branching vessel is the main artery, the **truncus arteriosus.**
□ Lift the ventricle to see the **posterior vena cava** and **anterior vena cavae,** which lead into the thin-walled **sinus venosus.**

In frogs there are two circuits of the circulatory system, the **systemic circuit** to the body and the **pulmonary circuit** to the lungs, but they are only partially separated. Deoxygenated blood from the body returns to the right atrium, and oxygenated blood from the lungs returns to the left atrium. The blood enters the ventricle

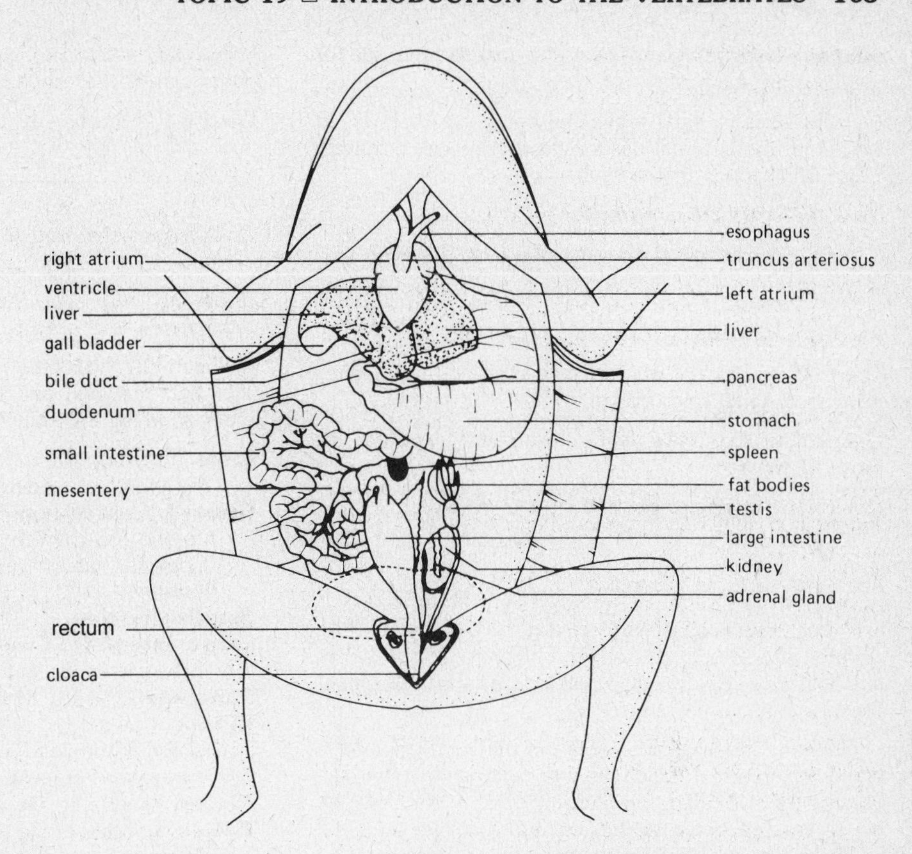

Figure 19–5. Internal anatomy of a male frog. The position of the bladder is shown by the dashed line. (Redrawn from W. H. Johnson et al., *Laboratory Manual for Biology*, 4th ed. New York: Holt, Rinehart and Winston, 1972.)

and is pumped out to the body and lungs. A spiral valve in the truncus arteriosus helps keep the oxygenated and deoxygenated blood separate. Some of the blood returning from the skin will be rich in oxygen, so there is mixing of oxygenated and deoxygenated blood in the ventricle to some extent. Thus the efficiency of the circulatory system is not as good as it would be if the circuits for oxygenated and deoxygenated blood were separated more completely. In reptiles, birds, and mammals, the skin is no longer used for respiration and the separation of the two circuits is more complete.

Be sure you understand how a frog heart differs from a fish heart and from your own mammalian heart.

□ Locate the small **lungs** by raising the liver.

Does the frog have a diaphragm, a partition separating the heart and lungs from the other organs?

____

The frog's brain has olfactory, visual, and equilibrium functions, but there are now cerebral hemispheres as well. The cerebral hemispheres help coordinate the activities of the other parts of the brain and allow the frog to take advantage of more complex behavior patterns.

CLASS AVES

Reptiles, birds, and mammals are classed together as the **amniotes** because they produce an **amniotic egg** with a protective shell. They do not depend on water for respiration or reproduction and so are the only truly terrestrial vertebrates. In birds, the amniotic egg is fertilized internally and a shelled egg is laid. Development of the young bird is **oviparous.**

Birds and mammals are the only two classes of vertebrates that are **endothermic** and closely regulate body temperature. They use physiological and anatomical adaptations to maintain a high body temperature. As a result they can live where reptiles and amphibians cannot, and they can remain active during very cold weather. Their respiration and circulation efficiency must be high because they are such active animals and must continuously meet the demands of thermoregulation.

Birds are the only vertebrates that have **feathers.**

□ Look at the feathers on demonstration.
□ Study a contour feather in the dissecting microscope and identify its parts:

Quill. The hard pointed end that attaches to the bird's skin.
Superior umbilicus. A small hole in the quill through which blood vessels pass.
Shaft. The hollow rod that supports the feather.
Barbs. The flexible filamentlike structures attached to the shaft. Notice how there is a system of tiny hooks, called **barbules,** that lock the adjacent barbs into position. The **down** at the base of the contour lacks these barbules.

If there are down and pin feathers also on demonstration, describe how a down feather differs from a pin feather. _______________________________

____

What are the two main functions of the feathers?

____

____

□ Examine a preserved pigeon or other bird. There is only a tiny **tail,** and the forelimbs have become the

wings. A special gland near the tail secretes oil for preening the feathers.

The jaws in birds move the horny bill or beak, in which are the external nares, and the brain is relatively large, filling much of the skull.

Does the bird have a **tongue**? ____________

Is there a **nictitating membrane** over the eye?

____

Are there **eyelids**? ____________

The flexible neck allows great mobility of the head. Birds have very good color vision.

□ Examine the bird's feet, which retain **scales** like those of reptiles.

How many digits are present on each foot?

____

Are there claws as in reptiles? ____________

One adaptation of birds for flight is that parts of the skeleton are fused for strength and many of the bones are hollow so that weight is reduced as much as possible.

□ Examine the internal structure of the bird skeleton on demonstration, and notice the internal struts that give the bones added strength. Some of the internal spaces in the bones are connected to the respiratory system.
□ Note the enlarged breastbone, the **sternum.** The powerful pectoral muscles used in flight are attached to the sternum.
□ Locate the **pectoral muscles** of the preserved bird.
□ Locate the **cloaca** which is the common exit for the urinary, genital, and digestive tracts. Birds have no bladder, so urine (mainly uric acid crystals) and feces are eliminated together in the bird's excrement, which in large deposits is called **guano** and makes excellent fertilizer.

The internal organs of the bird are shown in Figure 19–6.

□ Remove the skin and feathers over the region just below the neck, and expose the **crop,** which is an outgrowth of the esophagus.

What is the function of the crop?

____

□ Cut open the crop to see what it contains.

In pigeons, the crop of both males and females produces "crop milk" during the breeding season. It is regurgitated and used to feed the young birds.

□ Open the abdomen of the bird, first removing the skin and feathers, and locate the following structures, most of which are shown in Figure 19–6.

Liver. Empties into the small intestine directly through the bile ducts. There is no gall bladder.
Gizzard. Derived from the stomach and used to grind up the food which is often hard kernels of grain. Open the gizzard and see if any small pebbles are inside.
Small intestine.
Large intestine. Two small caeca (vestigial) are attached near the end of the large intestine.
Pancreas. Located in the folds of the small intestine.
Kidney. Long organ against the dorsal wall of the body cavity. Bird kidneys produce a concentrated urine containing uric acid.
Ureter. Leads into the cloaca. There is no bladder.
Testes. Located at the anterior end of the kidney; small unless the bird is preparing to breed.
Vas deferens. Leads from the testis to the cloaca. There is no penis in the pigeon nor in most birds. (A copulatory organ is present in primitive forms.)
Ovary. Only the left ovary is present; small unless the bird is breeding.
Oviduct. Again, only the left one is present. If the bird was breeding, there will be eggs in various stages of production within the oviduct. As the fertilized egg passes along, albumin, membranes, and a calcareous shell are added. The hard shell is necessary so that the eggs won't collapse when the parent birds sit on them during incubation. All birds lay eggs and development is always oviparous.

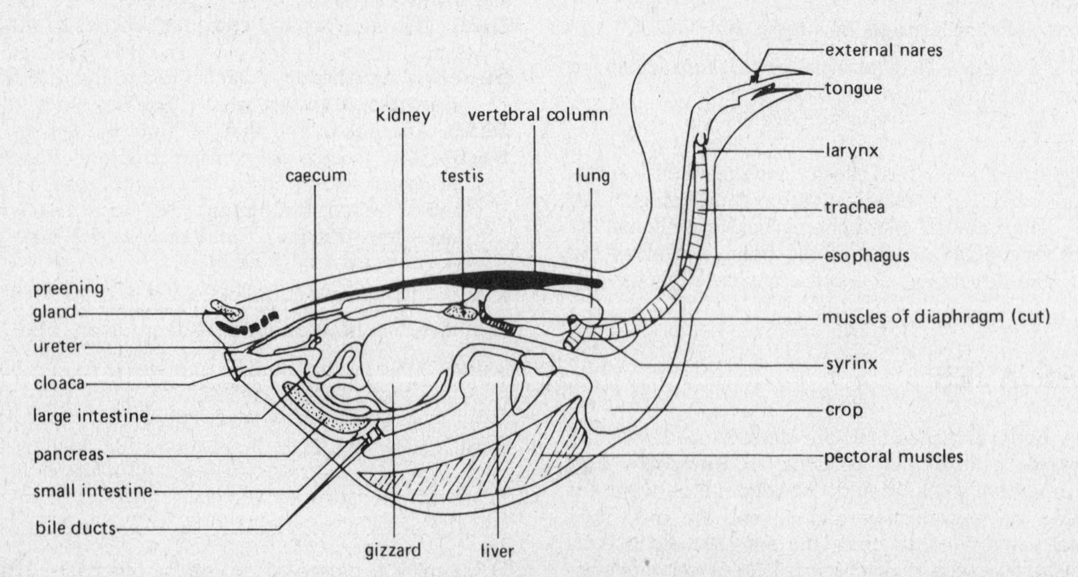

Figure 19–6. Internal anatomy of a male pigeon. There is no gall bladder or urinary bladder. The lungs connect to several air sacs within the abdominal and thoracic cavities and within the major bones (not shown).

□ Locate the **heart,** which is underneath the anterior portion of the sternum. You will have to cut away some of the bone. The four chambers are the **right atrium, left atrium, right ventricle,** and **left ventricle.** Birds require a very efficient circulatory system, and the systemic and pulmonary circuits are completely separated. The heart is fully divided into four chambers. Trace the path of a red blood cell from the right atrium to the main artery leaving the heart, using Figure 22–1 as your guide. In birds the main artery, the aorta, has retained only the right branch, which becomes the **dorsal aorta.**
□ Remove one of the wings, and cut through the ribs near the vertebral column to expose the underlying organs.
□ Locate the compact **lungs** in the most anterior and dorsal part of the abdominal cavity, pressed against the dorsal body wall.

The lungs of the bird are unique in that they are connected to a system of **air sacs.** (See text Figure 32–22.) You will not be able to identify air sacs in your dissection. During ventilation, air is drawn through the lungs and into the air sacs, so the lungs do not expand much. During expiration the air leaves the sacs and enters the lungs. Finally the air leaves after completing a circuit through the more-or-less one-way system. Most of the oxygenation of the blood takes place as the air flows through tiny air capillaries within the lungs. This respiratory system is very efficient and meets the stringent demands of the bird's circulatory system during flight.

The vocal sounds made by a bird are produced in a special organ, the **syrinx,** which is located at the posterior end of the trachea where it divides into the two bronchi. The **larynx** is the entrance to the trachea and is not involved in vocalization.

□ Remove one of the lungs and locate the **syrinx** of the pigeon.

The brain of the pigeon has greatly enlarged **cerebral hemispheres,** which permit the complex behavior patterns characteristic of birds such as courtship, pair bonding, mating, and migration. The **cerebellum** is well developed for the coordination required in flight.

CLASS MAMMALIA

Mammals are the only warm-blooded animals with **hair.** Their mammary glands produce **milk** which is usually released through the **nipples** to nourish the young during their development. Usually the young are retained within the body of the female during early development and are nourished through the blood vessels of the **placenta.** This organ allows the fetal and maternal circulations to exchange nutrients and wastes during the **viviparous** development of the young within the mother's body. Characteristics of the reptilian ancestors of mammals are sometimes retained, however: certain mammals lay eggs, some have scaly tails, and many have claws on their hands and feet.

The major orders of mammals are described in Table 30–4 in your text. Because the variation among the different orders is so great, it is difficult to pick one that is "typical." In this exercise you will concentrate on the features of the rat, but try to keep in mind that the organs and systems that you are studying may be greatly modified in other mammals.

□ If you are studying live rats, observe their behavior while they are being anesthetized by an overdose of a barbiturate. Notice the sniffing behavior and the gradual loss of coordination. Barbiturates and alcohol have a similar effect on the cerebellum, and the rat's behavior is similar to that of a human who is very drunk.
□ In the fully anesthetized or preserved rat, note the hair and nipples that characterize all mammals.
□ Note the elongated hair with **vibrissae** (whiskers) used for tactile sensing.

Why is it advantageous to have the vibrissae as far forward as possible?

____

□ Locate the **lips** and **nostrils.**

Does the **eye** have a nictitating membrane?

____

□ Find the external ears, or **pinnae** (singular: pinna). The **eardrums** are in a protected position within the skull.

□ Notice the **scales** on the rat's tail.

Are there scales on the legs, as in the bird?

____

Are there **claws** on the digits of the hands and feet?

____

□ Locate the **anus,** the posterior opening of the digestive tract.
□ If your specimen is a male, locate the **penis** in the fur-covered **prepuce,** or foreskin.

Is the penis directed anteriorly or posteriorly?

____

Anterior to the anus in the male is the **scrotum** containing the **testes.** The testes are located outside the abdominal cavity because they must be cooler than body temperature to produce sperm.

□ If your specimen is a female, locate the **vagina** anterior to the anus and the **urinary opening** anterior to the vagina. The urinary and genital systems have separate openings as in human females.
□ Open the rat's mouth and investigate the attachment of the **tongue.**
□ Locate the **glottis,** which opens into the trachea (leading to the lungs), and its protective covering, the epiglottis.

What is the function of the epiglottis?

____

____

____

□ On the roof of the mouth distinguish between the anterior **hard palate** and the posterior **soft palate.** The openings of the nasal passages are covered by the soft palate.
□ Locate the openings of the **Eustachian tubes** which lead into the middle ear.
□ Locate the large **incisors** which are the gnawing teeth characteristic of rodents. These teeth are constantly being worn down so they must grow at a fast rate.

The teeth of mammals show a lot of variation and

have become differentiated to serve different purposes. **Incisors** are front teeth adapted for cutting or cropping the food as in humans and deer or for gnawing as in rodents. The **canines** are farther back in the jaw and are for biting or holding prey and for tearing flesh as in a carnivore such as a wolf or a lion. In herbivores they are greatly reduced and also serve for cropping. In the rear of the jaw are the **premolars** and **molars,** which have a flattened surface and are used to crush and grind the food. Molars are especially prominent in herbivores which must spend a great deal of time chewing the food before it can be digested properly.

□ Examine the teeth of the skulls on demonstration and compare them to those of the rat.
□ Sometime during this lab, also examine and compare the mammalian skeletons on demonstration.
□ Cut open the skin of the rat over the abdomen from the pelvic girdle to the ribs, and cut sideways to form two flaps. Cut through the abdominal muscles to expose the internal organs.

The anatomy of the internal organs is shown in Figure 19–7. As you study your specimen, try to identify as many of the following organs as possible.

Esophagus. Leads to the stomach.
Stomach. Separated from the small intestine by the pyloric sphincter, a muscular valve.
Small intestine. Tightly coiled and very long compared to the size of the animal. A long intestine is characteristic of herbivores and omnivores like the rat.
Caecum. Enlarged sac at the beginning of the large intestine important in digestion in herbivores.
Large intestine. Ends in the muscular **rectum,** which probably contains some fecal pellets.
Liver. There is no gall bladder in the rat.
Pancreas. Located in the loop of the small intestine.
Spleen. Dark red organ that is located next to the stomach.
Kidneys. Compact organs in the dorsal part of the abdominal cavity. Urine travels through the **ureter** to the bladder where it is stored. Urine leaves the body through the **urethra,** which also carries sperm in the male. You will have to dissect below

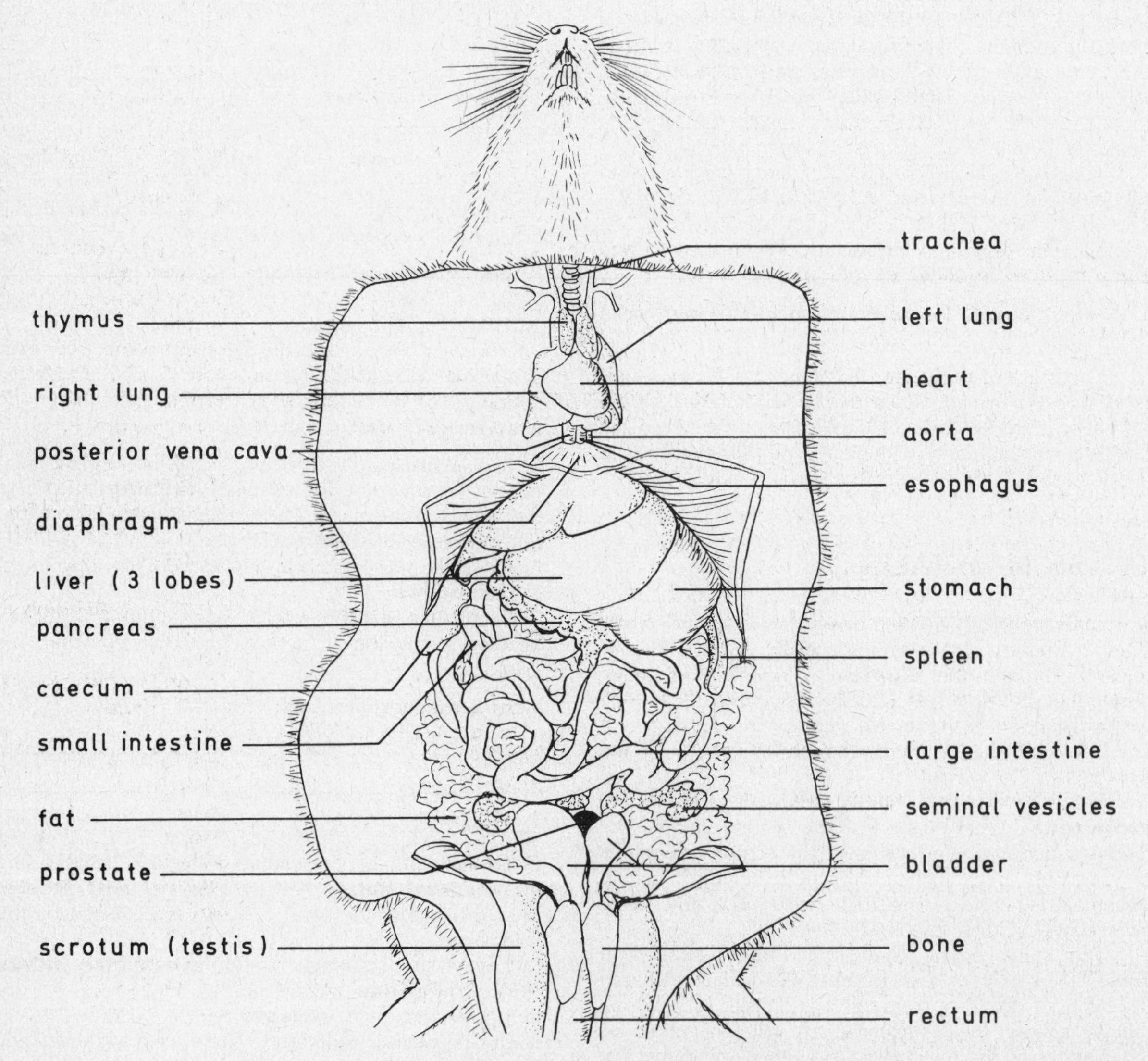

Figure 19–7. Internal anatomy of a male rat. Move the intestine aside to find the organs of the urogenital system.

the intestine and abdominal membranes to see these structures.

Adrenal glands. Small oval glands at the anterior end of the kidneys.

Testes. Carefully cut open the scrotal sacs and the tough membrane that surrounds the testis. Sperm are stored in the highly convoluted tubules of the **epididymis** and are carried in the **vas deferens** to the urethra.

Seminal vesicles and **prostate.** These add fluids to the semen before it passes into the urethra.

Ovaries. Pea-sized organs attached to the body wall.

Oviducts. Tightly coiled tiny tubes leading from the ovaries to the horns of the uterus.

Uterus. Consists of two tubes, or **horns,** in which the embryos develop. The horns join together at the posterior end and enter the **vagina.**

If your specimen is a female, is she pregnant? __________

□ Cut through the sternum and ribs to open the thorax and expose the **trachea;** it is kept open by its cartilaginous rings.

□ Locate the muscular **diaphragm,** which is a sheet separating the abdomen from the thorax.

What is the function of the diaphragm?

____

____

____

□ Remove the glandular material anterior to and covering the heart. This is the **thymus gland,** which is important during the development of the immune system.

The structure of the heart and circulatory system is like that of humans and other mammals. As in birds, the systemic and pulmonary circuits are completely separate because the chambers of the four-chambered heart are completely separate.

□ Identify the four chambers of the heart: **right atrium, left atrium, right ventricle,** and **left ventricle.**

□ Find the **pulmonary artery,** which is the most ventral vessel at the anterior end of the heart and carries blood from the right ventricle to the lungs.

□ Find the **aorta** which is dorsal to the pulmonary artery. It carries blood from the left ventricle to the body. The aorta is the main artery of the system, and it developed from the left systemic arch that you saw in the frog.

□ Lift the ventricle of the heart to see the **posterior vena cava,** which carries blood from the head and forelimbs to the right atrium.

□ Find the H-shaped **thyroid gland** covering the trachea in the neck region.

□ Locate the **right** and **left lungs.** Unlike those of the bird, they are closed sacs.

Within the lungs, gas exchange takes place across the thin membranes of the tiny air sacs (alveoli) located at the ends of the finely branching air tubes.

The brain of the rat has relatively large cerebral hemispheres. The enlarged olfactory bulbs show that smell is a very important sense in the rat. Vision is poor and unimportant. The visual functions are performed by the rear parts of the cerebral hemispheres.

REVIEW

Review what you have learned in this exercise by making sure that you have some understanding of each of the following structures or terms. If you are not sure of any, refer back to the text of this exercise. You should be able to locate the structures marked with an asterisk in any vertebrate in which they occur.

amnion
amniotic egg—cleidoic egg
anterior
anterior vena cava*
anus*
aorta*
aquatic
artery*
atrium*
auditory
bladder (urinary)*
branchial arteries*
caecum*
carnivore
cerebellum
cerebrum
claspers*
claws*
cloaca*
colon, or large intestine*
conus arteriosus*
crop*
ctenoid scale*
dermis
diaphragm*
digits*
dorsal
ear*
endothermic
epidermis*
esophagus*
Eustachian tubes
external fertilization
eye*
fins*
gall bladder*
gametes
gill slits*
gizzard*
glottis*
gut*
heart*
herbivore
internal fertilization
intestine (large or small)*
jaws*
kidney*
larva
larynx*
lateral line
liver*
lungs*
milk
mucus
nares, or nostril (singular: naris)*
nictitating membrane*
nipples*
notochord*
olfactory (lobes, bulbs, sacs)*
omnivorous
optic
oral (buccal) cavity*
ovary*
oviduct*
oviparous
ovoviviparous
palate (hard or soft)*
pancreas*
pectoral girdle*
penis*
pharynx*
pinna*
placenta
placoid scale*
posterior
posterior vena cava*
pulmonary vein*
rectal gland*
renal
scales (of epidermis)*
sinus venosus*
skull, or cranium*
spiracle*
spiral valve*
spleen*
stomach*
swim bladder*
syrinx*
tadpole
teeth (incisors, canines, premolars, molars)*
terrestrial
testis*
tetrapod, or quadruped
tongue*
trachea*
tympanic membrane, or eardrum*
ureter*
urethra*
uterus*
vagina*
vein*
ventral
ventricle*
vertebra*
vibrissa*
viviparous
vomerine teeth*

EXPLORING FURTHER

Romer, A. S. and T. S. Parsons. *The Vertebrate Body.* 5th ed. Philadelphia: W. B. Saunders, 1977.

McFarland, William N., F. Harvey Pough, Tom J. Cade and John B. Heiser. *Vertebrate Life.* New York: Macmillan, 1979.

Ostrom, John H. "Bird flight: How did it begin?" *American Scientist* 67:461, 1979.

Walker, W. F. *Vertebrate Dissection.* 5th ed. Philadelphia: W. B. Saunders, 1975.

Scientific American Articles

Baker, Mary Ann. "A brain-cooling system in mammals." May 1979 (#1428).

Buffetaut, Eric. "The evolution of the crocodilians." October 1979 (#1449).

Carey, Francis G. "Fish with warm bodies." February 1973 (#1266).

Gilbert, Perry W. "The behavior of sharks." July 1962 (#127).

Greenwaite, Crawford H. "How birds sing." November 1969 (#1162).

Heller, H. Craig, Larry I. Crawshaw and Harold T. Hammel. "The thermostat of vertebrate animals." August 1978 (#1398).

Johansen, Kjell. "Air-breathing fishes." October 1968 (#1125).

Schmidt-Nielsen, Knut. "How birds breathe." December 1971 (#1238).

TOPIC 20

Digestion

OBJECTIVES

When you have completed this topic, you should be able to:

1. Describe how a mammal breaks up its food mechanically.
2. Describe how food is broken up chemically.
3. Describe how food is absorbed.
4. Name the three major classes of macronutrients.
5. Name the parts of the mammalian digestive tract in order, and describe the function of each part in the digestion of each of the macronutrients.
6. List those organs that secrete digestive enzymes, and give the substances digested by the enzymes from each of the organs.
7. Describe where in the gut absorption takes place, what substances are absorbed in each place, and what happens to the food after it passes out of the gut lumen.
8. Explain the role of the liver in digestion.
9. Describe what happens to each type of macronutrient if it is eaten in excess of the body's needs.
10. Explain how the digestive system of a mammalian omnivore (human, pig) differs from that of an herbivore (cow) or carnivore (wolf).
11. Describe the role of symbiotic bacteria in the mammalian gut.

TEXT REFERENCES

Chapter 31 (especially C and G).
Macronutrients, mammalian digestive tract, digestive enzymes, liver, pancreas, omnivore adaptations, gut symbionts.

INTRODUCTION

You have already examined representative animals of the major animal phyla and looked at the different classes of the subphylum Vertebrata, which includes all animals with backbones. Now we will focus our attention on the Mammalia, the class of vertebrates to which we ourselves belong.

The mammals are classified into a number of different orders. Some of the more familiar ones are the Primata (primates), including humans, the Carnivora, including dogs and cats, and the Artiodactyla, including pigs, cattle, and deer. Although there is much diversity among the orders of mammals, there is enough similarity among the placental mammals that a study of the basic features of one representative mammal will be applicable to all. Consequently, even though pig and human belong to different orders, their basic mammalian structure is essentially the same. By carefully studying the anatomy of the fetal pig, you will learn much about the structure of your own body.

GENERAL PROCEDURES

Introduction to Dissection

The pigs you will be studying are unborn fetuses taken from the mother's uterus at the time of slaughter. Hog farmers often breed their sows before slaughter to increase their weight. Thus these fetal pigs are a by-product of the slaughtering process. The pigs have been prepared for dissection by being placed in a preservative such as formalin. (These preservatives are strong and irritating to the hands. Hand cream or lanolin applied to the hands before and after dissection can be helpful.) In addition, these pigs have had their circulatory system injected with colored latex to facilitate blood vessel identification. Veins should appear blue and arteries red, although exceptions occur.

YOU SHOULD NOT WEAR CONTACT LENSES DURING DISSECTION BECAUSE THE FUMES FROM THE PRESERVATIVES WILL IRRITATE YOUR EYES.

Your dissection must be planned and executed *carefully*. Thus it is especially important to read the instructions beforehand. Dissection is a procedure to help you learn the anatomy and structural relationships of the animal. If you do not follow the directions, you may remove, misplace, or lose parts before you have a chance to observe them. Remember that you are going to be using this animal for several laboratory periods, and parts must be preserved *intact* for future study. There are two good rules to keep in mind: do as little cutting as possible and *never* remove a part unless told to do so. If you use care, follow the directions in the manual, and pay attention to detail, the dissection of the fetal pig can be an excellent and interesting way to learn anatomy. You may be asked to work on your pig together with your lab partner. In that case, each person should do part of the actual dissection to get a feel for the procedures involved. Work slowly at first, and do not hesitate to ask your instructor for advice on how to proceed.

Dissecting Tools

For dissection you will need:

Razor blade (single edge) or **scalpel** to make the initial cuts through the skin. Use these sparingly. Most students tend to be heavy-handed with them and cut too deeply.
Scissors to use for most of the cutting. Use scissors instead of the scalpel whenever possible. Oil your scissors at the end of each lab to prevent corrosion.
Forceps to use for lifting and moving structures.
Blunt probe to use for lifting parts, separating tissues, and probing ducts.

Actually your fingers are your best dissecting tool. They can lift, separate, and expose organs far more efficiently than any of the above instruments.

Anatomical Terms

Many terms are necessary to describe the anatomical location of different parts of the animal. Some of the important ones are defined below. You should become familiar with these terms because they are used often in anatomical work.

Many of the terms are the same for four-legged animals and humans, but some are different because of our upright stance. For instance, the term **caudal** refers to a structure toward the tail end of a four-legged animal, whereas in human beings the comparable term would be **inferior.** Following are some terms to learn:

Left. The pig's left.
Right. The pig's right.
Anterior. Toward the head.
Posterior. Toward the tail.
Cranial. Toward the head (in humans, the term "superior" is used).
Caudal. Toward the tail (in humans, "inferior").
Dorsal. Toward the back (in humans, "posterior").
Ventral. Toward the belly (in humans, "anterior").
Lateral. Toward the side.
Medial. Toward the midline.
Proximal. Describes the part of a structure nearest a point of reference.
Distal. Describes the part of a structure that is farthest removed from a point of reference.
Pelvic. The hip region.
Pectoral. The shoulder region.

The terms **left** and **right** should be self-evident, but unfortunately these terms lead to more confusion than any of the other terms. It is important that you realize that left and right always refer to the *pig's* left or right, not your own! You will save yourself much confusion if you get this straight before you start.

MANY MORE TERMS ARE GIVEN IN THE FOLLOWING DIRECTIONS THAN YOU MAY BE EXPECTED TO LEARN.

Your instructor will give you some indication as to how much detail you will be expected to master, depending on the amount of laboratory time spent in dissection. To a certain extent, you will be expected to use your own judgment in deciding what is important. As a rough guide, you should be familiar with the location and function of those structures that are also mentioned in your text.

Care of Your Specimen

Although your specimen is preserved in embalming fluid, it will stay useful for dissection only if it is kept from drying out. It must be kept in a sealed plastic bag containing a little fluid at all times when you are not dissecting. During the laboratory period, you should add a little water or dissecting fluid from time to time to keep the tissues moist and flexible.

COVER THE PIG WITH WET PAPER TOWELS WHENEVER YOU ARE NOT ACTUALLY DISSECTING.

When you are ready to stop the dissection until the next period, make sure the organs are moist before you seal the specimen up in its plastic bag.

EXTERNAL ANATOMY

The body of a mammal is divided into a number of regions. The **head** is separated from the **trunk** by the movable neck. The trunk is generally divided into the anterior **thorax** and posterior **abdomen** and **pelvis.** Most mammals have a **tail.**

□ Examine the head and locate the **eyes,** the external ears, or **pinnae** (singular: pinna), the nostrils, and the mouth.
□ Note the paired legs and their various joints.
□ Compare the pig forelimb with your own arm, and try to figure out the corresponding parts.
□ Do the same for the hindlimb.
□ Pick up the pig and manipulate the joints.

Do they move the same way yours do?

yes

□ Notice the toes.

How many are there on each foot? 4

How many toes does the pig actually walk on?

2

The pig walks on the tip of his toenails, a method of locomotion known as "unguligrade."

How does this differ from the human method of walking?

Humans walk on their feet

□ Observe the **umbilical cord** by which the fetal pig was attached to the placenta during development within the uterus.
□ Observe the cut end of the cord, and notice the cut ends of the blood vessels that carried blood containing dissolved nutrients and wastes to and from the placenta of the mother. The human umbilical cord is normally cut at birth, and shrivels up, so all that remains of it is the **navel.**
□ On the ventral surface of the thorax and abdomen find the two rows of **mammary glands.** These glands are one of the distinctive features of mammals. The presence of **hair** is a second distinctive characteristic of all mammals.

Your next task will be to determine the sex of your pig.

□ Raise the tail and observe the **anus,** the posterior end of the digestive tract.
□ If your specimen is a female, there will be a second opening just ventral to the anus, **the urogenital opening.** This is the exit for both the reproductive and excretory systems. Protruding from the urogenital opening in the female is a small, fleshy **urogenital papilla.** In the male there is no second opening near the anus. Instead, the very small urogenital opening is located just posterior to the umbilical cord.

□ If your specimen is a male, find the scrotal sacs in the area ventral to the anus.

The scrotal sacs are not very apparent in the fetal pig. In the older pig the testes will be contained in the scrotal sacs.

Is your pig male or female? male

ANATOMY OF THE DIGESTIVE SYSTEM

□ Place your pig in the dissecting tray so that its ventral side is up.
□ Cut two pieces of string about 2 ft long.
□ Tie a piece of string securely around the ankle of each of the right legs.
□ Run the two pieces of string under the dissecting tray, and tie them tightly to the ankles of the left legs, thus holding the pig securely in place in the dissecting tray, as shown in Figure 20–1.

If you stretch the legs as far apart as possible at this point, it will be a lot easier to locate the inner organs during your dissection.

The digestive system of the fetal pig consists of the **mouth, oral cavity, pharynx, esophagus, stomach, small intestine,** and **large intestine.** The **liver** and **pancreas** are accessory organs. (See Figure 20–5.) As you study the anatomy of the digestive system, you should review the functions of each of these structures.

The Mouth and Pharynx

The mouth opens into the oral cavity.

□ Use bone scissors to cut through the corners of the mouth to the back of the jaw, so that the mouth opens widely. (See Figure 20–2.)
□ Note the **hard palate,** which makes up the anterior part of the roof of the mouth. It is made of bone and covered with folds of mucous membrane.

The hard palate separates the oral cavity from the nasal cavities.

□ Note that the membranes continue posteriorly, forming the **soft palate,** which does not contain bone.

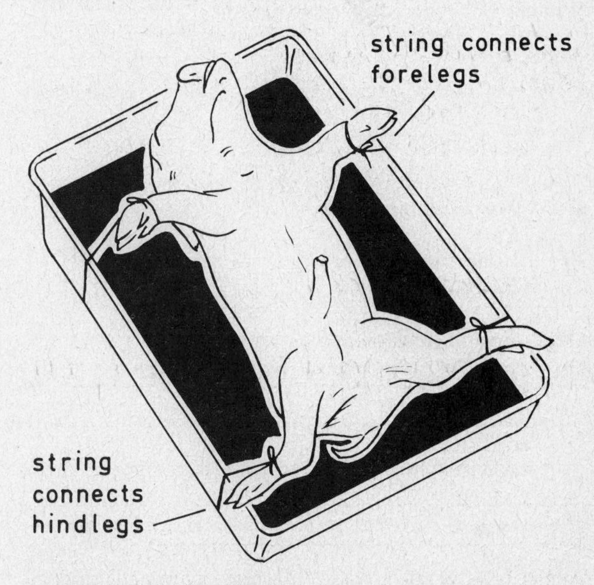

Figure 20–1. Fetal pig specimen. Stretch the legs as far apart as possible and tie them securely. Remove the pig for storage by slipping the strings out from under the dissecting pan.

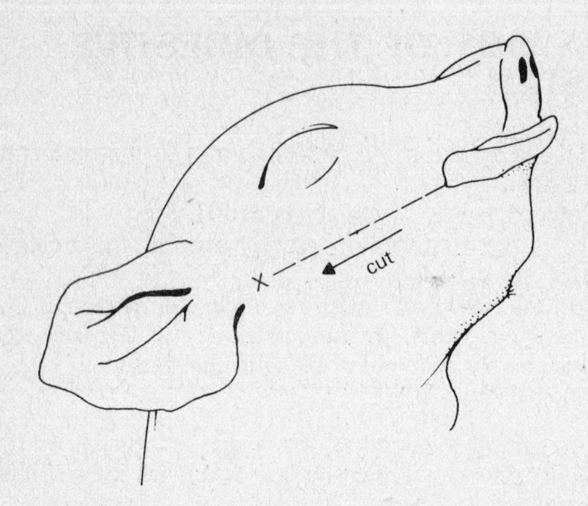

Figure 20–2. Cutting the jaw. Use heavy scissors to cut through the jaw cartilage so that the mouth can be fully opened.

In humans, there is a fleshy protuberance from the soft palate called the **uvula.**

□ Locate the uvula *on your neighbor.*
□ Notice the sensory papillae on the surface of the muscular **tongue.**

How is the tongue attached?

to lower jaw

Ducts from the salivary glands enter the rear of the oral cavity and produce **salivary amylase,** which breaks down starch enzymatically during mastication.

What are the products of starch digestion?

glucose

What substance other than starch is required for the reaction?

amylase

What is this type of reaction called?

hydrolysis

Have any **teeth** erupted from the gums in your specimen? yes

□ Cut into the gum of the pig along one side, and observe the developing teeth.

Are they all the same or are there different kinds?

there are different kinds

What are the functions of the teeth?

chewing + churning food

At the base of the tongue, the oral cavity ends and the pharynx begins. The pharynx is the common passageway for the digestive and respiratory tracts because both air and food must pass through this structure. You may have to extend the cuts you made earlier to open the mouth wide enough to see the pharynx and associated structures.

□ At the posterior end of the tongue find a trough-shaped fold of tissue, the **epiglottis.** You may have to reach deep into the pharynx with your probe to locate the epiglottis and pull it out into view. Dorsal to this is an opening, the **glottis,** which leads into the **larynx,** part of the respiratory system. (See Figure 21–3.)

During swallowing the epiglottis forms a kind of lid over the glottis and deflects the food away from the glottis. When the epiglottis fails to do this, the food enters the glottis and choking results.

□ Dorsal to the glottis find the second, wider opening into the **esophagus.** The esophagus is a collapsible muscular tube connecting the pharynx to the stomach.
□ Pass the end of your probe into the glottis and then into the esophagus. Be certain that you can differentiate them. Notice that air entering through the nose must cross the pathway of food during swallowing.
□ Relate Figure 20–3 to your pig to be sure you understand the anatomy of the pharynx.

At this time you will not follow the course of the esophagus as it proceeds through the thorax. It will be easier to study it when you dissect the respiratory system.

The Abdominal Cavity

□ Look at Figure 20–4 and study the cuts that you are going to make to open the abdominal cavity.
□ With the scalpel or razor blade make a shallow cut from a point about 5 mm anterior to the umbilical cord forward to the most anterior pair of mammary glands.
□ Cut around the umbilical cord on both sides, and continue the two parallel cuts posteriorly, keeping them *outside* the mammary glands.
□ Deepen these cuts carefully until you reach the body cavity. You will have cut deeply enough when you see the dark-colored abdominal organs underneath the transparent membrane called the peritoneum.

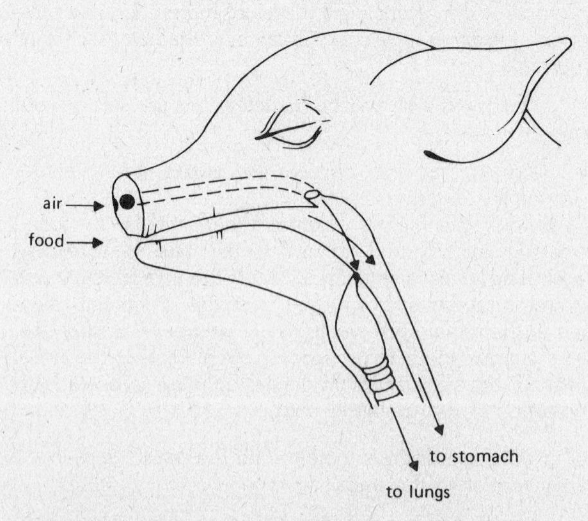

Figure 20–3. The pharynx in action. Air and food passages cross here, and there is some risk that food may enter or block the epiglottis, causing suffocation.

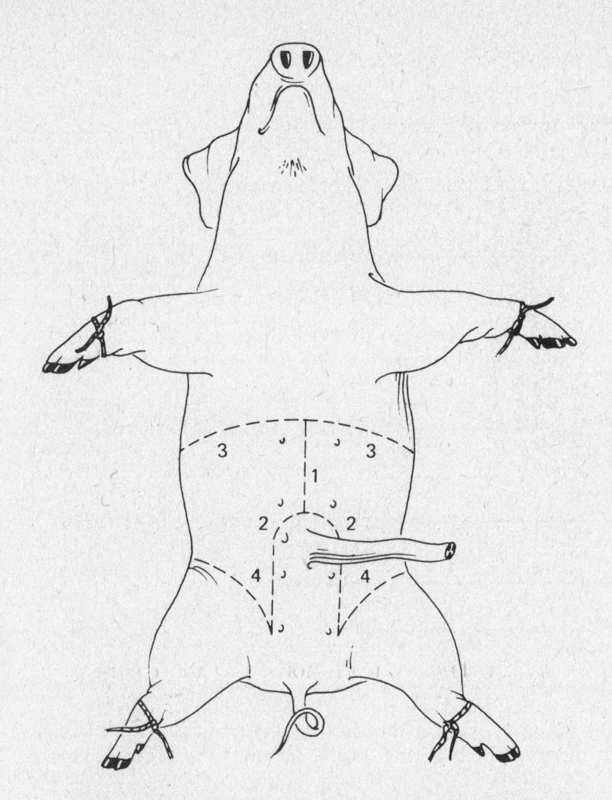

Figure 20–4. Dissection guide. Cut along the dotted lines to open the abdomen of your specimen.

☐ Pull up and back on the umbilical cord and you will see the **umbilical vein** going to the liver. This vein carries food and oxygen to the fetus from the placenta.
☐ Tie the ends of a piece of thread or string to the umbilical vein in two places to form a loop. The vein may now be cut between the ends of the string. The string will let you find the vein easily later on in the dissection.
☐ If the body cavity contains much fluid, flush it out with water. During the dissection, paper tissues can be used to soak up any remaining fluid in the body cavity.
☐ Locate the posterior end of the sternum, or breastbone, and cut laterally on each side following the posterior edge of the rib cage.
☐ Make similar cuts just anterior to the hind legs.
☐ Now fold out the flaps of the body wall, exposing the abdominal cavity. Its anterior end is bounded by a transverse partition, the **diaphragm.**
☐ Examine the lining of the abdominal cavity. This shiny membrane is called the **peritoneum.** The part of the membrane next to the body wall is called the **parietal peritoneum,** while that covering the surfaces of the internal organs is the **visceral peritoneum.** The thin, transparent sheets of tissue that connect the two are called **mesenteries.** The mesenteries suspend and support the visceral organs.
☐ Examine the large, dark brown **liver** located just posterior to the diaphragm.
☐ Notice that the liver consists of a number of lobes.
☐ Lift up the posterior edge of the liver and find the cut end of the umbilical vein anterior to the right side of the liver. (It should be attached to your string.) Just below the entrance of the umbilical vein into the liver you should see a greenish sac embedded in the surface of the liver. This is the **gall bladder.**

What substances are secreted by the liver?

bile
____

Does the liver secrete any digestive enzymes?

____

What is the role of the gall bladder?

The gall bladder stores bile
____

Use Figure 20–5 as your guide in locating the other abdominal organs.

☐ With the liver still elevated, identify the **stomach** on the left side of the abdominal cavity.
☐ Locate the point near the midline where the esophagus enters the stomach. This is the **cardiac** portion of the stomach. Food from the stomach leaves through the **pyloric** end of the stomach. The flow of food is governed by a circular sphincter muscle called the **pyloric valve.**
☐ Squeeze the pyloric valve and note how muscular it is.
☐ Cut the stomach open lengthwise, and notice the ridges on the inside of the stomach. They serve to increase the internal surface area.
☐ Observe the internal entrance of the esophagus and the pyloric valve, which marks the exit into the small intestine.

Are these valves normally open or closed?

____

☐ Notice that the stomach is not empty even though the fetus receives its nourishment from its mother through the umbilical vein. The fetus constantly drinks in the amniotic fluid surrounding it. The contents of the stomach consist of epithelial cells, hair, and other substances from the amniotic fluid. Cells lining the stomach secrete **hydrochloric acid** and **pepsin,** an enzyme that hydrolyzes proteins to amino acids under the highly acidic conditions found in the stomach.

Does any absorption take place from the stomach?

____

How would your own feeding habits have to be modified if your stomach were removed?

____
____
____

Attached to the stomach by a mesentery is the tongue-shaped **spleen.** The spleen is part of the circulatory system.

What color is the spleen? grayish brown

☐ Identify the anterior end of the **small intestine** leading from the stomach; it is the C-shaped **duodenum.** This is a very important part of the small intestine because the ducts from the liver, gall bladder, and pancreas enter here. Those from the liver and gall bladder join as the **common bile duct,** which can be seen as a white cord leading from the gall bladder to the duodenum. The **bile** produced by the liver and stored in the gall bladder enters through this duct.

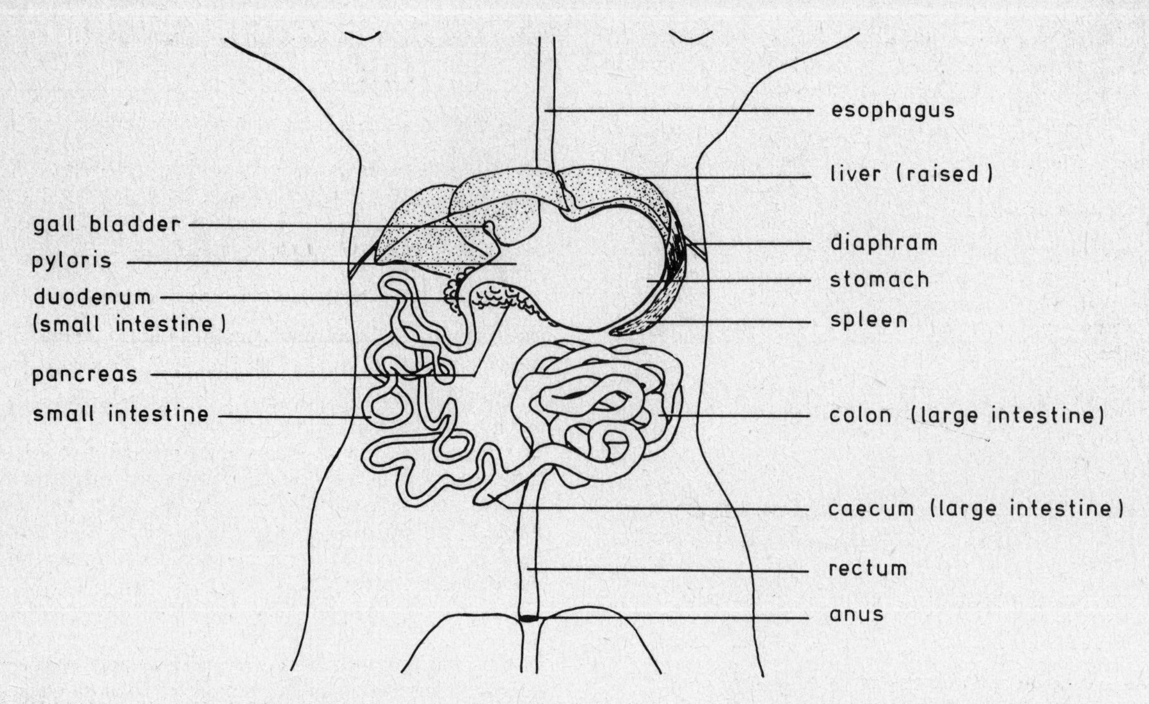

Figure 20–5. Abdominal organs. This diagram shows the approximate location of organs within the abdominal cavity. You will have to lift the liver to see the gall bladder underneath and move aside some of the other organs to see all of them.

What is the function of bile in digestion?

____

What is bile made up of?

____

□ If you lift the stomach and press the intestinal mass posteriorly, you will expose the light-colored glandular **pancreas.** (Glandular tissue has the appearance of used chewing gum.) The structural relationship between the duodenum and the pancreas has whimsically been called "the romance of the intestines" because the pancreas lies in the "arms" of the duodenum! The pancreas is connected to the duodenum by two tiny ducts that are too small to be easily seen.

What role does the pancreas play in digestion?

____

____

____

Which macronutrients (carbohydrate, protein, and fat) can be digested by enzymes from the pancreas?

____

What is the nondigestive function of the pancreas?

____

____

How does the pH of the small intestine differ from that of the stomach?

____

□ Trace the rest of the small intestine posteriorly from the duodenum. Notice the great length of the small intestine crowded into the small space.

What processes take place in the small intestine?

digestion capso absorption

____

Why is its length important?

____

____

____

Does the small intestine itself secrete enzymes?

____

□ Cut the small intestine next to the pyloris and next to the entrance to the large intestine. Dissect it free.
□ Measure its length and compare it with the pig's snout-to-rump length.

Humans, like pigs and rats, are omnivores and have a small intestine of intermediate length. If this length is comparable in pigs and humans, how long would the small intestine be in a person whose sitting height is 1 m?

	Pig	Human
Height (length)	cm	1 m
Small intestine	cm	m

□ Cut off a section of small intestine.
□ Slit it open lengthwise and examine it with a hand lens or under the dissecting microscope.

The velvetlike appearance is due to small projections called **villi,** which greatly increase the interior surface area. The villi contain capillaries and lymphatics that take up the end products of digestion and circulate them to the rest of the body, especially the liver.

□ Push the intestine to the right side, and relocate the point at which the small intestine enters the **large intestine.** It enters from the side, making a blind pouch called the **caecum.**

The caecum in plant-eating animals is usually large (up to 35 cm long in rabbits) and contains many microorganisms that will digest cellulose. The caecum is of little importance in the pig or human. In humans, the **appendix** is located at the end of the caecum, but it is absent in the pig. The main portion of the large intestine is called the **colon.** It forms a tight double coil and then goes to the dorsal abdominal wall and continues posteriorly along the midline.

This type of coiling in the colon is uncommon in mammals. The human colon is shaped like an upside-down U.

What is the function of the colon in digestion?

____

Does the colon secrete digestive enzymes?

No

□ Remove a short segment of the colon and cut it open.

How does the inner surface compare with that of the small intestine?

____

The colon contains a large number of bacteria that are continually being lost in the feces but that are also very important in human nutrition. They synthesize **vitamin K,** which is absorbed from the gut lumen and plays a vital role in blood clotting. Because they lack gut bacteria, newborn babies may be deficient in vitamin K and therefore especially susceptible to bleeding from small wounds.

HISTOLOGY OF THE INTESTINE

The intestinal wall is made up of four different layers. Review its structure by looking at a cross section of the intestine under low power of the compound microscope. (Refer to Figure 4–3.) The central opening, or **lumen,** has the villi projecting into it.

□ Examine one **villus.** Covering the villus is a layer of epithelial cells called the **mucosa.** Under the mucosa is an area of connective tissue, the **submucosa.** Identify the blood vessels running through the submucosa. The digested food must move from the intestinal lumen, through the mucosa and submucosa into these blood vessels. Within the villi there are also vessels of the **lymphatic system** that absorb food in the form of tiny droplets.

What kind of food molecules are absorbed by the lymphatics? lipids

____

What are the tiny droplets of this food called?

____

Outside the submucosa is the **muscularis,** which consists of two layers of muscle: an inner, circular layer and an outer longitudinal layer. This arrangement causes the peristaltic contractions of the intestine to move the food through the digestive tract. The outermost layer of tissue is the **peritoneum.**

Why does the intestine have villi?

____

What is the function of the goblet cells of the mucosa?

____

How does the shape of the intestine change when the inner layer of muscles contracts?

____

____

How does the shape of the intestine change when the outer layer of muscles contracts?

____

____

What tissue supports the blood vessels leading to and from the intestine? ______________

Reminder: When you are finished with your dissection for today, close up the abdomen of your pig and return it and any scraps of tissue to its plastic bag. Seal the bag, label it with your name, and store it properly. Clean your dissecting tools, oil the scissors to prevent rusting, and leave your workplace clean for the next student.

EXPLORING FURTHER

Deutsch, J. A., W. G. Young and T. J. Kalogeris. "The stomach signals satiety." *Science* 201: 165, 1978.

Grollman, S. *The Human Body. Its Structure and Physiology.* 4th ed. New York: Macmillan, 1978.

Morton, John. *Guts: The Form and Function of the Digestive System.* New York: St. Martin's Press, 1967.

Scientific American Articles

Davenport, H. W. "Why the stomach does not digest itself." January 1972 (#1240).

Fernstrom, John D., and Richard J. Wurtman. "Nutrition and the brain." February 1974 (#1291).

Kappas, Attallah, and Alvito P. Alvares. "How the liver metabolizes foreign substances." June 1975 (#1322).

Kretchmer, N. "Lactose and lactase." October 1972 (#1259).

Lieber, Charles S. "The metabolism of alcohol." March 1976 (#1336).

Scrimshaw, Nevin S., and Lance Taylor. "Food." September 1980 (#734).

Scrimshaw, Nevin S., and Vernon R. Young. "The requirements of human nutrition." September 1976.

TOPIC 21

Gas Exchange

OBJECTIVES

When you have completed this topic, you should be able to:

1. Describe the structures through which air must travel from the outside to the lungs of a mammal.
2. Explain the function of: the Eustachian tubes, turbinate bones, epiglottis, tracheal cartilage rings, alveoli, and diaphragm.
3. Distinguish between ventilation and respiration.
4. Describe how the rate of ventilation of human lungs is controlled.
5. Explain negative pressure breathing in a mammal, describing what happens during inspiration and expiration.
6. Explain positive pressure breathing in a frog.
7. Give two examples of groups of organisms that use the general body surface for gas exchange.
8. Cite one group of organisms that uses internal gills and one group that uses external gills for gas exchange.
9. Give the advantages and disadvantages of using air compared with water as a respiratory medium.
10. Explain what a tracheal breathing system is, and state the group of organisms in which it is found.
11. Name some organisms that do not need special structures for gas exchange.

TEXT REFERENCES

Chapter 32 (A–D, F–G).
Respiratory surface, ventilation, gas exchange, gills, tracheal system, positive and negative pressure breathing, control of ventilation.

INTRODUCTION

Unicellular organisms can obtain enough oxygen by simple diffusion and can dispose of their CO_2 and wastes in the same way. All larger and thicker organisms utilize a respiratory system of some sort to ensure that every one of their cells has an adequate oxygen supply. The respiratory system has a large, moist **respiratory surface** across which the actual gas exchange takes place, and a means of circulating air or water next to the respiratory surface. Animals whose respiratory system is in the form of lungs, gills, and sometimes the body surface, have a circulatory system intimately associated with the respiratory system. The circulatory system carries oxygen from the respiratory surface and distributes it throughout the body. Those organisms with a tracheal system, on the other hand, distribute air directly to all body cells and do not use the circulatory system for gas exchange.

THE RESPIRATORY SYSTEM OF THE FETAL PIG

The lining of the alveoli within the lungs is the respiratory surface where gases can be exchanged. Air is circulated to the alveoli through the nose, pharynx, larynx, trachea, and bronchi during ventilation of the lungs.

Nose

☐ Examine the external openings (**external nares**) of your pig's snout. Here air enters the nasal passages.
☐ Cut across the nose about 1 cm from the tip (see Figure 21–2) with your razor blade or scalpel, and open the cut, or look at the demonstration. Inside you see the

curved **turbinate bones,** which help increase the surface area of the passages.

Why is a large surface area needed?

A large surface area is needed for greater gas exchange

What is the role of the hairs, cilia, and mucous lining of the nasal passages?

To Filter out dirt and dust

Do comparable structures play à role in human nostrils?

Yes

Pharynx

□ Open the pig's mouth and review the location of the **glottis** and **epiglottis.** (See Figure 21–1.) They are located deep in the throat or oropharynx. Air enters the glottis, the opening to the lower respiratory tract, from the opening of the nasopharynx (pharynx above the soft palate). If the epiglottis is not evident, you may have to cut more deeply and fully open the pig's mouth. Use your blunt probe to fish for it, if necessary.
□ Slit the soft palate along its midline, and locate the opening of the **nasal passages.**

A pocket in this part of the pharynx has the openings of the **Eustachian tubes** which lead to the middle ear. They allow pressure within the middle ear to be equilibrated with the pressure outside so that there isn't any strain on the eardrum. Nose and throat infections can spread to the middle ear through the Eustachian tube.

Larynx

□ Touch the throat region of the pig to locate the hard, round larynx, or voice box.
□ Make the necessary cuts to expose the throat region, making sure the larynx is included within the exposed area. (See Figure 21–2.)

Beneath the skin of the throat are muscles and glands. The muscle tissue is mostly in ribbons with the fibers running lengthwise.

□ Note the large masses of light-colored glandular tissue, the **thymus glands.** These glands are part of the immune system, which produces antibodies as a defense against disease organisms. They are prominent early in development when the system is being established. Later in life the thymus decreases in size, and it is of little importance in the adult.
□ Near the lower jaws you may notice the **salivary glands** under the skin.
□ In the region of the larynx, use your probe or fingers to separate the muscles so that the **larynx** itself and the **trachea** posterior to it are exposed. Air passes from the pharynx into the glottis, through the larynx where the vocal cords are located, and through the trachea into the lungs themselves.
□ Notice that the larynx and trachea are hard; the trachea contains rings of cartilage that give it support.

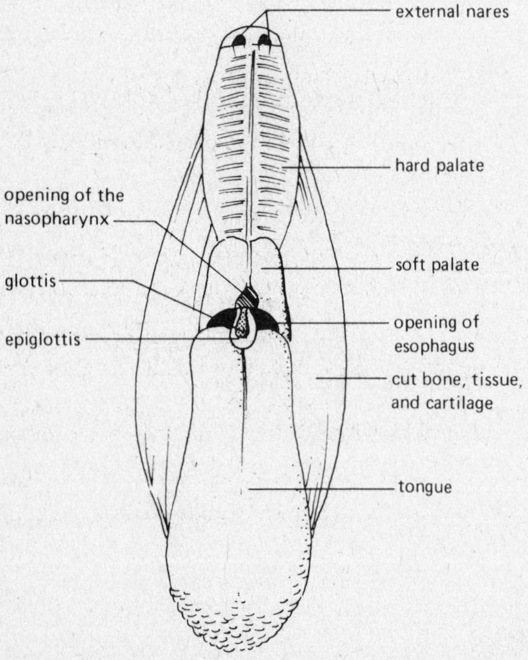

external nares
hard palate
opening of the nasopharynx
soft palate
glottis
opening of esophagus
epiglottis
cut bone, tissue, and cartilage
tongue

Figure 21–1. Inside the fetal pig's mouth. Cut the jaws open far enough to see the oral cavity and the oropharynx. The glottis is the opening within the folds of the epiglottis. The soft palate must be cut to find the openings of the Eustachian tubes and nasal passages.

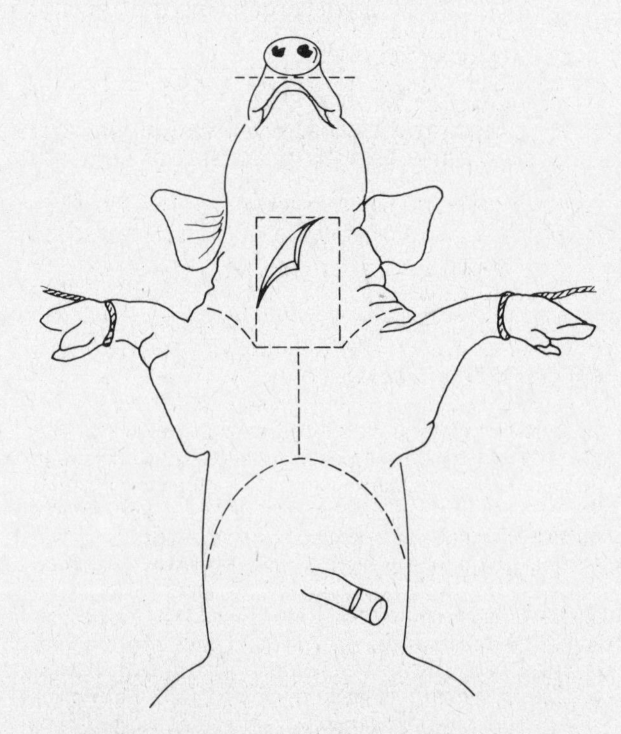

Figure 21–2. Dissection guide. Cut along the dotted lines to study the respiratory tract. Be careful not to damage underlying structures in the midline and shoulder regions.

Do the cartilage rings completely circle the trachea?

yes

What is the function of the cartilage?

The cartilage keeps the trachea open and strong

□ Ventral to the trachea near the joining of the neck and trunk, locate a smooth, brownish-colored gland, the **thyroid.** This is an endocrine gland important in regulating the body's metabolism.

□ Locate the **esophagus,** a collapsed flabby tube just dorsal to the trachea. To be sure you locate the right structure, pass a probe into the opening of the esophagus in the pharynx, and feel for it with your fingers beneath the trachea.

Following the dotted lines in Figure 21–2, open the thorax.

CUT CAREFULLY WITHOUT DAMAGING THE UNDERLYING ORGANS AND BLOOD VESSELS.

□ Use your scissors to cut through the bone and cartilage in the center. Be careful that you do not cut into the heart and blood vessels just underneath.

□ Cut along the margin of the rib cage as indicated by the curved line.

□ Cut towards the shoulders with care and only as far as necessary to open the thorax.

□ Locate the **diaphragm,** a sheet of muscle and tendon that separates the thoracic cavity from the abdominal cavity. The thoracic cavity is divided into three parts.

1. The **right pleural cavity,** lined by pleural membrane, contains the right lung.
2. The **left pleural cavity,** also lined by pleural membrane, contains the left lung.
3. The **pericardial cavity** is lined by the pericardial membrane and contains the heart.

□ Remove these membranes so that the heart and lungs are exposed.

The structure of the lower respiratory system is shown in Figure 21–3.

Air from the trachea enters the **bronchi,** which divide into smaller and smaller **bronchioles.** Finally the air ends up in the microscopic, thin-walled, moist sacs called **alveoli.** Their walls are the respiratory surface of the pig. Here gases in the air can be exchanged with gases in the blood of the capillary network because the wall of the alveoli are only one cell thick. Oxygen enters the capillaries by diffusion and facilitated diffusion and CO_2 crosses from the capillaries into the alveoli.

□ Examine the **lungs** of your pig.

Are the lungs equal in size? No

The right lung has four lobes and the left lung two or sometimes three. The lungs are filled with fluid in the fetal pig and are much more compact than they will be after they are inflated at birth. The point where the trachea divides into the right and left bronchi is hidden from view beneath the heart and blood vessels, but you will be able to see it later when you study the heart.

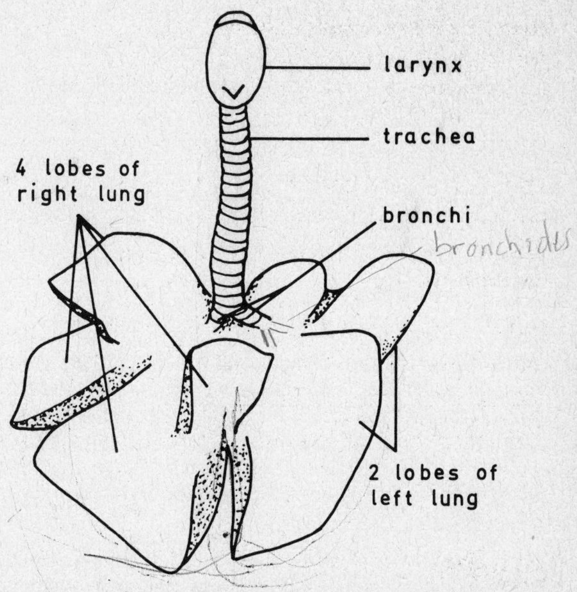

Figure 21–3. Lower respiratory tract of the fetal pig. The bronchi and bronchioles are surrounded by the spongy tissue of the lungs. In the fetal pig, the right lung has four lobes and the left has two (sometimes three).

How does the fetal pig carry on gas exchange?

Gas is exchanged by diffusion through the walls of the placenta

When you are finished with your dissection for today, store your fetal pig properly, clean your dissecting tools, and oil your scissors.

GENERAL BODY SURFACE

Frogs and many invertebrates use the skin of the body as a respiratory surface.

How is this surface "ventilated"?

gas is diffused through the skin - (blood vessels underneath) and carried to different parts of the body.

The frog uses its lungs as well as its skin for respiration.

What are some invertebrates that lack respiratory organs and use only the skin?

Earthworms.

Flatworms

GILLS

Gills consist of feathery outpocketings that greatly increase the area of the respiratory surface. In some aquatic animals they are external and unprotected, but usually they are protected in some way.

Necturus (Axolotl)

□ Examine the specimen on demonstration. In a living animal the exposed gills are pink.

Why are gills pink or red?

Because of blood vessels running through the gills

Perch

The gills are protected by a cover called the **operculum,** and water circulates from the mouth through the spaces between the gills and out through the opening of the operculum. The pattern of water circulation through the gills of a fish is shown in Figure 32–5 in your text. The gills themselves are so large that a **gill skeleton** is necessary to give support to the delicate filaments.

□ Examine the gill skeleton of a fish. The spiny anterior protrusions are called **gill rakers** and help keep large particles away from the delicate **gill filaments.**

TRACHEAL SYSTEM

Insects, centipedes, and millipedes obtain oxygen from air rather than from water. Their respiratory system, the **tracheal system,** is quite different from that of the lungs found in vertebrates: it does *not* depend on the circulatory system. Air enters the body through small openings called **spiracles,** which lead into the **tracheal ducts.** The ducts divide repeatedly into **tracheoles** and finally end in tiny, fluid-filled tubes where gas exchange takes place. The tiny tracheoles extend everywhere in the body, so that no cell is too far away from its air supply.

Figure 21–4 shows how air entering the spiracles reaches a muscle cell. The circulation of air is speeded up in larger, active insects by contraction of the abdominal muscles and flight muscles.

□ Examine specimens of a grasshopper or a cockroach. Locate the spiracles along each side of your specimen. The location of grasshopper spiracles is shown in Figure 18–5.

Why does a film of oil quickly kill an insect?

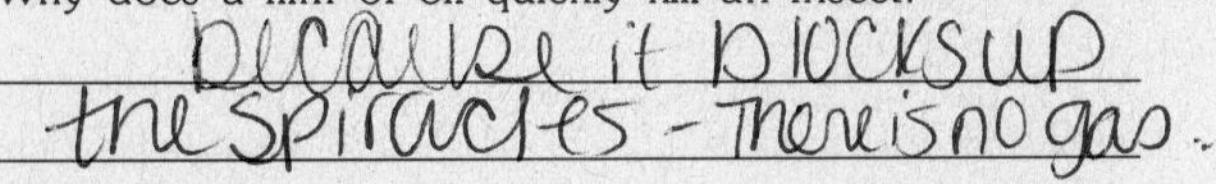

Why is a strong solution of a nontoxic detergent a good insecticide?

Because it can be brought in and clog up the spiracles

□ Remove the dorsal part of the exoskeleton by making cuts along its edges.

□ Locate a **trachea,** the tube leading from a spiracle into the insect's body.

□ Remove a bit of muscle, and tease it apart in a drop of water or Ringer's solution. Observe it under the compound microscope to see the fine tracheal tubes that carry oxygen to all of the muscle cells.

Why does the insect need a circulatory system if it isn't used for gas exchange?

Clean your slide and dissecting tools before going on to the next part of this exercise.

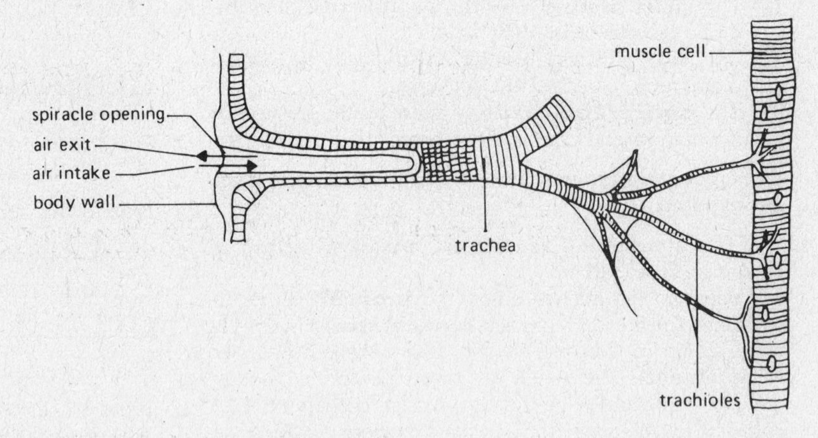

Figure 21–4. Trachea of an insect. Movements of the body cause the air within the trachea to circulate. (Redrawn from R. D. Barnes, *Invertebrate Zoology,* 3rd ed. Philadelphia: W. B. Saunders, 1974.)

EXPLORING FURTHER

Schmidt-Nielsen, Knut. *Animal Physiology: Adaptation and Environment.* New York: Cambridge University Press, 1975.

Scientific American Articles

Avery, M. E., N-S. Wang, and H. W. Taeusch. "The lung of the newborn infant." April 1973 (no offprint).

Comroe, Julius H., Jr. "The lung." February 1966 (#1034).

Hong, S. K., and H. Rahn. "The diving women of Korea and Japan." May 1967 (#1072).

Johansen, K. "Air-breathing fishes." October 1968 (#1125).

Kylstra, J. S. "Experiments in water breathing." August 1968 (#1123).

Naeye, Richard L. "Sudden infant death." April 1980 (#1467).

Student Name Kathleen Keenan **Date** Feb 6, 1983

HOW LUNGS ARE VENTILATED

Vertebrate lungs can be ventilated in two ways. In mammals the lungs are enclosed in the airtight thoracic cavity; enlargement of this cavity creates a **negative pressure** outside the lungs to that air rushes into them. The frog, on the other hand, uses **positive pressure** from contraction of the oral cavity to force air into the lungs. Negative pressure inflation of the lungs is like sucking in on your bubble gum so that it forms a bubble inside your mouth. Positive pressure inflation is like blowing a regular bubble.

Negative Pressure Breathing

Inspiration in mammals is due to expansion of the chest cavity when the rib cage lifts and the diaphragm contracts. Muscles between the ribs and tangential muscles in the diaphragm itself must contract for this to happen. The positions of the diaphragm and rib cage during inspiration and expiration are shown in Figure 32–12 in your text.

□ Place one hand on your stomach just below your ribs to feel the effect of your diaphragm contracting, and place the other hand on your rib cage.
□ Take a few normal breaths.

Does the rib cage move? yes

Does the stomach move? no

□ Try to breathe without moving the rib cage at all.
□ Then try not to move your diaphragm.

Can you get enough air using only one set of muscles? no

Positive Pressure Breathing

The frog is a good animal to observe using positive pressure breathing because it is easy to see how it is done. (Also see Figure 32–11 in your text.)

□ Observe the live frog on demonstration when it is out of the water.

What movement relating to breathing is most obvious? The movement of the floor of the frogs mouth

A frog can open and close the nostrils and can close off the glottis leading to the lungs. The floor of the mouth can be lowered to draw in air and raised to expel air. Complete the following chart to show what happens during inspiration in positive pressure breathing.

	Nostrils (Open/Closed)	Glottis (Open/Closed)	Mouth Floor (Raised/Lowered)
1. Air enters mouth.	open	closed	lowered
2. Air enters lungs.	closed	opened	Raised.

3. Finally air leaves the lungs under pressure from the stretched chest wall, and the breathing cycle is ready to start again.

HOW THE RATE OF BREATHING IS CONTROLLED

The rate at which you breathe is automatically controlled so that there will be enough oxygen and so that CO_2 will be promptly removed. **Sensors** in the brain stem (part of the spinal cord next to the brain) and in the aorta and carotid arteries (main arteries supplying the brain, which can be felt in the throat region) respond to the pH of the blood. Blood pH will decrease as the CO_2 level increases due to the formation of carbonic acid:

$$H_2O + CO_2 \rightleftharpoons H_2CO_3 \rightleftharpoons H^+ + HCO_3^-$$

As soon as the CO_2 level rises, the sensors detect carbonic acid and speed up the breathing rate to get rid of the CO_2. *Breathing rate is normally determined by the level of CO_2* rather than that of oxygen, probably because CO_2 is easier for the body to detect. (See Figure 32–20 in your text.)

IN THESE EXPERIMENTS, STOP AT ONCE IF YOU BEGIN TO FEEL FAINT. IF YOU HAVE ANY MEDICAL PROBLEM WITH YOUR HEART OR LUNGS, BE A TIMEKEEPER, NOT A SUBJECT.

Work with a partner to make the following measurements:

□ While sitting down and breathing normally, measure the number of breaths per minute three times. Record your results to the nearest half-breath in the chart below.
□ Hyperventilate by breathing as deeply and as fast as you can for 20 breaths. Then breathe normally.

Do you feel an urge to breathe right after hyperventilating? yes

Are your breaths deeper or shallower than normal? deeper

□ Measure your breathing rate for the first minute after you resume breathing.
□ Wait until you are breathing normally again before continuing.
□ Breathe into a plastic bag for 2.5 min, and record your breathing rate for each half-minute interval. Calculate breaths per minute for each interval.

STOP IF YOU FEEL FAINT. DO NOT EXCEED 3 MIN.

Condition	Time Interval	Breaths	Breaths per Min
Normal	1st min	13	13
	2nd min	13	13
	3rd min	13	13
After hyperventilation	1st min	16	16
Breaths into bag	1st half min		
	2nd half min		
	3rd half min		
	4th half min		
	5th half min		
	6th half min		

Did your breathing become more deep or more shallow? __________

There is still oxygen left in the bag. What would have happened if the CO_2 had been removed from each breath that you exhaled?

____

When you are finished with your work, return or dispose of all specimens and materials, clean your dissecting tools, and store your microscope properly.

TOPIC 22

The Transport System in Vertebrates

OBJECTIVES

When you have completed this topic, you should be able to:

1. Name four functions of the circulatory system.
2. Cite the function of each of the major parts of the circulatory system: heart, arteries, veins, and capillaries.
3. Describe the function of the placenta, umbilical arteries, and the umbilical vein.
4. Trace the path of a blood cell from the umbilical vein to the right atrium of the heart and back to the umbilical artery by the shortest route.
5. Describe the path of blood flow through the four chambers of the heart and the five major vessels connecting to it.
6. Name the parts of the circulatory system that have valves, and state the function of valves in regulating blood flow.
7. Describe the changes in structure and function that occur in these structures when the fetal pig is born: umbilical artery, umbilical vein, hepatic portal vein, ductus venosus, foramen ovale, pulmonary artery, pulmonary vein, ductus arteriosus.

TEXT REFERENCES

Chapter 33 (B,C,E).
Closed circulatory system, double circulation, mammalian heart, arteries, veins, capillaries, lymphatic system.

INTRODUCTION

The circulatory system of a mammal has four major functions:

1. The transportation of nutrients, wastes, hormones, and gases.
2. The regulation of the pH, water, and salt content of the tissues.
3. The regulation of body temperature.
4. The protection of the body by the immune system.

The **heart** is the pump that forces the blood through the pulmonary and systemic circuits of the circulatory system. The **arteries** are the main vessels carrying blood from the heart to the rest of the body, The arteries end in the tiny **capillaries,** where exchange of nutrients, gases, wastes, and other substances with the tissues can occur. Finally the blood returns from the capillaries of the body to the heart in the large **veins.** The blood and the capillaries are the physiologically active parts of the circulatory system. The main purpose of the heart and large arteries and veins is simply to move the blood quickly to the capillaries so that exchange can occur there.

THE HEART

Study the diagram of the adult heart in Figure 22–1. This view is from the ventral side and shows the internal structure of the heart.

Adult Heart

☐ Use the demonstration heart or a model of the heart to identify the major structures. The ventral side of the

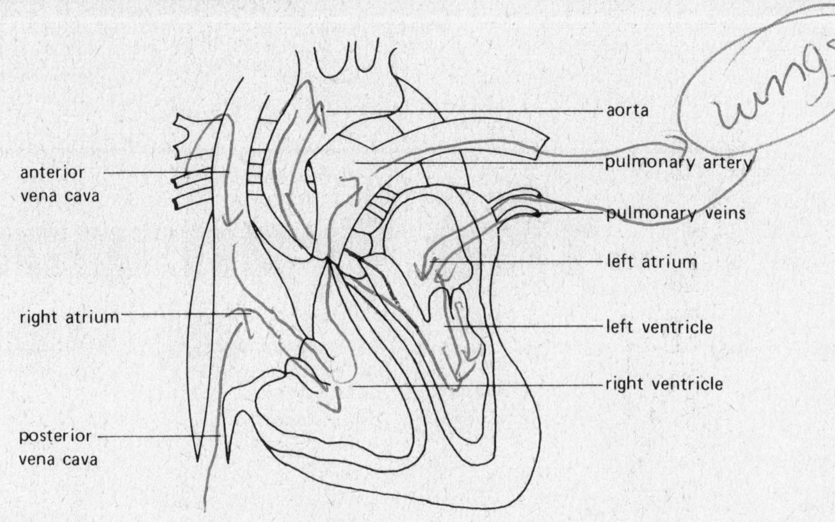

Figure 22–1. Adult heart. Blood from the anterior and posterior venae cavae flows into the right atrium, through the atrioventricular valves into the right ventricle, and leaves through the pulmonary artery to the lungs. Blood then returns from the lungs in the pulmonary veins, enters the left atrium, passes through the atrioventricular valves into the left ventricle, then leaves via the aorta.

heart can be identified easily because it has a diagonal line of fat. The dorsal surface is usually flatter and has a short, straight line of fat.

□ Find the ear-shaped **right atrium** and **left atrium** located on either side of the heart. The bulk of the heart consists of the **right** and **left ventricles.**

□ Examine the cut across the base or apex of the heart.

Which ventricle has the greatest muscle mass?

Left ______

□ Open the cuts in the heart so that you can see these valves:

1. The **tricuspid** valves are between the right atrium and right ventricle.
2. The **bicuspid** (mitral) valves are between the left atrium and left ventricle.
3. The **semilunar** valves are at the base of the pulmonary artery. There are also semilunar valves at the base of the aorta.

□ Note the two large arteries connecting to the heart. They are thick-walled and very muscular.

In which direction does blood flow in these arteries?

______

The more ventral artery is the **pulmonary artery.**

To which chamber of the heart is it connected?

right ventricle ______

The larger, more dorsal vessel is the **aorta.**

To which chamber of the heart is the aorta connected?

Left ventricle ______

There are three major veins connected to the heart. Because they are thin-walled, they are collapsed in this empty heart.

In which direction does blood flow in each of these veins?

into the heart ______

Is the blood flowing in these vessels under high pressure or under low pressure?

low ______

□ Find the two large vessels connecting to the right side of the heart. They are the **anterior vena cava** and **posterior vena cava.**

To which chamber of the heart do the anterior and the posterior vena cava connect?

Right Right ______

Finally, the **pulmonary veins** come together on the dorsal surface of the heart.

To which chamber do the pulmonary veins connect?

left Left ______

□ Trace the flow of blood from the right atrium to the aorta.

Draw arrows in Figure 22–1 to show the direction that blood flows in each vessel and chamber.

Fetal Pig Heart

The **pericardial cavity** contains the heart and is lined by the pericardial membrane.

□ Carefully remove the pericardial membrane to expose the heart.

□ Identify the major chambers and vessels of the heart, using Figure 22–2 as your guide; note the light-colored **ventricles** that form the posterior part of the heart and the dark-colored flaps that are the ventral parts of the **atria.** Larger portions of the ventricles are hidden from view.

□ You can cut off the tip of the heart with a razor blade to see the difference in size of the muscles making up the right and left ventricle.

Part of the blood entering the heart follows the pattern that you traced for the adult heart. In the fetal pig, however, there is an opening between the right and left atria called the **foramen ovale.** It allows some of the blood entering the right atrium to pass directly to the left atrium and out through the aorta, instead of first circulating to the fetal lungs.

□ Push the heart posteriorly and clean off pieces of thymus tissue and membrane so that you can see clearly the vessels that arise from the heart.

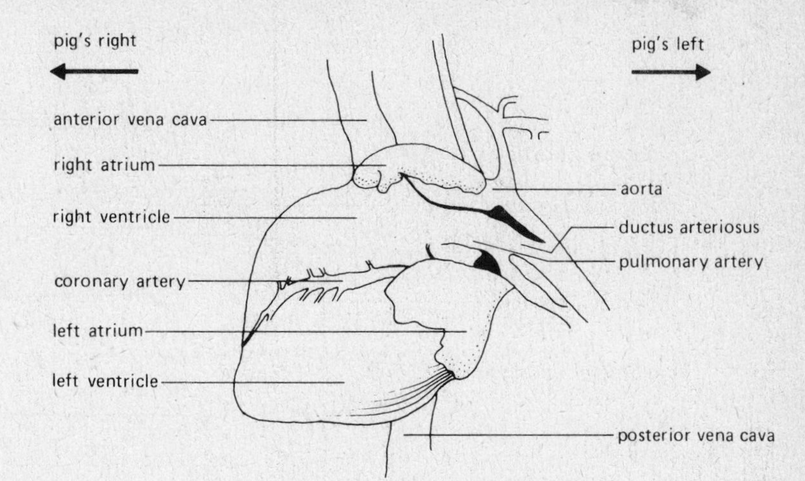

Figure 22–2. Fetal pig heart. Ventral view of its external structure. The pulmonary artery is the most prominent vessel leaving the heart.

1. The most conspicuous vessel is the thick-walled, light-colored pulmonary artery.
2. Beneath is the slightly larger aorta, which is also thick-walled and light-colored.
3. The pulmonary artery is connected to the aorta by the short, thick **ductus arteriosus.** In the fetal stage, most of the blood leaving the heart in the pulmonary artery by-passes the right and left branches of this artery and instead passes through the ductus arteriosus and into the aorta directly.

Why are the lungs by-passed before birth?

Because the mother is supplying all nutrient and gases through the placenta

4. Returning to the anterior part of the heart, the large, thin-walled, dark or blue-colored vessel on the right is the anterior vena cava. It brings blood from the anterior part of the body to the right atrium of the heart. Lift the posterior tip of the heart to see the pulmonary veins, which join together on the dorsal surface of the heart and enter the left atrium.
5. On the right side of the heart is a large blue vessel, which is the posterior vena cava and which joins the anterior vena cava in entering the right atrium.

ARTERIES

Arteries are all the vessels carrying blood away from the heart. If you have an injected pig, the arteries have been injected with pink latex so that they are easily seen and resistant to breakage. The veins have been injected with blue latex so that you can tell the difference, but sometimes the injection results are not perfect and a certain vessel, the umbilical vein or artery, for example, may be the "wrong color." Ask your instructor for help if you have trouble locating a certain artery or vein. Use Figure 22–3 to identify your pig's arteries.

In cross section, arteries have thick walls which are muscular and have elastic connective tissue. Arterial blood is under high pressure.

Pulmonary Artery

□ Locate the branches to the right and left lungs; lift the heart to see them.

What is the main vessel to which the pulmonary artery joins in the fetal pig?

~~capp~~ capillaries in lungs

Aorta

Know these

□ Note that this vessel arches as it leaves the heart, passes down in the thoracic cavity to the spine, and then runs posteriorly all the way to the umbilical arteries. It is the major artery in both the fetal and adult pig. The following vessels branch off from the aorta as it passes through the body (see Figure 22–3):

1. **Coronary artery.** The first vessel to branch from the aorta. It is difficult to see the branch point even when you lift the left atrium and look closely. The vessel itself is easy to see on the ventral surface of the heart. When this vessel constricts or becomes blocked by a blood clot (coronary thrombosis) during a heart attack, the oxygen supply to the active heart muscle is interrupted, and some of the muscle tissue dies. If the damage is extensive, the heart attack may be fatal.
2. **Brachiocephalic artery.** A large branch that divides into the **common carotid artery** and the **right subclavian.** Trace the common carotid to its branch point into the right and left carotids, and trace one of the carotids into the throat region. This artery serves the head and neck. The right subclavian sends blood to the right forelimb. The **internal thoracic arteries** send blood to the chest and ribs.
3. **Left subclavian artery.** The branch that leaves the thorax and enters the left forelimb.

□ Push the lungs to the pig's right, and locate the **dor-**

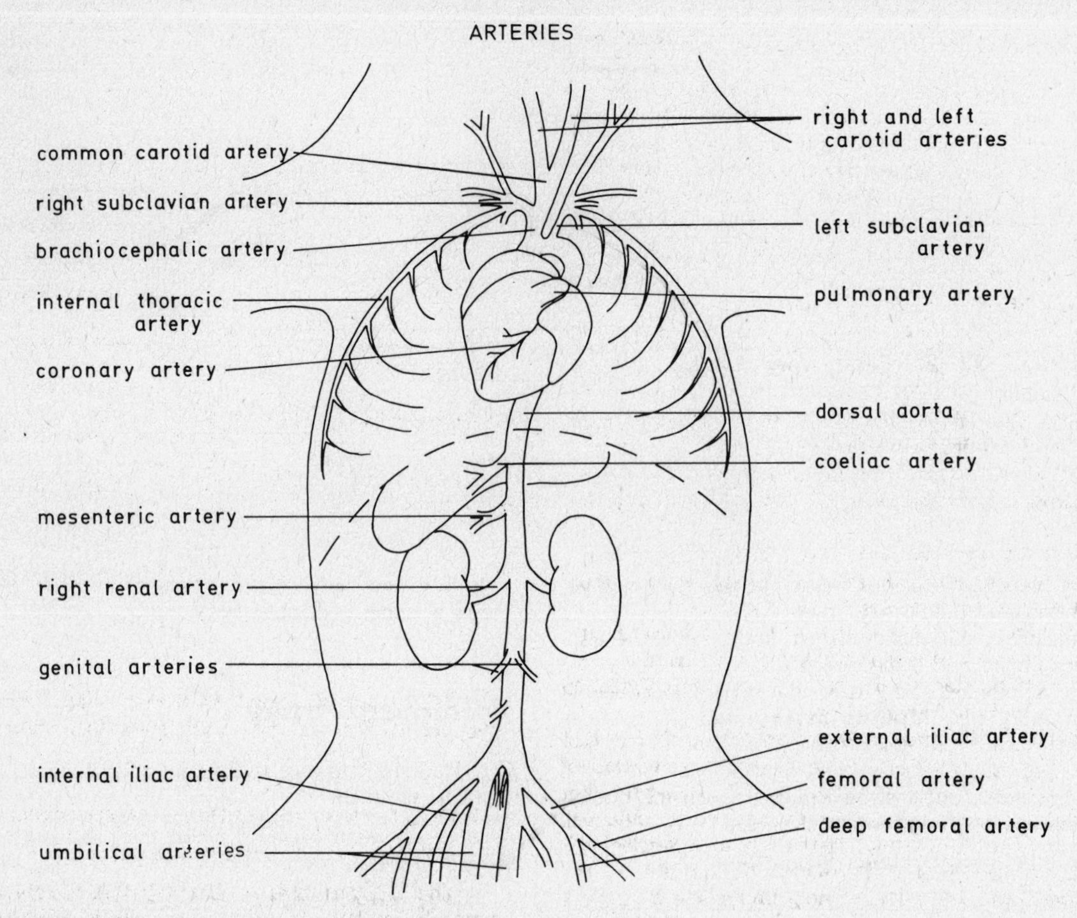

Figure 22–3. Arterial system of the fetal pig. In this view from the ventral side, the aorta arches dorsally, passes beneath the heart, and becomes the dorsal aorta deep within the thorax. The aorta divides to form the prominent umbilical arteries.

sal aorta deep in the thoracic cavity. Notice that it is still thick-walled and light-colored. Its walls are so thick that the color of the pink latex is not visible. Next to the aorta is the collapsed, flabby, tubular esophagus.

□ Trace the aorta and esophagus posteriorly as they pass through the diaphragm and enter the abdominal cavity.

4. **Coeliac artery.** Branches from the aorta just posterior to the diaphragm and serves the stomach, liver, spleen, and pancreas.
5. **Mesenteric artery.** The branches of this artery send blood to the small intestine and to the anterior part of the large intestine.
6. **Renal artery.** The right and left renal arteries send blood to the kidneys. Occasionally you might see a renal artery that is double instead of a single vessel.
7. **Genital artery.** A pair of small genital arteries that branches next to supply the testes in the male or the ovaries and uterus in the female.
8. **External iliac artery.** A pair of large arteries that supplies the hindlimbs. After each gives off a small branch, it branches into the **femoral artery** and **deep femoral artery.**
9. **Internal iliac artery.** The aorta ends at the spot where the external iliac branches off and it becomes the internal iliac artery for a short distance. The internal iliac in turn splits to form the umbilical arteries.
10. **Umbilical arteries.** These large arteries arise from the internal iliac and carry blood from the fetus to the placenta. Blood in these vessels would be very low in oxygen and nutrients but high in CO_2 and wastes.

You have traced the major arteries as an introduction to the complexity of the circulatory system anatomy. Each artery divides repeatedly into smaller **arterioles** and finally becomes a **capillary bed.** The capillaries join together into larger and larger **venules** and finally return blood to the large veins entering the heart.

COVER YOUR PIG WITH MOIST PAPER TOWELS DURING THE NEXT PART OF THE LABORATORY.

CAPILLARIES

The preserved capillaries in your fetal pig are much too small to be seen without special preparation for microscopy. Instead you should look at some living capillaries in which blood is actually circulating to get a feeling for the intense activity that goes on in your **microcirculation.**

Here in the capillaries gases and molecules are exchanged between the blood and the tissues. Because the blood is still under fairly high pressure when it enters the capillaries, some of the noncellular fluid (plasma) is forced out into the intracellular spaces around the capillaries so the area for exchange is increased even more. This fluid drains into the **lymphatic system** and is eventually returned to the circulatory system via large lymphatic vessels. Blood loses most of its pressure as it passes through the capillaries.

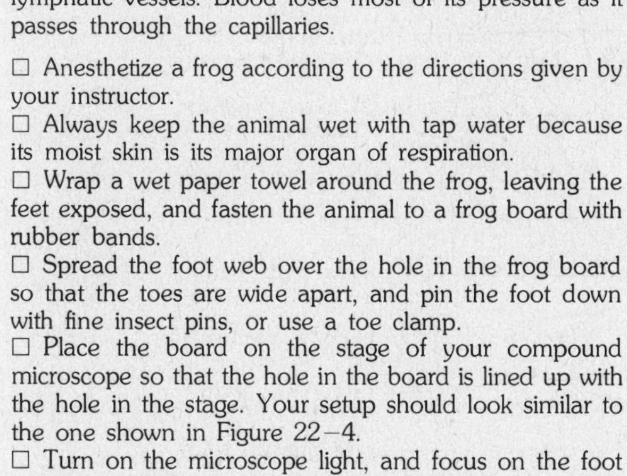

□ Anesthetize a frog according to the directions given by your instructor.
□ Always keep the animal wet with tap water because its moist skin is its major organ of respiration.
□ Wrap a wet paper towel around the frog, leaving the feet exposed, and fasten the animal to a frog board with rubber bands.
□ Spread the foot web over the hole in the frog board so that the toes are wide apart, and pin the foot down with fine insect pins, or use a toe clamp.
□ Place the board on the stage of your compound microscope so that the hole in the board is lined up with the hole in the stage. Your setup should look similar to the one shown in Figure 22–4.

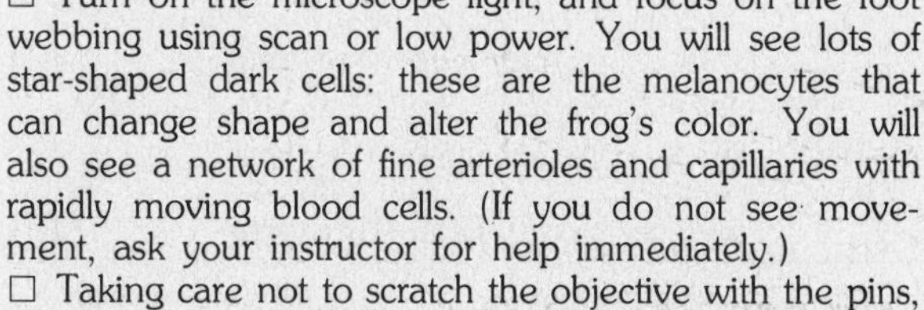

□ Turn on the microscope light, and focus on the foot webbing using scan or low power. You will see lots of star-shaped dark cells: these are the melanocytes that can change shape and alter the frog's color. You will also see a network of fine arterioles and capillaries with rapidly moving blood cells. (If you do not see movement, ask your instructor for help immediately.)
□ Taking care not to scratch the objective with the pins, switch to high power and observe the capillary bed more closely.

How wide is the smallest capillary compared with the width of a red blood cell?

About same size

□ Observe the small capillaries.

Does the blood always flow in the same general direction?

yes

Is the blood flowing in all of the capillaries or only in some?

all

□ When you are finished, free the frog and return it to the recovery tank.

VEINS

Veins are thin-walled and contain much of the blood. Because the blood in them is under very low pressure, veins have **valves** so that the blood can flow only towards the heart.

Pulmonary Veins

Blood from the right and left lungs drains into the branches of the pulmonary veins. You have already seen where the pulmonary veins join and enter the left atrium.

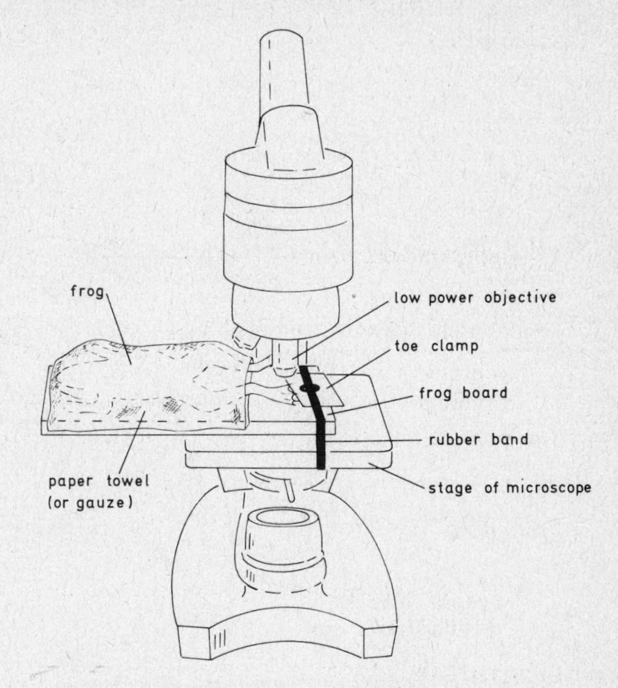

Figure 22–4. Frog capillary setup. Keep the frog and the delicate web between its toes moist.

Do the pulmonary veins carry as much blood before birth as they do after birth? No

The **azygos vein** is also beneath the heart and drains blood from the chest wall into the right atrium. The rest of the venous system is shown in Figure 22–5.

Anterior Vena Cava

Know these

Vessels leading into the anterior vena cava include:

1. **Coronary vein.**
2. **Internal jugular vein.** Parallel to the common carotid and drains the head and neck region.
3. **External jugular vein.** Joins the internal jugular close to the anterior vena cava.
4. **Brachial veins.** The main vessels draining the forelimbs. They join with the external jugulars before entering the anterior vena cava.
5. **Internal thoracic veins.** Carry blood from the chest and ribs.

Posterior Vena Cava

Vessels draining the posterior portion of the body join the posterior vena cava and include:

1. **External iliac vein.** Drains the anterior part of the hindlimb via the **femoral veins.**
2. **Internal iliac vein.** Drains the posterior part of the hindlimb and joins the external iliacs to form the common iliac. The two common iliacs join to form the beginning of the posterior vena cava.
3. **Genital veins.** Small veins that lead into the posterior vena cava and left renal vein.
4. **Renal veins.** Drain blood from the right and left kidneys.

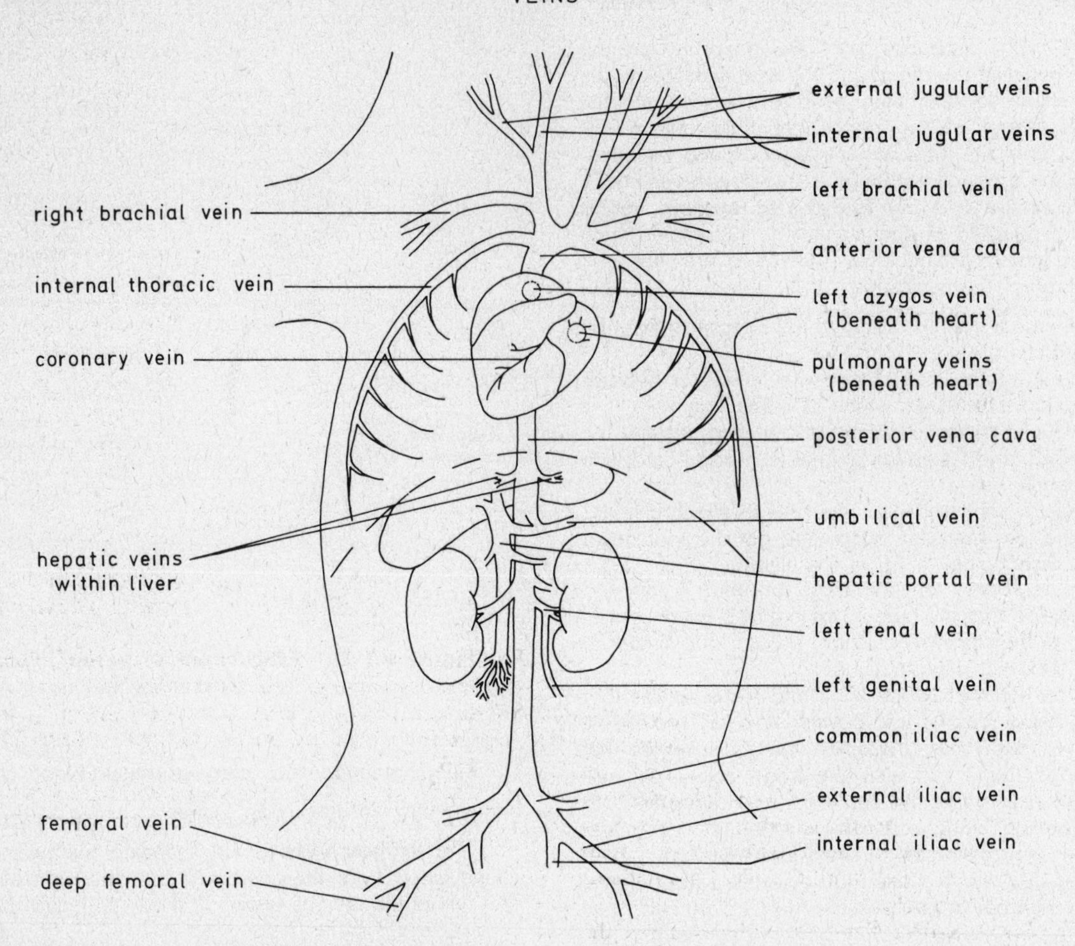

Figure 22–5. Venous system of the fetal pig. In this ventral view, the major veins leading to the heart are the venae cavae. The pulmonary veins are beneath the heart and the hepatic veins are within the liver.

5. **Hepatic veins.** Veins within the liver that drain into the posterior vena cava just below the diaphragm.
6. **Umbilical vein.** When you originally opened the abdomen, you cut this vessel, which leads from the umbilical cord into the liver (use the string to find the cut ends); one of its branches leads into the hepatic portal system below, and another branch, the ductus venosus, leads directly into the posterior vena cava.
7. **Ductus venosus.** In the fetal pig, part of the blood in the umbilical vein branches off in this vessel and goes directly into the posterior vena cava. This allows some blood rich in oxygen and nutrients to be pumped directly out to the body without passing through a capillary bed (see Figure 22–6).

Read

Hepatic Portal System

□ Gently separate the liver and stomach from the small intestine and locate the **hepatic portal vein** draining blood from within the mesentery of the small intestine into the liver. It may not be injected with blue latex in your pig—if not, find it in another specimen.

The structure of the portal system is shown in Figure 22–6.

Veins from the small intestine, large intestine, stomach, and spleen join to form the hepatic portal vein, which carries nutrient-rich blood to the liver after birth. During fetal development the gut has little function, so the hepatic portal system is not important in transport between gut and liver.

How do nutrients enter the fetal circulation?

____

____

____

____

Read this

TRANSFORMATION OF FETAL TO ADULT CIRCULATION AT BIRTH

At the time of birth the connection of the fetus with the placenta via the umbilical cord is broken. The umbilical vein no longer brings blood rich in oxygen and nutrients

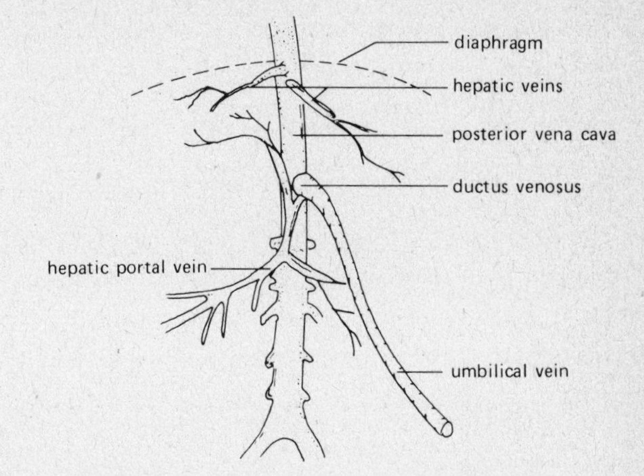

Figure 22–6. Hepatic portal system. The hepatic portal system drains from the intestine, stomach, and spleen into the liver. In the fetal pig, the umbilical vein empties partly into the portal system and partly through the ductus venosus directly into the posterior vena cava. (After C. A. Leone and P. W. Ogilvie, *Fetal Pig Manual.* Minneapolis: Burgess Publishing Co., 1960.)

to the liver, and the umbilical artery does not carry spent blood to the placenta to pick up more oxygen and nutrients and exchange wastes with the maternal circulation.

The fetal circulation responds to these changes in the following way:

1. **Lungs.** When the lungs are ventilated, the pulmonary circuit becomes functional for the first time. Blood leaves the right ventricle through the pulmonary artery, exchanges oxygen and CO_2 in the capillaries of the lungs, and returns to the left atrium via the pulmonary veins. It is now extremely harmful to by-pass the pulmonary circuit. Therefore the foramen ovale seals shut and the ductus arteriosus closes and degenerates.
2. **Digestive System.** Nutrients enter the gut as the piglet suckles and are carried to the liver by the hepatic portal system which thereby replaces the umbilical vein. The ductus venosus no longer has any function, and both it and the umbilical vein degenerate.
3. **Kidneys.** The kidneys now have to assume fully the function of removing wastes from the blood so the renal artery and renal vein become fully active. The urine* formed can no longer pass out through the umbilical cord (via the allantoic duct) and now must leave the body only through the urethra.

When you are satisfied with your understanding of the circulatory system, return your pig to its proper storage place or container. Clean your dissecting pan and instruments, and oil your scissors.

*The urinary system of the fetal pig is described in Topic 24.

EXPLORING FURTHER

Kaplan, Norman M. "The control of hypertension: A therapeutic breakthrough." *American Scientist* 68: 537, 1980.
Prosser. C. L. *Comparative Animal Physiology.* 3rd. ed. Philadelphia: W. B. Saunders, 1973.
Singer, C. *A Short History of Anatomy and Physiology: From the Greeks to Harvey.* New York: Dover, 1957.
Schmidt-Nielsen, Knut. *Animal Physiology: Adaptation and Environment.* New York: Cambridge University Press, 1975.

Scientific American Articles

Mayerson, H. S. "The lymphatic system." June 1963 (#158).
Stallones, Reuel A. "The rise and fall of ischemic heart disease." November 1980.
Wiggers, Carl J. "The heart." May 1957 (#62).
Wood, J. E. "The venous system." January 1968 (#1093).

TOPIC 23

Function of the Transport System

OBJECTIVES

When you have completed this topic, you should be able to:

1. List the major proteins present in blood plasma and give their functions.
2. List the major types of blood cells in human blood and give their functions.
3. Explain the function of hemoglobin and why it is found within cells rather than free in the blood.
4. Define blood pressure and demonstrate how to measure it.
5. Demonstrate how the beating of the frog heart is regulated by neurotransmitters.
6. Define pulse rate and describe how it is regulated by the nervous system.

TEXT REFERENCES

Chapter 33 (B–E); Chapter 34 (B); Chapter 37 (G,H,J); Chapter 38 (F,G).
Amphibian circulatory system, heart cycle, blood pressure, exercise, blood composition, blood clotting, lymphatic system, immune system, synapses, neurotransmitters, vagus nerve.

INTRODUCTION

In Topic 22 you studied the anatomy of the circulatory system and heart from the point of view of circulating nutrients, wastes, and gases around the body. The circulatory system actually has other important functions. Review its functions by listing the four major ones below:

1. Transportation of Nutrients, Wastes, gases & hormones
2. Regulation of Temperature
3. Regulation of pH, water and salt content of tissues
4. Protection of body by immune system

Several structures other than arteries, veins, capillaries, and the heart are important in the functioning of the circulatory system. The fluid that is lost in the capillaries drains back into the circulation through the **lymphatic system,** which also transports fats that are poorly soluble in water, and contains the lymph nodes and lymphocytes of the **immune system.** The **spleen** contributes cells to the immune system and stores red blood cells until they are needed. The **bone marrow** forms red and white blood cells, and the **liver** and **thymus** contribute cells to the immune system. The circulatory system does not act alone, but is closely involved with many other organs.

COMPOSITION OF THE BLOOD

Plasma

60% of blood volume
40% = blood cells

The liquid part of the blood, called the **plasma,** contains dissolved nutrients, wastes, salts, and the blood proteins, including albumin, globulins, and fibrinogen. **Albumin** helps maintain the osmotic strength of the blood, and contributes to its buffering capacity. The **gamma globulins** (one group of immunoglobulins) include the variety of antibodies present in the blood at all times. **Fibrinogen** is the structural protein from which clots are formed. It is split by another blood protein, the enzyme thrombin, during the chain reaction of the clotting process. Plasma from which the fibrinogen has been removed is called **serum.**

Blood plasma is a very good buffer.

What property of blood does a buffer control?

pH

Why is it important that this property be kept relatively constant?

Important because buffers maintain a neutral pH in the blood

The following experiment demonstrates the buffering capacity of blood. Record your results in Table 23–1.

□ Obtain two 100-mL beakers. Label them #1 and #2.
□ Add 10 mL of 0.9% NaCl to beaker #1.
□ Add 1 mL of blood plasma and 9 mL of 0.9% NaCl to beaker #2.

AVOID THE RISK OF SERUM HEPATITIS: DO NOT PIPET PLASMA BY MOUTH, AND AVOID CONTAMINATION IF HUMAN PLASMA IS USED.

□ Test the pH of each solution by wetting a glass rod in it, and touching a drop to a piece of pH paper; if pH sticks are used, place one stick into each beaker.
□ If the pH values do not agree, adjust the pH of the solution in beaker #1 by adding 0.05 M HCl or 0.05 M NaOH one drop at a time. Stir the solution and test its pH after adding each drop.

If the pH of the solution in the beaker is too high, should you add HCl or NaOH? ______________

□ When both solutions have the same pH, record the value in the table as **initial pH.**
□ Use a pipet to restore the volume of the solution in beaker #1 to 10 mL. Discard the excess solution.
□ Add five drops of 0.1% methyl orange indicator to each beaker.
□ *While counting the drops,* add 0.1 M HCl to beaker #1, one drop at a time, stirring the solution with a glass rod, until there is a definite and *permanent* color change from orange to pink. (This change occurs at a certain pH.)
□ Measure the pH of the solution and record the value under **final pH.**
□ Repeat the procedure for beaker #2, counting the drops needed to reduce the pH to the same value as you did for beaker #1, so that *both solutions are the same color.*

Which solution was more resistant to a change in pH upon addition of acid? ______________

How does a buffer react chemically upon the addition of acid or base?

____

____

____

Formed Elements

The three major types of blood cells are the **red blood cells** (erythrocytes), the **white blood cells,** and the **platelets.** Some of these types are illustrated in Figure 23–1. To view some of your own blood cells, use the following procedure, illustrated in Figure 23–2.

□ Obtain two clean slides, alcohol, sterile cotton, and a sterile blood lancet in an unopened wrapper.

AVOID THE RISK OF SERUM HEPATITIS: NEVER USE A PREVIOUSLY USED LANCET. DO NOT TOUCH MATERIALS CONTAMINATED WITH ANOTHER PERSON'S BLOOD.

If you are unable or unwilling to have your finger pricked, work with a partner who doesn't mind, or view a prepared, stained slide.

□ Massage a finger on your left hand if you are right-handed, and ask your partner to wipe it with alcohol and prick it for you or do it yourself.
□ Drip a drop of blood onto one slide.
□ Hold the second slide almost vertical and move it along until it just touches the drop of blood. The drop will spread out along the edge of the second slide.
□ Keep the second slide at an angle, and move it back to smear out the blood into a thin layer. Discard the slide that you used for spreading.
□ Allow the smear to dry, and place the slide on a staining rack.
□ Add enough drops of Wright's stain to cover the surface of the slide, and let it stand for 6 min.
□ Add 10 drops of distilled water, and let it stand for 5 min.
□ Finally rinse the slide with distilled water from a squeeze bottle to remove the dye. Rinse it with more distilled water at the sink, and stand it on end to dry.
□ Examine the dry slide under high power with the compound microscope.

Most of the cells are the pink-staining red blood cells, which have the form of biconcave discs (donuts without the holes).

Do human red blood cells have nuclei?

No

Where are these cells formed?

in stem cells of bone marrow

Where are they stored? (bone marrow) spleen

What is the respiratory pigment contained within them?

Hemoglobin

Red blood cells have a limited life span (about 120 days) because they have no nuclei. Old red blood cells are phagocytized in the spleen, and their hemoglobin is degraded in the liver and leaves the circulation as the bile pigments. Much of the iron can be recycled.

Less than 1% of the cells are white blood cells. The larger ones are the **granulocytes** and have a grainy cytoplasm. Cells with light pink granules are the **neutrophils.** Those with large deep red or orange granules are the **acidophils.** Cells with closely packed blue-violet granules are **basophils** (very rarely seen).

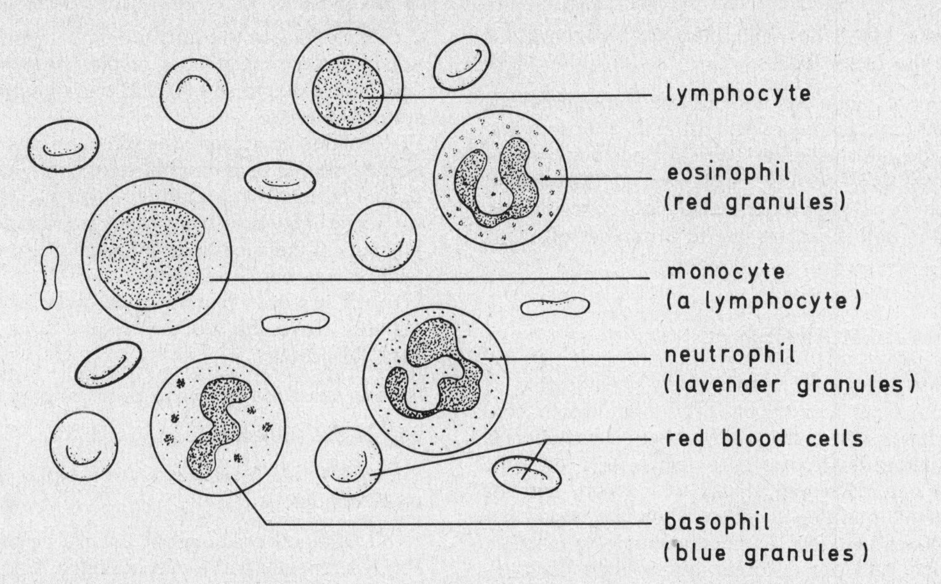

Figure 23–1. Human blood cells. White blood cells are very rare compared with red blood cells (1:500).

Some of the white blood cells are **agranulocytes** and have a clear cytoplasm. Small cells with only a thin ring of cytoplasm are the **lymphocytes.** If there is a larger amount of cytoplasm, the cells are **monocytes.** Both of these types are part of the immune system, and the monocytes are the cells that increase in number when a person has certain diseases, especially mononucleosis.

□ Locate 20 or 30 white blood cells, and note whether they are granulocytes (grainy cytoplasm) or agranulocytes (clear cytoplasm).
□ Record your results in Table 23–2 and calculate the percentage of each type.

Platelets are much smaller than white blood cells. They are very fragile, and were probably destroyed during the staining procedure.

Hemoglobin

The concentration of hemoglobin in red blood cells is very high. That the hemoglobin is packed into these cells is advantageous because otherwise the osmotic concentration of the blood would be extremely high, and the blood would be too viscous to move through the vessels of the circulatory system. As it is, the concentration of the respiratory pigment in blood can be greatly increased, and the efficiency of gas exchange is much higher than it would be if the pigment were merely free in solution.

HEMOGLOBIN CONTENT

Use the following procedure to measure the hemoglobin content of your blood.

□ Obtain two spectrophotometer tubes, and add 5 mL of distilled water to each one.
□ Prick your finger (or have your partner take a turn at being pricked), and draw up exactly 0.02 mL of blood in the capillary pipet provided for this purpose.
□ Blow the blood into the water in one of the tubes, and rinse three times with water from the tube.
□ Add 1 drop of concentrated ammonium hydroxide (NH_4OH) to the tube and to a second blank tube.
□ Wait 10 min.
□ Meanwhile, turn on a Spectronic 20 spectrophotometer to warm up, and set the wavelength at 545 nm.
□ Use the blank tube to adjust the **absorbance** to zero.
□ Determine the absorbance of the solution at 545 nm. Record the value in Table 23–3.
□ Calculate the approximate concentration of hemo-

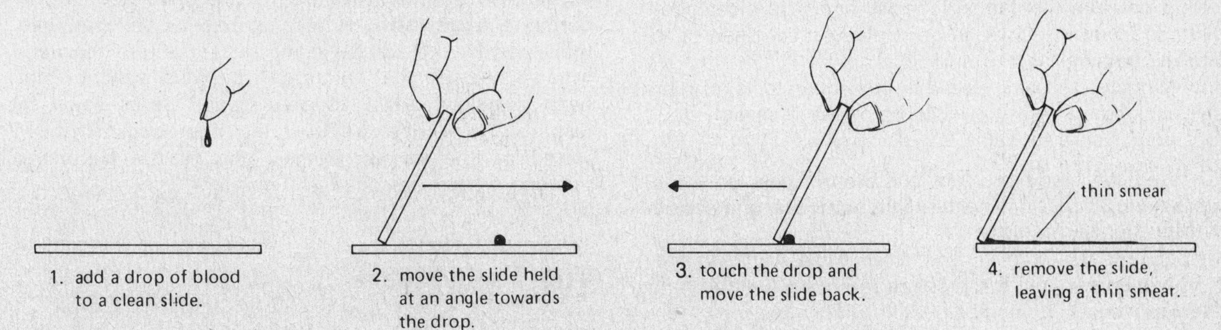

Figure 23–2. Blood smear preparation. After the smear has dried, stain the cells with Wright's stain.

globin in your blood by multiplying the absorbance at 545 nm by the factor 28.8.

The normal range for hemoglobin concentration is 16 ± 2 g/100 mL for males and 14 ± 2 g for females. Indicate in the table whether your hemoglobin concentration is in the low, normal, or high range. **Anemia** is the condition of having an abnormally low concentration of hemoglobin and it can be due to a number of different causes.

SICKLE-CELL HEMOGLOBIN

Persons with sickle-cell anemia have hemoglobin S rather than normal hemoglobin. This hemoglobin contains beta chains with a single amino acid substitution in which glutamic acid is replaced by valine. As a result the hemoglobin has a lowered affinity for oxygen, and the molecules tend to clump together when the concentration of oxygen is very low. During the clumping reaction, the red blood cells are distorted and assume "sickled" shapes.

If you did not study sickle-cell anemia in Topic 10, do so now. Read the section entitled SICKLE-CELL ANEMIA, and complete the exercise at the end of that section.

BLOOD PRESSURE

Blood pressure varies throughout the circulatory system, so it is usually measured in the upper arm for comparison of different individuals. The pressure is the pressure exerted on the walls of the blood vessels by the blood under the force of the contracting heart. It is greatest in the vessels adjacent to the heart, it drops off in the arterioles and capillaries due to the friction caused by the blood moving along the walls, and it is very low in the venules and veins leading back to the heart. Muscular contraction is required to squeeze the blood in the veins from the extremities to the heart because the pressure in these vessels is less than the force of gravity. **Valves** in the veins are extremely important in aiding the return of the blood to the heart, because its pressure is so low.

The blood pressure in the arteries follows a rhythmic cycle with each beat of the heart. When the ventricles contract, exerting the strongest pressure, the resulting high pressure phase is termed **systole.** While the heart relaxes the pressure in the arteries decreases in a low-pressure phase called **diastole.** Blood pressure is always measured at both systole and diastole, and the result is reported as the systolic value/diastolic value. The blood pressure reading is generally higher in older people and lower in people in good physical condition or in athletic training.

Working in pairs, use this procedure to take your partner's blood pressure while he or she is seated. Record your results in Table 23–4.

□ Wrap the blood pressure cuff around your partner's upper arm so that it is neither tight nor loose, and secure it with the Velcro fasteners.
□ Squeeze the bulb several times while watching the pressure gauge until the pressure reaches 180–200 mm Hg (760 mm is 1 atm).
□ Place the stethoscope in the hollow at the inner side of the elbow, and be ready to listen for the sound of the pulse.
□ Slowly release the pressure while watching the gauge, and make a mental note of the pressure at which you first hear a single sound with each heartbeat; this is the systolic pressure.
□ Continue to release the pressure until you hear the double sound of a normal heartbeat (lub-dup . . . lub-dup . . .); this is the diastolic pressure.
□ Record the pressure values for systole and diastole, and repeat the measurement twice more to be sure your value is reasonably accurate.
□ Next, ask your partner to lie down and rest for a few minutes. Take the blood pressure three times, and record the results.

Did the value for the blood pressure depend on the position of the subject? ______________

□ Repeat the measurement of blood pressure under other conditions.

The blood pressure will usually be higher if the person has recently had a cup of coffee, smoked a cigarette, or been subjected to stress. It is normally regulated by the activity of the autonomic nervous system. If the blood pressure is consistently too high, the condition of hypertension, a stroke or small hemorrhage in the brain, or kidney failure may result. Hypertension can be controlled with drugs or in some cases by conscious effort through biofeedback training.

CAPILLARIES

The active functioning of the blood as a transport tissue goes on only in the capillaries. Here the actual exchange of gases, nutrients, and wastes takes place, and white blood cells can enter and leave the circulatory system. The walls of the capillaries are thin enough so that the cells within them move along in single file.

If you did not see living capillaries in Topic 22 (section entitled CAPILLARIES), carry out the observation of capillaries described there.

HEART RATE

The beating pattern of the heart is not due to nervous stimulation, but is rather due to spontaneous electrical events starting with the pacemaker of the **sino-atrial (SA) node.** The rate of the heartbeat, however, is very responsive to changes in conditions within the body and to external stimuli. It speeds up when the demand for oxygen is high and also under the influence of the hormone **adrenalin.** Adrenalin may be released into the blood by the adrenal gland, and it is also the neurotransmitter of the autonomic nervous system. The heart rate slows down, especially when we lie down, in response to **acetylcholine,** the neurotransmitter released by the **vagus nerve** (part of the peripheral nervous system described in Topic 26).

The Frog Heart

The effects of different factors on the rate at which the frog heart beats will help you understand how the rate is normally adjusted to changing conditions.

Before starting this experiment, you and your partner should have access to all of the following:

Ice bath for the frog
Dissecting needle
Paper towels
Frog Ringer's solution
Scissors
Thread
Forceps
Pipets
5×10^{-5} M acetylcholine
5×10^{-5} M adrenalin
5×10^{-5} M atropine
Electrical stimulator (on *low voltage*)

PITHING THE FROG

□ Chill the frog in ice thoroughly. Remove and pat it dry, and carry out the following steps quickly.
□ Hold the frog firmly in your left hand with its front legs between your middle and ring fingers and the hind legs dangling.
□ Put your index finger over the head, and press down to bend the spinal cord.
□ Run your right index finger along the head until you locate the depression at the base of the skull where the spinal column begins. This is where you will insert the dissecting needle.
□ With a steady hand, insert the needle into the spinal cord, and sever it with a quick side-to-side movement. In this way you will be able to kill the frog as painlessly as possible.

If the frog shows signs of reviving, put it back into the ice bath and repeat the severing of the spinal cord.

□ Push the needle forward parallel with the external surface into the brain. Move it back and forth to sever all connections with the spinal cord so that the frog is neurologically dead.
□ Insert the needle into the spinal cord, and twist it to destroy the cord and the spinal reflexes. The legs will extend sharply as the needle contacts nerves within the cord.
□ When you remove the needle, the legs should be completely limp, and the pithing operation is finished.

EXPOSING THE HEART

□ Lay the frog on its back on a paper towel wet with Ringer's solution. The frog's skin must be kept moist at all times from this point on.
□ Use the forceps to pinch the skin over the abdomen, slightly to side of center, and snip a hole with the scissors.
□ Insert the point of the scissors carefully, being careful not to poke into the abdomen, and make an I-shaped incision by cutting in the order and direction shown in Figure 23–3.
□ Once you can see the underlying muscle, repeat the incision, cutting through the muscle as well. Fold back the flaps to expose the viscera.
□ Note that the beating heart is covered by the membrane of the **pericardium.**
□ Slit the pericardium *carefully* to expose the heart.

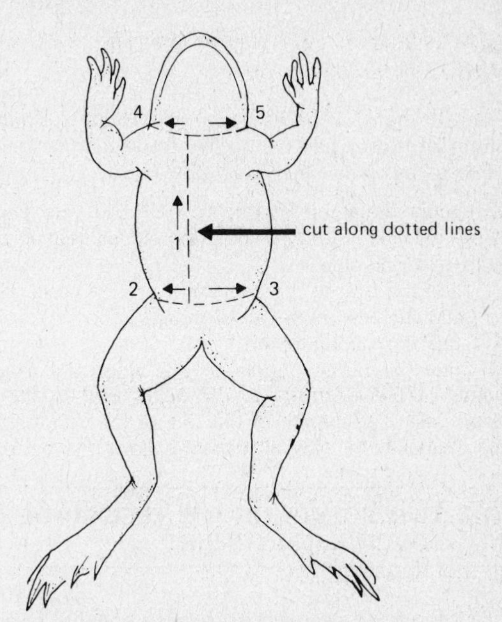

Figure 23–3. Frog dissection guide. Keep the frog moist inside and out at all times.

CHECK THAT THE FROG HAS WARMED UP TO ROOM TEMPERATURE BEFORE BEGINNING YOUR MEASUREMENTS.

□ Record the rate of beating for a 3-min interval, and calculate the beats per minute. Record the results in Table 23–5.

You are now ready to study the effect of the drugs and electrical stimulation on the heart.

ACETYLCHOLINE

This neurotransmitter is released by the vagus nerve, which is part of the parasympathetic nervous system. It should have the same effect as electrical stimulation of the vagus itself.

□ Add 3–4 drops of the acetylcholine solution to the fluid around the heart, not to the heart itself.
□ Observe and record the rate of beating for each 15-sec interval until there is a perceptible change.
□ Calculate the beats per minute by multiplying your final value by a factor of 4.
□ Rinse the heart with Ringer's two or three times, removing the fluid from around the heart with a pipet.
□ Give the heart a 5-min rest, and record a normal heart rate before going on.

ADRENALIN

This hormone is released into the blood by the adrenal glands and by the sympathetic nervous system. It prepares the entire body for an emergency response.

□ Add 2 drops of the adrenalin solution to the fluid around the heart, and observe the effect as you did for acetylcholine.
□ Record the results and allow the rinsed heart to rest.

ELECTRICAL STIMULATION OF THE VAGUS NERVE

Beneath the heart on its dorsal side, locate the bundle of stringlike nerves and connective tissue that join the sinus venosus as shown in Figure 23–4.

□ Gently draw the bundle to one side with your forceps, and tie a thread to it loosely so that it can be pulled to one side.
□ Take the frog to the stimulator or vice versa. Be sure to keep the frog moist at all times.
□ Turn the stimulator on.
□ Draw the nerve bundle to one side, and touch the points of the electrode to the nerve (*not* to the heart) until there is a change in the rate of the heartbeat. You may have to try several locations along the bundle.

THE TWO POINTS OF THE ELECTRODE MUST NOT TOUCH EACH OTHER.

□ Observe the new rate of beating and record it.

Did electrical stimulation of the vagus have an effect similar to that of acetylcholine? ______________

□ Allow the rinsed heart to rest until the beat is again close to the normal rate.

ATROPINE

This drug is an alkaloid poison that binds to and blocks the receptors for acetylcholine. Do not use this drug until you have completed your study of the effects of the neurotransmitters and electrical stimulation.

□ Add 3–4 drops of atropine to the fluid around the heart, and record the new beating rate in Table 23–5.
□ Apply electrical stimulation as you did before, and record the rate during electrical stimulation in the presence of atropine.

Did atropine change the effect of electrical stimulation?

____

How would atropine affect the functioning of the parasympathetic nervous system?

____

____

____

____

When you have finished with the frog, dispose of it properly, and clean your dissecting tools for the next student.

Human Heart Rate

□ Place the fingertips of your left hand in the hollow of your wrist at the base of your right thumb.
□ Count your heartbeats for 3 min, calculate the rate in beats per minute, and record the results in Table 23–6.
□ Jump up and down until you feel out of breath or tired, and again record your heart rate.
□ After your heart rate is back to normal, lie down for 5 min, and measure the rate while resting. Try to relax as fully as possible.

Exercise is supposed to benefit your circulatory system because the capillaries in the muscles, especially those of the heart, dilate in order to increase the supply of oxygen to the active tissues. After the exercise is over, the capillaries remain a little bit more open than before so that oxygenation of the tissues is improved all the time. An athlete in top condition will often have a very low heart rate of 50 beats/min or less. In order for the exercise to help significantly, however, the heart rate must reach 150 beats/min for at least a few minutes during each exercise session. Activities such as golf and gardening are just that: they do not increase the heart rate enough to be rated as exercise.

□ During the coming week, test your more strenuous activities to see whether they raise your heart rate to 150 beats/min for a significant period of time.

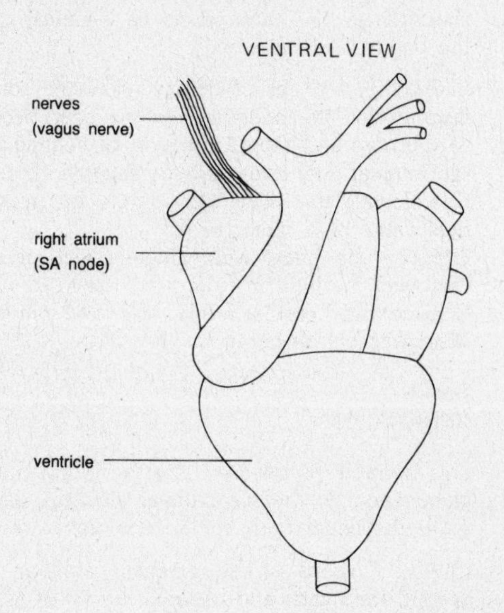

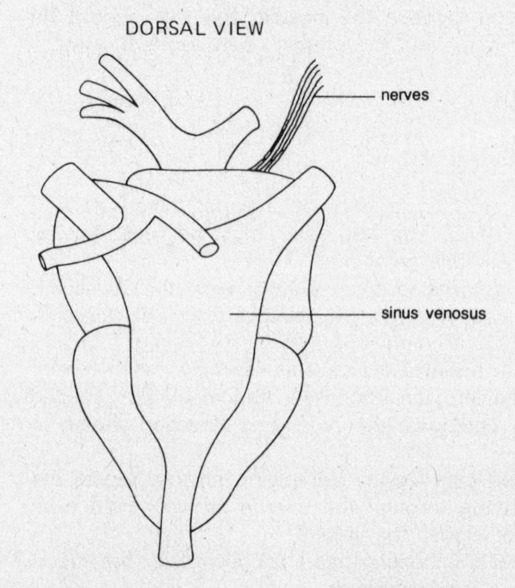

Figure 23–4. Frog heart. The vagus nerve is important in regulating the rate at which the heart beats.

EXPLORING FURTHER

Guyton, A. *Textbook of Medical Physiology*. 6th ed. Philadelphia: W. B. Saunders, 1981.
MacFarlane, R. G., and A. Robb-Smith. *Functions of the Blood*. New York: Academic Press, 1961.
Wessel, Morris A., and A. Dominski. "Our children's daily lead." *American Scientist* 65: 294, 1977.

Scientific American Articles

Allison, A. C. "Sickle cells and evolution." August 1956 (#1065).
Adolph, E. F. "The heart's pacemaker." March 1967 (#1067).
Axelrod, J. "Neurotransmitters." June 1974 (#1297).
Benditt, E. P. "The origin of atherosclerosis." March 1967 (#1067).
Cerami, A., and C. M. Peterson. "Cyanate and sickle cell disease." April 1975 (#1319).
Chapman, C. B., and J. H. Mitchell. "The physiology of exercise." May 1965 (#1011).
Cooper, M. D., and A. R. Lawton III. "The development of the immune system." November 1974 (#1306).
Friedman, Milton J., and William Trager. "The biochemistry of resistance to malaria." March 1981.
Hock, Raymond J. "The physiology of high altitude." February 1970 (#1168).
Mayerson, H. S. "The lymphatic system." June 1963 (#158).
Perutz, M. F. "Hemoglobin structure and respiratory transport." December 1978 (#1413).
Rose, Noel R. "Autoimmune diseases." February 1981.
Schmidt-Nielsen, Knut. "Countercurrent systems in animals." May 1981.
Zucker, Marjorie B. "The functioning of blood platelets." June 1980 (#1472).
Zweifach, B. W. "The microcirculation of the blood." January 1959 (#64).

Student Name ______________________ **Date** ______________

TABLE 23–1. PLASMA BUFFERING

Solution	Initial pH	Final pH	Drops HCl Added
Beaker #1 (control)			
Beaker #2 (plasma)			

TABLE 23–2. WHITE BLOOD CELLS

	Number	Percentage of Total
Total White Blood Cells		100%
Granulocytes		
Agranulocytes		

TABLE 23–3. DETERMINATION OF HEMOGLOBIN CONCENTRATION

Absorbance at 545 nm	.73
Hemoglobin concentration (Absorbance × 28.8)	g/100 mL
Sex (Male or Female)	21.0 16.7
Range (Low, Normal or High)	High

2/18/86
.58
.58 × 28.8
17.0
Normal

TABLE 23-4. BLOOD PRESSURE

Condition	Reading	Blood pressure Systolic/diastolic
Initial pressure	1	/
	2	/
	3	/
Resting pressure	1	/
	2	/
	3	/
Other condition: ______________________	1	/
	2	/
	3	/
______________________	1	/
	2	/
	3	/

Student Name ______________________ **Date** ______________

TABLE 23–5. FROG HEART RATE

Condition	Beats/time interval		Beats/min
Initial rate			
Acetylcholine			
After rest			
Adrenalin			
After rest			
Vagus nerve			
After rest			
Atropine			
Atropine + Electrical Stimulation			

TABLE 23–6. HUMAN HEART RATE

Condition	Beats/time interval		Beats/min
Initial normal			
Jumping			
Normal			
Lying down			
Other activities: ______________			
______________			
______________			
______________			

TOPIC 24

Excretion and Reproduction

OBJECTIVES

When you have completed this topic, you should be able to:

1. Define the process of excretion in mammals; name the major nitrogenous waste product and the chemical molecules which give rise to it.
2. Trace the path of urine from the kidney to the exterior.
3. Describe the nephron, and relate its structure to the internal structure of the kidney.
4. Name and contrast the position and functions of primary sex organs in the male and female.
5. Trace the path an egg follows from its formation to the exterior of the body, assuming no fertilization.
6. Trace the path of sperm cells from their site of origin to the exterior, and give the function of each accessory organ along the way.
7. If copulation and fertilization occur in the pig, state where the semen is deposited, where fertilization occurs, and where development of the young piglets takes place.
8. Describe how fertilization and early development in the human differ from those processes in the pig.
9. For each of the major types of contraception, explain how the method works, give the exact step in conception that is blocked, and specify the anatomical site where the method has its effect.

TEXT REFERENCES

Chapter 35 (A,B,F,G); Chapter 36 (A–F).
Nitrogenous wastes, male urinary system, female urinary system, kidney function, nephron, male reproductive system, female reproductive system, menstrual cycle, gamete formation, fertilization, birth control.

INTRODUCTION

Wastes from within cells are removed from the body in the process of **excretion.** The wastes are chemicals that are formed by the metabolic activity of cells and cannot be gotten rid of except as dissolved substances. In mammals, the main nitrogenous waste is **urea,** a simple molecule formed from CO_2 and ammonia. (See Figure 35–2 in your text.) Nitrogenous wastes are formed whenever proteins or amino acids in the diet or in cells are broken down to carbon skeletons and ammonia in the process of **deamination.** Urea is soluble and relatively nontoxic, so it can be eliminated in urine in a fairly concentrated form.

Although the kidney is the main excretory organ, some nitrogenous wastes are also excreted in the sweat and from the lungs.

Food remains that are not and have never been inside of cells are eliminated from the body through the rectum as feces. This process is **egestion** and is basically different from excretion. Excretory wastes from cells of the liver and large intestine do enter the gut, but they make only a small contribution to the fecal wastes.

The reproductive system is anatomically associated with the excretory system and, in fact, complete separation of the two is rare among vertebrates. In female and male fetal pigs both systems have a common opening to the exterior, so you will study them both at the same time. You will be responsible for understanding both the male and female systems, so exchange your pig for one of the opposite sex when you are finished to complete your study of the reproductive system.

As you study the anatomy of reproduction, you will also be learning how it is possible to prevent conception, or use **contraception,** in several different ways. Permanent contraception, usually called sterilization, prevents the egg or sperm cells from ever reaching their intended destination. Other methods are temporary and can be started and stopped again whenever desired. The common methods are mechanical, chemical, or hormonal. In many vertebrates other than humans, contraception of some sort is an important part of reproduction. These animals have behavioral or physiological mechanisms for limiting reproduction, whereas human beings must spend millions of dollars to achieve the same end through contraception.

EXCRETORY SYSTEM

□ Push aside or remove the intestines of your pig and locate the large paired **kidneys** (see Figure 24–2 or 24–3). These are reddish organs located deep within the abdominal cavity, underneath the peritoneum on either side of the aorta. They are not enclosed within the peritoneum.

□ Before you clear away the membrane from one of the kidneys, locate the **adrenal gland** pressed against the inner side of its anterior part. The gland is a narrow strip of light brown or yellow tissue and is easily destroyed. The **adrenal medulla** (inner part) is a neuroendocrine organ that secretes adrenalin into the bloodstream, whereas the **adrenal cortex,** or outer part, secretes a variety of hormones.

□ Now, expose the kidney and study its external structure. Identify the following parts:

Renal hilum. Where the blood vessels and **ureter** are attached.

Renal artery. Carries blood from the aorta to the kidney.

Renal vein. Carries blood from the kidney to the posterior vena cava.

Ureter. Hollow tube which carries urine from the kidney to the **urinary bladder.** The ureter is also covered by the peritoneum.

□ Slice through the kidney from right to left with a razor blade, and study the internal structure. (See Figure 24–1.)

Within the kidney are about one million **nephrons.** The capsules of the nephrons, which receive fluid from blood within the capillaries, are located in the outer part of the kidney, the **cortex.** For each nephron a tubule then passes down into the inner region, or **medulla,** back to the cortex, and finally again into the medulla, where the tubules now join larger **collecting tubules.** The collecting tubules can be seen as rays radiating from the central part of the kidney. They empty into the hollow space, the pelvis, which leads into the beginning of the ureter. The structure of a nephron is shown in Figure 35–13 in your text.

In humans, kidney stones may form within the pelvis of the kidney. During a kidney stone attack, a small stone may become stuck in and block the ureter. Extreme pain results as urine pressure builds up behind the stone. In extreme cases, the stone may completely fill the pelvis of the kidney and take the form of a "staghorn." Draw a "staghorn" stone within the kidney in Figure 24–1.

□ View the section of kidney cortex on demonstration in the microscope. The larger dark structures surrounded by membranes are the **glomeruli** surrounded by the capsules of the tubules. The round hollow tubes are the **proximal tubules,** and the irregularly shaped tubes are the **distal tubules.**

□ Follow the ureter posteriorly to where it joins the **urinary bladder,** which lies between the umbilical arteries. The structure of the urinary tract is shown in Figures 24–2 and 24–3. Any urine formed while the pig is in the uterus can pass out of the bladder via the **allantoic duct,** which forms part of the umbilical cord. After birth urine leaves the bladder by the **urethra.** Locate the urethra, and clear away enough tissue so that you can see how it disappears deep within the girdle of pelvic bones. You will be able to follow it all the way to the

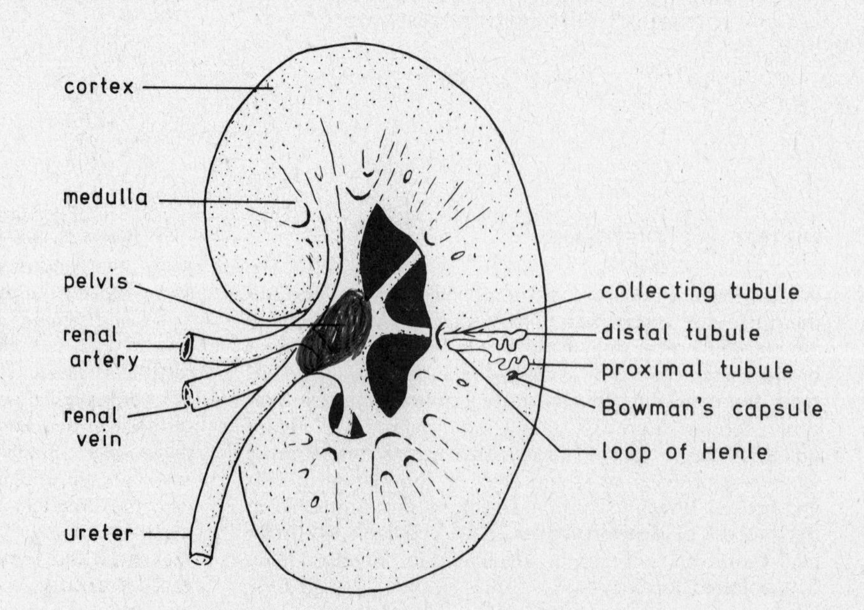

Figure 24–1. Internal structure of a kidney. The detail shows the position of the parts of a single nephron. Each kidney will have about one million nephrons. Urine flows out of the collecting tubule into the hollow pelvis of the kidney and then exits through the ureter to the bladder. Capillaries permeate the kidney and are closely associated with each individual nephron.

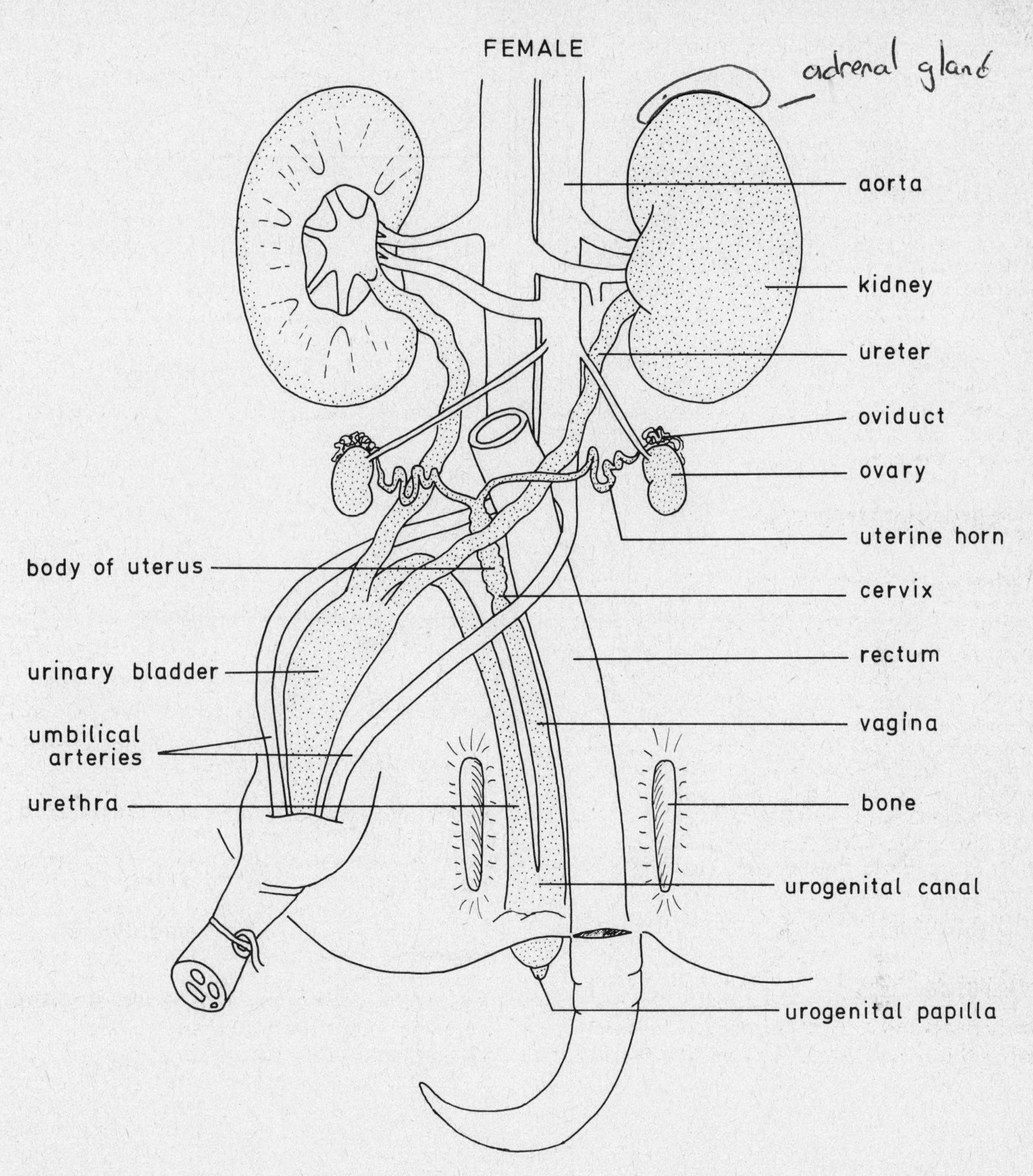

Figure 24–2. Female urogenital system. Dissection of a female fetal pig to show the structures of the urogenital system.

outside when you have finished studying the reproductive system.

FEMALE REPRODUCTIVE SYSTEM

The gonads, or primary sex organs, of the female are the **ovaries.**

□ Locate these light-colored, small spheres, which are posterior to the kidneys and are held in place by mesenteries.

Ovaries have a double function: they contain all the **egg cells** that the female will ever produce, and they are the endocrine glands that release **female sex hormones** into the bloodstream. Closely attached to the top of the ovary is a *tiny,* twisted white tube.

□ Locate this tube, which is the **oviduct,** or **Fallopian tube.**

An egg released into the abdomen by the ovary is swept into the open end of the oviduct (ostium) and on down the duct to the uterus by ciliary action. If sperm are present, the egg will be fertilized along the way while it is traveling through the oviduct. The pig's **uterus** is quite different from that of the human. It consists of two long **horns,** in which the developing eggs implant, and a small **body.**

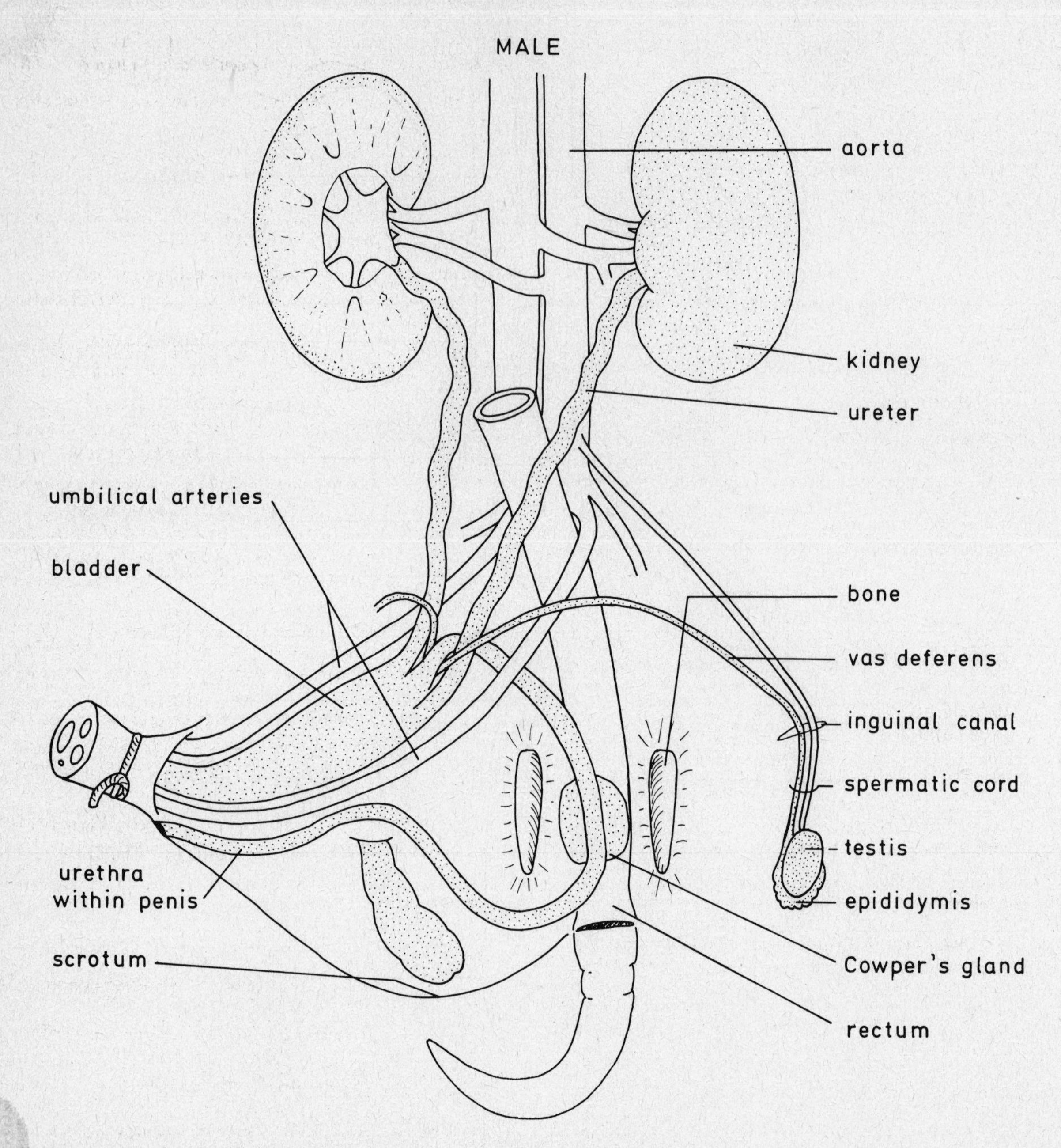

Figure 24–3. Male urogenital system. Dissection of a male fetal pig to show the structures of the urogenital system.

□ Locate the long horns, which are larger than the oviducts, and begin close to the ovaries. Locate the body of the uterus where the two horns join. The body ends at the **cervix.**

The uterus leads into the **vagina,** or birth canal, and the beginning of the vagina can be seen as a constriction soon after the uterine horns join.

□ Follow the vagina posteriorly until it disappears into the pelvic girdle.

At this point you should be able to show the rectum, urethra, and vagina all entering the pelvic girdle. It is now necessary to cut through the muscle and the bone of the pelvic girdle to see what happens next.

□ Carefully make a cut with your razor blade along the midline of the pig.
□ Cut almost all the way back to the urogenital papilla.
□ Deepen the cut to the bone, and then cut through the bone, trying not to cut the structures located underneath it.
□ Spread the legs apart, and force the bones of the girdle open to expose the structures within it. Follow the vagina and urethra posteriorly, and see where they join to form the **urogenital canal.**

This canal leads to the urogenital opening of the female pig. As the pig develops, the vagina and urethra become more and more separated so that finally there will be two openings: a ventral one for the urethra and a more dorsal one for the vagina. The rectal opening is most dorsal, just beneath the tail.

On each side of the urogenital opening are the rounded lips, or **labia.** They would have developed into the scrotum if the pig had been male.

□ Slit the side of the urogenital canal, and look on the ventral surface of the canal for the **clitoris,** the female organ that is the counterpart of the male penis.

MALE REPRODUCTIVE SYSTEM

Before starting to dissect, study Figure 24–3 to familiarize yourself with the overall anatomy. Notice that the pathways for urine and sperm join in a common urethra, which passes down through the pelvic girdle, then ventrally and anteriorly along the abdominal wall until it finally exits through the urogenital opening just below the umbilical cord. You will need to dissect quite a bit to see the entire pathway.

□ Raise the flap of abdominal wall where the umbilical cord enters, and relocate the **bladder** between the umbilical arteries.
□ Use your probe to separate the **ureters,** which drain from the kidney into the bladder.
□ If it helps, cut through the middle of the bladder and umbilical arteries, and separate them from the abdominal wall so that you can locate the beginning of the **urethra,** which leads out of the bladder. This area is called the neck of the bladder.
□ Notice the two small tubes that are looped over the ureters and which join the urethra at the bladder neck. They are the **vasa deferentia,** or "tubes," of the male, and carry sperm from the testes to the urethra.
□ To find the **testes,** or male gonads, investigate the scrotum of the pig to see whether they have already moved down from the abdomen, or descended. If not, as is usually the case, they can be found in the groin area, on the inner side of the upper part of the hind leg.
□ Cut through the skin and muscle to locate each testis, which will be enclosed in a tough, white, opaque membrane. Be careful not to cut between the testis and the vas deferens that leads from it.
□ Cut open the outer membrane so that you can see the reddish testis itself.

Like the ovary, the testis has two functions: it will produce the **sperm cells** of the male that enter the vas deferens, and it produces **male sex hormones** that determine the sex of the pig and allow it to function sexually. Covering part of the testis is a whitish mass of tightly coiled tubules, the **epididymis.** Sperm cells produced in the testis pass into the coils of the epididymis where they are stored before entering the vas deferens.

□ Dissect as much as you need in this area to show the path of the vas deferens leading from the testis to the urethra.

Each vas deferens passes from outside the abdomen through the abdominal wall to the inside through the **inguinal canal.** In humans this canal is a weak spot and may rupture or split open, forming a hernia. Testes, like kidneys, are outside the peritoneum. When the testis is within the scrotum, it is supported by a ligament and has its own blood vessels leading to and from the abdomen: the vessels, ligament, and vas deferens form the **spermatic cord** from which the testis is suspended.

□ Follow the vas deferens to the bladder, and observe where it enters the urethra.

The situation here is very complicated and is diagrammed in Figure 24–4.

□ Pull the bladder toward you, and try to locate the following structures on the *dorsal* side:

Don't need to know

Seminal vesicles. Small, paired, light-colored glands that branch off from the urethra just posterior to the vasa deferentia.

Prostate. Very small, light-colored, smooth, round gland at the junction of the seminal vesicles. It is sometimes too small to be seen.

As sperm cells pass along the urethra, fluids from these glands are released to help make up the **semen.**

□ Push the flap of abdominal wall containing the umbilical cord to one side: you will need it momentarily.
□ To expose the urethra as it leaves the abdomen, cut through the skin, muscle, and bone of the pelvic girdle slightly to one side of the midline. Try not to cut beneath the bone.
□ Locate the urethra just beneath the bone: it is now quite large and well developed, and is the combined pathway for both the excretory and reproductive systems.
□ Pry apart the bones of the pelvic girdle, and locate the bean-shaped, white **Cowper's glands** (Figure 24–3) on either side of the urethra. They also secrete fluids into the semen.

The urethra leaves the abdomen, and enters the long, muscular **penis** within the abdominal wall.

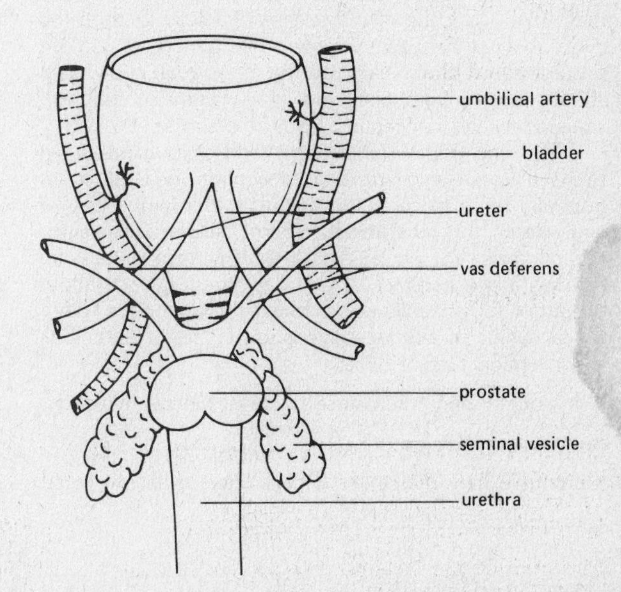

Figure 24–4. Male reproductive system. Dorsal view showing the structures in the region of the neck of the bladder. If you dissect carefully, you should be able to see all of these structures except the prostate, which may be too small to see. (Redrawn from Odlaug, Theron O., *Laboratory Anatomy of the Fetal Pig*, 6th Ed. © 1969, 1975, 1980 Wm. C. Brown Company Publishers, Dubuque, Iowa. Reprinted by permission.)

□ Feel that free strip of abdominal wall to find the penis and dissect it free with your probe. You can now trace the excretory system from the kidney and the reproductive system from the testis through the common urethra to the external urogenital opening.

□ Finally, follow the digestive tract as it exits through the pelvic girdle. The **rectum** lies just dorsal to the urethra and ends in the external opening, the **anus.**

What color is the rectum? ______________

Varicose veins within the human rectum are called hemorrhoids.

When you are finished, seal the pig and any extra tissue into the bag, and dispose of it properly. Clean your pan and tools, and oil your scissors for the next student.

HUMAN REPRODUCTION: CONCEPTION

Each human female has about 300,000 eggs that are set aside very early in development, about 3 weeks after conception. Some of these potential eggs mature, one at a time, once the woman is sexually mature. Thus any damage to the eggs or ovaries of a woman, such as from X-rays or toxic chemicals, is permanent. No new eggs can ever be formed. (In the testes, on the other hand, sperm are formed continuously and damaged sperm are quickly replaced.) The physiological conditions that cause eggs to be released and make the reproductive tract receptive to fertilization and implantation of the resulting embryo are under hormonal control. Regular changes in the levels of these hormones cause the events of the menstrual cycle: the egg is released halfway through the cycle, midway between two menstrual periods. If you are unsure about the details of the menstrual cycle, you should turn to your text, and review them so that you can understand conception. (See text section 36–C.)

The egg cell is rather short-lived once it has been released from the ovary, and conception, or fertilization, normally takes place in the oviduct within half a day of egg release, or **ovulation.** Sperm cells are viable for several days, so regular sexual activity means that the egg has a good chance of being fertilized. In fact, about 80 out of 100 sexually active couples will conceive within a year when no contraceptive is used. This is defined as a conception rate of 80%.

If a couple had intercourse 2 days before ovulation, could conception occur? Yes

If a couple had intercourse 2 days after ovulation, could conception occur? Yes

HUMAN REPRODUCTION: CONTRACEPTION

If natural events were allowed to take their course, most women would be pregnant most of the time. In our society, personal freedom, good medical care, improved general health and nutrition, and widespread preference for bottle rather than breast feeding all help contribute to this potentially high pregnancy rate. Of the many ways that pregnancy can be prevented, the only one that is 100% sure is sexual abstinence. Although this method is used to a degree by many societies, it has almost no role in our own. All the other methods of birth control or contraception, including sterilization, do have some finite risk of pregnancy. This risk is given as the chance of pregnancy in a group of couples using the method to prevent pregnancy during the first year of use.* On the average, the couples had a pregnancy rate of 3.7%; they would have expected a rate of about 80% if they had not used any contraception.

Some common methods of contraception are shown in Figure 24–5.

Mechanical Methods

CONDOM

□ Look at the various types of condoms on demonstration.

The rubber sheath is rolled over the penis before intercourse. It prevents sperm from entering the vagina. Thinner condoms are supposed to be more "natural," but are also more likely to break, releasing sperm into the vagina. Lubricants such as Vaseline cause the rubber of the condom to deteriorate and weaken. Condoms are available without prescription and are the only method that gives protection against venereal disease. This method is coitus-related, and requires some practice, care, and self-control when the condom is put on and removed.

Conception rate is 7%. If the condom is used carelessly, the rate will be much higher than if the condom is used carefully with foam.

DIAPHRAGM

□ Look at the one size of diaphragm on demonstration.

This rubber shield is used with spermicidal jelly and inserted into the vagina before intercourse. To be effective, it must fit properly over the cervix (tip of the uterus), so each woman must be individually fitted for her diaphragm by a physician. Diaphragms are available by prescription only and must be used with a spermicidal agent. They also have to be used at the time of intercourse and require practice in learning to insert them properly. Conception rate is 10%.

IUD (INTRAUTERINE DEVICE)

□ Look at the various types on display.

IUDs come in all shapes and sizes. The device is inserted into the uterus through the cervix, and so long as it is in place, it will block pregnancy by preventing implantation of any eggs that happen to be fertilized. An IUD must be

*Table 2 in B. Vaughan, J. Trussell, J. Menken, and E. F. Jones, "Contraceptive failure among married women in the United States, 1970–73," *Family Planning Perspectives* 9:6, 1977.

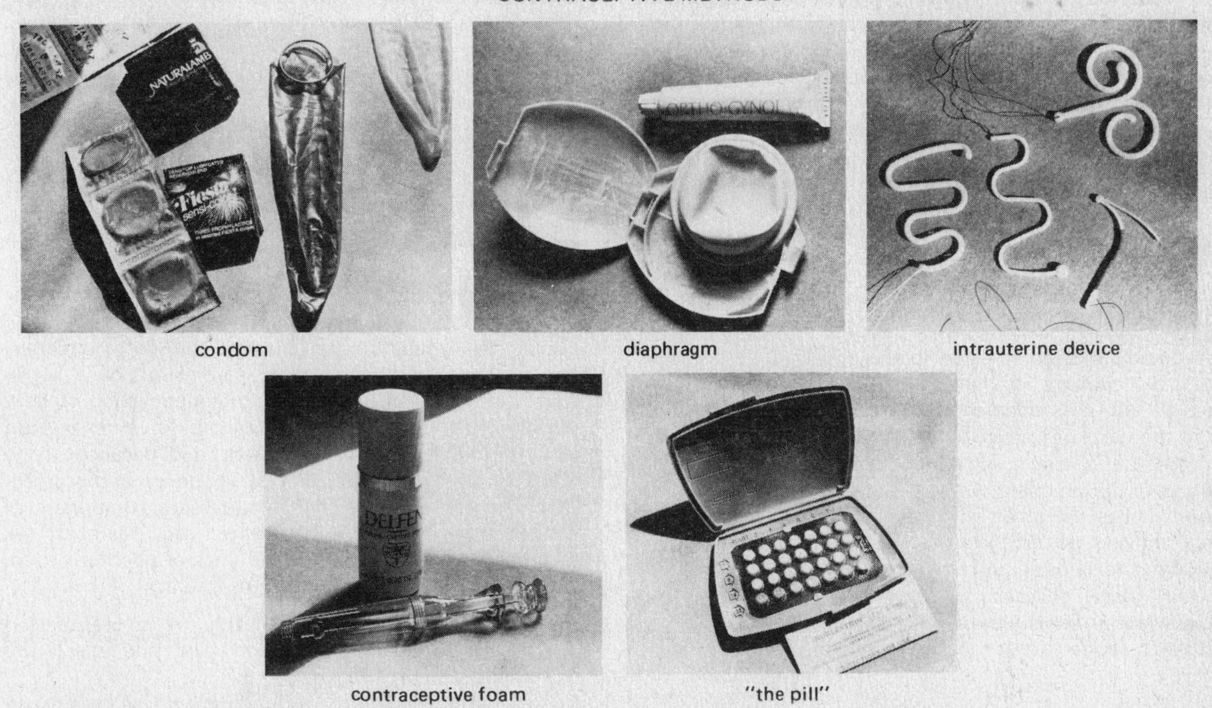

Figure 24–5. Contraceptive methods. These methods are widely used in the United States. (Courtesy of Planned Parenthood Federation of America, Inc.)

inserted by a physician or nurse-clinician, and it can be removed any time the woman desires. It has the great advantage that its insertion is not related to intercourse, but many women experience undesirable side effects such as cramps, excessive bleeding, and an increased susceptibility to infection. The woman must regularly check the IUD to make sure that it has not been expelled. Some IUDs contain copper or progesterone to make them even more effective in preventing implantation. The mechanism by which copper inhibits implantation is not known. Conception rate is 3% if the device is not expelled accidentally.

Chemical Methods

FOAMS, CREAMS, AND JELLIES

Foams from pressurized cans are put into the vagina with an applicator something like a tampon. Creams are applied in the same way. Foams, as well as creams and jellies, all contain chemicals that effectively kill sperm cells without irritating the vagina. They are available without prescription and must be used each time before intercourse. Some products can be used without a diaphragm, but the effectiveness of all such products can be increased if a diaphragm or condom is used in addition. Jellies are usually designed only for use with a diaphragm. Conception rate is 13%.

FEMININE HYGIENE PRODUCTS (INCLUDING DOUCHES) ARE NOT EFFECTIVE CONTRACEPTIVES.

Hormones

"THE PILL"

The most commonly used birth control pill is a combination of synthetic estrogen and progesterone. It works by blocking the release of anterior pituitary hormones (FSH and LH) that are needed for the egg to mature, be released, and implant properly. (See text Figure 36–7.) The pills are taken for 3 weeks to prevent ovulation, and then stopped for 1 week so that a menstrual period occurs. The pill is popular because it is cheap, effective, not used at the time of intercourse, and often makes the menstrual periods less bothersome. On the other hand, some women experience annoying side effects, it must be taken every day, and older women will have an increased chance of blood clots (thromboembolism). The pill may also increase the risk of minor vaginal infections, and it does not prevent venereal disease. Conception rate is less than 2%, and probably 0% if a woman never forgets to take the pills exactly as prescribed.

IMPLANTS

If there is a problem in taking the pill every day, the same effect can be achieved by implanting a capsule under the skin of the arm or thigh. The capsule will slowly release hormones into the body and thereby prevent pregnancy. This method is reversible, but the implant must be inserted and removed by a physician. Implants are less widely used than the other methods.

BREAST FEEDING IS NOT AN EFFECTIVE BIRTH CONTROL MEASURE.

After birth it often takes the woman's hormones and menstrual cycle some time to get back to normal, and ovulation may be inhibited for a few months. This process is so variable, however, that it cannot be relied upon to prevent pregnancy. Some women ovulate very soon after giving birth, and many mothers have become pregnant while they are still nursing.

Surgical Methods

FEMALE STERILIZATION

A surgical incision is made in the abdomen (or vagina) and the oviducts, or "tubes," are cut and tied off. As a result, any eggs released will just stay in the far end of the tube and will never be able to reach the uterus. This procedure is easy to do after childbirth but can also be done in nonpregnant women through a very small incision. It has the great advantage of being very effective and having no side effects. On the other hand, it is relatively expensive and is limited to women who do not want (more) children. It can only rarely be reversed. Conception rate is almost 0%. Once in a great while the tubes manage to grow back together.

MALE STERILIZATION

The vasa deferentia, or "tubes," are cut and tied off through a small incision in the scrotum. This procedure can usually be done in the doctor's office and is much cheaper and safer than female sterilization because the testes are outside the abdominal cavity. The other advantages and disadvantages of the two procedures are the same. After sterilization, the man's semen eventually will contain no sperm cells (the stored sperm must first be used up), but otherwise the semen will be the same as before the operation. Sexual functioning is just the same except that the man will not be able to make the woman pregnant. In some cases, the tubes have been surgically rejoined so that the man is again fertile, but this method is at present limited to men who do not plan to have children in the future. Conception rate is almost 0%. The male tubes have an even smaller chance of rejoining than do the female tubes.

Biological Methods

RHYTHM

The couple using this method tries to predict when the woman will ovulate, and avoids having intercourse around that time. Ovulation is supposed to occur just about 14 days before the next menstrual period begins, but the time between the end of the preceding menstrual period and the time of ovulation tends to be rather variable. Hints about the timing of ovulation are sometimes obtained by noting an increase in body temperature and changes in stickiness of cervical mucus. Even under the best of conditions sickness might raise the body temperature, and all sorts of physical and psychological changes will affect the timing of ovulation, so that it will never be possible to predict the correct time with absolute certainty. This method cannot be used by the many women with irregular or somewhat variable cycles, and it always involves periods of sexual abstinence. Conception rate is 10% under the most favorable conditions. Usually the rate is higher than with any other method. Reported success depends strongly on the physiology and motivation of the women involved.

Pregnancy Termination (Abortion)

Many pregnancies, perhaps 30–60%, end in spontaneous abortions (miscarriages) soon after conception. Usually this results from a defective embryo or imperfect implantation. Induced abortion, carried out by a physician, is not a satisfactory means of contraception by itself because of the strain being pregnant for even a short time puts on the woman's system, and because of the finite risk of infection or other consequence of the procedure. Abortion is, however, a necessary component of contraception because every method used has some risk of pregnancy. There are several types of abortions used, depending on the stage of the pregnancy:

1. A shot of estrogen within 24 hours of unprotected sex will prevent implantation. The side effects are unpleasant.
2. Vacuum aspiration, an outpatient procedure in which the cervix is dilated and the embryo removed under suction, can be used for up to 12 weeks.
3. Dilation and curettage (D and C) is used during the first 12 weeks and sometimes later. The cervix is dilated, and the contents of the uterus are scraped out.
4. A saline solution can be injected into the amniotic sac, and as a result, the uterus expels the embryo in a few hours. This more complicated procedure can be used through up to 6 months of pregnancy.
5. Under unusual circumstances, a pregnancy can be ended even later by an operation (hysterotomy) in which the abdomen is surgically opened and the fetus removed. This is a major operation, but the uterus is not removed as it is in a hysterectomy.

Risk of Complication

Many people do not realize that for women under 40 the overall risk to health of any form of contraception is much smaller than the risk involved if the woman used no contraception and carried each pregnancy that occurred to term. Unprotected women under 25 have a risk of pregnancy-related death of 5 in 100,000 per year, whereas any combination of birth control methods, including IUD, pills, abortion, and so on, has a risk of 1–2 in 100,000 per year. Once a woman has become pregnant, the risk of serious illness or death is much smaller for having an abortion compared with carrying out the pregnancy.

Review what you have learned about human reproduction by tracing the path of an egg and a sperm cell from formation to the site of fertilization.

□ Locate each of the following sites:

Where egg and sperm cells are formed.
Where sperm are deposited.
Where fertilization normally occurs.

Where each of the following acts to interfere with fertilization:

Male sterilization
Female sterilization
Condom
Foam
Diaphragm
IUD

EXPLORING FURTHER

Boolootian, R. A. *Human Reproduction.* New York: John Wiley and Sons, 1971.

Boston Women's Health Book Collective. *Our Bodies, Ourselves.* Rev. 2nd ed. New York: Simon and Schuster, 1976.

Demarest, Robert, and John Sciarra. *Conception, Birth and Contraception.* 2nd ed. New York: McGraw-Hill, 1976.

Hatcher, R. A. et al. *Contraceptive Technology 1980–81.* 10th rev. ed. New York: Irvington Publishers, 1980.

Ganong, W. F. "Renal function." *Review of Medical Physiology.* 8th ed. Los Altos, California: Lange Medical Publishers, 1977.

Peel, J., and D. M. Potts. *Textbook of Contraceptive Practice.* Cambridge: Cambridge University Press, 1969.

Turner, C. D., and J. T. Bagnara. *General Endocrinology.* 6th ed. Philadelphia: W. B. Saunders, 1976.

Vaughan, B., J. Trussell, J. Menken and E. F. Jones. "Contraceptive failure among married women in the United States 1970–73." *Family Planning Perspectives* 9: 6, 1977; *Contraceptive Efficacy among Married Women in the United States.* National Center for Health Statistics (ser. 23, no. 5). Ed. by Klaudia Cox, 1979.

VD Handbook and Birth Control Handbook. Montreal: Montreal Health Press, P.O. Box 1000, Station G.

Scientific American Articles

Beaconsfield, Peter, George Birdwood and Rebecca Beaconsfield. "The placenta." August 1980 (#1478).

Fuchs, Fritz. "Genetic amniocentesis." June 1980 (#1471).

Grobstein, Clifford. "External human fertilizaion." June 1979 (#1429).

Smith, Homer W. "The kidney." January 1953 (#37).

Tietze, C., and S. Lewit. "Legal abortion." January 1977 (#1348).

TOPIC 25

Animal Development

OBJECTIVES

When you have completed this topic, you should be able to:

1. Describe the changes that take place in an invertebrate egg during fertilization and that distinguish an unfertilized from a fertilized egg.
2. List the four main stages of embryonic development following fertilization.
3. Correctly explain and use the following words: blastocoel, blastopore, blastula, cleavage, gastrocoel (archenteron), gastrula, gastrulation, neurula, neurulation, zygote.
4. Draw a diagram of an embryo containing the three primary embryonic germ layers; label the germ layers, blastocoel, blastopore, and gastrocoel (archenteron); and label the embryo with its correct embryological stage.
5. Define embryonic induction and give an example.
6. List or recognize the body parts formed from each primary embryonic germ layer.
7. Define metamorphosis and explain the selective advantage to animals of going through metamorphosis at some stage of the life history.
8. Given a developing chicken egg, open it properly, and decide whether it is 24, 48, or 72 hr old by identifying the features or structure characteristic of that age.

TEXT REFERENCES

Chapter 12 (C,D,E,G); Chapter 29 (A); Chapter 36 (F–H).
Echinoderms, fertilization, embryonic development, birth, metamorphosis, amniote egg.

INTRODUCTION

You have already studied **meiosis,** the process in which a diploid organism prepares for reproduction by packaging its genetic material into haploid gametes, each of which contains a complete genome, but only part of the parents' genetic material. The reproductive cycle is continued by the process of **fertilization,** in which haploid gametes join to make a diploid **zygote,** thereby restoring the original number of chromosomes. The reproductive cycle is completed when the zygote undergoes **development** into the many-celled adult organism, which becomes sexually mature, undergoes meiosis, and then reproduces in its turn. Today's lab will focus on embryonic development, which begins with fertilization of the egg and ends in the production of a young organism similar in form to the adult. Many developing organisms, including invertebrates such as the sea urchin, insects, and some amphibians, pass through a developmental stage called the larva, which is different from the adult, but usually exists independently of the parents.

STRUCTURE AND FUNCTION IN THE SEA URCHIN

If live sea urchins are available, your instructor may ask you to complete this section on the sea urchin as a representative echinoderm. Although echinoderm adults are not at all like humans, early development in these organisms is similar to that in humans in several important ways.

☐ Observe the sea urchins in the aquarium or supply dish. The spines that cover the top (aboral) surface of the body are mobile, and can pivot on little bumps at the base of each spine.

If the urchin is healthy, the spines will stick straight out.

Can you tell which sea urchins are male and which are female? ______________

If the urchins are crawling on the surface of the aquarium, you will be able to see the **tube feet** sticking to the glass and the **mouth,** which is located on the flat underside. (See Figure 25–1.)

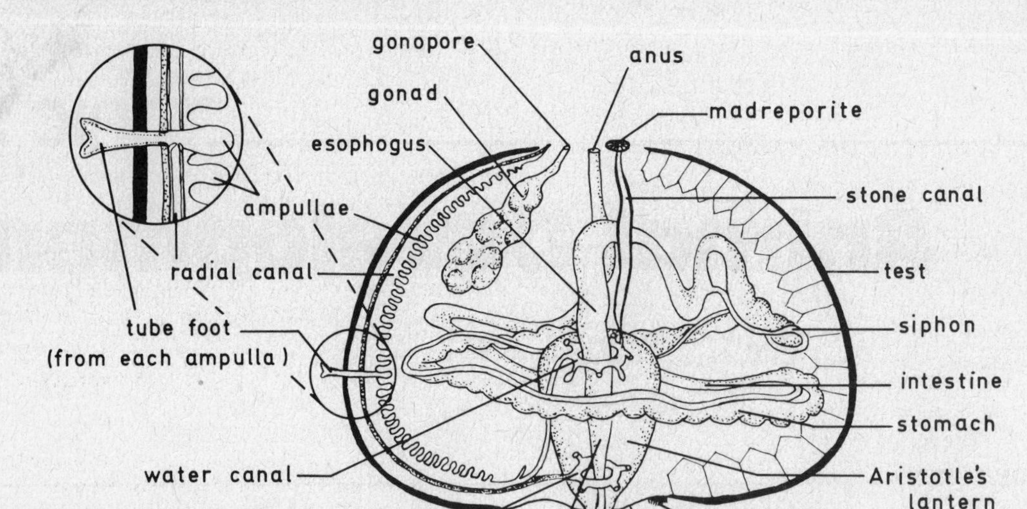

Figure 25–1. A sea urchin. Aristotle's lantern is the feeding apparatus used for ingesting kelp and seaweed. The gonads, test (skeleton), and water-vascular system show pentaradial, or five-fold, symmetry, although diagrammed only once. The many moveable spines that cover the surface of the sea urchin are not shown.

□ Use large forceps or wear rubber gloves when you are ready to take one of the animals

---

DO NOT TOUCH SEA URCHINS WITH BARE HANDS.

□ Place the urchin in a beaker of sea water or on a Petri dish of sea water with the flat (oral) side down.
□ Observe the animal for a few minutes, and notice the tiny, clear tube feet that extend from between the spines and stick to the surface of the glass.

About how many tube feet does your sea urchin seem to have? ______________

□ Place a piece of lettuce or a small object on the surface of the urchin, and wait 5 or 10 min. Then gently try to remove the object.

Does the sea urchin attach its tube feet to the object?

____

Study Figure 25–1, and note how the **water-vascular system** of radial canals and ampullae connects to all of the tube feet. This is a hydraulic system that allows the feet to be extended by increasing the pressure within the ampulla at the base of each foot. When the foot contacts something, suction can be exerted so that the foot sticks to the object like a tiny suction cup. Each foot can be operated individually. Water circulates into and out of the water-vascular system through a porous plate called the **madreporite,** which is located on the aboral surface. Also located on the aboral surface is the **anus** and the five genital pores through which gametes are released from the gonads into the sea.

The shape of the sea urchin is due to a hard endoskeleton called the **test.** Both the test and the spines are covered by a delicate epidermis. Study the sea urchin test on demonstration, using the dissecting microscope when appropriate.

□ Locate the five **gonopores** on the aboral surface.
□ Notice the tubercles where the spines were once attached.
□ Find the numerous double perforations where the tube feet extended through the test. These occur in five regions that are separated from each other by five sectors in which there were no feet and thus no perforations.

By now you should be able to describe the type of symmetry that the sea urchin shows (twofold/threefold/fourfold/fivefold/sixfold) (bilateral/radial):

____

□ On the oral surface notice the large opening in the test. This is where the feeding apparatus, called **Aristotle's lantern,** originally protruded through the test, but the membranes holding it together and in position disintegrated when the test was prepared. You can see this structure in your living animal.
□ Again using forceps or gloves, invert the animal so that the flat side is on top. The animal is now upside down from its normal position.
□ Note the feeding structure, Aristotle's lantern, in the center of the oral surface. It contains five teeth that are used to tear the kelp on which the animal feeds, and it is held in place by a membrane that attaches it to the skeleton.

The mouth is the opening in the center of the lantern.

□ Take a single tooth from those on demonstration, and study its structure in the dissecting microscope.
□ When you are finished observing the structure of the oral surface, leave your animal upside down for awhile, and observe its response from time to time.

How did your sea urchin react to an upside down situation?

______

What structures were used in the response?

______

When you are finished with your sea urchin, return it to the aquarium or supply pan.

SEA URCHIN DEVELOPMENT

Mature sea urchins are male or female. When the gametes are ripe, they are released through pores on the top (aboral) side of the animal, and fertilization takes place in the sea water.

Is this type of fertilization internal or external?

______

In the laboratory, the urchins can be induced to release their gametes separately so that fertilization can be carried out and observed under the microscope. If the fertilized eggs are carefully washed and incubated in sea water, they will develop for many days, and will eventually form a larval stage known as the **pluteus.** The pluteus is a free-swimming, feeding organism which eventually attaches to the bottom, and metamorphoses into a tiny sea urchin. Depending on the time of the year, you will use an East Coast sea urchin (*Arbacia*) or a West Coast urchin (*Strongylocentrotus*) to study early development. The schedule for the development of these species is shown in Table 25–1.

Cleavage

Sea urchin eggs are very large cells, so you will have to use a depression slide, or omit the cover slip while you are observing them.

□ Place a drop of stirred sea urchin egg suspension on your slide, and observe it under low power.

What is the diameter of the eggs in micrometres?

______

What is the color and texture of the cytoplasm?

______

What does the cell membrane look like?

______

□ If you are studying *Arbacia* eggs, add a drop of India ink so that you can see the clear jelly coat surrounding the egg.

About how thick is the jelly coat in micrometres?

______

□ Clean the slide and dry it before adding a fresh drop of eggs. Focus on the eggs under low power.

□ Add a drop of *diluted* sperm cells provided by your instructor. Mix gently with a toothpick while observing in the microscope, and record the time and temperature in Table 25–2 at the end of this topic.

When the sperm cells penetrate the jelly coat, they eventually come into contact with the membrane of the cell. Then one lucky sperm will enter the cytoplasm, leaving its tail behind, and its nucleus will combine with the egg's nucleus to form the new diploid nucleus of the zygote.

You can barely see the individual sperm cells under low power, but you can see right away that something is happening.

What change occurs within 1 or 2 min of adding sperm cells to the eggs?

______

When the egg cell is stimulated by the arrival of the sperm, a chain of events begins to unfold that is very precise and almost inevitably ends in the production of a normal sea urchin larva. First an electrical reaction begins at the site where the sperm penetrates, and spreads all over the surface of the cell. Then Ca^{++} ions released from this region of the membrane induce changes in the egg cytoplasm, signaling that development should begin. The **vitelline membrane,** which lies between the cell

TABLE 25–1. DEVELOPMENT OF SEA URCHIN EMBRYOS

Developmental Stage	*Arbacia punctulata* at 23°C	*Strongylocentrotus purpuratus* at 15°C
Fertilization	0	0
Fertilization membrane forms	1–2 min	1–2 min
First cleavage	50–70 min	90–110 min
Second cleavage	80–110 min	140–170 min
Blastula	5–6 hr	8–10 hr
Hatching of blastula	7–8 hr	18–20 hr
Gastrula	12–15 hr	22–26 hr
Pluteus	24 hr	72 hr
Metamorphosis	2 weeks	2–8 weeks

membrane and the jelly coat, separates from the cell membrane and enlarges so that there is a wide space between two membranes. It is now called the **fertilization membrane.** The cell "membrane" now looks double; this is the obvious change that you see whenever fertilization occurs. Any sperm still struggling through the jelly coat are lifted away from their goal and the fertilization membrane provides a protected environment for the developing embryo.

□ Seal your slide of fertilized eggs by applying Vaseline with a toothpick around the coverslip, and set it aside, but observe it every 15 min for the rest of the period. You should see signs of cleavage in about 1 hr. Your instructor may ask you to sample a culture of developing eggs instead of saving the fertilized egg on your slide. Record the age, stage, and appearance of the embryos in Table 25–2 each time you observe them.
□ Take an ordinary slide, and prepare a wet mount of sperm cells from the suspension indicated by your instructor. Observe under high power or oil immersion. Cut down the light by closing the diaphragm.

Are the sperm cells in motion? ______________

Can you see the flagella of the sperm cells?

____

In addition to the nucleus and the flagellum, what organelle is important for sperm cells to carry out their reproductive role? (*Hint:* This organelle provides energy needed in large amounts for sperm motility.)

____

Once the first division has taken place, cleavage will continue for many cell divisions. In sea urchins cleavage is **indeterminate:** each of the cells formed early in development has the potential to develop into a complete organism. If the daughter cells should happen to separate after the first division, the result would be two identical organisms. When this happens in human development, the result is identical twins. Cleavage is **radial** because each cleavage plane is at right angles to or parallel to the original axis of the zygote. (See Figure 36–14 in your text.) In the sea urchin the yolk is in the form of cytoplasmic granules, so the cleavage furrow can pass completely through the egg without difficulty. Study the later stages in cleavage that are available.

How large is the embryo compared to the egg after several divisions?

____

The **morula** is a solid ball of cells that gradually develops a hollow interior called the **blastocoel.** The embryo is then at the **blastula** stage, and the outer cells are ciliated so that the blastula is motile, and can move about within the fertilization membrane. After several hours of cleavage, the blastula is fully developed, and "hatches" as the fertilization membrane finally breaks open and releases the embryo. The blastula now enters the next stage of embryonic development.

Gastrulation

During gastrulation the embryo changes from a blastula with one layer of cells into a **gastrula,** which has three layers. In this process cells migrate into the interior of the blastula at a point called the **blastopore,** and form a lining so that the embryo now has two layers: an outer ectoderm and an inner endoderm. The ectodermal cells will form the skin and its derivatives, and give rise to the whole nervous system. The endoderm develops into the lining of the gut and of certain digestive organs. The space inside both layers is called the **gastrocoel,** or **archenteron,** and will eventually form the lumen of the organism's gut. The cells in the dorsal part of the endoderm form pouches, and bud off to become the cells of the mesoderm. Mesoderm gives rise to the muscle, bone, and some internal organs of the adult. In humans, mesoderm forms in the same way.

□ Examine a drop of the embryos that have reached the gastrula stage.
□ Record the age of the embryos and their appearance in the table.

Can you see the primitive gut? ______________

Neurulation

In vertebrates such as yourself and the chick, the gastrula soon develops a dorsal groove, which pinches off and becomes part of the dorsal nerve cord and the brain. An embryo undergoing this process is at the neurula stage.

Organogenesis

The process of organogenesis in which organs and organ systems are developed will be discussed in the section on the chick embryo where it can be studied more easily.

If older sea urchin embryos are available, you will be able to see a simple digestive system and skeletal system in this case as well.

Metamorphosis

The sea urchin embryo develops skeletal elements, and begins feeding as the **pluteus larva.** (See Figure 25–2.) It continues this existence for up to four months until it finally settles down on the substrate, attaches, and changes into a tiny adult sea urchin. The whole process of **metamorphosis** may take only an hour, and the result is a miniature adult only 1 mm in diameter.

□ Observe the embryos that have reached the pluteus stage.
□ Draw an arrow in Figure 25–2 to indicate the direction in which the pluteus swims.
□ Note the skeletal spines that support the body and arms.
□ Note whether colored pigment granules are present.
□ Identify the complete digestive tract visible through the transparent body of the pluteus.

What is the advantage of a motile larval stage?

____

____

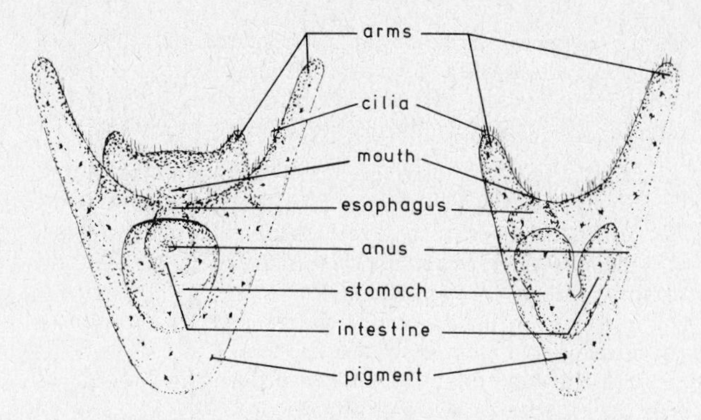

Figure 25–2. Pluteus larva. The sea urchin larva develops from the zygote within a few days. The motile larva feeds and swims about independently until it settles down and metamorphoses into a tiny adult.

CHICK DEVELOPMENT

Fertilization and Early Development

The egg cell of a chicken consists of a yellow mass of **yolk** and a small area of yolk-free cytoplasm, the **germinal disc,** that will develop into the embryo proper. **Cleavage** and development occur only within the disc, and the yolk supplies required nutrients and energy to support the embryo.

As the egg cell passes through the reproductive tract of the female chicken, it is modified in several ways. A solution of the protein albumin is added to make up the white of the egg. This layer provides physical protection for the delicate embryo, and antimicrobial substances in it protect the embryo from bacterial infection. The albumin itself will provide the chick with a store of amino acids to draw on for protein synthesis, and the liquid is the chick's only source of water. Later on, two membranes and then a porous shell are added. The hard shell protects the embryo and its food supply but also allows gas exchange with the environment.

□ Examine an unfertilized egg on demonstration. Identify the germinal disc, which is normally on the upper surface of the yolk.

Is fertilization in the chicken internal or external?

internal

If the egg cell has been fertilized, it will begin cleavage and start to develop right in the hen's oviduct. Since eggs are held there for various lengths of time before being laid, the exact stage of development at the time of laying is different for each individual egg, but usually cleavage, blastula formation, and gastrulation have already been accomplished.

How would an unfertilized egg differ from one that has been fertilized?

Nothing developing in Yolk

Is there any nutritional advantage in eating fertilized eggs?

Cleavage is imcomplete in the chick because the yolk is so massive. Early in development, the cells of the **blastoderm** split into two layers: the upper layer is the **ectoderm** and the lower layer is the **endoderm.** A thickening, called the **primitive streak,** then develops in the center of the blastoderm. Here cells migrate into the space between ectoderm and endoderm, and form the embryo's **mesoderm.** Mesodermal tissue near the center of the embryo pinches off to form the **notochord,** and the pairs of **somites** that lie in rows along each side of the notochord. The notochord induces the ectoderm above it to form a **neural plate.** The plate develops folds, which eventually meet and fuse to form the **neural tube.** The neural tube in turn differentiates into the brain and the spinal cord. Finally the embryo lifts, and the ectoderm folds beneath it so that the embryo is completely covered by ectoderm and the primitive gut is enclosed by the endoderm. These changes are shown in Figure 25–3. When you have time, mark the ectoderm, mesoderm, and endoderm

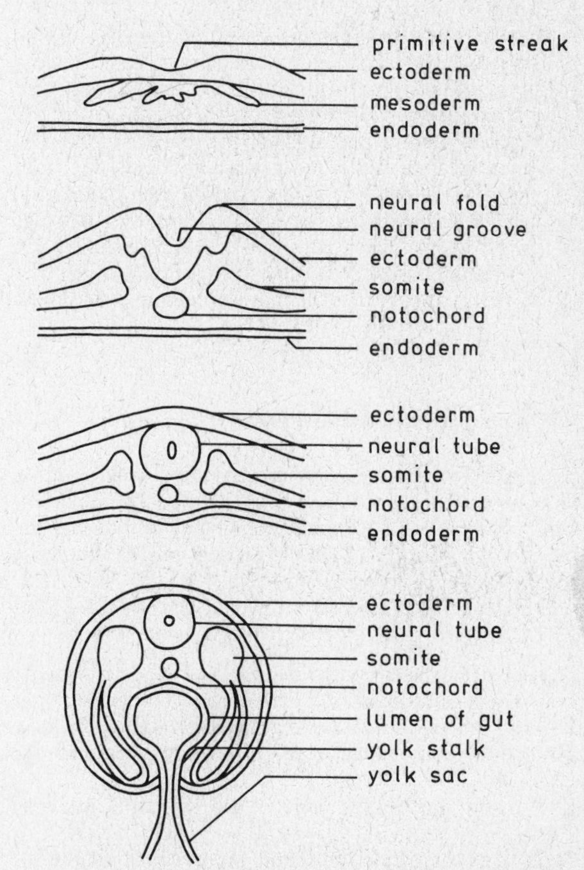

Figure 25–3. Chick embryo in cross section. The notochord of the gastrula induces part of the ectoderm to invaginate and form the neural tube. The brain and nervous system will eventually differentiate from the the neural tube. You can color-code structures derived from the primary germ layers: ectoderm = blue, mesoderm = red, and endoderm = yellow.

with different colored pencils (blue, red, and yellow, respectively) so that you can follow the developmental process more clearly

Meanwhile, specialized cells in the area of the blastoderm surrounding the embryo form **blood islands,** which give rise to extraembryonic blood vessels and blood cells. The embryo develops a primitive heart, and blood soon begins to circulate between the embryo and the surface of the yolk.

Other cells in the blastoderm grow into protective **extraembryonic membranes.** Although these membranes are not part of the embryo proper, they are vital for its nourishment and development.

1. The **amnion** develops as a fold of tissue growing down over the embryo's head. It meets with another fold growing up over the tail, and fuses with it so that the embryo is completely enclosed in an **amniotic sac,** which is filled with amniotic fluid. The fluid acts as a protective cushion.
2. The **yolk sac** grows out from the primitive gut, and eventually surrounds the whole yolk with a double membrane. Blood vessels in its walls are important for the nutrition of the growing embryo.
3. The **chorion** grows outside of the amnion and the yolk sac, and completely surrounds the whole embryo, including the yolk. It protects the embryo, and holds it in contact with the yolk.
4. Finally, the **allantois** grows out as a saclike extension of the primitive gut, expands up over the embryo, and eventually surrounds the embryo and yolk with a double membrane: the outer layer of the allantois presses up against the chorion, and the inner layer against the amnion and yolk sac. Because the allantois is richly supplied with blood vessels, it allows gas exchange with the external environment. The wastes that accumulate during development, mostly in the form of uric acid, are stored in the interior of the allantois until the chick hatches.

Living Chick Embryos

Each pair of students will be provided with a chick embryo that has been incubated for 33, 48, or 72 hr. You are responsible for understanding all of these stages, so study the prepared slides or another student's embryo to see all the structures.

Follow the procedure shown at the top of Figure 25–4 to open your incubated egg:

□ Add a little warmed saline (0.9% NaCl) to a finger bowl so that the bottom is just barely covered.
□ Take an egg from the incubator or tray, hold it in exactly the same orientation, and carry it carefully to your desk.
□ Tap the egg on the edge of the dish until the shell cracks.
□ Holding the egg in the saline, pry the shell apart at its bottom without disturbing the upper surface of the yolk.

If your embryo is 72 hr old, you should see the embryo and its beating heart immediately. If you do not see your embryo, ask your instructor for help. Either the egg failed to develop, or the yolk is turned upside down so that the embryo is hidden.

After you have observed as much structure as you can in the dissecting microscope, remove the embryo from the yolk, and transfer it to a watch glass so that you can see it more clearly. This process (shown at the bottom of Figure 25–4) is not difficult if you work carefully:*

□ Fold a piece of filter paper in half, and cut out a spoon with a hole just larger than the embryo.
□ Add a teaspoonful of warm saline to a watch glass.
□ Lay the filter paper spoon on the yolk so that the embryo shows through the hole.
□ Press down gently on the paper with your probe, if necessary, so that it is wet all around.
□ Cut the membrane in a circle around the outside of the filter paper to free the embryo.
□ Lift the spoon slowly so that the embryo lifts from the surface of the yolk.
□ Lay the spoon in the watch glass, and observe in the dissecting microscope.

33-Hour Chick (Figure 25–5)

This embryo is a streak in the center of the germinal disc, and you will have to transfer it to a watch glass to see much. The brain has begun to differentiate, and the **optic vesicles** are developing.

The **vitelline veins** have joined to form the primitive **heart** and there are about 12 somites along the sides of the **notochord.** The heart might be feebly contracting, but there is no blood flow yet. If you are looking at a stained slide, notice the **blood islands** in the germinal disc outside of the embryo itself. They will give rise to the blood cells and some of the vessels. Because the tail region develops more slowly than the head region, it may still show the primitive streak that was the very first sign of development. The amnion is beginning to develop as a fold of tissue growing down over the head region. It will eventually contact the posterior fold to enclose the embryo.

Finally, you may be able to see the foregut, which is growing toward the head region from underneath the embryo as an outpocketing of the endoderm. It will later extend toward the tail region as well.

What structure does the notochord give rise to?

vertebral column
____

Examine the demonstration of a vertebra to be sure that you understand the relationship of the notochord and the spinal cord.

What structures arise from the somites?

muscles and bones
____

____

□ Quiz your partner on the structures in Figure 25–5.

48-Hour Chick (Figure 25–6)

This embryo has started to twist to the right so that the head region is now lying on the chick's left side. This is the process of **torsion,** which will continue until the whole embryo is lying on its left side. At the same time,

*Suggested by Professor A. W. Blackler, Cornell University.

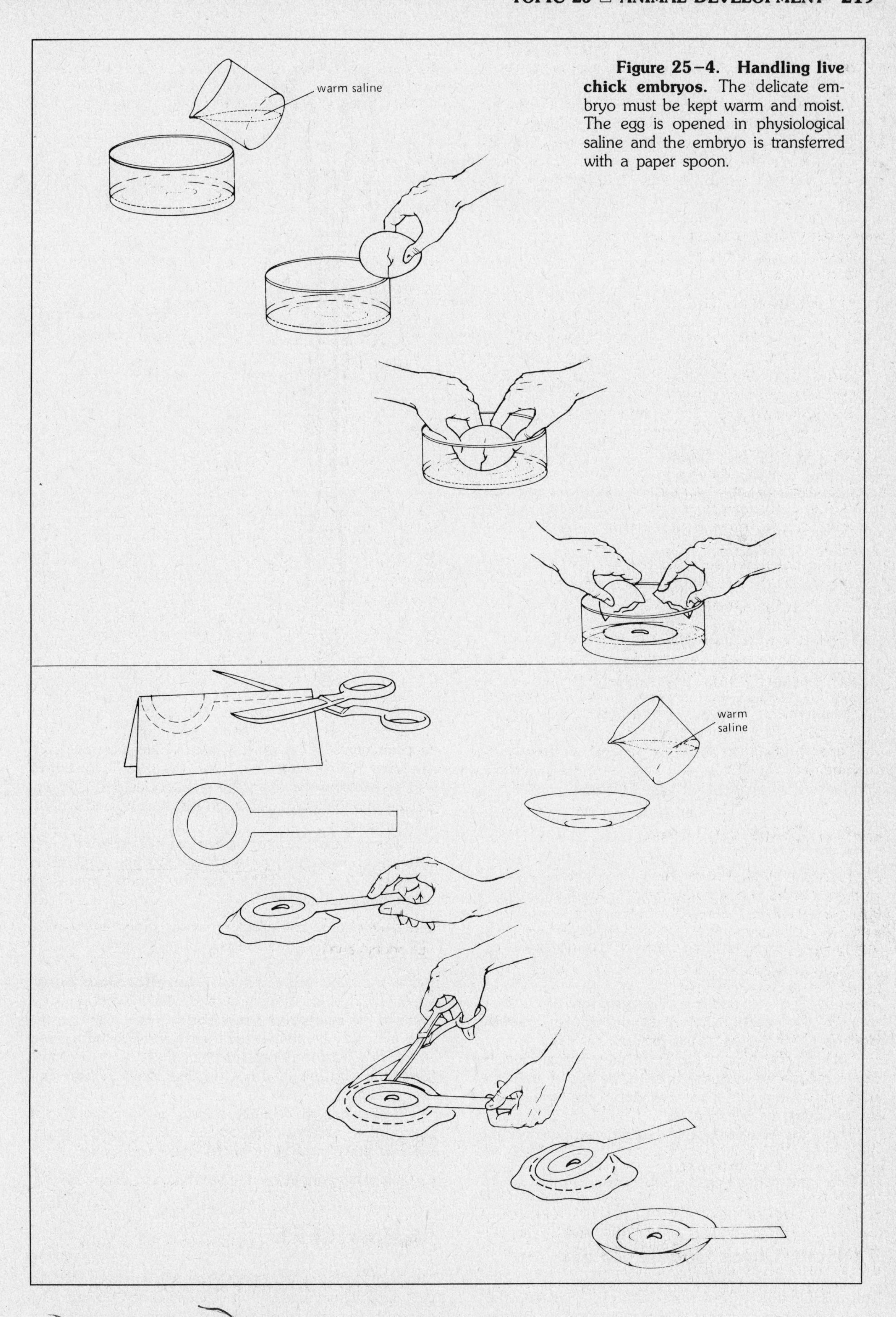

Figure 25–4. Handling live chick embryos. The delicate embryo must be kept warm and moist. The egg is opened in physiological saline and the embryo is transferred with a paper spoon.

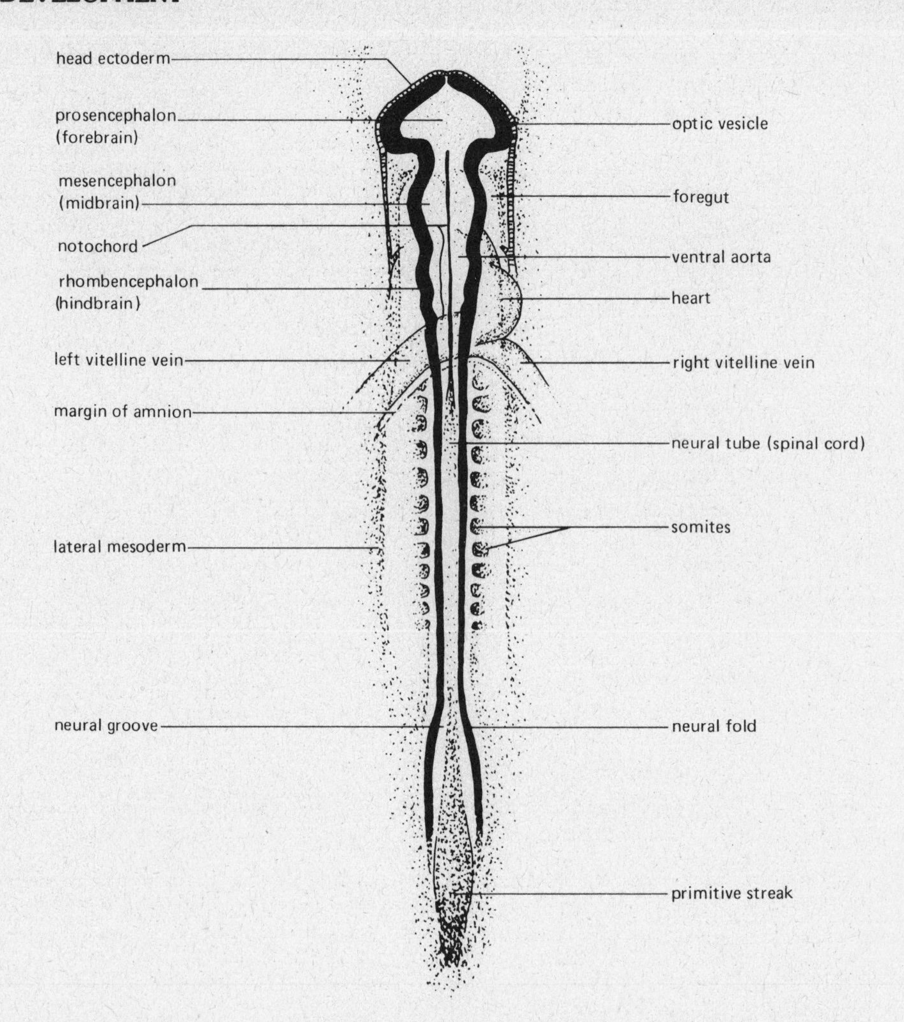

Figure 25–5. Whole mount of a 33-hour chick embryo. The anterior end of the neural tube has begun to differentiate while the posterior end has not yet formed from the neural groove. (Redrawn from Wodsedalek, J. E. and Charles F. Lytle, *General Zoology Laboratory Guide,* 8th Ed. © 1971, 1977, 1981 Wm. C. Brown Company Publishers, Dubuque, Iowa. Reprinted by permission.)

the embryo is curling up in the process of **flexure** so that the neural tube takes the shape of a question mark, with a bend in the region of the hindbrain.

How many somites are still showing their dorsal sides?

____

How many somites does your embryo have? ________

Blood circulation usually begins at about the 16-somite stage.

If you are viewing a living embryo, is there blood circulating? yes

Trace the flow of blood from the **ventricle** of the heart through the **aorta,** and then through the **aortic arches** in the region of the pharynx.

Identify the parts of the brain: the optic cup has induced the formation of the **lens** of the eye by the ectoderm, and the neural tube has induced the formation of the **otocyst,** or primitive ear.

Note the **vitelline arteries** growing out over the yolk.

□ Quiz your partner on the structures in Figure 25–6.

72-Hour Chick (Figure 25–7)

The chick is now almost completely on its left side (torsion). Flexure of the brain is prominent, and it is starting to differentiate into distinct regions. Note the optic cup, the lens, and the otocyst, or primitive ear. The **atrium** and ventricle of the heart are greatly enlarged, and are actively pumping blood through the vitelline and embryonic arteries.

There are several **pharyngeal clefts** and **pouches.** These structures appear at some stage in all vertebrate embryos. There are too many somites to count easily (about 36).

What membrane completely encloses the embryo itself?

amnion

Below the heart you can see the **anterior limb buds,** which will develop into wings, and near the embryo's tail you see the **posterior limb buds,** which will form the legs. You may be able to see the small saclike **allantois** in the region of the posterior limb buds, but it is probably hidden by the tail, and is better seen in later embryos.

If you can successfully transfer your embryo to a watch glass, observe circulation in the blood vessels under low power with your compound microscope.

□ Quiz your partner on the structures in Figure 25–7.

96-Hour Chick

This embryo has developed about 41 somites, and the yolk is partly covered with prominent blood vessels. The flexure of the body posterior to the brain is now obvious,

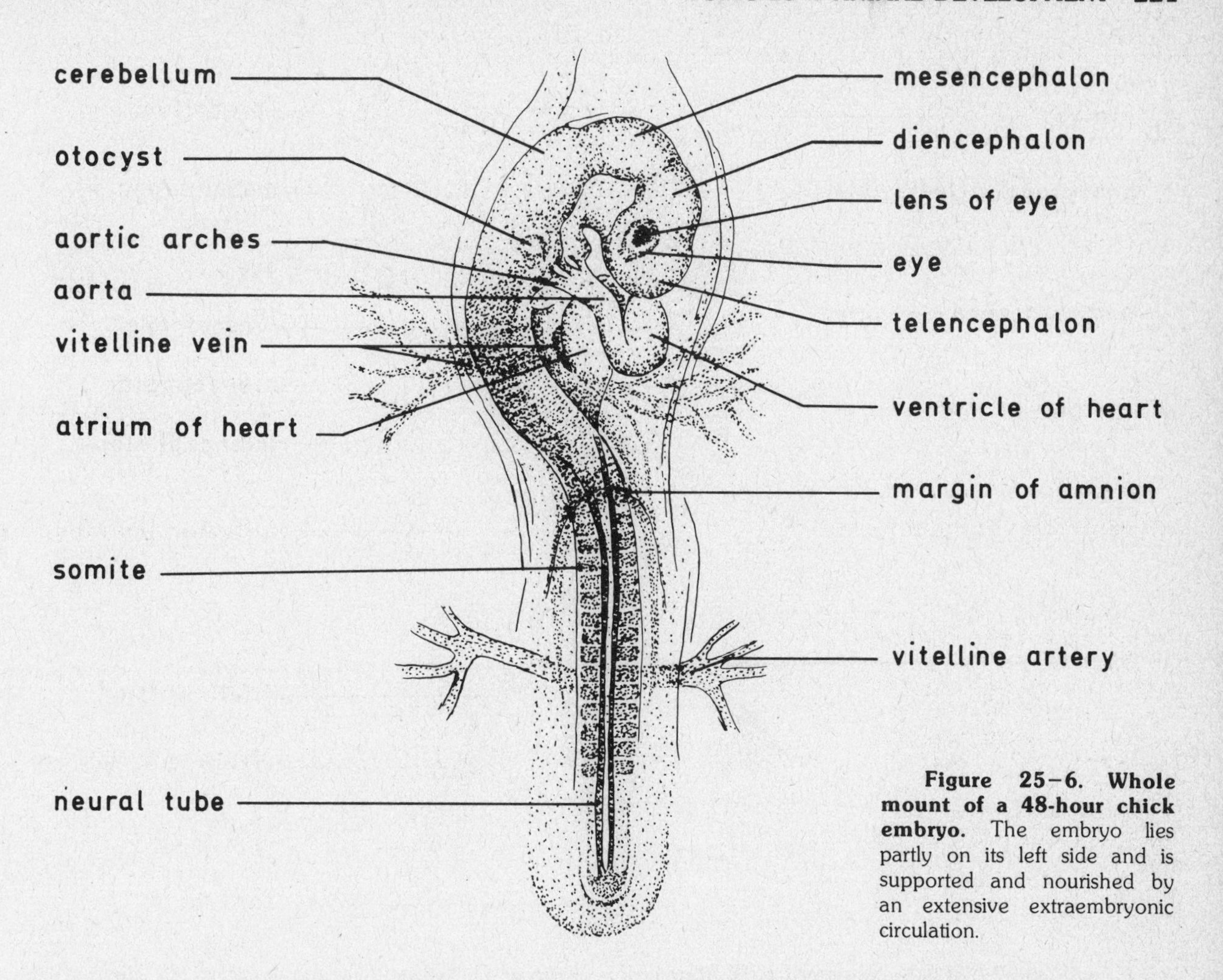

Figure 25–6. Whole mount of a 48-hour chick embryo. The embryo lies partly on its left side and is supported and nourished by an extensive extraembryonic circulation.

and the brain is divided into three distinct regions. The eye is easier to see because it is pigmented, but the ear is less distinct because the tissues have become much thicker.

□ Remove the membranes surrounding the embryo so that you can see the heart and limb buds. You can use a probe or teasing needle to straighten out the embryo so that you see the pharyngeal clefts and pouches.

The fluid-filled sac at the posterior end is the allantois. It will expand and eventually fuse with the chorion, and become highly vascularized.

What are the two functions of the allantois?

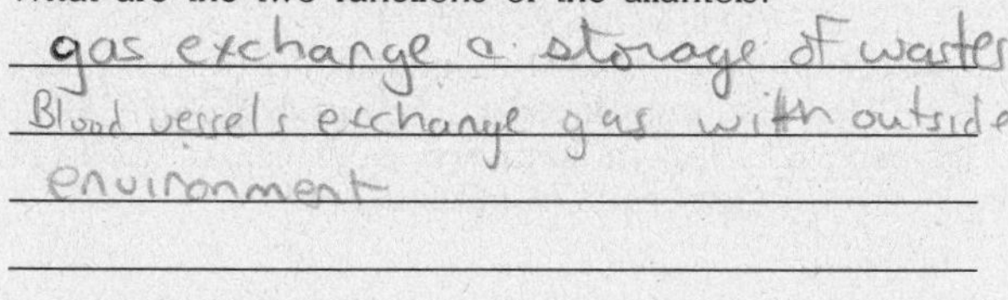

When the chick hatches, the allantois will remain behind as a waste-filled membrane stuck to the shell.

Later Development

Your instructor may demonstrate some older living embryos that are even further along in development. All of the major developmental processes happen during the first 6 days of development. The embryo then grows and matures during the rest of the incubation period until it hatches at 21 days, and starts its independent existence as a wobbly, fluffy chick.

When you are finished with your work, discard your debris in the EGG WASTES container, clean your dissecting instruments, return the prepared slides, and store your microscopes properly.

EXPLORING FURTHER

Browder, L. W. *Developmental Biology.* Philadelphia: W. B. Saunders, 1980.

Harvey, E. B. *The American Arbacia and Other Sea Urchins.* Princeton, New Jersey: Princeton University Press, 1956.

Patten, B. M. *Early Embryology of the Chick.* 5th ed. New York: McGraw-Hill, 1971.

Romanoff, A. J. *The Avian Embryo.* New York: Macmillan, 1960.

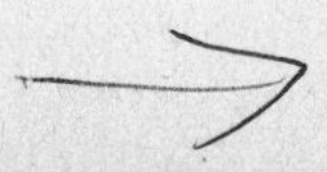

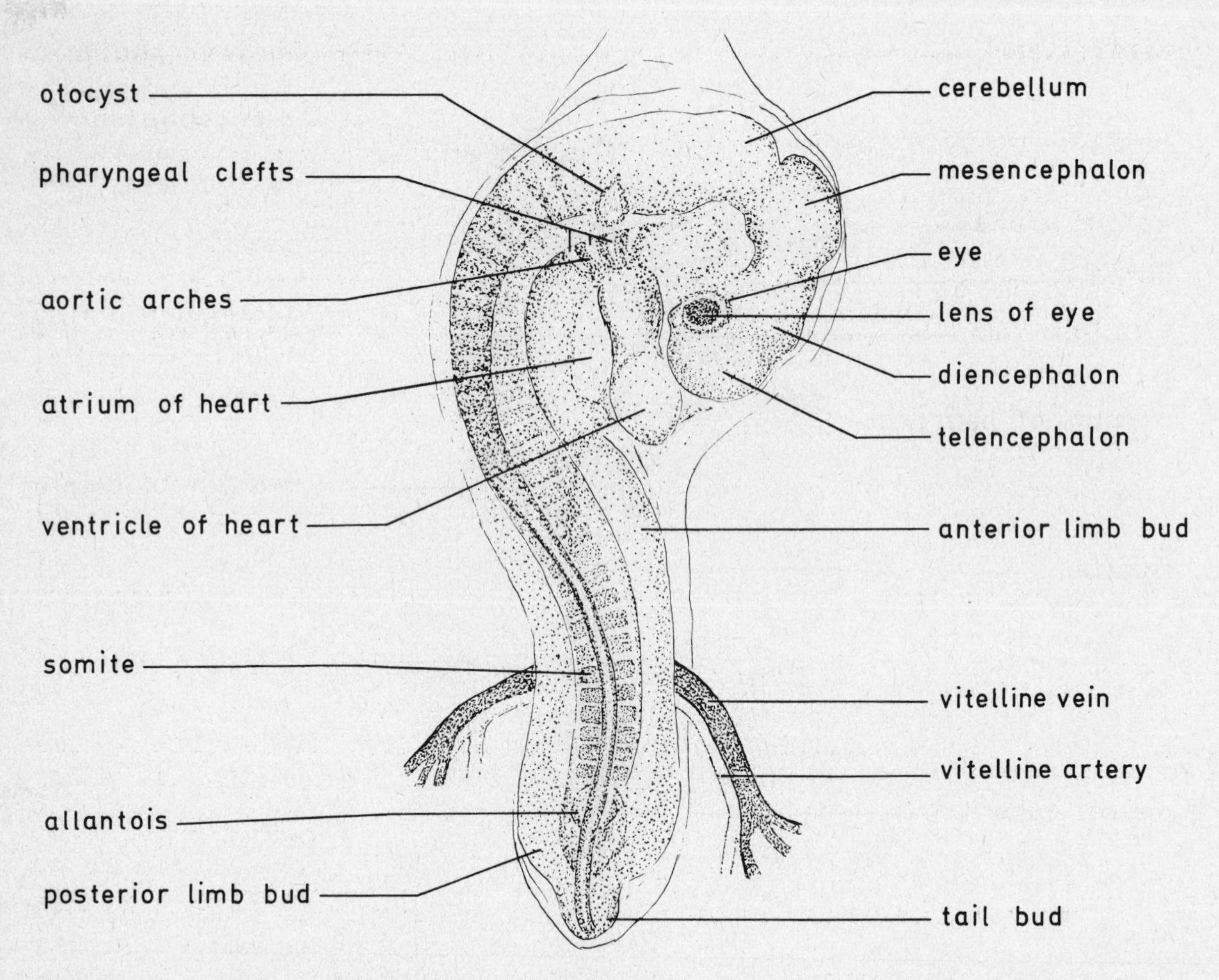

Figure 25–7. Whole mount of a 72-hour chick embryo. The embryo lies mostly on its left side and the allantois and limb buds are beginning to form.

Scientific American Articles

Bryant, P. J., S. V. Bryant and V. French. "Biological regeneration and pattern formation." July 1977 (#1363).

DeRobertis, E. M. and J. B. Gurdon. "Gene transplantation and the analysis of development." December 1979 (#1454).

Epel, David. "The program of fertilization." November 1977 (#1372).

Garcia-Bellido, Antonio, Peter A. Lawrence and Gines Morata. "Compartments in animal development." July 1979 (#1432).

Gordon, R. and A. G. Jacobson. "The shaping of tissues in embryos." June 1978 (#1391).

Lane, Charles. "Rabbit hemoglobin from frog eggs." August 1976 (#1343).

Levi-Montalcini, Rita, and Pietro Calissano. "The nerve-growth factor." June 1979 (#1430).

Peakall, David B. "Pesticides and the reproduction of birds." April 1970 (#1174).

Rahn, Hermann, Amos Ar and Charles V. Paganelli. "How bird eggs breathe." February 1979 (#1420).

Wolpert, L. "Pattern formation in biological development." October 1978 (#1409).

Student Name ______________________________ **Date** ____________________

TABLE 25–2. OBSERVATIONS ON SEA URCHIN DEVELOPMENT AT ________ °C

Genus ____________________

Species ____________________

Time	Time after Fertilization	Stage	Appearance
	0 min		
	min		
	min		
	min		
	min		
	min		
	min		
	min		
	min		
	min		
	hr		
	hr		
	hr		
	hr		
	hr		
	hr		

TOPIC 26

The Nervous System

OBJECTIVES

When you have completed this topic, you should be able to:

1. Describe the basic structure of a neuron.
2. Show in which direction(s) an impulse travels if the neuron is stimulated at any point.
3. Describe the function of a myelin sheath, and give the location of myelinated neurons in the nervous system.
4. Describe the structure of the mammalian brain, and give the functions controlled by the medulla, cerebellum, hypothalamus, pituitary, thalamus, reticular formation, colliculi, and cortex.
5. Describe the structure of the spinal cord, and give the main types of neurons entering, leaving, or functioning within it.
6. Distinguish between the sympathetic and parasympathetic nervous systems.
7. Describe the five main types of receptor cells, and name some sense organs in which each is found.
8. Describe the structure of the vertebrate eye and the types of receptors found in it.
9. Explain how the transducers in the inner ear work.
10. Describe where proprioceptors are found in the body, and explain their function.

TEXT REFERENCES

Chapter 37 (A,B, G–J); Chapter 38 (A–G); Chapter 39.
Neurons, synapses, central nervous system, peripheral nervous system, vision, hearing, taste, touch.

INTRODUCTION

The function of the nervous system is to coordinate the activities of the different parts of the body. **Receptor cells** located in internal and external sense organs send it information, allowing it to respond by means of its **effectors,** which are usually muscles or glands. The nervous system maintains equilibrium, or **homeostasis,** within the body and allows the organism to respond to external stimuli with appropriate **behavior** patterns.

The **neurons,** or nerve cells, of the vertebrate nervous system are organized into the **central nervous system** and the **peripheral nervous system.** The brain and spinal cord make up the central nervous system (CNS), which receives input from the major sense organs and coordinates most of the body's activities. Many types of neurotransmitters are active in the synapses of the CNS.

The **somatic** part of the peripheral nervous system includes the cranial and spinal nerves. These nerves control voluntary activities and use acetylcholine as the neurotransmitter. The **autonomic** part of the peripheral nervous system controls those internal functions that are usually carried out at the subconscious level. It includes the **sympathetic** nervous system, which uses noradrenalin as a neurotransmitter, and the **parasympathetic** nervous system, which uses acetylcholine.

NEURONS

The neurons that make up the nervous system conduct electrochemical impulses from one site to another through long cell processes. Those processes leading toward the cell body are called **dendrites,** and those leading away are called **axons.** Both types of neurons usually connect with adjacent neurons, receptor cells, or effector cells by means of **synapses.** Dendrites receive stimuli from their synapses, whereas axons stimulate their synapses with adjacent cells.

□ Examine a prepared slide of a neuron.
□ Locate the cell body, nucleus, axon, and dendrites.

Where do impulses reach this cell in the form of neurotransmitter molecules?

Dendrites
____

Where does a generator potential occur?

~~Axon~~ synapse dendrite
____

Where is an all-or-none action potential generated?

nucleus, cell body synapse
____

In vertebrates, some axons are surrounded by special Schwann cells, which are filled with a fatty insulating material called **myelin.** The effect of the myelin sheath is to speed up the rate of impulse transmission along the axon. At intervals along the length of the axon, there are interruptions in the myelin sheath known as the nodes of Ranvier. Instead of traveling at the normal rate, the impulse skips from node to node at a much faster rate. (See text Figure 37–12.) Myelinated nervous tissue appears white, whereas unmyelinated tissue is gray.

THE CENTRAL NERVOUS SYSTEM

The Brain

The vertebrate brain is divided into the forebrain, the midbrain, and the hindbrain. In mammals the forebrain is the most prominent part, and it is considerably more developed than it is in lower vertebrates. The external features of a sheep brain are shown in Figure 26–1.

The brain is covered by a triple layer of membranes called the **meninges.** The hollow spaces within the brain are called the **ventricles.** The ventricles and the hollow interior of the spinal cord are filled with **cerebrospinal fluid.**

The **forebrain** consists of the cerebral hemispheres and the underlying parts. The **midbrain** in mammals is a small region near the medulla, which is concerned with certain responses of the eye and ear. The **hindbrain** is the cerebellum and the medulla.

□ Examine a preserved sheep brain, and note the two **cerebral hemispheres,** which are the brain's largest structures. Although the hemispheres appear to be symmetrical here, in humans the two halves differ in size, and some functions are carried out only in one or the other hemisphere.
□ Locate the longitudinal **cerebral fissure** between the two hemispheres.
□ Spread the two hemispheres apart, and find the thick, white **corpus callosum.** It consists of bundles of myelinated axons, and connects the two hemispheres to each other.

The cortex of the highly convoluted cerebrum contains the **gray matter** of the brain, consisting of cell bodies and unmyelinated axons. Associations, memory, and higher brain functions such as abstract thinking are carried on by the cerebrum. The corpus callosum contains only myelinated and unmyelinated axons, and thus consists of **white matter** due to the presence of myelin.

Know these ✓

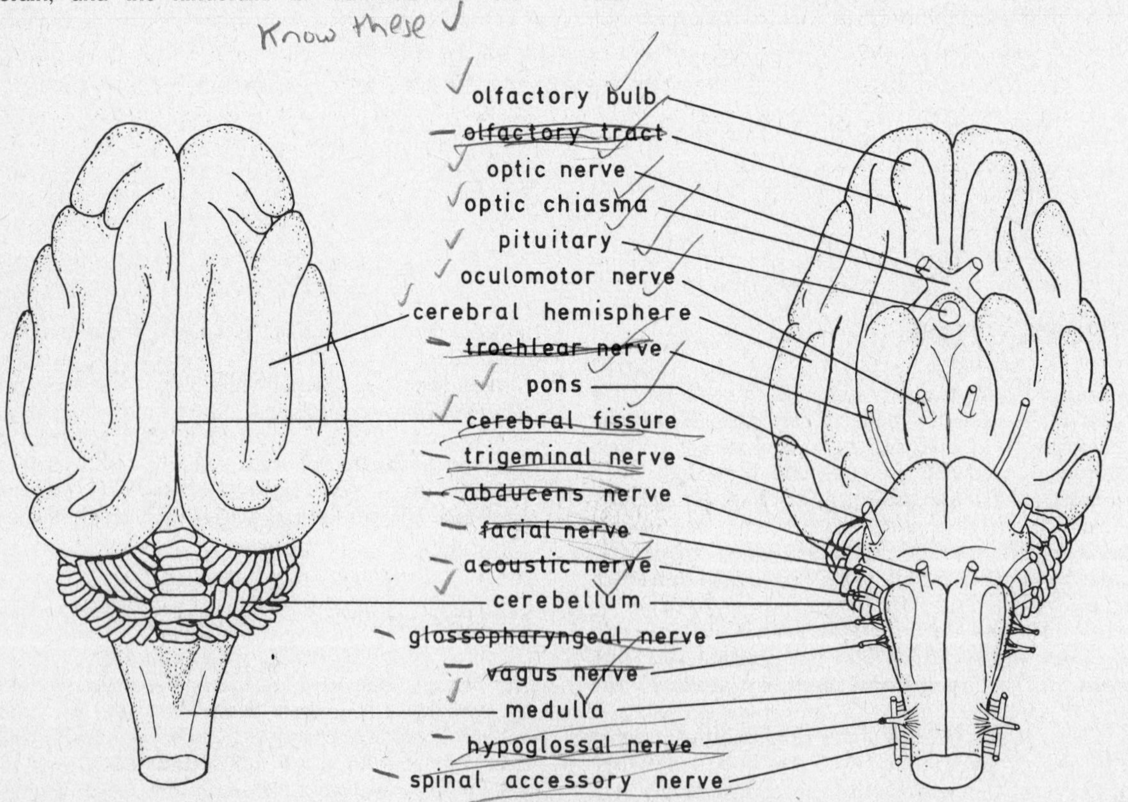

Figure 26–1. Sheep brain. The cerebral hemispheres are much smaller relative to the rest of the brain than in humans. The pituitary may be missing in your specimen.

□ Examine the ventral surface of the brain, and find the **olfactory tracts,** which lead to the chambers of the skull near the nasal passages.
□ Find the **optic nerves,** which cross in the **optic chiasma** before entering the brain.
□ Posterior to the optic chiasma, look for the **pituitary,** which is suspended from the part of the brain called the hypothalamus. Secretions from the hypothalamus regulate the release of hormones from the pituitary. (The pituitary may be lost or damaged in your sheep brain.)
□ Find the **cerebellum,** which is a rounded structure with a convoluted surface just posterior to the cerebrum.
□ Identify the **medulla** which is connected to and continuous with the anterior part of the spinal cord. The medulla and cerebellum both develop from the hindbrain of the sheep embryo.

The internal structures of the brain are shown in Figure 26–2.

□ Examine a sagittal section of the sheep brain, and find the location of the pituitary.
□ The dense **hypothalamus,** just dorsal to the pituitary, can now be seen better. In addition to controlling the pituitary, it has centers that control hunger, thirst, emotional states, sleep, osmoregulation, sex drive, and body temperature.
□ Just dorsal to the hypothalamus, find the white matter of the **thalamus.** This is an important area for sorting sensory messages before they enter the cerebrum.
□ Locate the **midbrain,** posterior to the thalamus and anterior to the cerebellum. Its two anterior protuberances are the **anterior colliculi,** which control visual reflexes. The posterior protuberances are the **posterior colliculi**, which receive auditory input. The ventral part of the midbrain, the cerebral peduncles, contains tracts of fibers that conduct impulses anteriorly and posteriorly.

What part of the brain does the medulla lead to?

midbrain

What is the internal appearance of the cerebellum?

convoluted structure?

The cerebellum coordinates body movements and is important in maintaining equilibrium. The medulla transmits impulses between the spinal cord and the rest of the brain. The **reticular formation** is a network of neurons that connects the medulla and thalamus. It receives various kinds of sensory input. It also controls the general arousal level of the animal and how strongly it responds to external stimuli. The **reticular activating system** is located between the reticular formation and the cerebral cortex. It filters all sensory input before it can reach the cortex and be consciously perceived.

The Spinal Cord

A cross section of the spinal cord is illustrated in Figure 26–3.

□ Examine a slide of the spinal cord in cross section in your dissecting microscope, and orient it so that the **ventral fissure** is toward you.
□ Locate the H-shaped area of **gray matter** around the central canal; it consists mainly of cell bodies.
□ Notice the **white matter** that makes up the rest of the spinal cord; it contains axons and dendrites.

How does the arrangement of gray and white matter here compare with the arrangement in the brain?

White matter on inside of Brain
Gray matter in outer part opposite
in spinal cord

□ Find the right and left **dorsal roots** through which sensory nerve fibers enter the spinal cord.
□ Near the dorsal root, find the **dorsal root ganglion,** which contains the cell bodies of the **sensory neurons.**

Figure 26–2. Sheep brain in sagittal section. The ventricles of the brain are continuous with the interior of the spinal cord and contain cerebrospinal fluid.

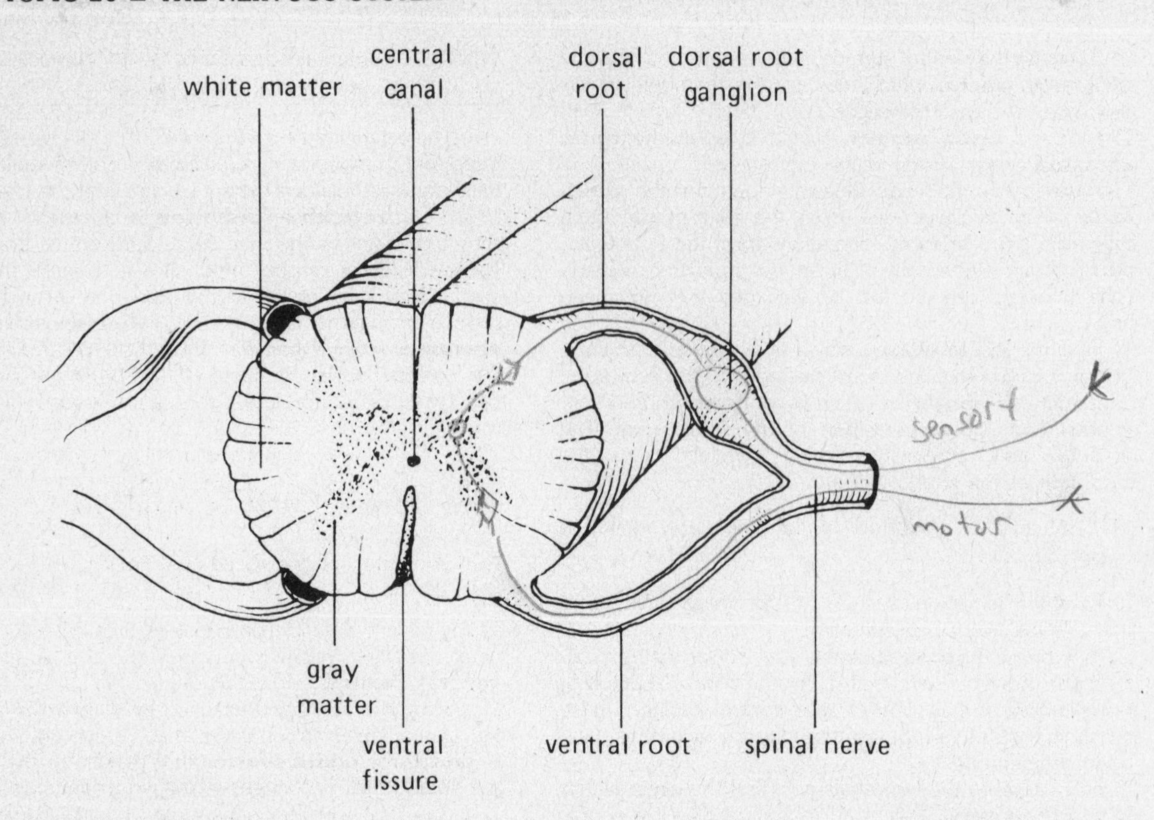

Figure 26–3. Spinal cord. This cross section is at the level of a spinal nerve and shows the dorsal and ventral roots. The white matter containing the axons surrounds the gray matter, which consists of cell bodies.

□ Locate the right and left **ventral roots** through which **motor neurons** carry impulses away from the spinal cord.

□ Examine the ventral part of the gray matter in the compound microscope. Locate the cell body of a motor neuron, a large, triangular cell that is darkly stained.

What is the function of the motor neurons?

Return impulse to effected organ or muscle

□ Locate an **association neuron** in the dorsal portion of the gray matter; it will have a more rounded cell body.

What is the function of an association neuron?

connect sensory impulse with motor Neuron

□ Find the rounded body of a **sensory neuron** within the dorsal root ganglion.

In which direction does this cell conduct impulses?

toward spinal cord

□ Examine the white matter to see cross sections of nerve fibers. The dark central part of each fiber is the axon of the neuron, and the space around it is where the myelin is located. (Myelin may be absent as a result of the process of making the slide.) The myelin sheath is surrounded by the membrane of the **neurilemma.**

□ Use the diagram in Figure 26–3 to sketch in a **reflex arc.** Include the sensory neuron, the association neuron, and the motor neuron in your sketch.

THE PERIPHERAL NERVOUS SYSTEM

The dorsal and ventral roots join to form a **spinal nerve.** There are 32 pairs of spinal nerves branching from the human spinal cord. The spinal nerves and spinal cord can control reflex behavior without any input from the brain. Normally most responses other than very basic reflexes are modified by impulses from the brain. Only some of the sensory input to the spinal cord will reach the brain.

The major sense organs and heart are served by 12 pairs of **cranial nerves,** which lead directly into the brain rather than first passing through the spinal cord.

The location of these cranial nerves is shown in Figure 26–1:

I	Olfactory	sensory	smell
II	Optic	sensory	sight
III	Oculomotor	motor	eye
IV	Trochlear	motor	eye
V	Trigeminal	mixed	face, mouth
VI	Abducens	motor	eye
VII	Facial	mixed	face, tongue
VIII	Acoustic or Auditory	sensory	hearing, balance
IX	Glossopharyngeal	mixed	mouth, throat
X	Vagus	mixed	neck, internal organs
XI	Spinal accessory	mixed	larynx, shoulder muscles
XII	Hypoglossal	motor	tongue

□ Locate as many of the cranial nerves as you can on the sheep brain.

The autonomic nervous system consists of chains of ganglia alongside the spinal cord and nerves leading from the ganglia to major organs, blood vessels, and skin. The **sympathetic** division triggers appropriate physiological responses to an emergency situation. The **parasympathetic** division generally has the opposite effect and stimulates routine activities such as digestion. The autonomic nervous system usually works without conscious control but can be influenced by the central nervous system. An emergency response, for instance, is often initiated by a terrifying visual or auditory stimulus. In conditioning by biofeedback, the responses of the autonomic nervous system are relayed to the conscious level of the CNS, which then allows the brain to exert a positive or negative influence on the response. In this way "autonomic functions" can be brought under conscious control to a certain extent.

SENSE ORGANS

Sense organs contain specialized neurons called receptors and accessory structures that increase the efficiency of the receptors in perceiving stimuli. The five types of receptors are:

1. *Photoreceptors,* which respond to light.
2. *Mechanoreceptors,* which respond to movement or physical stress.
3. *Chemoreceptors,* which respond to chemical substances.
4. *Thermoreceptors,* which respond to heat or cold.
5. *Electroreceptors,* which respond to electrical fields.

In this exercise you will be studying the first four types. The receptors synapse with sensory neurons that carry impulses through nerves to the CNS or to ganglia of the autonomic nervous system.

The Eye

In the eye, photons of a visual stimulus are **transduced** into nerve impulses by photoreceptors in the **retina.** In mammals there are two types of receptors: **rods** for perceiving the intensity of light and **cones** for detecting color or wavelength. The cones are concentrated in the **fovea** of the retina, the area of most acute vision. The rods and cones synapse with neurons in the retina, the neurons interact in the initial processing of the visual input, and finally the information is transmitted to the brain via nerve impulses in the optic nerves. The actual transduction process occurs when the photon is absorbed by the vitamin-A-derived part of the visual pigment **rhodopsin** and undergoes a structural change. This change causes a depolarization of the receptor cell membrane and sets up a generator potential. If enough photons strike the cell, an action potential will result.

The structure of the vertebrate eye is shown in Figure 26–4.

□ Take a preserved sheep eye and note the outer covering, the **sclera,** to which muscles are attached to move the eye.
□ Find the **optic nerve** on the posterior surface.
□ The transparent part of the sclera over the pupil is the **cornea.** The cornea is thicker in the center and acts as a lens to help magnify the image.
□ Beneath the cornea, find the **iris,** which surrounds the opening into the eye, or **pupil.** The amount of light entering the pupil is regulated by the contraction of muscles controlling the iris.
□ Orient the eye so that the optic nerve is to your left and the pupil is horizontal.
□ Cut through the eye with your scalpel from front to back leaving part of the sclera at the bottom intact to hold the halves together.
□ Find the transparent **lens** just behind the pupil. It is hard in a preserved eye, but in life it is flexible and can be thickened or flattened by its associated muscles to focus the image on the retina. In most people over 40, the lens is no longer flexible enough to change shape, and corrective lenses must be worn for close vision.
□ Note the **aqueous humor,** the fluid material that was contained between the cornea and lens.
□ The main part of the eye is filled with the gel-like **vitreous humor.**
□ Print something on a scrap of paper.
□ Place a blob of vitreous humor on the printing.

Does the vitreous humor contribute to magnification of the image? ___Yes___

□ Note the black lining of the sides of the eye. The pigment is located in the **choroid** layer.
□ Locate the **retina,** which lines the interior part of the eye opposite the pupil. The retina contains the rod and cone receptor cells, nerve cells, and blood vessels.
□ Locate the white area in the retina where the optic nerve leaves the eye. This **blind spot** is filled with nerve fibers and has no receptor cells.

When would the iris be fully open?

___dark___

When would the iris be tightly closed?

___a lot of light___

Why is the black lining important?

___helps absorb light so it is not reflected out___

When light rays pass through the lens of the eye, they are bent so that the image of an object falls on the retina upside down. A thicker lens bends the rays more strongly. The lens of the eye thickens in order to bring nearby objects into focus. Sketch an object in front of the eye shown in Figure 26–4, and show how its image would fall on the retina.

If the object were moved farther from the eye, would the lens have to thicken or flatten to keep the image in focus? ___flatten___

Why don't you see objects upside down?

____

____

The Ear

The structure of the human ear is shown in Figure 26–5. The outer ear is the part you are already familiar with: it consists of the **pinna** and the **external canal.** Water in the external canal gives you the sensation of

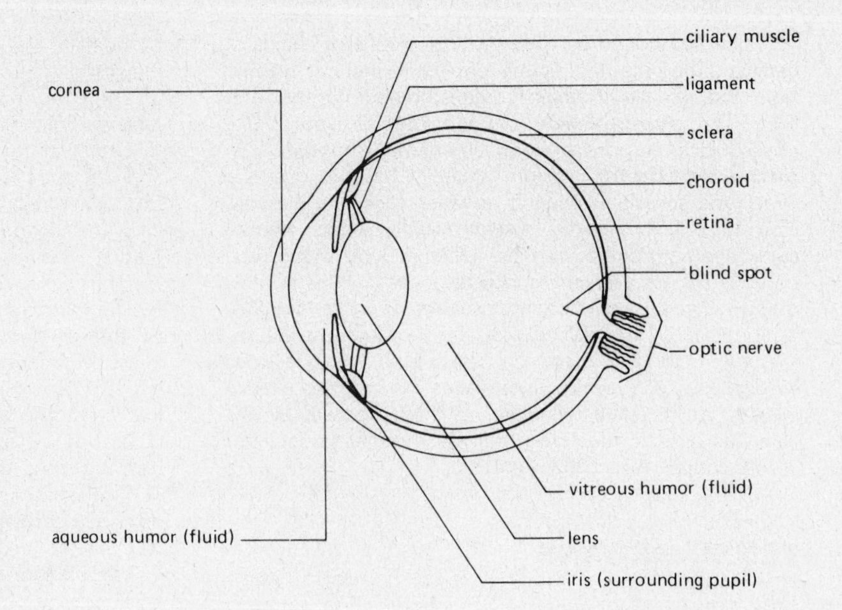

Figure 26–4. The vertebrate eye. Light passes through the transparent cornea and the pupil to the cells of the retina, where its energy is transduced into nerve impulses. The iris regulates the amount of light entering the eye by changing the pupil size.

"water in the ear," and too much ear wax in this canal will interfere with the transmission of sound to the **eardrum** (tympanum) at its inner end.

What is the function of the pinna?

captures sound

Study the features of the middle ear and inner ear in a cutaway model. The middle ear consists of the space between the eardrum, or tympanum, and the cochlea of the inner ear, and three small bones which conduct the sound. It is connected to the pharynx by the **Eustachian tube.**

Why is a connection of the middle ear with the exterior necessary?

regulate pressure

Respiratory infections can travel from the throat up the Eustachian tubes and cause painful inflammation and infection of the middle ear.

□ Locate the **cochlea** in the inner ear. The cochlea contains the organ of Corti, which is the organ of hearing.

If the cochlea were uncoiled, a cross section of it would appear as shown in Figure 26–6.

Within the cochlea there are three fluid-filled canals: the **vestibular canal,** which connects to the oval window, the **cochlear canal,** and the **tympanic canal.** Sound waves reaching the oval window set up vibrations within the vestibular canal which then travel back through the tympanic canal to the round window. Indirect vibrations are set up in the cochlear canal which contains the mechanoreceptor **hair cells** of the **organ of Corti.** Vibrations of the tectorial membrane above these cells cause movements of the gel-like matrix around the hairs and excite the cells by moving the hairs. Impulses from the receptors are transmitted along the basilar membrane on which the cells are located and reach the auditory nerve which carries them to the brain. Vibrations of different frequencies (pitches) will excite cells at different positions along the membrane, so that pitch as well as intensity can be sensed by the ear.

□ Examine a microscope slide of the **cochlea;** several sections through the three canals will be visible.
□ Identify the **vestibular, cochlear,** and **tympanic canals** in one section.

A second part of the inner ear, the **labyrinth,** has nothing to do with hearing, but also contains hair cells, which are mechanoreceptors. This organ of equilibrium consists of fluid-filled canals, which are stimulated when the fluid flows through them in either direction. The **semicircular canals** are in three planes and allow the organ to sense **acceleration** (change in speed or direction) of the head.

The base of the semicircular canals consists of two sacs, the **sacculus** and the **utriculus,** which are also fluid-filled and contain crystals of calcium carbonate. The crystals rest on a bed of sensory hair cells and change position slightly as the head is moved. The pattern of cells being stimulated while the head is at rest in any position allows the organ to sense the **resting position** of the head relative to the force of gravity.

The sense of resting position in humans is very weak, as any scuba diver can testify, and is easily overridden by visual cues. The sense of movement is much stronger. When it conflicts with the visual cues, as during a boat trip or plane flight, the result is the nausea of motion sickness.

HUMAN SENSORY PERCEPTION

The Blind Spot

□ Obtain a sheet of white paper, and mark on it a heavy black cross and a heavy black dot about 8 cm apart.
□ Cover your left eye, and hold the paper at arm's length with the dot at the left and the cross at the right.

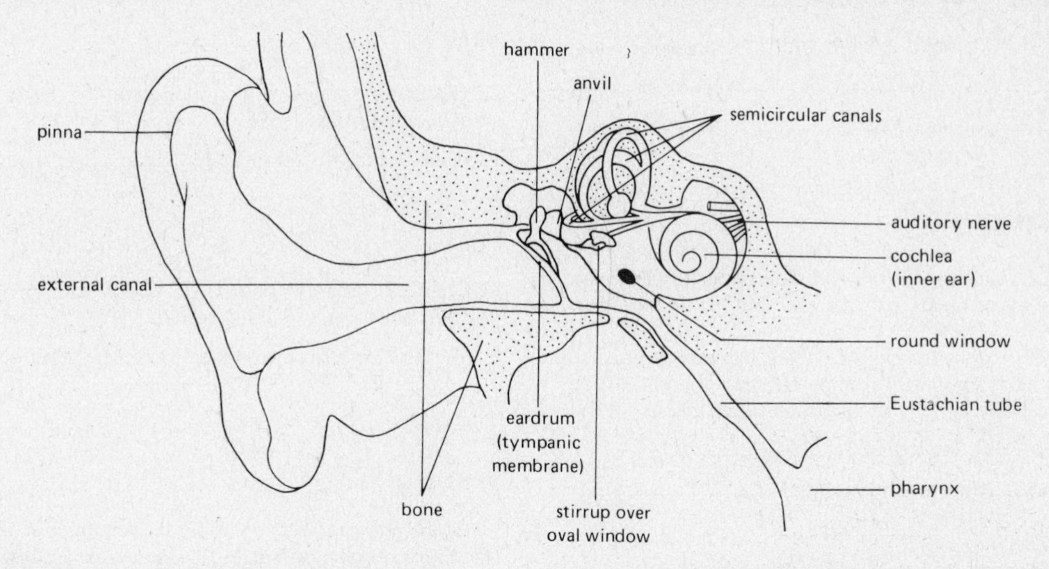

Figure 26–5. The human ear. The mechanical energy of sound waves is transduced into nerve impulses by cells within the cochlea. The semicircular canals sense the movement and position of the head and are the organ of equilibrium.

□ Stare hard at the dot while you slowly move the paper toward you.
□ Note that the cross is visible at first, disappears completely, and then reappears.

What do you see instead when the cross disappears?

the plain paper

Map your blind spot as follows:

□ Turn the paper sideways so that there is a clear space to the right of the black dot. Lay it flat on the desk.
□ Cover your left eye, and place your right eye about 25 cm from the black dot. Stare at the dot intently while your partner moves a small, brightly colored object (such as a bead on a white pipecleaner) toward the dot from your right side.
□ As soon as the object disappears, tell your partner to stop and mark the position of that edge of the blind spot.
□ Next ask your partner to move the object from the black dot toward the blind spot, and then to stop and mark the spot as soon as the object disappears.
□ Repeat several times, moving the object toward the blind spot from different directions, until your blind spot is well mapped, and then draw in its outline.

Does your blind spot have a regular or irregular shape?

Compare the size and shape of your blind spot with those of other members of the class. The distance from the dot to the blind spot will depend partly on the distance between your eye and the dot.

Why don't you have an empty space in your field of vision all the time?

binocular vision

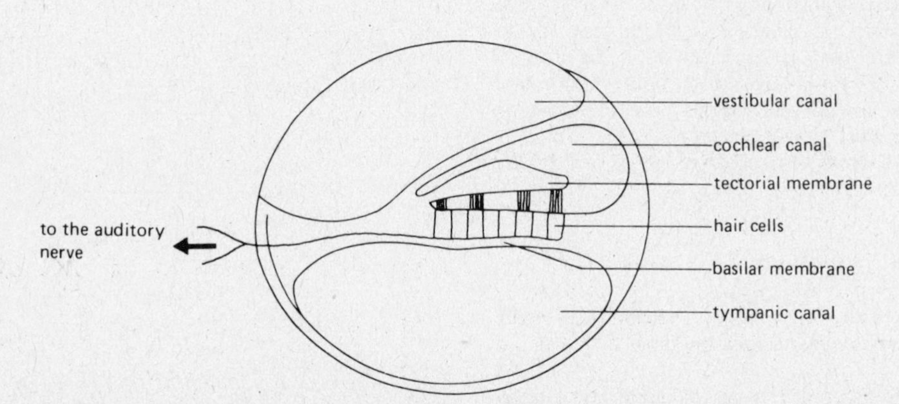

Figure 26–6. Cross section of one spiral of the cochlea. Vibrations from the bones of the middle ear travel through the vestibular canal and back through the tympanic canal. The mechanoreceptors that detect the vibrations are located on the basilar membrane of the organ of Corti.

Explain why we have a blind spot in each eye. Place where optic nerves are on retina

Afterimages

□ Place a small square of solid blue paper or cardboard on a sheet of white paper, and put it under strong illumination.
□ Working with your partner, stare intently at the blue paper while your partner times 60 sec. Then shift your gaze and stare intently at the plain white paper.

Do you see an afterimage? Yes

How long did the afterimage last? Few seconds

What color is it? yellow

□ Place a small square of solid yellow paper on a white sheet, and repeat your observation.

What color is the afterimage when the object is a yellow square? blue

While you are staring at the blue square, the cones that are sensitive to the blue light become **adapted** (less responsive), whereas those sensitive to red and green light are unstimulated. Then when you suddenly switch to white light, which contains all three colors, the blue receptors give a response much lower than normal, whereas the red and green receptors give a response stronger than normal. This is exactly the response that occurs with yellow light, and your brain interprets the plain white paper as being yellow. As the receptors return to their normal response levels, the afterimage slowly fades.

Chemoreceptors

Record your observations in Table 26–1.

TEST A

□ Obtain slices of apple and potato cut to the same size.
□ Close your eyes, pinch your nose shut, and have your partner place one slice on your tongue. Try to detect whether the slice is from an apple or from a potato without moving it around in your mouth and without breathing for 10 sec. Record your identification of the slice as correct (+) or incorrect (–).
□ Repeat several times with slices of apple and potato in random order.

TEST B

□ Suck on the slice, and move it around on your tongue. Record your observations in Table 26–1 at the end of this topic.
□ Repeat several times with slices of apple and potato in random order.

Can you distinguish apple and potato by taste alone? ____________

TEST C

□ Blot your tongue dry, and repeat the tests. Record your results in the table.

Is saliva necessary for tasting? ____________

TEST D

□ Chew a slice of potato or apple with your nose pinched shut. Record your results.

Can you distinguish apple from potato without olfaction?

____

TEST E

□ Again chew a slice of apple or potato, but this time breathe through your nose. Record your results.

Is it easier to distinguish them using olfaction?

____

TEST F

Proceed as follows to map the location of your taste receptors.

□ Obtain dropping bottles of the following solutions:

5.0% sucrose	(sweet)
10% NaCl	(salty)
0.1% quinine sulfate	(bitter)
1.0% acetic acid	(sour)

□ Open your mouth wide, without sticking out your tongue, and have your partner place a drop of one solution on spot #1 on the tip of your tongue. Repeat on the side and finally the back as shown in Figure 26–7. Rinse after each drop with plain water.
□ Rate the taste for each drop as absent (0), faint (+), or strong (++), and record the results in the appropriate column in Table 26–2.
□ Repeat for the other three solutions.

When you finish testing, map your areas sensitive to sweet, salty, bitter, and sour tastes on the outlines in Figure 26–8.

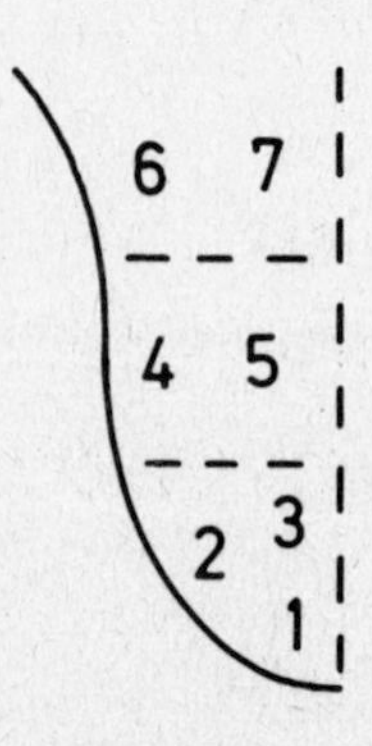

Figure 26–7. Tongue and taste. Numbers label the spots to be tested.

Proprioceptors

Proprioceptors in your muscles and joints inform your nervous system about the location of your body and its appendages without visual input. A certain posture can be maintained by reflexes involving antagonistic muscle pairs and their proprioceptors. Test the effectiveness of proprioception alone by performing these actions with your eyes closed. Record the results in Table 26–3.

TEST A

□ Stand with your feet together and arms at your sides. Have your partner measure the extent of swaying and your success in attempting to compensate for it.

TEST B

□ Stand on one leg with your arms extended. Have your partner time how long you can stand on one leg with your eyes closed, then repeat with your eyes open.

How important is visual input in supplementing your proprioceptors?

____

Other proprioceptors (kinesthetic receptors) allow detection of movement of body parts without the benefit of visual input. Test your kinesthetic sense as follows.

TEST C

□ Close your eyes, and extend your arms fully to the sides at shoulder height.
□ Move your arms together without bending your elbows, and attempt to touch your index fingers side to side.
□ Have your partner measure how far apart your fingers are if your fingers fail to touch.

TEST D

□ Close your eyes, and extend your arms sideways at shoulder height, but this time bend your elbows sharply so that your index fingers point at each other.
□ Move the fingers together until they touch.
□ Have your partner measure how far apart your fingers are if they fail to touch.

Thermoreceptors

MAPPING THERMORECEPTORS

□ Stamp a grid on a hairless part of your forearm, or draw one 3 × 3 cm marked off with a grid 0.5 × 0.5 cm.
□ Have your partner hold a probe in a boiling water bath, dry it, and quickly touch it *lightly* to one point on the grid while you look the other way.
□ Indicate whether you felt a sensation of warmth, and if so, record the result on the grid in Table 26–4.
□ Repeat for the other points on the grid, and calculate the number of heat-sensitive points per square centimeter in this area of skin. Next, place a probe in a bath of ice water, dry, and touch it to one point on the grid. Record whether there was a sensation of cold.
□ Repeat for the other points on the grid, and calculate the number of cold-sensitive receptors per square centimeter.

Which type of receptor is more concentrated in this area of skin? ______________

HABITUATION

□ Obtain three beakers; fill one with cold water, one with warm water, and one with lukewarm water.
□ Place the index finger of your left hand in the beaker of warm water and the index finger of your right hand in the beaker of cold water. Then count to 60.
□ Now place both fingers in the beaker of lukewarm water.

What sensation do you feel with your left finger?

warm

What sensation do you feel with your right finger?

cool

Explain your sensations.

____

____

When you are finished with this lab, return all materials and store your microscope properly.

EXPLORING FURTHER

Arnold, A. P. "Sexual differences in the brain." *American Scientist* 68: 165, 1980.

Grollman, S. *A Laboratory Manual of Mammalian Anatomy and Physiology.* 4th ed. New York: Macmillan, 1978.

Jacobs, B. L., and M. E. Trulson. "Mechanisms of action of LSD." *American Scientist* 67: 396, 1979.

Katz, B. *Nerve, Muscle and Synapse.* New York: McGraw-Hill, 1966.

Oscar-Berman, M. "The neuropsychological consequences of long-term chronic alcoholism." *American Scientist* 68: 410, 1980.

Scientific American Articles

September 1979: This issue is devoted to the brain.

Axelrod, J. "Neurotransmitters." June 1974 (#1297).

DiCara, L. V. "Learning in the autonomic nervous system." January 1970 (#525).

Favreau, Olga Eizner, and Michael C. Corballis. "Negative aftereffects in visual perception." December 1976 (#574).

Hodgson, E. S. "Taste receptors." May 1961 (no offprint).

Hubel, David H., and Torsten N. Wiesel. "Brain mechanisms of vision." September 1979 (#1442).

Iverson, Leslie L. "The chemistry of the brain." September 1979 (#1441).
Jouvet, Michel. "The states of sleep." February 1967 (#504).
Kandel, Eric R. "Small systems of neurons." September 1979 (#1438).
Kety, Seymour S. "Disorders of the human brain." September 1979 (#1445).
Land, Edwin H. "The retinex theory of color vision." December 1977 (#392).
Lassen, Niels A., David H. Invar and Erik Skinhoj. "Brain function and blood flow." October 1978 (#1410).
Morell, Pierre, and William T. Norton, "Myelin." May 1980 (#1469).
Nathanson, J. A., and P. Greengard. "Second messengers in the brain." August 1977 (#1368).
Parker, Donald E. "The vestibular apparatus." November 1980.
Pribram, K. H. "The neurophysiology of remembering." January 1969 (#520).
Rushton, W. A. H. "Visual pigments and color blindness." March 1975 (#1317).
Snyder, S. H. "Opiate receptors and internal opiates." March 1977 (#1354).
Stevens, Charles F. "The neuron." September 1979 (#1437).
Tosteson, Daniel C. "Lithium and mania." April 1981.
von Bekesy, G. "The ear." August 1957 (#44).

Student Name ______________________ **Date** ______________

TABLE 26–1. DATA ON TASTE CHEMORECEPTION

Record (+) if the identification is correct, (−) if the identification is incorrect, and (0) if no identification can be made. Space has been provided in each test for up to six sets of observations.

Test		1	2	3	4	5	6
A. No sucking	Apple						
	Potato						
B. Sucking	Apple						
	Potato						
C. No saliva	Apple						
	Potato						
D. Chewing	Apple						
	Potato						
E. Breathing and chewing	Apple						
	Potato						

TABLE 26–2. MAPPING DATA FOR TASTE CHEMORECEPTORS

Rate each drop of solution as either no taste (0), mild taste (+), or strong taste (++). Map the results in Figure 26–8.

		POSITION ON TONGUE BEING TESTED						
		Front			*Side*		*Back*	
RECEPTOR	**SOLUTION**	*1*	*2*	*3*	*4*	*5*	*6*	*7*
Sweet	5.0% sucrose	++			+		0	0
Salty	10% NaCl	++			+		0	0
Bitter	0.1% quinine sulfate	0			+		++	++
Sour	1.0% acetic acid	+			++			0

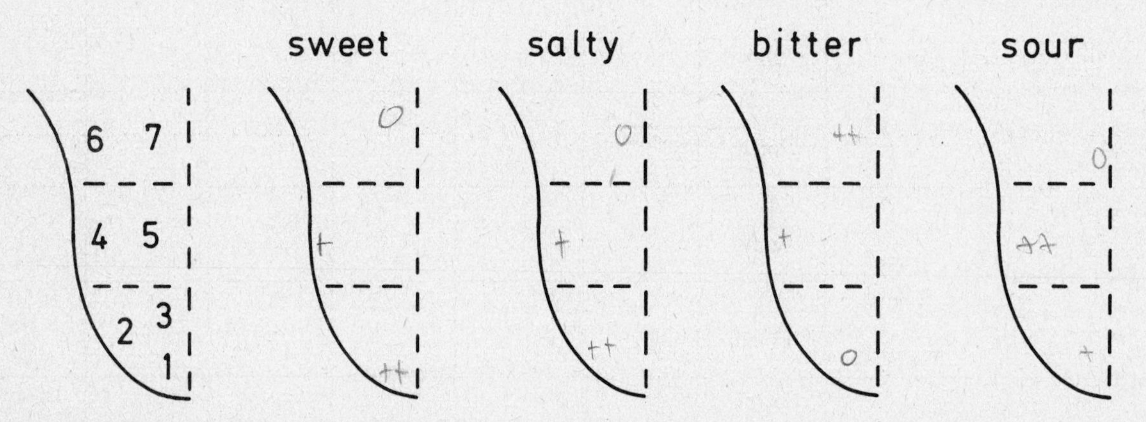

Figure 26–8. Results of taste test. Mark the positive responses for each type of taste receptor where they occurred on the tongue.

Student Name ______________________ **Date** ______________

TABLE 26–3. DATA FOR PROPRIOCEPTION OF STATIC AND DYNAMIC EQUILIBRIUM

Test	Observations
A. Feet together	
B. One leg, eyes closed One leg, eyes open	
C. Touch fingers	Fingers were ______ cm apart
D. Touch fingers	Fingers were ______ cm apart

TABLE 26–4. DATA ON THERMORECEPTION IN THE SKIN

Warmth Receptors ____________________

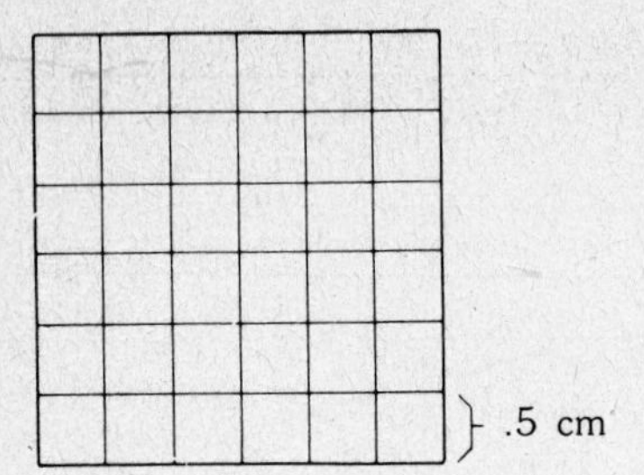

concentration: ______/cm^2

Cold Receptors ____________________

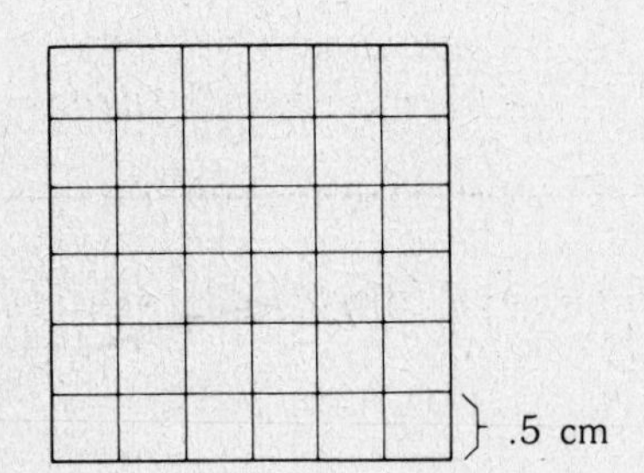

concentration: ______/cm^2

TOPIC 27

Muscles and Skeletons

OBJECTIVES

When you have completed this topic, you should be able to:

1. Describe the type of movement that is carried out by striated muscle, smooth muscle, and cardiac muscle; give the advantages of each muscle type.
2. Identify the major bones of the human skeleton and distinguish whether they belong to the axial skeleton or the appendicular skeleton.
3. Describe the major structural differences between bone and cartilage and give the advantageous properties of each tissue.
4. Identify the major muscles of a vertebrate.
5. Explain what happens when a muscle is given an increasingly intense stimulus.
6. Explain what happens when a muscle is given an increasingly frequent stimulus.
7. Define flexor, extensor, antagonist, reciprocal inhibition, tetany, fatigue, and twitch.

TEXT REFERENCES

Chapter 40.
Smooth muscle, cardiac muscle, striated muscle, muscle physiology, antagonistic muscles, skeletons, joints.

INTRODUCTION

Muscles are **effectors.** They respond to stimuli detected by other cells. Glands, light organs, and electric organs are also effectors.

The major function of muscles is to move things by changing chemical energy to kinetic energy (motion). **Locomotion** is the most obvious form of movement in which the muscles in cooperation with the skeleton move the whole organism, but muscles also move blood, food, air, and eyeballs, regulate body temperature, and cause countless other physiological changes.

The skeleton is clearly necessary for muscles to be effective for locomotion in vertebrates, but it also serves other functions. It cradles and protects internal organs, especially the heart and brain, and it supports the body as a whole; antlers help some animals to protect themselves. The long bones are the main blood-generating organs and help maintain the immune system. Bones also serve as a depot for fat and calcium.

Because muscles can react only by contracting and not by "expanding," they must work in pairs and must be attached to the part to be moved, usually a bone. Consider the two bones joined by the fibrous connective tissue of a **ligament** to form a **joint,** as shown in the sketch in Figure 27–6 at the end of this topic.

☐ Draw in a muscle attached to both bones that will draw them together so that the angle of the joint becomes smaller; this is the **flexor** of the joint. Label the upper attachment point the **origin** (nearest the body) and the lower, the **insertion.** Now draw in the antagonistic muscle which widens the angle of the joint: this is the **extensor.** Label its origin and insertion also.

Are flexors or extensors important when you hit a baseball? Extensors

Are flexors or extensors important when you pick up a book? flexors

Muscles can also work without being attached to bones. When they are in the form of a ring they can contract to close off a tube (pyloric sphincter), or when they are in the form of a tube they can contract in a ring or parallel to the length of the tube to change its shape and force the tube contents to move, as in peristalsis. Cardiac muscle contracts rhythmically for a lifetime even without being stimulated to contract by individual nerve endings. Most of the activities of **smooth** and **cardiac** muscle are not under our conscious control, so we are not as aware of them as we are of **skeletal** muscle.

SKELETON

The vertebrate skeleton is divided into the **axial skeleton** consisting of the skull, vertebral column, and sternum, and the **appendicular skeleton,** which consists of the pectoral and pelvic girdles and their appendages. Locate the axial and appendicular skeletons in Figure 27–1, which shows the bones of a human being.

The connection between any two bones is a joint. These are some different types of joints:

1. A **suture** is a joint where the bones don't move (example: skull).
2. A **hinge joint** is limited to movement in one plane (example: knee, finger joints).
3. A **ball and socket joint** allows rotation of the bones as well as flexion (example: hip, shoulder).
4. A **condyloid joint** allows movement in any direction but not rotation (example: joints between fingers and the hand).
5. A **plane joint** allows some slipping of two bones but very little flexion or rotation (example: joints between vertebrae).

□ Study the major bones in Figure 27–1. Then complete Table 27–1 at the end of this topic by matching the technical names of the bones with their common names and deciding whether they are part of the axial or appendicular skeleton.

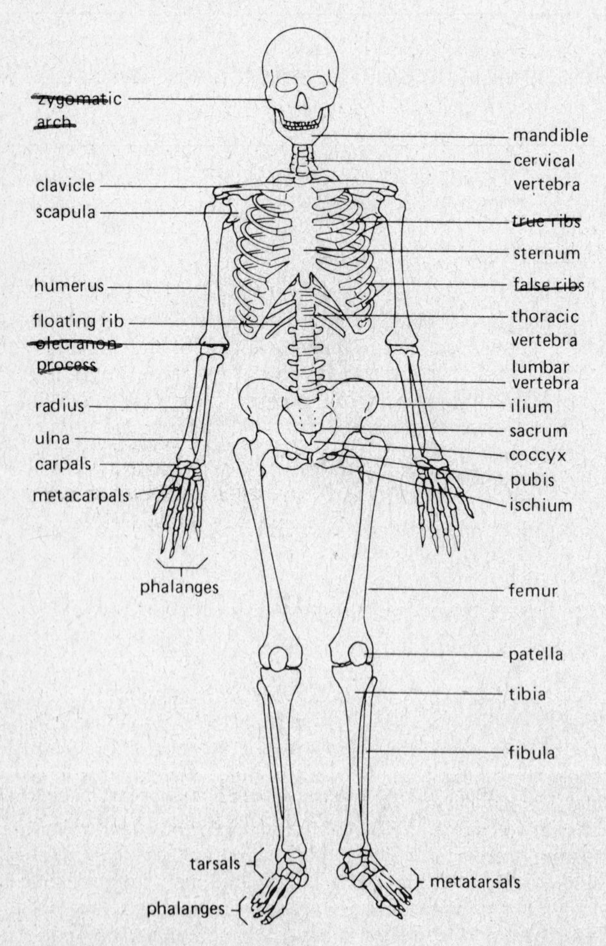

Figure 27–1. Human skeleton. Some of the 206 bones (29 in the skull alone).

□ If there are other vertebrate skeletons on demonstration in your laboratory, compare them to each other and to the human skeleton. Some good examples to use are a frog, cat, bat, pigeon, fish, snake, or crocodile.
□ Compare a bipedal animal with one that is quadrupedal.
□ Notice the adaptations of the skeleton for the type of locomotion of the animal.
□ Consider a particular bone such as the femur, ischium, or mandible, and see how it differs in different organisms. Pay particular attention to these features:

Articulation of the spine and skull
Structure of the vertebrae
Ribs
Scapula
Clavicles
Pelvic girdle
Carpals and metacarpals
Tarsals and metatarsals
Phalanges of fore- and hindlimb

Your instructor may ask you to make a table showing how these skeletal parts show different adaptations in different organisms.

A typical long bone is shown in Figure 27–2. The ends are covered with smooth **cartilage** (hyaline cartilage) and rub against the cartilage of the adjacent bones. Ligaments hold the bones in place at either end and the muscles are attached to the bone by **tendons.** During development, there are sections of cartilage within the end of the bone that grow as the bone lengthens. (The very ends of the bone, the **epiphyses,** have to be hard and bony or they would wear away.) When the growing cartilage becomes converted to bone so that the epiphyses and **shaft** (diaphysis) unite, growth in the length of the bone stops.

The hollow interior of the bone is the **marrow** and is filled with blood-forming cells and fat. In mammals, only the very outer shell of the bone is hard, or **com-**

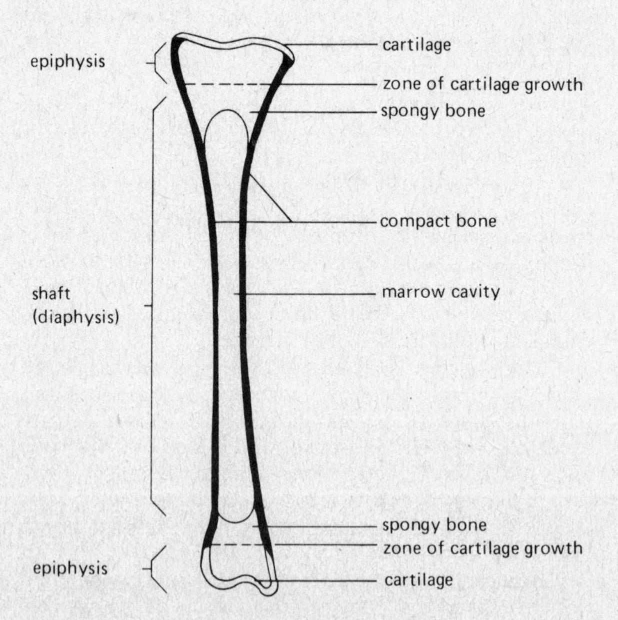

Figure 27–2. A long bone. During development, the bone grows longer in the regions marked with dashes. When the epiphyses and shaft unite, growth stops.

pact bone. The rest is **spongy bone,** which is rich in blood vessels.

MICROSTRUCTURE OF BONE AND CARTILAGE

Compact Bone

□ Examine a slide of compact bone; this is bone ground thin enough so that light can penetrate. The concentric circles are the **Haversian systems.** Each contains a **Haversian canal** in which there are one or two blood vessels (a large capillary or a small arteriole and a venule) as well as a nerve. The rings are the lamellae, which are rings of the calcium phosphate **matrix** that makes up the substance of the bone. The tiny open spaces around the edges of each circle are the **lacunae,** which contain the bone cells that produced all of the matrix. Tiny canals (canalicules) connect each lacuna with a Haversian canal, allowing extracellular fluid to reach and nourish the living cells.

□ Draw a section of the bone in the space provided in Figure 27–6 to show that you understand its structure.

Cartilage

Cartilage differs from bone in that it does not have mineralization in its matrix and it lacks a blood supply.

□ View a slide of cartilage in the microscope. The homogeneous area is the elastic **matrix,** which contains collagen and elastic fibers. The individual cells that secrete the matrix are scattered throughout the matrix in holes called **lacunae.** Each hole contains one or more cells.

□ Examine your own head and the front of your neck, and list four sites where you find cartilage.

1. ______________________________
2. ______________________________
3. ______________________________
4. ______________________________

Where does the cartilage occur in the spinal column?

____

Where does it occur in the thoracic cage?

____

What parts of the pelvic girdle are made of cartilage?

____

MICROSTRUCTURE OF MUSCLE

Striated Muscle – Skeletal muscle

□ Examine a slide of striated muscle in the microscope. Each strand is a muscle **fiber,** a large cell with many peripheral nuclei.

What is this kind of cell called? ______________

The membrane of the muscle fiber is the **sarcolemma.** Within the fiber are the many longitudinal **myofibrils,** or bundles of thick and thin filaments.

□ Locate a muscle fiber in the slide, and identify some of its nuclei.

The fine striations that give striated muscle its name are the result of the lining up of thick and thin filaments in exactly the same register in many adjacent myofibrils. Striated muscle is found in the voluntary muscles that move the parts of the body and that are under conscious control.

Smooth Muscle

□ View a slide of smooth muscle.

This type of muscle consists of bundles of spindle-shaped cells. Each cell has a central nucleus and is filled with myofibrils. There is no pattern of cross striations. Smooth muscle is designed for slow, sustained contractions.

Where is smooth muscle found?

____

____

Cardiac Muscle

□ Examine a slide of cardiac muscle.

This type of muscle is found only in the walls of the vertebrate heart.

Is cardiac muscle striated? ______________

Does it have nuclei? ______________

What conspicuous features does it show that are not present in smooth or striated muscle?

____

____

□ Locate the position of the **sarcolemma.** (The membrane is too thin to see.) The dark bands scattered throughout the tissue are the **intercalated discs,** which are junctions between individual muscle cells.

MUSCLE ANATOMY

Movement of the body parts and locomotion are accomplished by pairs of antagonistic muscles. Because the forces developed are quite large, there must be reflex circuits that inhibit the contraction of a given flexor while the extensor is contracting, and vice versa. This set of reflexes is called **reciprocal inhibition.** Tension on a muscle triggers nerve impulses from **stretch receptors** in the muscle tendons, which provide the brain with information needed to coordinate contraction of a muscle and relaxation of its antagonist.

Study the human muscle pairs shown in Figure 27–3. If the **rectus femoris** muscle of the thigh contracted, the leg would be straight at the knee. If the **biceps femoris** contracted, the leg would be bent.

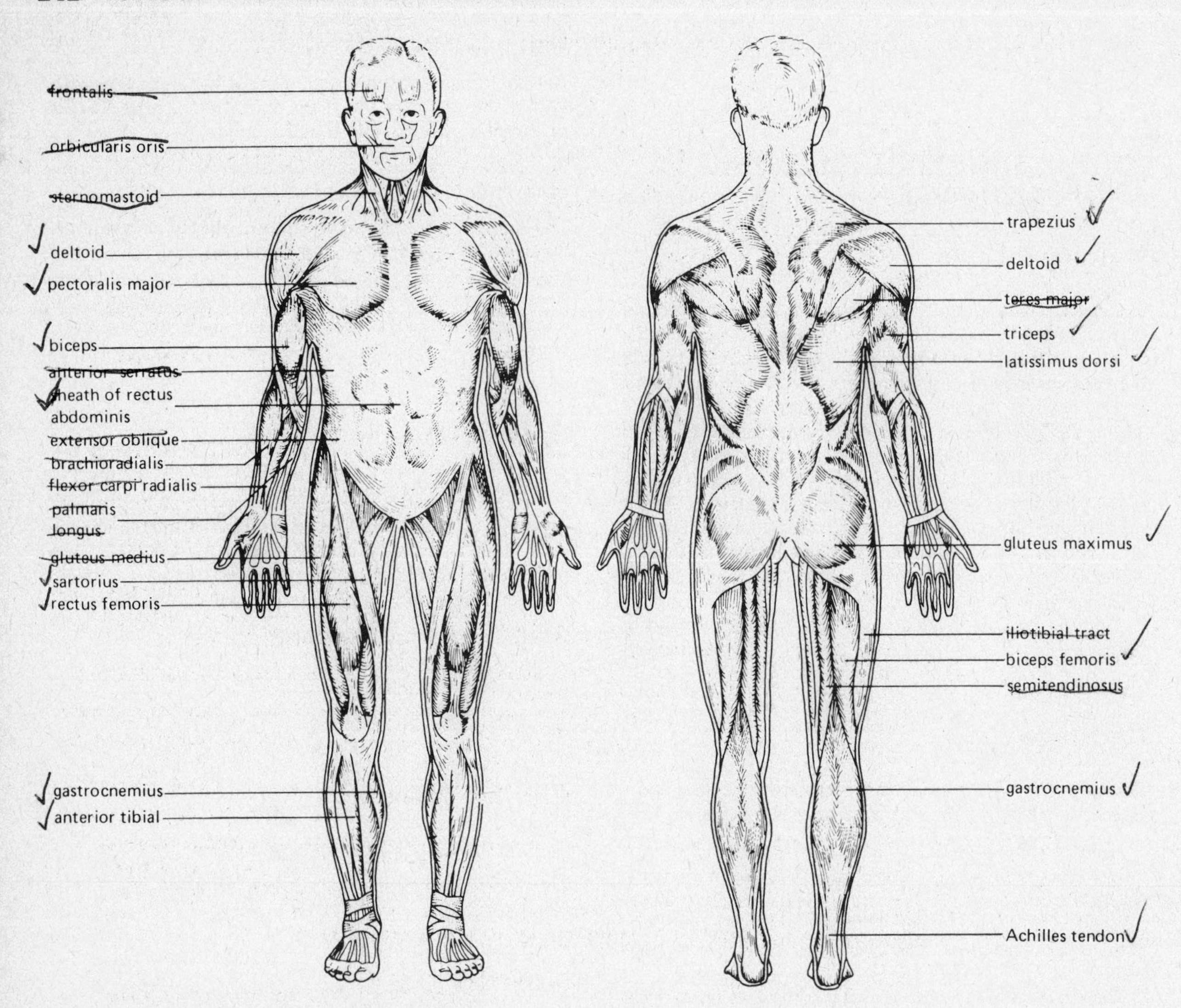

Figure 27–3. Human muscles. (Reprinted with permission of Macmillan Publishing Co., Inc., from S. Grollman, *A Laboratory Manual of Mammalian Anatomy and Physiology*, 4th ed., New York, 1978.)

What pair of antagonists control the bending and straightening of the elbow?

Tricep and Bicep

What muscle is involved in each of the following activities?

Frowning:

Puckering the mouth:

Clenching the fist:

Lifting a cup of coffee: bicep

Pushing a door shut: tricep

Rising from a deep knee bend: rectus femoris

Shrugging your shoulders: trapezius

Kicking a ball: rectus femoris

The muscles of a frog are shown in Figure 27–4 for comparison. Use this information for your dissection of the leg in the next part of the laboratory.

MUSCLE PHYSIOLOGY

Preparation

SET UP YOUR EXPERIMENTAL APPARATUS *BEFORE* BEGINNING YOUR FROG DISSECTION.

The **kymograph** is a drum that revolves at a constant rate. Your instructor will give you directions for operating it.

□ Put a clean recording paper around the drum with the shiny side exposed. Lap the left edge of the paper over the right so that the writing pen steps down when the drum is rotating, rather than catching on the edge of the paper.

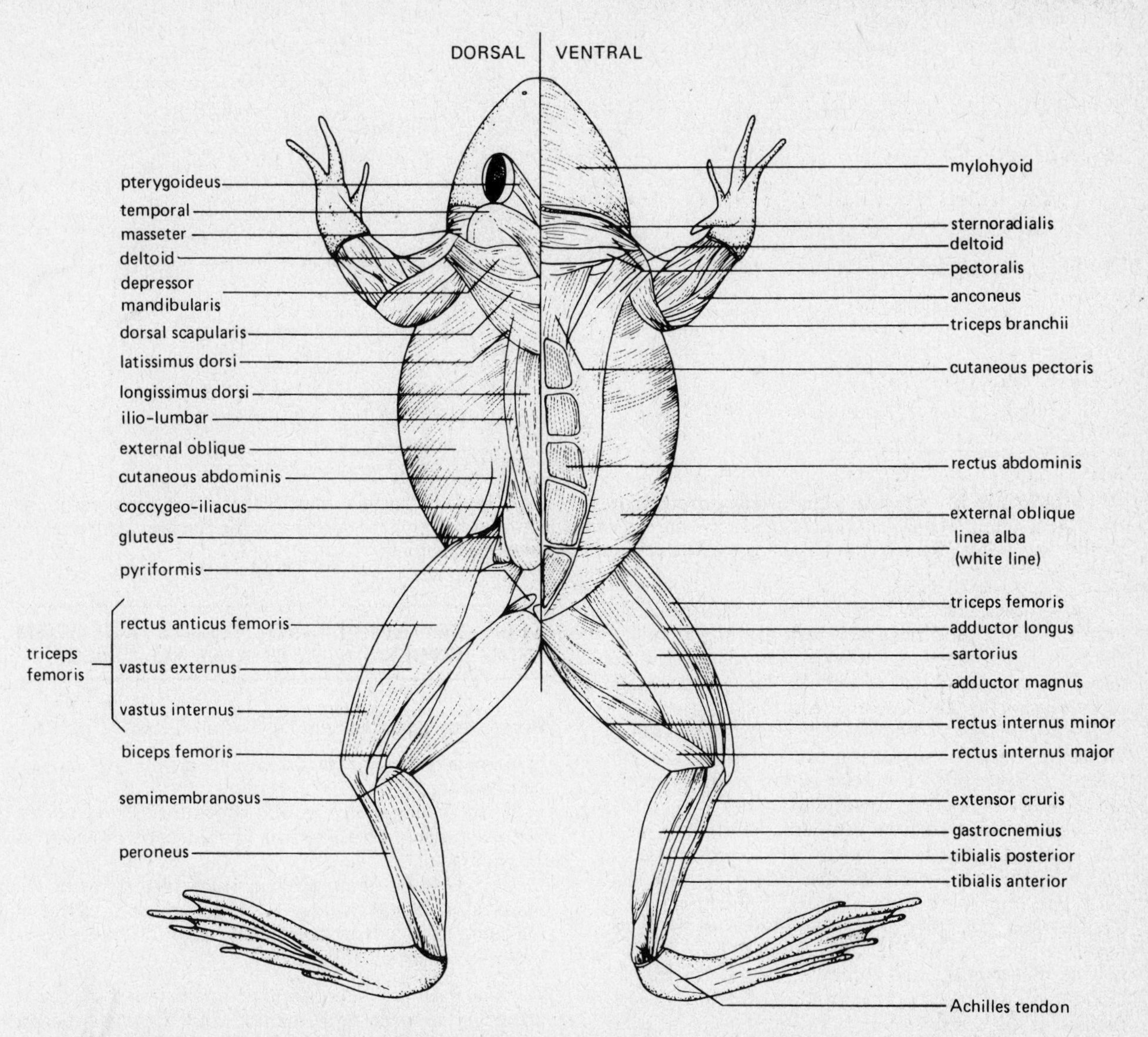

Figure 27–4. Frog muscles. (Redrawn from Wodsedalek, J. E. and Charles F. Lytle, *General Zoology Laboratory Guide,* 8th Ed. © 1971, 1977, 1981 Wm. C. Brown Company Publishers, Dubuque, Iowa. Copyright © by Kendall/Hunt Publishing Co. Reprinted by permission of Kendall/Hunt Publishing Co.)

□ Set up the ring stand as shown in Figure 27–5. Attach a **femur clamp,** a utility clamp with a *clean* slide, and a muscle lever. The nerve leading to the frog's muscle will lie on the slide and *must be kept moist with frog Ringer's solution at all times,* or the sensitive nerve cells will quickly die.

□ Set up the signal magnet and the time marker if they are provided with your experimental apparatus.

□ Attach leads to the stimulator or inductorium as shown in Figure 27–5 or follow the directions given for your equipment.

□ Lead 1 connects the signal marker with the 6-V battery and then goes to the MAG terminal of the stimulator.

□ Lead 2 connects the signal marker with the MAG terminal directly.

□ Lead 3 connects the heart clip clamp to the MOD IN terminal.

□ Lead 4 connects the slide and utility clamp with the SYNC IN terminal. It should be switched to the femur clamp if you wish to stimulate the muscle directly.

WHEN YOUR INSTRUCTOR HAS APPROVED YOUR SETUP, PROCEED WITH YOUR FROG DISSECTION.

□ Locate the **gastrocnemius** muscle and the Achilles tendon that attaches it to the bones of the foot as shown in Figure 27–4. This is the muscle that you will study. Locate the semimembranosus and the vastus externus muscles. Between them is the **sciatic nerve** (not visible in Figure 27–4), which you will stimulate to excite the gastrocnemius muscle.

□ Obtain a pithed frog or pith your own according to your instructor's directions. The procedure for pithing a frog is given in Topic 23.

□ Cut the skin completely around the "waist" of the frog, and pull it off, freeing it from the legs.

□ Using a probe, *carefully* separate the large muscles (vastus externus and semimembranosus) on the *dorsal,* medial side of the upper leg. Locate the femur, along which you will find blood vessels and a white or light yellow fiber, which is the sciatic nerve.

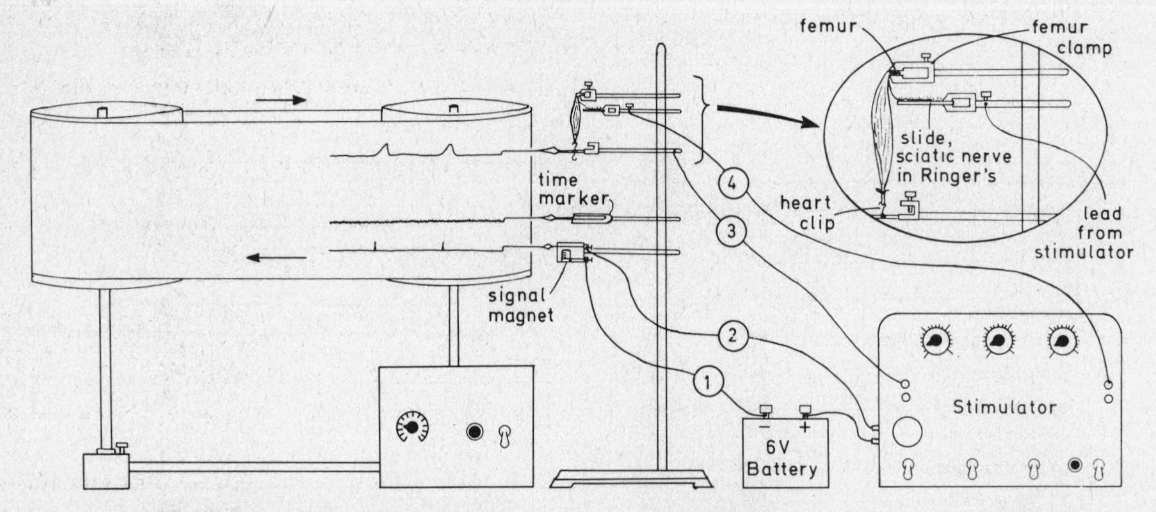

Figure 27–5. Muscle physiology experiment. Apparatus for stimulating a muscle and recording its response on a kymograph. The signal magnet marks the time of stimulation and the time marker shows the duration of the latent period and twitch. Set up the 4 leads according to your instructor's directions.

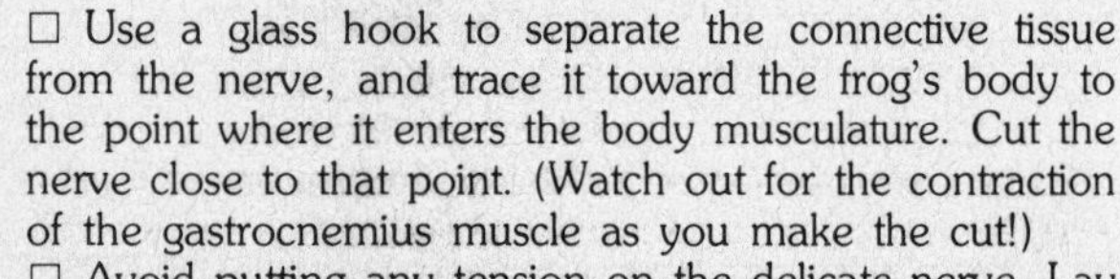

□ Use a glass hook to separate the connective tissue from the nerve, and trace it toward the frog's body to the point where it enters the body musculature. Cut the nerve close to that point. (Watch out for the contraction of the gastrocnemius muscle as you make the cut!)
□ Avoid putting any tension on the delicate nerve. Lay the nerve over the lower leg so that it is out of the way.
□ Pull the muscles of the upper leg away from the femur, and cut them off close to the knee.
□ Cut the femur halfway between the hip and knee. Leave enough of a stump to grip with the femur clamp.
□ Trace the broad Achilles tendon around the heel, and cut it free from the bottom of the foot. Free the gastrocnemius from the other muscles and bones of the lower leg.
□ Cut the muscles and bones of the lower leg just below the knee, taking care not to damage the gastrocnemius or the sciatic nerve.
□ Mount your preparation in the apparatus by attaching the femur to the femur clamp and attaching the Achilles tendon to the muscle lever with a heart clip and a piece of thread.
□ Make these adjustments as necessary to get a good recording:

Pen. The ink should be flowing freely to give a good tracing. If not, close the cap of the reservoir bottle, pinch it until ink flows from the pen, and then loosen the cap slightly while pinching. Adjust the height of the bottle so that ink can flow out.

Heart clip string. Align the muscle and heart clip string so that they are vertical to the muscle lever.

Muscle tension. Raise or lower the muscle lever on the ring stand so that the Achilles tendon is under slight tension.

Pen arm. Move the position of the thread so that the arm responds freely to movement of the muscle.

Kymograph speed. Adjust the rotation rate so that a single twitch gives a record with a recognizable form.

SINGLE MUSCLE TWITCH

Your instructor will give you directions for stimulating the nerve.

KEEP THE NERVE AND MUSCLE WET WITH FROG RINGER'S SOLUTION AT ALL TIMES.

Record the stimulus intensity used in Table 27–2.

□ Stimulate the muscle through its nerve, wait 15 sec, and repeat.
□ If the tracings are similar, copy the tracing of the **single twitch contraction** in the space provided in Table 27–2.
□ Compare the length of the twitch, beginning at the mark of the signal marker, with the number of cycles in the same interval. Calculate the duration of the twitch in msec according to Table 27–2.

For example, suppose your twitch lasted 12 cycles according to your time marker, and the time marker vibrated at 100 cycles/sec. Then the time that the twitch took would be:

$$\frac{12 \text{ cycles}}{100 \text{ cycles/sec}} = 0.12 \text{ sec} = 120 \text{ msec}$$

If you do not have a time marker, measure the circumference of the drum in cm. Then measure the time in sec for a complete rotation of the kymograph drum. Divide the length by the time in order to calculate the rate of rotation in cm/sec. The time that the twitch took would be:

$$\frac{\text{length of twitch in cm}}{\text{rate of rotation in cm/sec}}$$

The twitch consists of three parts: the **latent period** is the time from stimulus to the beginning of contraction; **contraction** is the period when the muscle is shortening; and **relaxation** is the time during which the muscle is lengthening.

□ Calculate the duration of the latent period, contraction period, and relaxation period for your twitch, and enter the results in Table 27–2.

What chemical events go on during the latent period?

____

____

____

____

What molecules move to cause the contraction?

____

Is energy required for the relaxation process?

____

Did every muscle fiber in the muscle contract during the single twitch? ______________

□ Set the stimulator for a higher intensity, and repeat the single twitch.

Did a more intense stimulus cause more intense contraction? ______________

Why or why not?

____

____

____

* FATIGUE

A muscle can contract indefinitely so long as it has an adequate rest period between twitches. If the stimuli are close together, however, the circulation won't be able to keep pace with the demands of contraction, and the muscle will show **fatigue.** Fatigue could occur at the neuromuscular junction or in the muscle.

What substances could run out and thereby cause muscle fatigue?

glucose

What substance builds up in the muscle when it is showing fatigue? lactic acid

□ Set the stimulator for multiple shocks at a frequency of 4/sec and an intensity of 3 V (0.3 V × 10 on the multiplier control). Set the kymograph for slow recording.

□ Stimulate the nerve and record until the contractions are 2 mm high, and then turn off the stimulator.

□ Record the result in Table 27–2.

Was the decrease in the strength of contraction gradual or sudden? ______________

□ Test whether the fatigue really occurred in the muscle, or whether it was due to inadequate stimulation of muscle by the nerve by stimulating the muscle itself directly through the femur clamp using the same intensity of stimulus.

Was the muscle itself fatigued? ______________

What causes fatigue at the neuromuscular junction?

____

____

□ Stimulate the muscle repeatedly with the same settings that you used for the nerve.

Does the muscle eventually show fatigue at this rate of stimulation? ______________

Rinse the muscle with Ringer's, and give it a 5-min rest.

Temporal Summation

If the muscle is stimulated several times in a row, the intensity of contraction during the twitches may increase. This is called **temporal summation.**

How can the muscle show more contraction on a second or later twitch?

____

□ Stimulate the muscle repeatedly at intervals shorter than the length of a single twitch.

Do you observe temporal summation?

____

□ Record the results in Table 27–2.

* Tetany

Tetany occurs when the stimuli are so close together that the muscle doesn't have time to relax between twitches and stays contracted.

□ Set the stimulator so that the frequency of stimuli is the maximum possible.

□ First give the muscle a single stimulus so you have a basis for comparison.

□ Then give the muscle the same strength stimulus at the high frequency, and record the muscle response for several seconds.

□ Record the results in Table 27–2.

Does the muscle show temporal summation at first?

____

Does the strength of contraction during tetany gradually decrease during the sustained contraction?

____

When might a tetanic contraction occur in your own body?

____

When you are finished with your experiment, dispose of the animal remains in the "animal wastes" container, then clean and store your equipment and microscope properly.

EXPLORING FURTHER

Grollman, Sigmund. *The Human Body: Its Structure and Physiology.* 4th ed. New York: Macmillan, 1978.

Schmidt-Nielsen, Knut. *Animal Physiology: Adaptation and Environment.* Cambridge, England: Cambridge University Press, 1975.

Wilkie, D. R. *Muscle.* New York: St. Martin's Press, 1968.

Scientific American Articles

Cohen, Carolyn. "The protein switch of muscle contraction." November 1975 (#1329).
Evarts, Edward V. "Brain mechanisms of movement." September 1979 (#1443).
Feld, Michael S., Ronald E. McNair and Stephen R. Wilk. "The physics of karate." April 1979 (#3042).
Hildebrand, M. "How animals run." May 1960 (no offprint).
Lester, Henry A. "The response of acetylcholine." February 1977 (#1352).
McLean, F. C. "Bone." February 1955 (no offprint).
Napier, John. "The antiquity of human walking." April 1967 (#1070).
Ross, R., and P. Bornstein. "Elastic fibers in the body." June 1971 (#1225).
Smith, D. S. "The flight muscles of insects." June 1965 (no offprint).
Sonstegard, David A., Larry S. Matthews and Herbert Kaufer. "The surgical replacement of the human knee joint." January 1978 (#1378).

Student Name ______________________ **Date** ____________

TABLE 27–1. COMMON AND SCIENTIFIC NAMES OF SOME HUMAN BONES

Common Names	Scientific Names	Axial	Appendicular
Toe bones	phlanges		✓
Neck bones	cervical vertebra	✓	
Hip bone	ilium		✓
Backbone	thoracic vertebra		
Ankle bones	tarsals		✓
Collar bone	clavicle		✓
Shoulder blade	scapula		✓
Wrist bones	carpals		✓
Knee cap	platella		✓
Jawbone	mandible	✓	
Shin bone	tibia		✓
Thigh bone	femur		✓
Foot bones	metatarsals		✓
Upper arm bone	humerus		✓
Forearm bones	radius, ulna		✓
Bone of calf	fibula		✓
Hand bones	metacarpals		✓
Finger bones	phlanges		✓
Funny bone	declanon process		✓
Breast bone	sternum	✓	
Tail bone	coccyx	✓	
Cheek bone	zygomatic arch	✓	

Figure 27–6. A joint. Add the extensor and flexor muscles with their origins and insertions.

DIAGRAM OF A JOINT

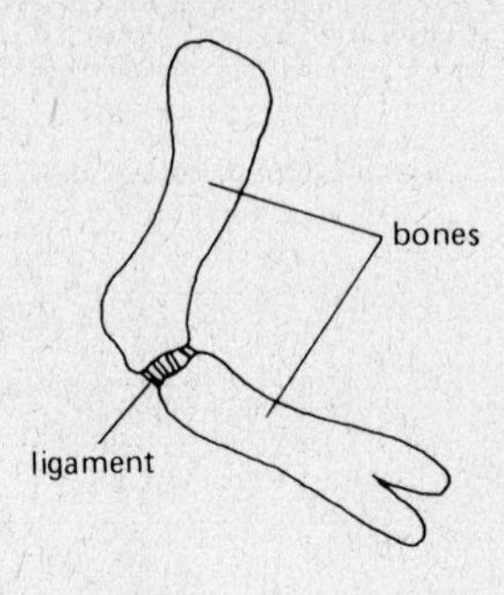

DRAWING OF COMPACT BONE

Student Name ____________________ **Date** ____________

TABLE 27–2. DATA FROM THE FROG MUSCLE PHYSIOLOGY EXPERIMENT

Stimulus intensity was: __________

A. DIAGRAM OF A SINGLE TWITCH CONTRACTION:

B. DATA FOR A SINGLE MUSCLE TWITCH:

TIME MARKER CALCULATION

cycles/twitch: __________ cycles

Frequency of time marker: __________ cycles/sec

Duration of twitch = #/frequency: __________ sec = __________ msec

ALTERNATIVE CALCULATION

Rate of drum rotation = $\frac{\text{drum circumference: \_\_\_\_\_\_\_\_\_\_ cm}}{\text{time for one rotation: \_\_\_\_\_\_\_\_\_\_ sec}}$ = __________ cm/sec

Duration of twitch = length of twitch/rate = __________ sec = __________ msec

Phase	Length of Tracing	Time (msec)
Latent period		
Contraction		
Relaxation		

TABLE 27–2 (Cont.). MORE DATA FROM THE FROG PHYSIOLOGY EXPERIMENT

C. DIAGRAM OF FATIGUE:

D. DIAGRAM OF TEMPORAL SUMMATION:

E. DIAGRAM OF TETANY:

TOPIC 28

Behavior

OBJECTIVES

When you have completed this topic, you should be able to:

1. Define orientation, kinesis, sign stimulus, fixed action pattern, ritual, communication.
2. Define taxis and state the three major types of stimuli that result in the response of taxis.
3. Describe an experiment to test for the effectiveness of chemical stimuli in chemotaxis.
4. List the factors involved in fish schooling and describe an experiment to test for each factor.
5. Explain the basis for the stimulus used in the optomotor response experiment.
6. Define agonistic behavior and explain why it is important in the survival of a species.
7. Describe the fight and flight components of agonistic displays in the Siamese fighting fish.

TEXT REFERENCES

Chapter 42.
Instinct, learning, territoriality, agonistic displays, courtship rituals, communication.

INTRODUCTION

Behavior is the response of an organism to any internal or external stimulus. In the broadest sense it includes just about anything that the animal or plant does. It is convenient to describe behavior as **innate** when it seems to be an automatic and consistent response to a certain stimulus and **learned** when the response tends to change with experience, so that it becomes more and more appropriate or elaborate when the stimulus has been given many times. Both types of behavior are important in animals with at least moderately complex nervous systems, including humans. In this exercise, you will be studying mainly innate behavior.

Orientation is a type of behavior pattern in which an organism turns its body or moves either toward or away from a certain stimulus. This type of behavior is basic to the success of the animal. Whatever it does, whether it is seeking food, water, sunlight, a mate, interacting with other members of a group, or avoiding predators, there must be some sort of orientation.

Kinesis is a simple type of orientation behavior in which the organism changes its locomotion only indirectly in response to the stimulus. If the response is positive kinesis, the animal will slow down its locomotion in the presence of the stimulus and speed up if the stimulus is absent. In a way, this is a primitive form of seeking behavior: if a necessary environmental condition or factor is missing, the animal moves about quickly, thereby increasing its chances of finding better conditions. If the conditions are favorable, it stays close to that spot by slowing down. For example, an organism that must have a moist environment, such as a sowbug, will show positive kinesis toward water: it will move faster under dry conditions than under wet conditions.

If the individuals in a culture of *Paramecium* tend to accumulate near a bubble of CO_2 as a result of kinesis, does lower pH make them swim slower or faster?

slower

In organisms that can detect stimuli at a distance, the orientation behavior will be a form of **taxis,** in which there will be a sustained movement either toward or away from the stimulus. This kind of response may be shown for light, heat, cold, moisture, gravity, sound, or chemicals. The response will be termed **phototaxis** if the stimulus is light, **geotaxis** if the stimulus is gravity, and **chemotaxis** if the response involves a chemical stimulus.

One way in which an organism can show taxis is to try to balance the stimuli reaching a pair of sense organs, such as its photoreceptors or chemoreceptors. If less stimulus reaches the right side, the animal turns to the left until the stimulus on its right increases to the level of the left side (or vice versa). Even if the receptors are very crude, this strategy (called tropotaxis) will allow movement toward or away from the stimulus. If one of the paired receptors is destroyed, the animal will receive no stimulus from the damaged receptor and will continuously move in futile circles *away* from that side. It is trying to increase the stimulus to the damaged side. These futile circular movements are called "circus movements" and are diagnostic for organisms that use tropotaxis.

Many stimuli cause very predictable responses. These stimuli are called **sign stimuli,** or releasers, and the response is called a **fixed action pattern.** The advantage of a fixed action pattern is that an appropriate response can be made quickly and accurately the first time without the need for learning. The disadvantage is that it will occur even if the animal has been tricked by nature or a biologist into making the response under inappropriate conditions. For example, a robin, whose red breast is a sign stimulus warning other males away from its territory, will peck at red feathers that bear no resemblance to another bird.

In addition to responding to environmental conditions, animals must usually interact with other animals of the same or different species. Interactions between members of the same species, usually by visual, auditory, or chemical cues, are called **communication.** Many of the signals used in communication are sign stimuli, which cause very predictable responses in the other individuals.

Sign stimuli are important in courtship, in mating, and in **agonistic,** or **conflict, displays** over territory, food, or mates. The stimuli are often built up into chain reactions called **rituals,** in which the response to a sign stimulus acts as a second sign stimulus, that response as a third stimulus, and so forth. Courtship rituals, for example, ensure coordination of physiological readiness to mate, mating within the same species, and sometimes care of the young.

Agonistic displays allow members of the same species to settle disputes over territory, food, or mates without direct conflict and usually without damage to either party in the dispute. The presence of a competitor will release responses that show the readiness of the opponent to fight. Signs of the intention to fight will cause the opponent either to show its own fight intention or to withdraw. Signs of timidity or **appeasement** will tend to reduce the aggressiveness of the opponent.

The agonistic display will be a mixture of "fight" and "flee" signals, and the proportion of each will be influenced by the **motivation** of the individual. Males will be more motivated to fight when they are in their own territory and when it is the breeding season. They will be less motivated to fight when the female approaches during the breeding season and will therefore be able to begin the courtship ritual. In many species courtship is a means of overcoming the agonistic behavior of the male to the extent that mating is possible.

Sign stimuli often cause agonistic behavior and sometimes the animal will respond even more strongly to an artificial stimulus than to another organism. In today's experiments you will study the effects of some sign stimuli on agonistic behavior.

TAXIS IN *DROSOPHILA*

You may have noticed during the genetics experiment that the flies had the annoying habit of sitting on the cotton plug of the food vial. This behavior may have been due either to positive phototaxis, negative geotaxis, or both types of taxis.

Phototaxis

□ Place five flies in a clean food vial (without the food), cover it with another clean vial, and tape the vials together as shown in Figure 28–1.
□ Cover the vials with a black paper shield so that only one end is open to the light, and lay it on a horizontal surface.
□ Place a light source near the open end of the tube, and wait 3 min.
□ Quickly remove the shield, and count the flies in each half of the tube; record the results on a data sheet at the end of this topic.
□ Tap the flies back into the center of the tube and reverse the tube so that the former right end is now the left end. Replace the shield and light source, which should be exactly as before.

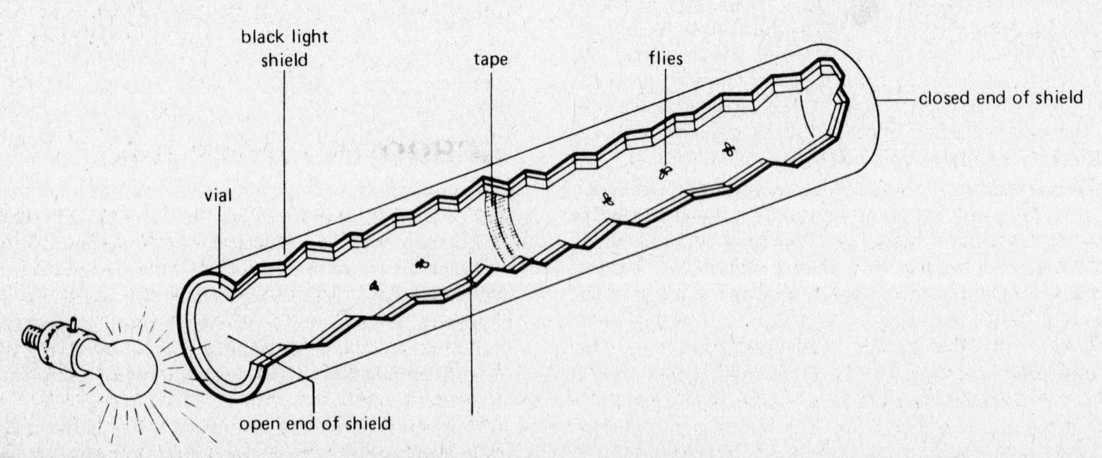

Figure 28–1. Phototaxis in *Drosophila*. In this experimental setup, the open ends of the vials are taped together, and the vials are covered with a black shield open to the light at one end only.

□ Wait 3 min, remove the shield, and count the flies once again.
□ Repeat once more unless your results agreed very well in the first two trials.

Did the flies show phototaxis? ______________

If present, was it positive or negative? ______________

Why did you have to reverse the vials between trials?

____

____

____

____

Why did you have to keep the vials horizontal?

____

Geotaxis

□ Use the same vials or take five new flies, put them in a clean food vial, and tape it end to end to another clean food vial.
□ Let the flies attach to the glass, and then cover the vials completely with a black shield.
□ Hold the vials in a vertical position for 3 min.
□ Quickly remove the shield, and count the flies in each vial. Record the results on your data sheet.
□ Tap the flies to the middle of the tube, cover it with the shield, and hold it in a vertical position, reversing the vials so the one that was held up at first is now down.
□ Wait 3 min, remove the shield, and count the flies in each vial. Record the result.
□ Repeat once more unless your results in the first two trials matched very well.

Did the flies show positive geotaxis or negative geotaxis?

____

Why did you have to reverse the position of the vials between each trial?

____

____

Phototaxis and Geotaxis

Test the relative strengths of the response to light and gravity by opposing these stimuli.

□ Enclose the vials with five flies in the black shield that is open at one end.
□ Hold the vials in a vertical position so that the covered end is the end where you would expect the flies to be under the influence of gravity alone.
□ Hold a light source near the open end of the tube, and wait 3 min.
□ Remove the shield, and count the flies in each vial. Record the results.
□ Repeat two or three more times until you see a definite trend.

Was phototaxis strong enough to overcome geotaxis?

____

Chemotaxis

Orientation toward volatile substances helps *Drosophila* locate food and a mate and avoid unpleasant or harmful substances. A simple T-maze for testing chemotaxis is shown in Figure 28–2. The substance to be tested is placed in one vial, the second vial acts as the control, and the flies are placed in the third vial. The progress of the flies can be viewed through the glass.

□ Place some fermenting *Drosophila* food or fermenting fruit on a small piece of aluminum foil, and put it into vial a. Attach it to the T-tube.
□ Place a piece of moist filter paper in vial b, and attach it to the T-tube.
□ Place 12–15 flies in vial c, and attach it to the T-tube.
□ Watch the flies for 5 min. Record the number of flies in each vial.
□ Record the numbers of flies again after 10 min.

Do you see any trend in the movements of the flies?

____

□ Put the maze in a dark place for an hour, and record the number of flies in each vial after 60 min.

Did the flies show chemotaxis toward the food?

____

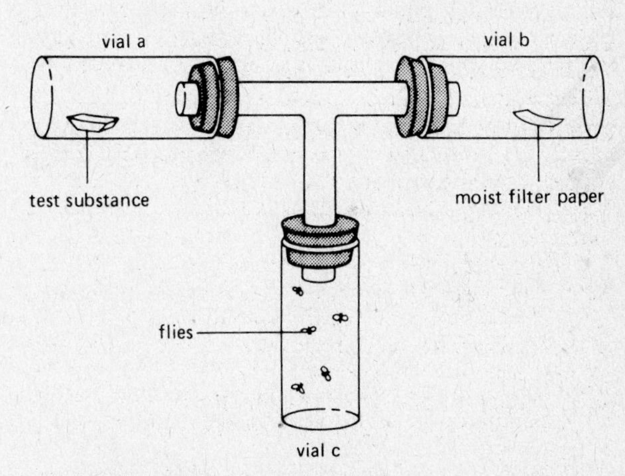

Figure 28–2. Chemotaxis in *Drosophila*. In the T-maze, vial (a) contains the substance to be tested, and the control, vial (b), contains moist filter paper. The flies are introduced into vial (c).

SCHOOLING BEHAVIOR

Animals may gather in a group for different reasons. If they are attracted to the same spot by the presence of food, light, or some external stimulus, the group is called an **aggregate.** If, however, they form a group because they are mutually attracted to one another, the group is called a **school** in the case of fish, and a **herd,** or **flock,** in the case of mammals or birds. In fish, vision, swimming movements in the water, and olfaction may all contribute to keeping the members of the school together, but vision is the most important cue. Schooling fish seem to be attracted to each other mainly by their appearance, and the attraction is strongest for other members of the same species. Schooling is a form of communication between members of the same species, but fish of differing species do sometimes school together as well.

Schooling is very prevalent among all sorts of fish, from very primitive ones to the more advanced species. The members of the school may be better protected from predators than single fish, and they seem to be able to swim more efficiently. A predator in a school has a greater chance of locating food than does a lone predator. For plankton-feeding fish, however, food is always present, and there will be less food for each fish in a school than there would be for a single fish. Members of a school can learn from each other more quickly than a single fish can learn, and they do not have to spend energy in locating mates for reproduction.

However, one big disadvantage of schooling behavior is that it has made the human fishing industry very successful and efficient and may ultimately lead to the decline or even extinction of many fish species.

Schooling of Fish with Visual Markings

In these experiments you will be testing the role of the visual component in schooling behavior. Because the fish will be separated by a glass barrier, there can be no communication by sound or chemical signals. In many fish, vision is the prime factor in the attraction of schooling fish toward each other, but olfaction and sound seem to help maintain the cohesion of an established school.

BEHAVIOR EXPERIMENTS WILL NOT WORK UNLESS THE ANIMALS ARE TREATED WITH CARE AND PATIENCE.

DO:

1. Always use a net to transfer the fish.
2. Allow time for the fish to adjust to new conditions before beginning your observations.
3. Wash and rinse your hands before reaching into the experimental tanks to place beakers, and so on.
4. Treat the fish with care.
5. Report any sick-looking or dead fish to the instructor.
6. Return all fish to the proper tanks when you are finished.

DON'T:

1. Disturb the fish more than necessary for the experiments.
2. Expect the fish to respond instantaneously to a new stimulus.
3. Reach into the stock tanks where the fish are kept.
4. Leave fish in the experimental apparatus or tanks.

Work with a partner and try these tests of schooling behavior. The experimental setups are shown in Figure 28–3 and Figure 28–4.

TEST 1 (Figure 28–3)

□ Place several fish with markings, such as zebra danios, in a beaker. Cover the beaker with the screen, and immerse it slowly in the aquarium. Place it into position on side A.

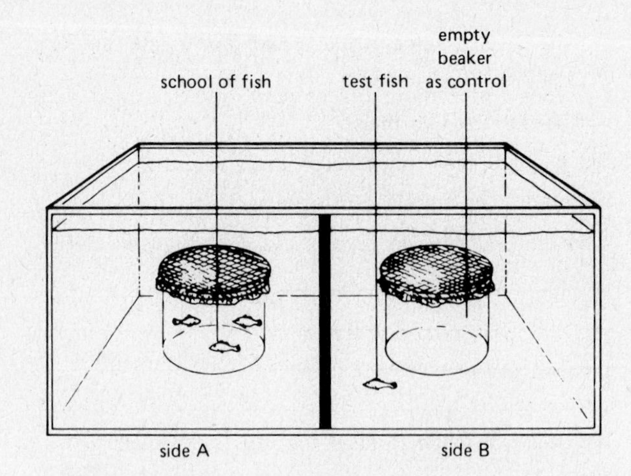

Figure 28–3. Schooling behavior. In the aquarium setup for testing with one school, a beaker with fish is placed on side A and an empty beaker on side B to serve as the control. The test fish is free to swim throughout the aquarium.

□ Place an empty beaker on side B to serve as the control.

□ Place a single test fish of the same species in the tank.

□ Use a stopwatch to time the number of seconds spent by the test fish in each half of the tank during a 5-min period (300 sec). Record the results on your data sheet.

□ Calculate the percentage of the time the test fish spent on side A and on side B.

Did the fish spend more time on side A or side B?

____

TEST 2 (Figure 28–3)

□ Place two fish in the beaker, and repeat the experiment using a different single fish of the same species. Record the results.

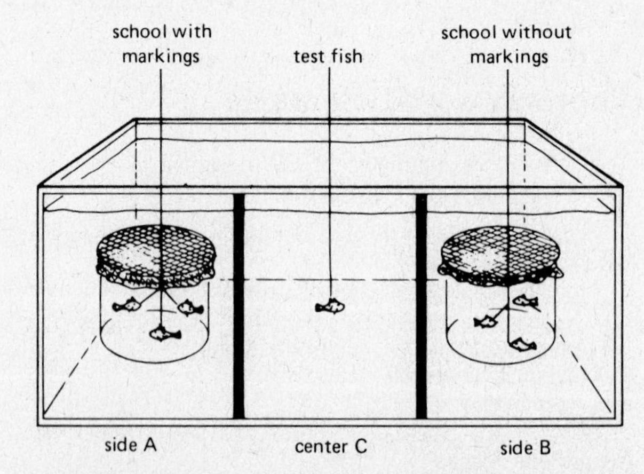

Figure 28–4. Schooling behavior. When two schools are used, the aquarium is divided into three parts. A school of one species of fish is placed on side A and a school of another species on side B. The free test fish can choose school A, school B, or neither school.

□ Calculate the percentage of the test period that the fish spent on each side.

If you have time, repeat the test using different numbers of fish in the beaker, and using a different test fish each time.

□ Calculate the percentage of the test period that the test fish spent with the school and the percentage of time that it spent alone.

Did the number of fish in the school affect the tendency of the single test fish to spend time with the school?

____

TEST 3 (Figure 28–4)

□ Place three fish with markings, such as zebra danios, in a beaker on side A of the aquarium, and three fish of another species without markings, such as white clouds, in a beaker on side B. Be sure to put the *same number* of fish in each beaker.
□ Place a single test fish *with markings* in the center of the aquarium.
□ Use a stopwatch to time how many seconds in a 5-min period the test fish spends with its own species (section A), with the other species (section B), and alone (section C).
□ Calculate the percentage of time that the test fish spent in each section of the tank.

In which of the three sections did the fish spend the most time?

____

Is there a big difference in the percentage of time spent with the two species? ______________

Schooling Fish without Visual Markings

Visual markings may help fish to identify other members of the same species, but not all schooling fish have prominent markings. Test the schooling tendency of fish without prominent markings in the following experiments, and compare your results to those you obtained for the fish with markings.

TEST 4 (Figure 28–3)

□ Place several fish without markings in a beaker on side A of the aquarium, an empty beaker on side B, and a single test fish of the same species free in the aquarium.
□ Use a stopwatch to time the number of seconds spent in each half of the aquarium during a 5-min period. Record the results.
□ Calculate the percentage of the time the test fish spent on side A and on side B.

On which side did the test fish spend more time—side A or side B? ______________

TEST 5 (Figure 28–3)

□ Place two fish in the beaker, and repeat the experiment using a different single fish of the same species. Record the results.
□ Calculate the percentage of the test period that the fish spent on each side.

If you have time, repeat the test using a different number of fish in the beaker and using a different single fish each time.

□ Calculate the percentage of time the fish spent with the school and alone for each test.

Did the number of fish in the school affect the tendency of the single test fish to swim with the school?

____

TEST 6 (Figure 28–4)

□ Place three fish with markings in a beaker on side A of the aquarium and three fish without markings in a beaker on side B. Be sure to put the *same number* of fish in each beaker.
□ Place a single test fish *without markings* in the center of the tank.
□ Use a stopwatch to time how many seconds in a 5-min period the test fish spends with its own species (section B), with the other species (section A), and alone (section C). Record your results.
□ Calculate the percentage of time that the test fish spent in each section of the tank.

In which of the three sections did the fish spend the most time? ______________

Is there a big difference in the percentage of time spent with each of the two species? ______________

Was the test fish more attracted to visual markings or to members of its own species?

____

Return all fish to the proper tanks when you are finished.

THE OPTOMOTOR RESPONSE

In this experiment, the fish is presented with a pattern of moving stripes in the apparatus shown in Figure 28–5.

A response to the moving stripes has been found in both schooling and nonschooling fish, and it may be present even before young fish have learned how to school. Test for the presence of the optomotor response in both the marked and the unmarked fish.

□ Place a fish with markings in the beaker, and observe its behavior with no special stimulus. Record the results.
□ Attach a paper cylinder with vertical stripes to the drum of the kymograph, and observe the reaction of the fish to stationary vertical stripes.
□ Start the drum rotating at about 0.5 revolutions per second, and observe the behavior of the fish for 3 min.

Did the rotating stripes change the behavior of the fish?

____

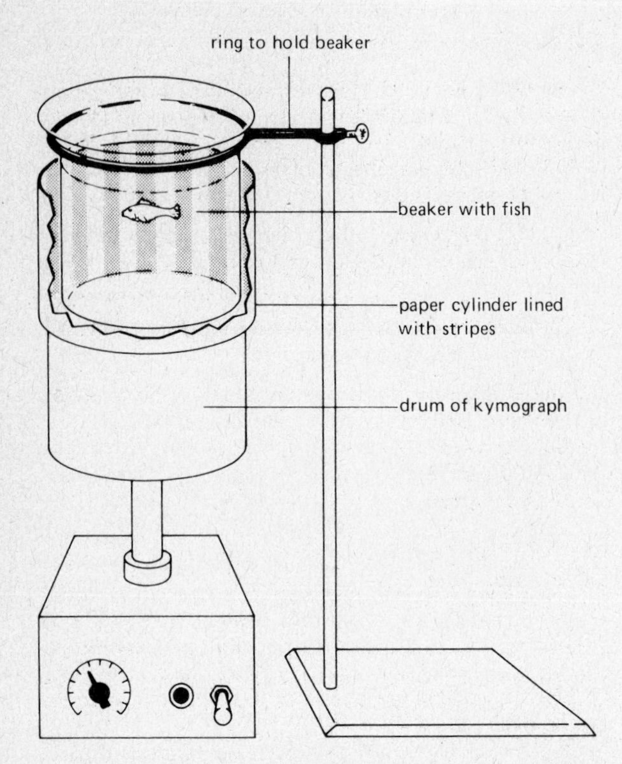

Figure 28–5. Optomotor response. A striped paper cylinder is the visual stimulus. It is attached to the drum of a kymograph and rotated around the outside of the beaker containing the test fish.

□ Change the paper cylinder to one with horizontal stripes, and observe the response of the fish to stationary horizontal stripes.
□ Start the drum rotating at about 0.5 revolutions per second, and observe the behavior of the fish for 3 min.

Did the rotating stripes change the behavior of the fish?

____

Be sure to return the fish to the aquarium when you are finished.

□ Place a fish without markings in the beaker, and repeat your observations on the fish's response to no stimulus, stationary and moving vertical stripes, and stationary and moving horizontal stripes.
□ Record the results, and return the fish to its tank.

Did the fish without markings show the optomotor response? ______________

AGONISTIC BEHAVIOR IN SIAMESE FIGHTING FISH

These fish (*Betta splendens*) live in the freshwater tropical streams of Indochina and have been bred for exaggerated agonistic behavior. The male fish shows its intention to attack another male or to flee by visual displays. One of the male fish eventually will get the upper hand and bluff its competitor into leaving.

When the fish are confined within an aquarium, however, the winner will chase and bite the losing fish.

YOU MUST BE READY TO REMOVE ONE OF THE FISH AS SOON AS THE WINNER IS CLEARLY APPARENT.

The balance between the tendency of the fish to fight or flee is strongly dependent on whether it is in its own territory and whether it has just defeated another fish or has just been defeated itself. The most impressive displays are observed with closely matched fish.

The components of the agonistic display are shown in Figure 28–6. The normal resting fish (a) is drab or only slightly colored and keeps the dorsal and ventral fins folded. When another male appears, the fish will probably challenge the intruder with a **broadside display** (b). Two fish may swim side by side for several minutes in this display, each trying to intimidate the other. Finally, one will get a slight edge, and the subordinate fish will gradually give up the broadside display, and lose some of its brilliant color. The dominant fish will then probably attempt a **frontal attack** (c) and will pursue and thrust at the fins of the submissive fish.

(a) RESTING OR SUBMISSIVE

(b) BROADSIDE DISPLAY

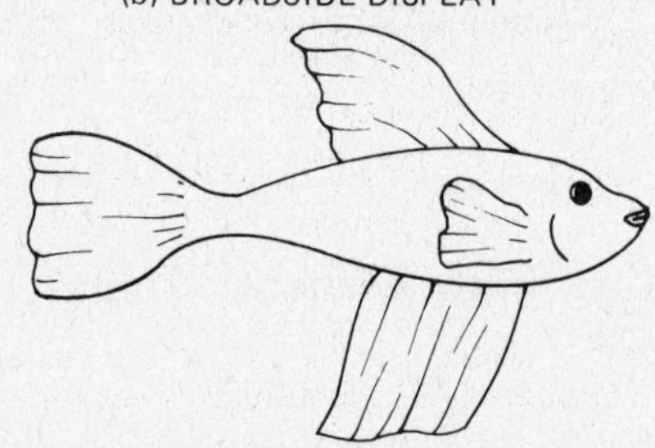

(c) FRONTAL ATTACK

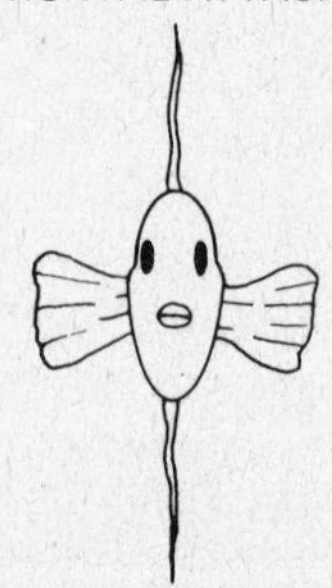

Figure 28–6. Agonistic displays in a Siamese fighting fish. The display in (a) is the least aggressive and is shown by the resting fish. Its coloring is dull or moderately bright. The broadside display in (b) is commonly seen between closely matched equals, brilliant in color. The fish in (c) is dominant and about to drive the submissive fish away with its brilliant coloring and fierce attack.

The display of agonistic behavior will also be shown in response to artificial stimuli. The most effective is a mirror image of the fish itself. A fish-colored model in the shape of the fish or even in a different shape will usually also cause a response. Models of different colors are generally ignored. Artificial stimuli are most effective if the fish has recently defeated another fish and is in its own territory so that its motivation to display is high. The broadside display of the male is also used during the courtship ritual to attract the female to the bubble nest. The male maintains the display during mating, and the female must make herself as inconspicuous as possible to avoid being attacked.

Record your observations on the fighting fish on a data sheet at the end of this topic.

TEST 1

□ Place a male fighting fish (fish #1, "Red Raider" or whatever) in an aquarium, and observe its normal behavior.
□ Place a second fish (fish #2) in the tank, and observe the behavior of both. Note their orientation to each other, their color, the position of their fins, and the position of the gill covers (opercula). Record what you observe.

If the fish were equally matched before the experiment, which one theoretically has an advantage now?

#1 ___

Why?

He got to the territory first ___

□ Continue your observations for 10 min or until one of the fish starts pursuing the other. *Immediately remove one of the fish in this case.*

TEST 2

□ Replace the dominant fish with one provided by the instructor, which was submissive in a previous experiment.
□ Observe the two submissive fish for 10 min.

Did either of the submissive fish show agonistic behavior? Yes ___

TEST 3

□ Place a dominant fish in the aquarium, and allow it 5 min to adjust.
□ Hold a mirror alongside the tank close to the fish, and move it slightly so that the image appears to move. It may take the fish a few minutes to respond.
□ Continue until there is a response or until 5 min have passed.

Did the fish respond to its mirror image?

Yes ___

Did it try to attack the image?

Yes ___

TEST 4

□ Choose one of the models provided, and move it slowly close to a dominant fish. If your models are plastic, they can be held in the water near the fish.
□ If the fish moves away, follow it, but avoid moving the model very close to the fish or touching the fish with it.
□ Continue presenting the model until there is a response or until 5 min have passed. Record your observations by checking for each of the following reactions shown by the fish:

no response
swimming ✓
swimming away
pectoral fin movement ✓
raising of dorsal fin ✓
lowering of ventral fin
expansion of tail ✓
increased coloring ✓
extension of gill covers ✓
facing the model ✓
biting the model ✓

Which of the following general response patterns did the fish show?

no response
fleeing from the model
broadside display ✓
frontal attack ✓

___

Was the response greater or less than the response to another fish?

Just as great as response to real fish ___

□ Obtain the results of other class members using different models, and record them alongside your own results for comparison.

Did shape strongly influence the response of the fish to the models?

No ___

Did color strongly influence the response?

Yes ___

When you are finished with these experiments, return all the fish to their proper containers as directed by your instructor.

EXPLORING FURTHER

Alcock, J. *Animal Behavior.* Stamford, Conn.: Sinauer, 1975.
Kalmus, H. *101 Simple Experiments with Insects.* Garden City, N.Y.: Doubleday, 1960.
Lorenz, K. Z. *King Solomon's Ring.* London: Methuen, 1942.
Shaw, Evelyn. "Schooling fishes." *American Scientist* 66: 166, March 1978.
Tinbergen, N. "On war and peace in animals and man." *Science* 160: 1411, 1968.

Scientific American Articles

Benzer, S. "Genetic dissection of behavior." December 1973 (#1285).
Crews, David. "The hormonal control of behavior in a lizard." August 1979 (#1435).
Eibl-Eibesfeldt, Irenäus. "The fighting behavior of animals." December 1961 (#470).
Hallman, Jack P. "How an instinct is learned." December 1969 (#1165).
Hess, Eckhard H. "'Imprinting' in a natural laboratory." August 1972 (#546).
Keeton, William T. "The mystery of pigeon homing." December 1974 (#1311).
Mykytowycz, Roman. "Territorial marking by rabbits." May 1968 (#1108).
Saunders, D. S. "The biological clock of insects." February 1976 (#1335).
Tinbergen, N. "The curious behavior of the stickleback." December 1952 (#414).
Todd, John H. "The chemical language of fishes." May 1971 (#1222).
Wilson, E. O. "Pheromones." May 1963 (#157).

Student Name ______________________________ **Date** ____________________

BEHAVIOR DATA SHEET

Student Name ______________________________ **Date** ______________

BEHAVIOR DATA SHEET

Student Name ______________________ **Date** ______________

BEHAVIOR DATA SHEET

TOPIC 29

Plant Structure and Function

OBJECTIVES

When you have completed this topic, you should be able to:

1. Explain why the division of labor between roots, stem, and leaves is necessary for the success of a flowering plant.
2. List four functions of a root and compare the advantages and disadvantages of a fibrous root system and a tap root system in carrying out those functions.
3. Describe the differences between the external coverings and the internal structures of a stem and a root.
4. List five main functions of a stem.
5. Explain the difference between primary and secondary growth in vascular plants; describe the process of primary growth in a root and primary and secondary growth in a stem.
6. Trace the movement of a molecule of water from the soil to a guard cell, listing all the structures through which it passes.
7. Define girdling, list the layers that are removed, and describe the consequences to the tree.
8. Define meristem and list the locations in a plant where it is found.
9. Explain the term apical dominance and describe its role in the growth of the plant.
10. Compare the way that branches form in a stem and in a root.
11. Distinguish between simple and compound leaves, and between alternate, opposite, or whorled leaf arrangements.
12. Define node, internode, axillary bud, intercalary meristem, pericycle, vascular cambium, and cork cambium.
13. Given a twig, tell its age and identify the external features, including the bud scales, axillary buds, leaf scars, bud scale scars, and lenticels.
14. Given a cross section of a leaf, stem, or root, identify the following structures if they are present: epidermis, root hairs, cuticle, stomata, guard cells, endodermis, pericycle, xylem, phloem, pith, cortex, collenchyma, palisade layer, spongy mesophyll, and air spaces.
15. List the important features that distinguish monocots and dicots.

TEXT REFERENCES

Chapter 43; Chapter 44; Chapter 45 (C–E).
Roots, stems, woody stems, leaves, xylem, phloem, plant nutrition.

INTRODUCTION

Angiosperms are characterized by their **flowers** and their highly efficient **vascular tissues.** The rapid uptake of water and minerals in the roots and their distribution to the rest of the plant allow an angiosperm to maintain large leaves with broad surface areas. The higher rate of photosynthesis carried on by these leaves allows the plant to grow much more quickly than a gymnosperm, and it may reach reproductive maturity in a few weeks. The details of reproduction will be considered later. Here you will study the structure of flowering plants and the important role played by their vascular tissues.

Plants may survive for one, two, or many years. The **annuals** die after a season's growth and allow their seeds to continue the species the following spring. **Biennials** reproduce after two seasons and then die. **Perennials** live on for many years. The body of an annual grows mainly in length, from the tips of its shoot system and from the tips of its roots. This type of growth is called **primary growth.** Some annuals and most perennials also increase in size sideways. This growth in girth is called **secondary growth.** The stem of some perennials—the trunk of a tree, for example—continues to increase in diameter year after year. Other perennials, such as the chrysanthemum and bulb-forming plants, die back each year and produce little or no secondary growth.

When secondary growth occurs, large amounts of **secondary xylem** called **wood** may be formed. This tissue gives great strength to the stems and roots of **woody plants.** Plants that do not form wood are called **herbs. Secondary phloem** is also produced.

Angiosperm plants show great diversity, but they fall into two main groups: the **monocots** and the **dicots.** These names are derived from the structure of the seed. A monocot has one seed leaf (cotyledon), whereas a dicot has two. The characteristics of the roots, stems, and leaves are also different. Monocots are usually herbs, whereas dicots include both herbs and woody plants. When you finish with the laboratory work in this exercise, state the main characteristics of monocots and dicots in Table 29–2 at the end of this topic.

TRANSPIRATION

When plants are actively photosynthesizing, they need to use water and carbon dioxide. Air must be allowed to enter the leaf interior by opening the **stomata,** and water reaches the leaf by rising through the **xylem** from the roots. When the stomata are open, however, a great amount of water is lost from the leaves, especially on a hot, sunny day with the breeze blowing. This lost water must be replaced by rapid uptake from the soil around the roots, or the plant will wilt. The process of losing water by evaporation from the leaves is called **transpiration.**

Demonstrate transpiration by carrying out this simple experiment:

□ Add a drop of dye to a small amount of water in a beaker or flask.
□ Make a clean cut with a sharp razor in the stem of a white flower, such as a daisy or carnation.
□ Place the flower in the beaker, set the beaker in a well-illuminated location, and check the petals from time to time during the laboratory for the presence of dye.
□ Wait 1–2 hr.
□ After the dye has reached the flower, note which part of the petals stained nore darkly.
□ Make a sketch in the space provided in Figure 29–11 at the end of this topic to show the pattern of staining of the petal. Label the part of the petal where transpiration was most rapid.
□ Make a thin cross section of the stem with a sharp razor blade, and examine it under the dissecting microscope.
□ Note the xylem vessels through which the dye rose.
□ Make a sketch of your cross section in Figure 29–11 and label the location of the xylem.

You can also measure the amount of water lost by a plant quantitatively by measuring the rate at which it is taken up by the stem. Record your results in Table 29–3.

□ Assemble the materials shown in Figure 29–1 and a sharp razor blade.
□ Hold the stem of a plant such as part of a geranium plant beneath the water, and cut the stem off without crushing it.

DO NOT WET THE LEAVES.

□ Hold the piece of rubber tubing under the water, and insert the cut stem into it. (The tubing must be just a bit narrower than the stem.)
□ Immerse the pipet in the pan of water, and attach it to the other end of the tube.
□ Gently squeeze the tubing to expel any air bubbles that might have been trapped below the stem.

IF THE EXPERIMENT IS TO WORK PROPERLY, ALL AIR BUBBLES MUST BE REMOVED.

□ Holding the pipet in a horizontal position, lift the plant and the pipet out of the water, and set the pipet on a flat surface in a good source of light.
□ Record the position of the water column in the pipet every 15 min until transpiration is well established.
□ Calculate the rate of water uptake by the stem in mL/hr.
□ Place the plant in darkness, and continue to follow the uptake of water.
□ Calculate the rate of water uptake by the stem in mL/hr in the dark.

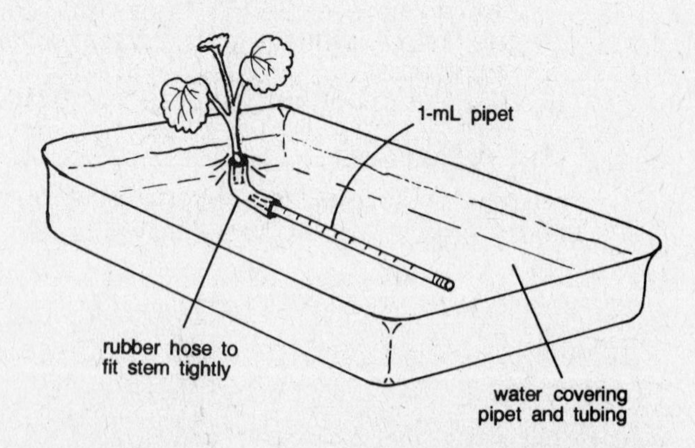

Figure 29–1. Transpiration experiment. Once all air has been excluded from the system by immersing it in water, the plant cutting is lifted out and placed in strong light so that the uptake of water by the stem can be measured.

How did the absence of light affect the rate of transpiration?

____

ROOTS

Types of Root Systems

When a seed germinates, it sends a single **primary root,** or **radicle,** down into the soil. This root sends out side branches called **lateral roots,** and these branch out in turn until a **root system** is formed.

Why does the root develop before the rest of the plant?

____

____

If the primary root continues to be the largest and most important part of the system, the plant is said to have a **tap root** system. If many main roots are formed, the plant has a **fibrous** root system. Most monocots, such as grass, and some dicots have a fibrous root system. Some trees, dandelions, and carrots have a tap root.

□ Examine the demonstrations of a fibrous root system and a tap root system.

In Table 29–1, describe how well each type of root system is adapted for carrying out the major functions of a root.

Root Tips

Most plant growth is **indeterminate** and stops only when the plant dies. Not all parts of the plant undergo cell division, however. Only certain tissue, called **meristem,** contains actively dividing cells. In a germinating seed, the meristematic tissue of the seedling is found only at the tip of the shoot (the part above the ground) and the tip of the root (the part underground).

□ Examine a longitudinal section of a root tip from a plant such as an onion under low power in your compound microscope.
□ Referring to Figure 29–2, identify the **root cap,** the zone of **cell division,** the **quiescent center,** the zone of **elongation,** and the zone of **maturation.**
□ Note the **root hairs** in the zone of maturation. They increase the surface area of the root.
□ Switch to high power, and examine the zone of cell division more carefully.
□ Note the many mitotic figures in the columns of **meristem** cells along the sides of the root and the absence of mitosis in the region of the quiescent center.

The new cells formed by mitosis elongate and push the root tip away from the older section of the root. The mature cells remain fixed in position relative to the soil while the tip of the root moves down continuously.

Why are the delicate root hairs found only in the mature part of the root tip?

____

____

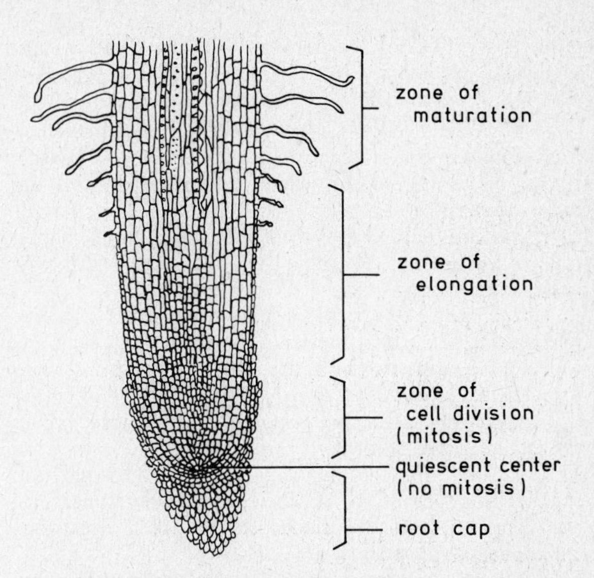

Figure 29–2. A growing root tip. Mitosis occurs mainly in the outer columns of the zone of cell division. Cells of the root tip are sloughed off as the root grows through soil.

The inner layer of root cap cells constantly divides to produce more cells. These replace the cells sloughed off from the outer layers of the root cap as the growing root forces its way through the soil.

What is the purpose of having a root cap?

____

Certain cells of the root cap sense gravity and aid the **geotropic** response so that the root grows downward.

□ Examine a 2-day-old radish seedling that has sprouted.
□ Note the fuzzy appearance of the radicle, which is covered with hundreds of root hairs.

Are the root hairs longer at the top or bottom of the root?

____

How does the growth of the root cause this pattern?

____

____

Root Structure

Because one function of the root is absorption, the surface of young roots is not covered with a waterproof layer but is somewhat permeable to the water and dissolved nutrients with which the root is in contact. Roots also need oxygen in order to survive and must usually also be in contact with air pockets in the soil so that they can carry on gas exchange. The mineral content of the water that is sent to the rest of the plant is controlled to some extent by having a waterproof substance between the endodermal cells that surround the central part of the root. The vascular tissue is located within the central region, so that it does not passively take up whatever is in the water around the plant. Thus, the fluid leaving the root in the vascular tissue has passed through living cells that concentrate certain things needed by the plant.

□ Examine a cross section of a root under low power. Locate the inner **stele,** which is separated from the outer **cortex** by the layer of endodermis.
□ If you are looking at a dicot root, note the X-shaped core of **xylem** vessels within the stele. Xylem in monocots forms a ring.
□ Use Figure 29–3 to locate the xylem and the **phloem** tissue which make up the vascular tissue.
□ Find the **endodermis,** the layer of cells surrounding the stele and regulating the fluid entering it. Just inside the endodermis locate the cells of the **pericycle,** which retain the ability to divide.
□ Note that the cells making up the cortex all look similar. This type of cell is known as **parenchyma** and may contain grains of stored starch within organelles called **amyloplasts.**
□ Study the outermost layer of cells, the **epidermis,** and note that **root hairs** are actually extensions of the epidermal cells.

During primary growth, the root may send out side branches, or **lateral roots.** These begin growing from the meristematic tissue of the pericycle and push their way out through the cells of the cortex into the soil.

□ Examine a cross section of a root, showing a developing lateral.

Does the new root have a root cap?

____

Does it have vascular tissue yet?

____

If the root is from a woody plant, it will form a layer of lateral meristem between the xylem and phloem, called the **vascular cambium.** Division of cells within the cambium will allow the formation of secondary xylem and seconday phloem, so that the root can grow in diameter as well as in length.

All roots contain food for their own needs, and the roots of overwintering plants must contain the nutrients to support the growth of new shoots the following spring. Some roots are especially modified to store large amounts of food and, as a result, are popular human food as well.

□ Obtain a piece of a carrot and make a very thin cross section with a razor blade.
□ Place the slice on a slide and add a few drops of iodine stain.
□ After the stain has soaked in for a few minutes, examine the slice under the microscope.

Where are the grains of starch located?

____

____

□ While you are waiting for the results of the starch test, make a longitudinal cut through the section of carrot.
□ Note the light-colored ring that divides the inner phloem from the outer xylem.
□ Locate the light-colored, horizontal streaks from the inner part of the carrot to the edge. These are the connections to the lateral roots that were present along the tap root before it was pulled up out of the ground.

Although nutrients are required for a plant's healthy growth, too much fertilizer can be harmful.

□ Place a piece of carrot root in a solution of 2% ammonium phosphate or ammonium nitrate for 1 hr, and place another piece in distilled water.
□ Describe the condition of the two pieces after 1 hr:

Carrot in 2% fertilizer:

____

Carrot in distilled water:

____

If fertilization of plants is overdone, they may be "burned" by the high concentration of salts and die as a result.

STEMS

The Pattern of Stem Growth

When a seed germinates, the part that grows upward is called the **shoot.** Growth occurs at the tip of the shoot where the dividing cells are found in the **apical meristem.** As the shoot grows, it produces not only the lengthening stem, but also **leaves** that branch from the

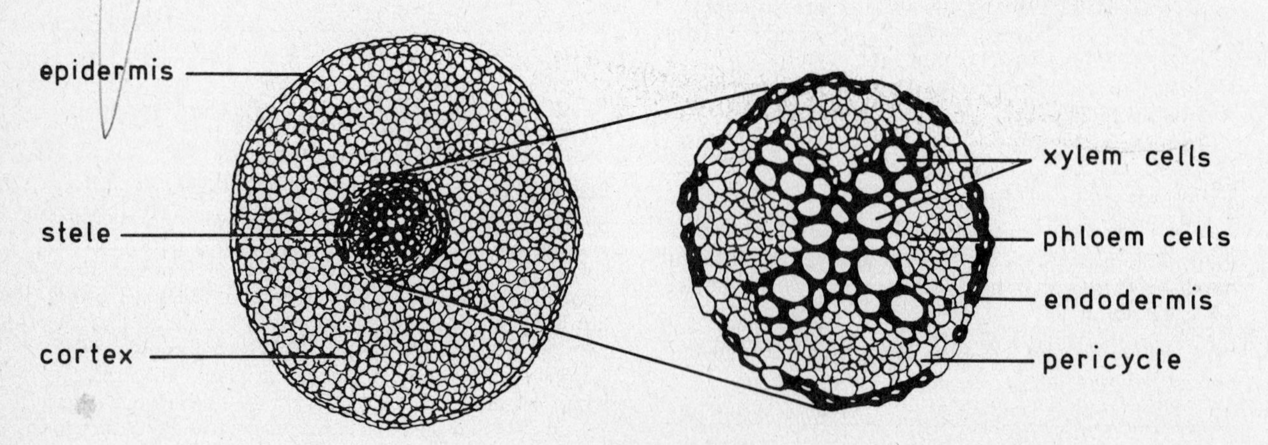

Figure 29–3. A dicot root. The vascular tissue is located within the stele.

stem and side or **lateral branches.** The point where a leaf or branch joins the stem is called a **node,** and the section of stem between the nodes is called an **internode.**

□ Examine a young plant, such as a bean plant, and locate the nodes and internodes.
□ Examine the junction of a leaf or the lateral branch with the stem closely, and note the **lateral** or **axillary bud** that is found in the axil of the leaf or lateral.
□ Locate the **apical bud** at the tip of the growing shoot.

The Shoot Apex

The apical bud produces hormones that keep the adjacent lateral buds dormant for a time. This inhibition of the lateral bud development is called **apical dominance.** Eventually the apical bud will grow far enough away so that its influence lessens, and the lateral buds will develop into lateral branches. If the apical bud is removed by a grazing animal or for experimental purposes, it will no longer inhibit the lateral buds, and they will immediately begin to grow.

□ Examine a prepared slide showing the apical bud of a plant such as a lilac or *Coleus* in longitudinal section. The internal structure is shown in Figure 43–8 in your text.
□ Identify the **apical meristem,** where the most actively dividing cells are found.
□ Note the **leaf primordia** adjacent to the meristem; each one (or pair) is attached to the stem at a node, and the small spaces between them are the internodes.
□ Look for **vascular tissue** in some of the older leaf primordia.
□ Lumps of tissue above the leaf primordia are **bud primordia,** which will form the buds that will eventually give rise to new laterals or to flowers.

As the apical meristem grows, the stem between the leaf primordia lengthens and the new leaves become fully developed. The process is something like pulling out the sections of a collapsible telescope, except that the number of sections keeps growing and can be extended indefinitely.

Internal Structure of a Stem

The organization of the stem is somewhat different from that of the root. Instead of being at the center of the stem, the vascular tissues are arranged in bundles that may be located around the edge of the stem (in dicots), or scattered throughout the interior of the stem (in monocots). The structure of a dicot stem is shown in Figure 29–4. A monocot stem is shown in Figures 43–23 and 44–18 in your text.

□ Examine a cross section of the stem of an herb under low power, and locate the **vascular bundles,** the cells of the vascular tissue.

Is the stem from a monocot or dicot?

____

□ Note the central **pith,** which may have been pulled apart during growth to make the interior of the stem hollow. It is composed of **parenchyma** cells.
□ Note the layer of **epidermis** around the outside of the stem. The epidermis is only a single cell layer thick.
□ Switch to high power, and examine the edge of the stem closely.
□ Note the transparent layer of the **cuticle.** The stem and leaves but not the root are covered with a cuticle formed from the substance **cutin,** which is secreted by the cells of the epidermis.

What is the function of the cuticle?

____

____

____

□ Look for **epidermal hairs** that may be present on the surface of the stem. Hairs usually have a protective function.
□ Study the layer of cells within the epidermis. It consists of **collenchyma** cells, which have characteristically thickened corners and act as supporting columns for the stem.
□ Note the inner part of the cortex, which is composed of parenchyma cells.

Some of the stem cells are photosynthetic, and some may contain amyloplasts for storing starch.

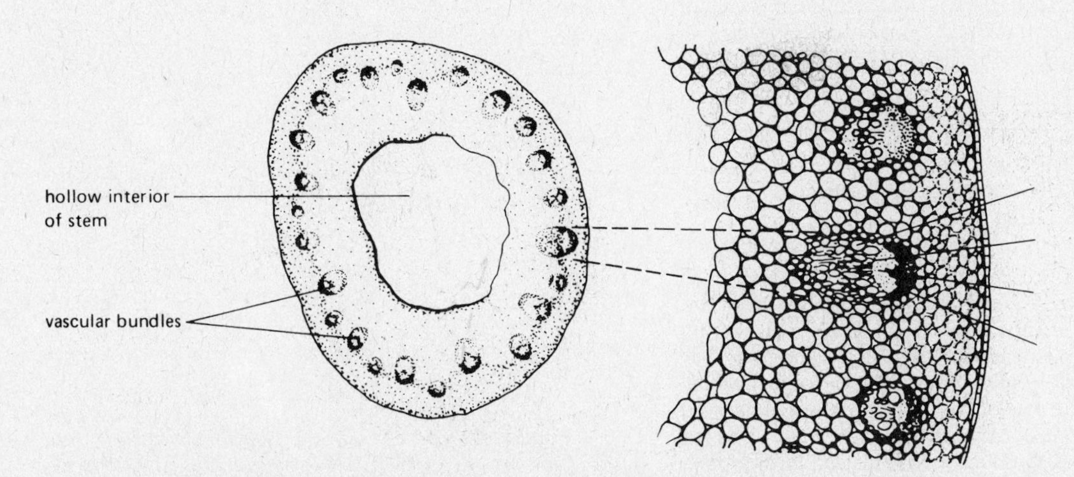

Figure 29–4. A dicot stem. The many vascular bundles form a ring.

□ Examine a single vascular bundle. Locate the larger thick-walled cells of the **xylem** vessels in the inner part and the smaller thin-walled **phloem** tubes in the outer part of the bundle.
□ Locate the cells of the **cambium** between the xylem and the phloem.
□ Locate the thickly packed, darkly staining **fibers** along the outer edge of the vascular bundle. They give added strength to the stem.
□ Label the parts of a vascular bundle in Figure 29–4.

The conducting cells of the phloem tissue are the **sieve tube members.** These cells contain living cytoplasm but lack nuclei. They are closely associated with **companion cells** which do have nuclei. Individual sieve tube members are connected end-to-end across perforated plates known as sieve plates. The structure of phloem tissue is shown in Figure 44–17 in your text.

The major function of the phloem is to transport nutrients, often in the form of sucrose, from the leaves to the roots for their nourishment and for storage. It can conduct dissolved substances in both directions, however: some sieve tubes may be conducting downward while others are conducting different substances upward. In this way, minerals and nutrients can be continuously redistributed among the organs of the plant to meet their changing needs.

□ Locate the small **companion** cells adjacent to the sieve tube members in your slide.

Storage Stems

In some plants, the stem rather than the root is used to store food for the next season. In the potato, for example, part of the stem is a storage organ and also functions in asexual reproduction.

□ Examine a cross section of a potato through one of the "eyes."
□ Note that the eyes are really buds that will send up shoots if the potato is planted in moist soil or exposed to the light.

You should now be able to list five important functions of stems:

1. ______
2. ______
3. ______
4. ______
5. ______

The Woody Twig

In woody plants, unlike in herbs, the apical meristem has to survive the winter, and the green shoot must be protected from drying out or freezing. The young branches of the plant are called **twigs,** and they become covered with a protective layer of corky material called **bark.** At the end of the growing season, changing environmental conditions cause the terminal bud to cease vegetative leaf production and instead form protective **bud scales.** Primary growth of the plant for that season comes to an end. In the spring, the bud scales will fall off, the terminal bud will grow into a new shoot of primary growth, and the "old" twig will now begin secondary growth. It will continue cycle after cycle of secondary growth as long as the plant is alive.

□ Examine the twig of a woody plant such as a hickory tree, and use Figure 29–5 as your guide to its structure.
□ Note the **bud scales** covering the apical bud.
□ Identify the **lateral buds** along the sides of the twig. Those in the older portions of the twig have become latent and will not develop unless the terminal bud is damaged.
□ Just below each bud is a **leaf scar** where the old leaf was once attached. The bud was thus formed in the leaf axil.
□ Look at the leaf scar under the dissecting microscope. The dots within it plugged vascular bundles that once led into the leaf before the leaf was dropped from the tree.
□ Locate the **bud scale scars,** which are a ring of

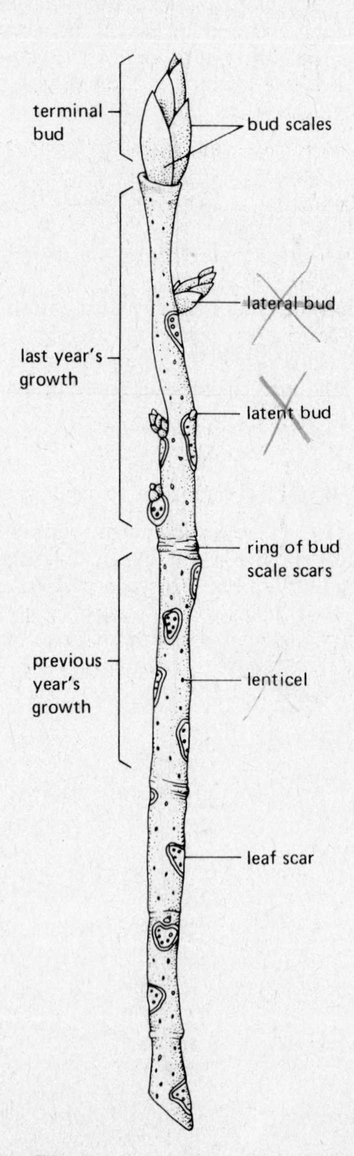

Figure 29–5. A woody twig. This hickory tree grew the length of twig between the ring of bud scale scars and the terminal bud in one season.

wrinkled tissue. This was the position of the terminal bud 1 year before its present position.
□ Look further down the stem to find the next ring of bud scale scars. This was the position of the terminal bud 2 years before its present position.

The point where the twig joins the next larger branch marks the place where it was a lateral bud before it started developing into a new branch.

How many seasons has a twig been growing if it has four rings of bud scale scars between the terminal bud and the branch that it grows from? 5 years

□ Peel off a bit of the bark from the twig, and note that it is much too thick to allow gas exchange.
□ Look for small rough spots on the surface of the bark called **lenticels,** and examine one with the dissecting microscope.
□ Note that the surface is spongy and not smooth like that of the bark. Its porous structure allows gas exchange between the cells of the stem and the exterior.

The internal structure of a woody twig that has grown for only part of 1 year is not much different from that of a herbaceous dicot. As the twig continues to grow, however, important changes must take place. The meristematic cells of the vascular cambium, located between the primary xylem and phloem in the vascular bundles, continue to divide and produce more vascular cells. As they do so, the vascular bundles become fused into a ring with phloem on the outside, xylem on the inside, and a layer of dividing cambium cells in between. A simplified diagram of this stage is shown in the left part of Figure 29–6.

Secondary growth begins when the vascular cambium becomes active and produces secondary phloem and secondary xylem. The layer of primary phloem is split apart as it is forced outward, and eventually it degenerates. The secondary phloem now functions to transport food. The primary xylem is left behind as the cambium moves outward in a slowly expanding cylinder, leaving the newly formed cells behind. The oldest xylem is in the center of the stem nearest the pith, and the newest is right next to the cambium. A diagram showing the structure of a twig that has grown for two seasons, with both primary and secondary growth, is shown on the right in Figure 29–6. The layer of secondary xylem is called **wood,** and the twig now has an annual ring. After another season, it will have another **annual ring** of secondary xylem. The older stems, which you will examine next, may have dozens of annual rings.

The Woody Stem

Just as the layer of phloem split apart as the vascular cambium grew outward, so the epidermis split and a mechanism was necessary to keep the outer covering of the stem intact. Some of the cells outside the phloem became meristematic tissue, and formed the cylinder of **cork cambium.** Cells produced by the cork cambium move outward and secrete the substance **suberin,** which acts to waterproof the surface of the stem. As the layers accumulate, the cells die and form much of the bark of the branch or tree trunk. The cork cambium and its derivatives are known as the periderm of the stem and provide an actively growing protective covering.

□ Examine a woody stem that is several years old under low power in your microscope.
□ Referring to Figure 29–7, identify the **vascular cambium,** which consists of the flattened cells along the outer edge of the largest annual ring.
□ Note the annual rings of **secondary xylem** inside the vascular cambium.
□ Counting inward from the vascular cambium, note the total number of layers of xylem. (The inner layer nearest the pith is primary xylem.)

The number of layers is equal to the age of this part of the stem in growing seasons.

How old is the stem you are studying?

____

□ Note the scalloped layer of **secondary phloem** growing from the outer layer of the vascular cambium. This layer is always being split apart as the cambium enlarges.
□ Note the pyramids of tissue within the splits of the phloem. These are called phloem rays.
□ Outside the phloem, locate the **cortex** and the thin layer of **cork cambium** that surrounds it.
□ Note the dead and dying cells on the outside of the cork cambium that will make up the **cork,** or outer bark.

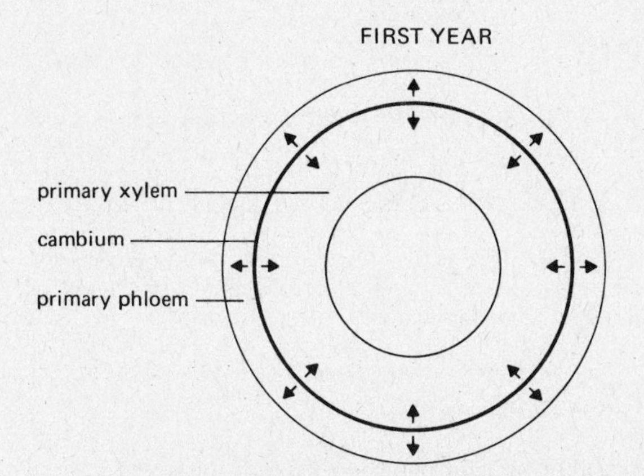

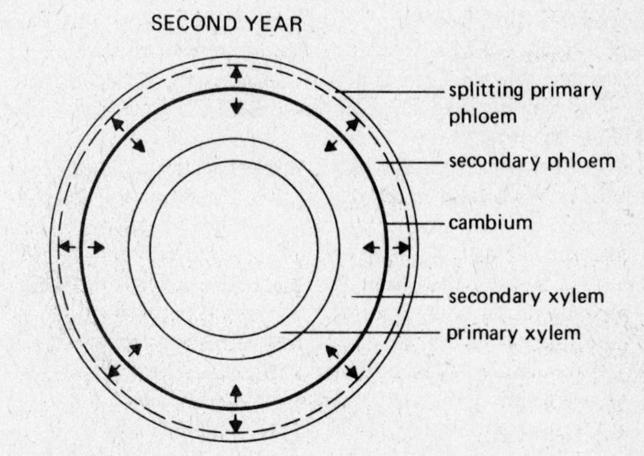

Figure 29–6. New tissue growth in a woody stem. Dividing cells are found only in the cambium.

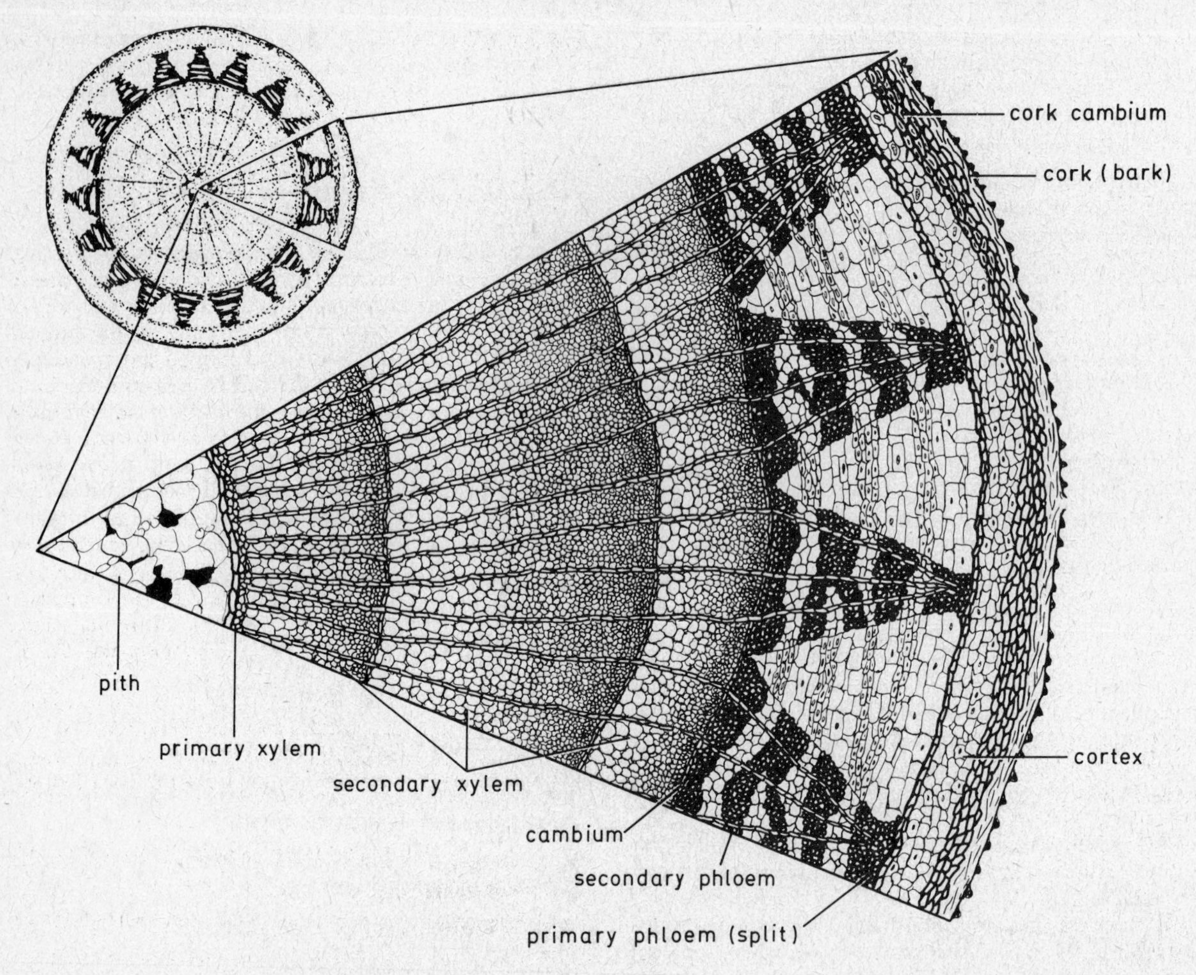

Figure 29–7. A woody twig in cross section. This twig has grown for three seasons. It has one ring of primary xylem and two rings of secondary xylem.

□ The remains of the original epidermis might still be seen on the outermost surface of the bark.

If you were to peel the bark from this twig, it would separate from the stem at the level of the vascular cambium. The bark includes not only cork but also the cork cambium, cortex, and secondary phloem. Removing the bark in this way cuts the supply lines through which the roots receive nourishment from the leaves. If the tree is **girdled** by removing a ring of bark that completely encircles the tree, the roots will starve, and the tree eventually will die. If even a thin strip of bark remains to connect the top and bottom of the stem, however, the roots may receive enough food and the tree may continue to grow.

As the stem grows older, the secondary xylem in the interior no longer conducts sap. It becomes the **heartwood** of the tree. The cells of the heartwood may become filled with metabolic wastes and resins and may turn a darker color than the still-active xylem, the **sapwood.** Heartwood is often the most beautiful and aromatic wood in a tree. Sometimes it starts to rot, or it may be eaten away by insects to make the tree hollow. The physiology is not upset as a result, but the tree is weakened and may split or break under stress.

□ Examine a cross section of an older tree, such as you might come across while splitting firewood.

□ Identify the **bark, sapwood,** and **heartwood,** and indicate the position of the **vascular cambium.**
□ Label these structures in Figure 29–11.

How old was your tree when it was cut down?

____

Wood

What is the botanical term for the tissue that we call wood?

Secondary Xylem Heartwood

Although the cells of wood are mostly dead, the structure of wood is rather complicated and involves interesting cell adaptations.

GYMNOSPERM WOOD

Tracheids are the only type of conducting cells found in gymnosperms. These cells are primitive because they have a small diameter, and because there are only small pores or pits connecting the cells end-to-end and sideways.

□ Examine a cross section of pine wood or a prepared slide showing tracheids in the microscope.
□ Note the uniform appearance of the tracheids and their small diameter.
□ Note also that the tracheids in each annual ring are somewhat larger toward the center of the tree and smaller and more closely packed toward the outside of each annual ring.
□ Note the very large **resin canals** scattered through the wood.

The inner layer of each ring, the **springwood,** is formed during the spring when the tree is growing faster. The outer layer of **summerwood** is more compact and was formed during the hot, dry months.

ANGIOSPERM WOOD

Angiosperms have developed more efficient conducting cells called **vessel members,** which are stacked vertically to form **vessels.** They have a larger internal diameter and have sometimes completely lost the end walls between the ends of adjacent cells. The wood of these trees is a mixture of fibers, tracheids, and vessel cells.

□ Examine a cross section of angiosperm wood such as maple, elm, or basswood under low power in your microscope.
□ Note the very large vessel members characteristic of angiosperms.

Are the vessels scattered uniformly through the annual ring, or are they concentrated mainly in the springwood?

____

____

□ Identify the **tracheids,** which have a much smaller diameter but do conduct some water.
□ Locate the **fibers,** which are very long, thin cells with heavy walls. The presence of fibers makes the wood strong, or "hard."

Technically (according to the U.S. Department of Agriculture) gymnosperms are softwood and angiosperms are hardwood. Actually, the wood of many gymnosperms is harder than that of many angiosperms: balsawood is from an angiosperm but lacks fibers and is extremely soft. The concentration of fibers determines the strength of the wood.

Within the sapwood are two structures containing living cells. Vertical sheets, or **rays,** of living parenchyma cells run out from the center of the tree to connect the living cells of the cambium with the conduits of the xylem in the sapwood.

□ Find the **rays** in the cross section of wood.

There are also columns of living parenchymal cells that secrete **resin,** a sticky, pungent substance that may help protect the wood against physical trauma or damage by insects. Vertical resin canals are very prominent in the wood of gymnosperms, as you saw in the cross section of pine wood.

LEAVES

Types of Leaves

A leaf consists of a flat **blade** (lamina) and a stalk, the **petiole,** which attaches the leaf to the stem. There will be a lateral bud in the leaf axil where the petiole joins the stem. Some leaves are **simple** and consist of just one blade and its petiole, but others are **compound** and consist of several **leaflets** that share a single petiole. The types of leaves are shown in Figure 29–8.

If it is unclear whether a leaf is a compound leaf or several simple leaves, look for buds. Because the buds are always found on stems, leaves that join without buds in their axils are part of a single compound leaf.

□ Examine the leaves on demonstration, and decide whether each is simple or compound.

The vascular tissue within the leaf is in large bundles called veins. If the veins run from the petiole more or less parallel out to the tip of the leaf, the leaf has **parallel venation.** If there is one main vein with smaller veins branching off it and running parallel to each other, the venation is **pinnate.** If several main veins radiate out from the petiole and have many smaller branches, the leaf has **palmate venation.** Most monocots have parallel venation, and dicots tend to have pinnate or palmate venation. These types of venation are shown in Figure 29–8.

□ Identify the three types of venation among the leaves on demonstration.

The arrangement of the leaves on the stem may be

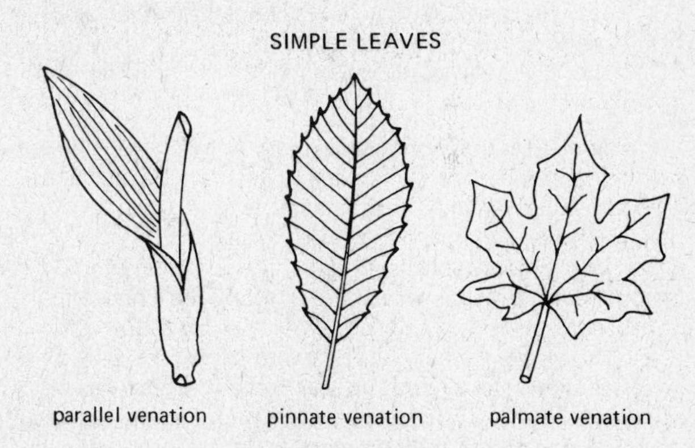

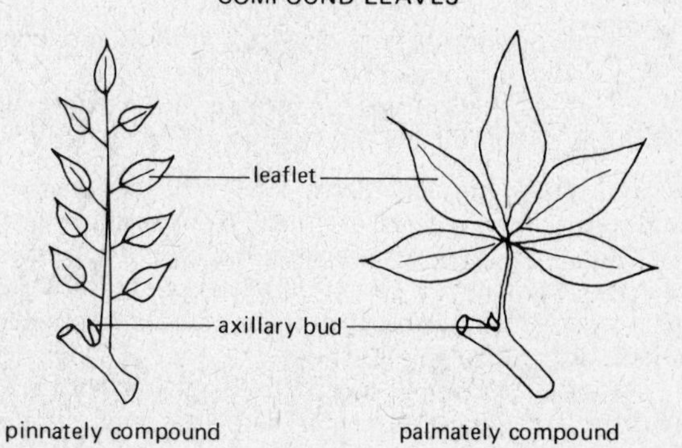

Figure 29–8. Leaves. A compound leaf has a bud only at its base, not next to its individual leaflets.

opposite, with two leaves at a node on the opposite sides of the stem; **alternate,** with one leaf per node and leaves appearing first on one side and then on the other; or **whorled,** with three or more leaves clustered at a node. If the leaves are opposite, the twigs and branches will also be opposite because they arise from buds in the leaf axils.

□ Find an example of opposite, alternate, and whorled leaf arrangements among the leaves on demonstration.

Structure of the Leaf

Because leaves are covered by an impermeable cuticle, access to the leaf interior is through the pores called **stomata.**

□ Obtain a dicot leaf, and study its upper and lower surface under high power in the microscope to locate the stomata.

On which surface are the stomata located?

lower

Are they open or closed? ______

What are the adjacent cells that open and close the stomata?

Guard cells

□ Review the internal structure of the dicot leaf by examining a cross section of a leaf such as a privet leaf.
□ Locate the **palisade mesophyll,** the **spongy mesophyll,** the **air spaces,** and the stomata leading into them.
□ Identify the **cuticle** that covers the leaf and the cells of the **epidermis** that secrete it.
□ Locate a bundle of vascular tissue, and identify the **xylem** and **phloem.**

The leaves of monocots tend to be long and pointed and oriented vertically. They may not have a palisade layer because the sunlight strikes them at an oblique angle. Often the leaf will be adapted to curl up when the amount of water lost through transpiration becomes too high. The curled leaf keeps the concentration of water vapor inside high, but also cuts down the efficiency of gas exchange. The cross section of a blade of grass that has curled up to prevent water loss is shown in Figure 29–9. Notice that the stomata are all on the inside where there is more water vapor, and that they are located in sunken pits.

□ Examine a cross section of a leaf from a monocot such as a lily or oleander.
□ Note where the stomata are located and whether air spaces are present or absent.

Sometimes monocots have a layer of **intercalary meristem** at the base of each leaf. The meristem continues to divide and can replace the leaf tissue as it becomes old, damaged, or cut off by a lawnmower. This is the process that allows grass to be mowed frequently without killing it.

Turn to Table 29–2 and fill in the characteristics of monocots and dicots that you have observed. If the potted plants that you studied were not in bloom, consult your text for the difference between monocot and dicot flowers.

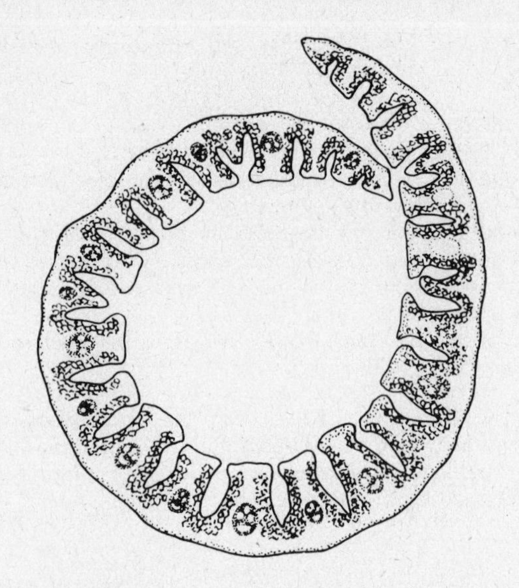

Figure 29–9. A blade of grass in cross section. The leaf is curled and the stomata are located in deep pits to minimize the loss of water.

Note the results of the transpiration experiments, and return all materials and equipment before leaving the laboratory.

EXPLORING FURTHER

Balbach, M. K., L. Bliss and H. J. Fuller. *A Laboratory Manual for General Botany.* 5th ed. New York: Holt, Rinehart and Winston, 1977.
Bold, H. C. et al. *Morphology of Plants and Fungi.* 4th ed. New York: Harper and Row, 1980.
Emboden, William A. *Bizarre Plants: Magical, Monstrous and Mythical.* New York: Macmillan, 1974.
Galston, A. W. et al. *The Life of the Green Plant.* Englewood Cliffs, N.J.: Prentice-Hall, 1980.
Galston, A. W., and C. L. Slayman. "The not-so-secret life of plants." *American Scientist* 67: 337, 1979.
Jensen, W. A., and F. B. Salisbury. *Botany: An Ecological Approach.* Belmont, Calif.: Wadsworth, 1972.
Lee, A. E., and C. Heimsch. *Development and Structure of Plants: A Photographic Study.* New York: Holt, Rinehart and Winston, 1962.
Torrey, J. G., and L. J. Feldman. "The organization and function of the root apex." *American Scientist* 65: 334, 1977.
Wilson, C., W. Loomis and T. Steeves. *Botany.* 5th ed. New York: Holt, Rinehart and Winston, 1972.

Scientific American Articles

Albersheim, Peter. "The walls of growing plant cells." April 1975 (#1320).
Altshul, Siri von Reis. "Exploring the herbarium." May 1977 (#1359).

Biddulph, Susann and Orlin. "The circulatory system of plants." February 1959 (#53).
Elias, T. S. and H. S. Irwin. "Urban trees." November 1976 (no offprint).
Epstein, Emanuel. "Roots." May 1973 (#1271).
Erickson, Ralph O., and Wendy Kuhn Silk. "The kinematics of plant growth." May 1980 (#1470).
Fritts, Harold C. "Tree rings and climate." May 1972 (#1250).
Heslop-Harrison, Yolande. "Carnivorous plants." February 1978 (#1382).
Overbeek, Johannes. "The control of plant growth." July 1968 (#1111).
Rick, Charles M. "The tomato." August 1978 (#1397).
Zimmerman, Martin H. "How sap moves in trees." March 1963 (#154).

Student Name ______________________ **Date** ______________

TABLE 29–1. ROOTS

Function	Tap Root	Fibrous Roots
1. Anchoring the plant		
2. Absorbing water and minerals (surface area)		
3. Transport to the stem		
4. Storage of nutrients		

TABLE 29–2. CHARACTERISTICS OF MONOCOT AND DICOT ANGIOSPERMS

Characteristic	Monocots	Dicots
Type of seed leaf (cotyledon)		
Type of leaf venation (vein pattern)		
Arrangement of vascular tissue		
Formation of wood	No	Yes Some
Symmetry of flower parts (multiples of 3, 4, and so on)	3	4, 5
Type of root system	Fiberous	Tap

Figure 29–10. Transpiration. Sketch the results from the dye experiment.

Drawing of Petal After Transpiration	Drawing of Xylem Cross Section After Transpiration

TABLE 29–3. RESULTS FROM THE TRANSPIRATION EXPERIMENT

Transpiration in Light			Transpiration in Dark		
Time (min)	*Reading*	*Change (mL)*	*Time (min)*	*Reading*	*Change (mL)*
0					
15					
30					
45					
60					
75					
90					
105					
120					
Average rate = _______ mL/hr			Average rate = _______ mL/hr		

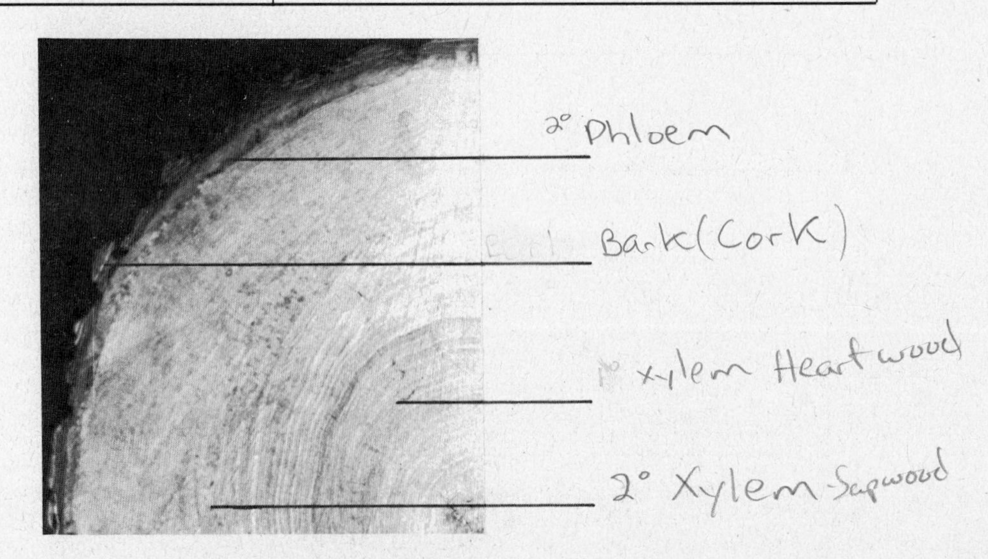

Figure 29–11. A tree trunk in cross section. Label the parts with their correct names.

TOPIC 30

Plant Reproduction

OBJECTIVES

When you have completed this topic, you should be able to:

1. List the reproductive characteristics that distinguish angiosperms from other seed plants.
2. Given a diagram of a flower or an actual flower, identify it as a monocot or dicot and point out the following parts: sepal, petal, stamen, anther, carpel, ovary, style, stigma, ovule, pollen.
3. Define pollination and fertilization, and show where they occur in a flower.
4. Explain the role or origin of the following structures and identify them in a microscope slide: microspore, male gametophyte, pollen grain, pollen tube, megaspore, female gametophyte, embryo sac, micropyle.
5. Give the botanical definition of a fruit, and name some "vegetables" that are botanically fruits.
6. Given an actual fruit or a diagram, identify the following parts: ovule, locule, endocarp, mesocarp, exocarp, attachment sites of old sepals and petals.
7. List or identify the parts of a bean seed and of a corn kernel; state the function or fate of each part.
8. Describe the role of animals in pollination and seed dispersal.
9. State four factors that are important in seed germination.
10. Explain the role of plant hormones in germination, tropisms, and apical dominance.
11. Explain the advantages and disadvantages of vegetative reproduction; give an example of vegetative reproduction from a root, from a stem, from a leaf, and from an infertile fruit.

TEXT REFERENCES

Chapter 43 (A,F,G); Chapter 46.
Flowers, pollen, seeds, germination, development, plant hormones, vegetative reproduction.

INTRODUCTION

As we have already seen in surveying the plant kingdom, most seed plants have dispensed with a free flagellated sperm in their reproductive cycle and have instead developed the pollen grain as a carrier of the male gamete. The process of fertilization leads to the formation of a seed that contains a small embryo and also a source of nutrition within a tough seed coat. The seed is usually resistant to harsh environmental conditions and begins growing only when conditions become favorable.

Angiosperms differ from the other seed plants in that the seeds are formed within **flowers.** In many plants the flowers are showy, or have an odor so that animals are attracted to them and become involved in pollination. The seeds are enclosed within a fruit that may be a source of food for animals. The seeds are then scattered about when the animal feeds or are carried away in its digestive tract and are later expelled far from the parent plant. Thus animals are an important factor in seed dispersal for many angiosperms.

BEFORE BEGINNING THE NEXT SECTION, TURN TO THE SECTIONS ENTITLED "POLLEN" AND "GERMINATION," AND SET UP THE DEMONSTRATIONS OF POLLEN TUBES AND SEEDLING GREENING, UNLESS YOUR INSTRUCTOR GIVES OTHER DIRECTIONS.

FLOWERS

Angiosperms that are pollinated by animals have several strategies for attracting attention. Their flowers may have large, showy petals, which are brightly colored to attract the notice of potential pollinators. They may even have color patterns that are not visible to us but that attract insects such as bees because some insects can see into the ultraviolet range further than we can. Within the flower there may be nectar, which is an important food source for the pollinators. Some flowers produce volatile chemicals that give the flower a very attractive odor. (Putrid odors that are disgusting to humans may be very attractive to pollinators such as flies.) Many angiosperms have a sort of symbiotic relationship with humans, in which they are protected and propagated in exchange for their production of flowers.

Flowers are modified shoots in which the leaves have become modified to form the flower parts.

□ Obtain a large, simple flower, such as a gladiolus. Cut it from the stem with a razor blade far enough down so that you take all of the green part at the base in addition to the petals.
□ Identify the green **receptacle** at the base of the flower.
□ Carefully dissect the flower from its outer whorls inward. The flower may be enclosed in a green bract, or leaflike part.
□ Note the **sepals,** which make up the outer whorl, called the calyx. They may be green or the same color as the petals.

What color are the sepals?

green ______________

□ Note the colored **petals,** which make up the **corolla** of the flower.
□ Locate the **stamens,** the male reproductive structures that form the next whorl.
□ Locate the **pistil.** It is made up of one or more **carpels,** which are the female reproductive organs. You will be able to determine the number of carpels later, when you dissect the ovary.

How many petals, sepals, stamens, and carpels does the flower have?

Petals: ________

Sepals: ________

Stamens: ________

Carpels: ________

Monocots tend to have the flower parts in threes, whereas dicots tend to have four or five of each part.

Considering the other characteristics that distinguish monocots and dicots, as well as the structure of their flowers, do you think this flower is from a monocot or from a dicot? ________________

Pollen is formed in the **anther,** which is the top part of the stamen supported by the stalklike **filament.** It is then deposited on the sticky surface of the **stigma** at the top of the pistil during **pollination.** The germinating pollen grain must grow a tube through the **style** of the pistil to the ovary at its base. The ovules (prospective seeds) within the ovary are fertilized when the pollen tubes reach them and sperm nuclei enter to unite with nuclei within each ovule.

When you have finished this part of the dissection, label the structures indicated in Figure 30–5 at the end of this topic.

□ Make a cut across the base of the carpels where the green **ovary** is the widest, as shown in Figure 30–6 at the end of this topic.
□ Make a second cut and remove the slice from the flower.
□ Note the small **ovules** that will mature into seeds; each ovule will be fertilized by one pollen tube.
□ Note the spaces containing the seeds, called **locules.** The number of locules is equal to the number of carpels that make up the pistil.

How many locules (carpels) does the ovary contain?

____

□ Identify the wall of the **ovary,** which surrounds the locules. It will develop into part of the fruit.
□ Make a sketch through the ovary in the space provided in Figure 30–6, and label its internal structures.

There are many variations in flower structure. The flower that you have just studied is **simple** because it contains a single ovary, and it is **complete** because it contains both stamens and carpels. A flower such as a daisy is a **composite** flower and consists of many tiny individual flowers. Each white petal is formed by a separate ray flower, and the central disc consists of many other tiny flowers. Sometimes the carpels and stamens are in separate flowers, which are termed **incomplete.**

Some flowers are self-fertilizing and have petals that close over the reproductive organs. The anther fertilizes the carpel right next to it, and the closed petals prevent the entry of pollen from another flower carried by wind or insects. Flowers that are normally pollinated by the wind rather than by animals are often inconspicuous and may lack petals altogether. The flowers of many trees and grasses lack petals.

□ Examine other types of flowers that are on demonstration.

In the flower that you studied, the sepals may have been green, or they may have been colored and almost indistinguishable from the petals. The sepals usually form a whorl of their own outside of the true petals. In some flowers the sepals are large and showy, while the rest of the flower is very inconspicuous. In other flowers, such as the poinsettia, the red "petals" are modified leaves called bracts that surround the tiny flowers in the center.

POLLEN

Note: At the beginning of the laboratory session, set up your demonstration in the following manner.

□ Obtain a Petri plate of agar or a slide with a drop of 10% sucrose.
□ Shake some pollen from the anther of a flower such as gladiolus or *Tradescantia* onto the agar or the drop.
□ Cover the pollen, and set it aside in a warm, moist place to allow pollen tubes to develop from the grains of pollen.

The male reproductive organ, the stamen, consists of the anther (pollen sac) which is supported by the filament.

□ Locate the filament and anther of your flower.
□ Shake a few grains of pollen onto a slide, make a wet mount, and observe the pollen under high power in your microscope.
□ Note the shape of the pollen grains.
□ If the pollen from other sources is available, study it also, and compare the shapes of pollen grains from different sources.

Is all pollen alike? ______________

The cells within the anther that undergo meiosis are the **microspore mother cells,** and the haploid cells that are produced are the microspores. This tetrad of microspores will produce four male gametophytes which are parasitic within the anther as they mature into pollen grains. The formation of pollen is shown in Figure 30–1.

□ Examine a prepared slide of an anther of the lily containing pollen tetrads.
□ Look for groups of four cells, which are the tetrads.
□ Examine a slide showing a mature anther that is almost ready to shed pollen.
□ Note that many grains have two functional cells, and there are some mature pollen grains that are dark and opaque.

A single grain of mature pollen contains a generative nucleus and a tube nucleus. The generative nucleus will participate in fertilization. The pollen is deposited on the stigma during pollination and the **tube nucleus** then directs the growth of the **pollen tube,** into which it moves. The generative nucleus divides to form two **sperm nuclei,** which also move out into the tube. Only the sperm nuclei will eventually enter the ovule and contribute genetic material to the new embryo through fertilization. In angiosperms, pollination and fertilization may be separated by a few hours or whatever time it takes for the pollen tube to reach its destination.

□ Examine your demonstration of germinating pollen to see if any pollen tubes have grown.

Are the tubes straight or irregular? ______________

If there are no tubes yet, examine the grains again later.

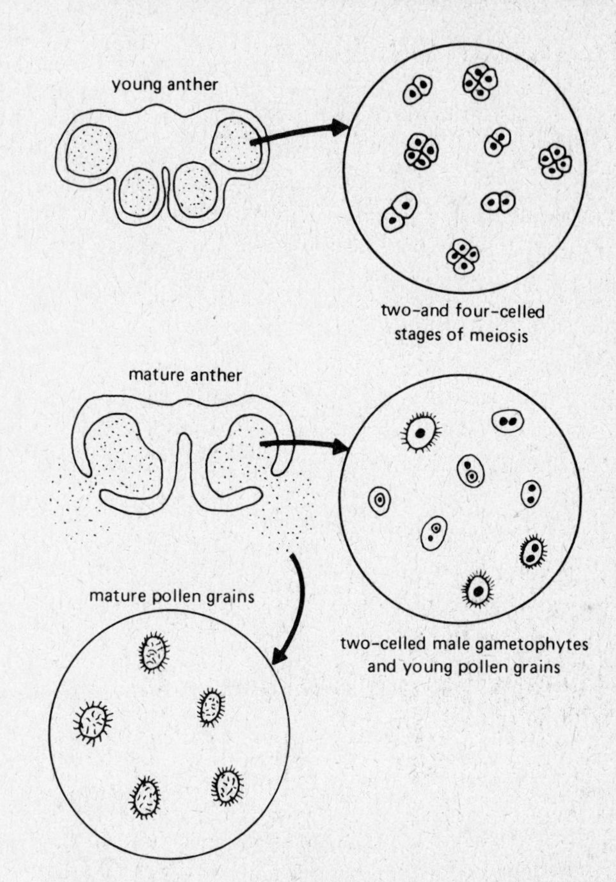

Figure 30–1. Production of pollen grains. A microspore mother cell produces four microspore daughters in a tetrad. Each of the haploid cells gives rise to a pollen grain with its generative and tube nuclei. The mature pollen grain, the male gametophyte, has a characteristic spiny covering.

OVARY

The ovary at the base of the pistil contains one or more ovules or **megasporangia.** The reproductive cell within the ovule, called the **megaspore mother cell,** undergoes meiosis to form four haploid cells; only one survives to form the female gametophyte. The exact number of cells forming the gametophyte varies, depending on the species of flower, but there will be one **egg nucleus** and two other nuclei that fuse to form a **central nucleus.** The development of the female gametophyte, or **embryo sac,** is shown in Figure 30–2.

□ Examine a prepared slide showing the cross section of the ovary of a lily under low power.
□ Referring to Figure 30–2, scan the six ovules of one cross section to locate an **embryo sac,** a large structure with granular cytoplasm. (Many of the ovules will not show an embryo sac because the section was at the wrong level.)
□ When you have located an embryo sac, switch to high power, and note the nuclei within it. Developing sacs will have two or four nuclei, and mature ones will have eight (rarely seen).

Are any of the cells in the embryo sac in the process of dividing? ______________

Fertilization occurs when the sperm nuclei of the pollen tube enter a pore at the bottom of the embryo sac, called the **micropyle,** and fuse with the nuclei within. One sperm nucleus fuses with the egg nucleus to form the **embryo,** and the second sperm nucleus fuses with the central nucleus to form the **endosperm.** The zygote will be diploid, whereas the endosperm will be triploid or pentaploid. Both patterns are shown in your text in Figure 46–9. Mitosis then produces the embryo and the tissues of the endosperm. This **double fertilization** occurs only in angiosperms.

Label the important structures in the diagram of fertilization shown in Figure 30–7 at the end of this topic, and give the function or fate of each part listed in Table 30–1.

SKIP

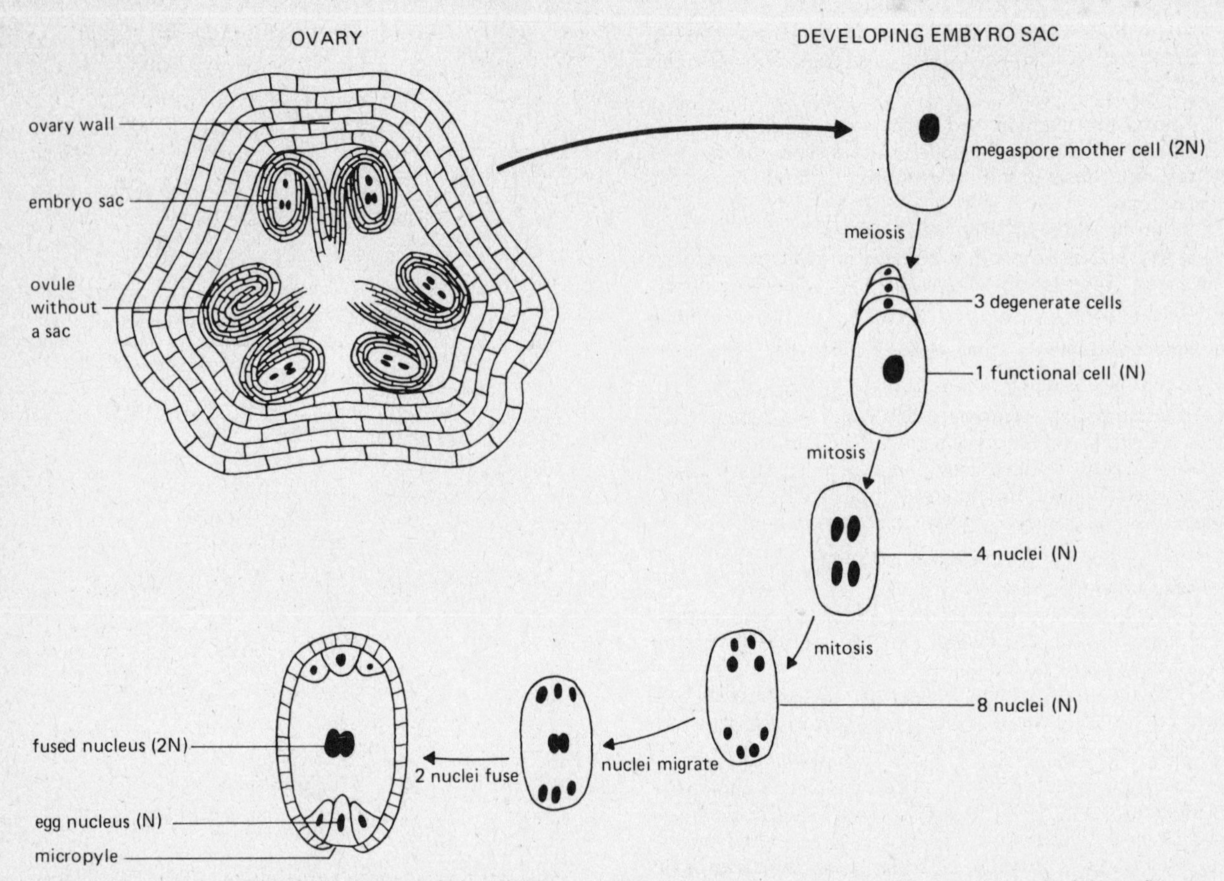

Figure 30–2. Development of the embryo sac. The developmental pattern shown is typical of most flowers. (Development in the lily is more complex.) Note that only some of the ovules will have a visible embryo sac. The embryo sac represents the female gametophyte and develops from a haploid nucleus.

TABLE 30–1. FERTILIZATION

Structure	Function or Fate
Egg nucleus	
Central nucleus	
Ovule wall	
1st sperm nucleus	
2nd sperm nucleus	
Micropyle	

FRUIT

After fertilization, the ovule develops into a seed, and the entire ovary matures into the **fruit.** Botanically speaking, a fruit is a ripened ovary or an ovary fused with nearby tissue. Ethylene, a plant hormone, promotes fruit maturation or ripening.

□ Obtain a fruit such as a bean pod. If the fruit was recently picked, you may see old petals and sepals clinging to the end of the pod.

□ Split the pod open and find the seeds. Each one represents an **ovule.**

□ Note that each seed is attached to the pod at a single point.

□ Identify the space surrounding the seeds, known as the **locule.**

How many locules does the bean have?

____

□ Note the pod, which represents the wall of the ovary and is called the **pericarp.**

While a bean pod is not very "fruitlike," it is a perfectly good example of a fruit. The pod itself is a compressed version of the kind of fruits you are used to: it has an inner layer, the **endocarp**; a middle, juicy layer, the **mesocarp**; and an outer layer, the **exocarp.**

□ Make a cut through the pod of the bean, and exam-

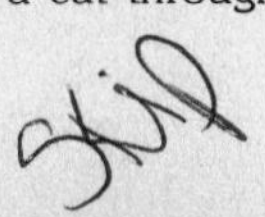

ine a slice of pod material in the dissecting microscope.
□ Note the lining of the pod, the endocarp, the juicy mesocarp, and the tough outer exocarp.
□ Label the parts of the bean pod shown in Figure 30–8 at the end of this topic. (One ovule failed to get fertilized.)

What was the fate of the unfertilized ovule?

____

____

A more familiar fruit such as a peach has the same layered structure. The seed is enclosed within a space surrounded by the stony endocarp that we call the pit. The juicy flesh is the mesocarp, and the skin is the exocarp. Label the parts of a peach fruit in Figure 30–8 at the end of this topic.

Sometimes the whole pericarp is reduced to a dry skin that surrounds the seed. This is seen in a wheat grain or in the fruit of a dandelion.

SEEDS

Anatomy

Take a closer look at the structure of a seed by dissecting a bean seed that has been soaked to loosen the skin and soften it. The external structure is shown in Figure 30–3.

□ Note the rough whitish area called the **raphe.** This is where the seed was once attached to the pod.
□ Examine the seed with your dissecting microscope, and find the tiny opening of the **micropyle.** Pollen entered the ovule through the micropyle.
□ Carefully split the seed into two halves, and refer to Figure 30–3 to identify the parts: **cotyledon, plumule** or **epicotyl, hypocotyl, radicle, foliage leaves.**

The bean is a dicot, and its seed has two cotyledons. The seeds of monocots have only one. The endosperm tissue of the bean has been completely absorbed by the cotyledons.

□ Obtain a soaked corn kernel, and cut it in half as shown in Figure 30–4.
□ Identify the single **cotyledon,** the **endosperm,** the **hypocotyl, epicotyl,** and the **foliage leaves** of the embryo. The shoot is covered by a sheath called the **coleoptile.**

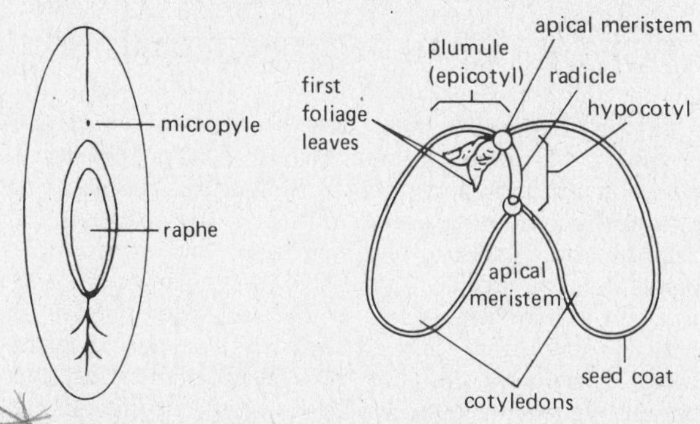

Figure 30–3. A bean seed. The bean on the right has been split open to reveal the embryo (epicotyl and hypocotyl). The apical meristem of the shoot is hidden within the first foliage leaves.

In the corn kernel, the cotyledon has absorbed only part of the endosperm tissue, and there is still a substantial amount of endosperm left. The ovary wall has become the pericarp.

□ Add a few drops of iodine stain to the cut surface of the corn kernel to stain the stored starch.

Does the endosperm or cotyledon contain more stored starch?

____

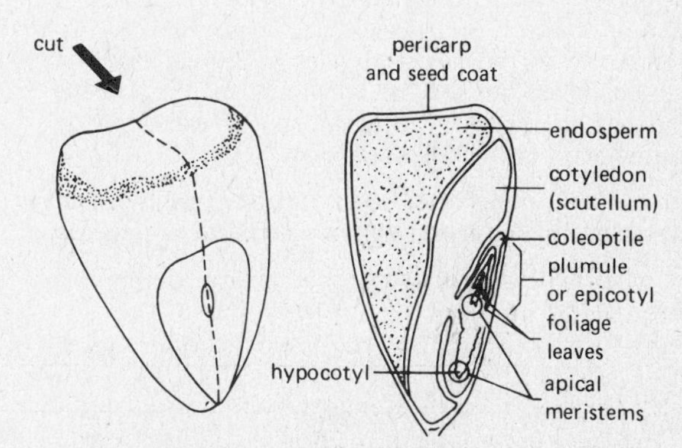

Figure 30–4. A corn kernel. The shoot is covered by the sheath of the coleoptile, and the pericarp has fused with the seed coat.

Metabolism

The seed has lost most of its water, and carries on metabolism at a very slow rate. This dormant state is maintained by the hormone abscisic acid. Some seeds will be able to germinate under the right conditions, and are **viable** whereas other are **inviable.** It is easy to test for metabolic activity in seeds by soaking them and observing the reduction of a dye by the living tissue. The dye is tetrazolium, which is normally colorless, but turns pink when it is reduced by living tissue.

□ Cut several soaked corn kernels in half, and put them into a test tube.
□ Add enough 0.5% tetrazolium solution to cover the kernels, and set them aside.
□ Observe the cut surfaces from time to time for the production of a pink color.
□ After the pink color is well developed, remove the kernels, and examine them to see which part of the seed has the darkest color and therefore the greatest metabolic activity.

Where was there the greatest metabolic activity?

____

Where any of the seeds totally inactive with the dye (dead)? ______________

GERMINATION

Note: At the beginning of the laboratory session, set up your demonstration in the following manner.

□ Obtain a few alfalfa sprouts, a Petri dish, and some filter paper or paper toweling.
□ Cut a circle of toweling to fit the Petri dish, and moisten it with tap water.
□ Place the sprouts on the toweling, and cover the dish.
□ Note the color of the cotyledons and hypocotyls of the sprouts.
□ Place the dish in strong sunlight or artificial light.

Several factors may be important in the process of germination: a period of cold, absorption of water, abrasion or digestion of the seed coat, and an increase in the content of the hormone gibberellin. Seeds differ in their requirements, but they all require a supply of water in order to germinate. The absorption of water that starts germination is called **imbibition.**

□ Examine some bean seeds that have been soaked in water overnight, and compare them with dry seeds.

In what two ways does the seed change during imbibition?

____

____

During the first days of development, the seed must rely on its own stored food present in its cotyledons. It must reach sunlight and a source of water within a limited time span or it will not survive. In order to increase its chance for success, its growth is closely controlled by hormones.

The level of abscisic acid is high in the seed and maintains dormancy. After imbibition has taken place, gibberellin production increases and stimulates elongation of the stem. Abscisic acid and inhibitory concentrations of auxin in the cells of the root cap cause the root to show positive geotropism and to grow downward toward the water. Auxin helps the shoot to show positive phototropism and grow up toward the sunlight.

□ Examine some young bean seedlings that have been allowed to sprout on agar plates or paper towels.
□ Note that the hypocotyl has grown downward, whereas the plumule has turned up toward the light.
□ Look at the oldest plants, and note how the cotyledons have shriveled up as their food was removed by the embryo. They will soon fall off the plant.

An interesting example of the vital importance of hormone balance can be seen in seedlings that have been kept in complete darkness so that they have become **etiolated.**

□ Examine some etiolated seedlings, and compare them with normal seedlings of the same age.

What color are the seedlings? ______________

What is their appearance compared with that of normal seedlings?

____

Gibberellin secreted by a fungus has long been known to cause "foolish seedling" disease of rice, in which the seedlings grow tall and spindly and then fall over and die. Excessive gibberellin is also responsible for the abnormal growth of etiolated seedlings. During the first days of growth, a seedling must reach the light or it will die. The concentration of gibberellin is high to stimulate rapid growth of the stem while the seedling is in the darkness underground. As soon as it reaches the light, the red wavelengths of light transform phytochrome P_r to P_{fr}, and the production of gibberellin falls off as a result. The stem's growth slows down, and normal development follows. If, however, the shoot is unlucky and never reaches the sunlight, the phytochrome stays in the form of P_r and continues to stimulate the production of large amounts of gibberellin. The seedling grows and grows until it becomes severely etiolated and dies. A seedling that is being kept in complete darkness can be saved by irradiating it with red light to reduce the amount of P_r and decrease gibberellin production.

What other process in angiosperms is under the control of P_r and P_{fr}? ________________________

As the seedling emerges into the light, it has the potential for carrying on photosynthesis, and will turn green within a short time. One of the steps in the production of chlorophyll is a photoreaction, which requires light. As long as the seedling is in the dark, the reaction is blocked, and there is no chlorophyll. There is, however, a large amount of the compound preceding the light-sensitive step. When light becomes available, the production of chlorophyll proceeds rapidly, and the seedling turns green, as the chlorophyll builds up.

□ Examine the alfalfa sprouts that you set up at the beginning of the period for signs of greening.
□ If they are not yet green, examine them once more before you leave the laboratory.

How long did it take the sprouts to turn visibly green?

____

Why is it advantageous for the seedling to postpone formation of chlorophyll until it reaches the light?

____

____

____

____

VEGETATIVE REPRODUCTION

Although there is a great deal of concern about the possibility of cloning human beings, growing clones of plants in the laboratory and in agricultural practice has already become widespread: one or a few meristematic cells can be used to grow plant tissue and even whole plants that are genetically identical to the parent cell and to each other. Vegetative, or asexual, reproduction is basically the same process: cells of the parent plant produce progeny that are genetically identical to the parent. Vegetative reproduction is a way of life among plants, and often it is the only means by which a plant can reproduce at all. When a successful plant reproduces asexually, the progeny will most likely also be successful because they will be well adapted to local conditions.

Sometimes plants have it both ways: they produce reproductive organs and seeds so that they have good dispersal and a lot of offspring, but the seeds are genetically identical to the parent and their formation does not involve meiosis. This process of **agamospermy** happens in dandelions, some grasses, and well-adapted plants that have given up sexual reproduction but have retained their reproductive organs. Other plants produce fruits, but have no fertile seeds, and the plants must be propagated vegetatively. Some examples are the pineapple, banana, and tangelo.

□ Examine the interior of a banana.
□ Note the tiny black dots. These are degenerative ovules that failed to form viable seeds.
□ Examine the various examples of vegetative reproduction on demonstration in the laboratory.
□ Match the examples that you see with the corresponding type of vegetative reproduction shown in Figure 30–9 at the end of this topic.

Check your demonstrations of pollen tubes and alfalfa sprouts, if you have not already done so, and return all materials. Clean up your work area, and store your microscope properly before leaving the laboratory.

EXPLORING FURTHER

Hutchinson, J. *The Genera of Flowering Plants.* Vols. I and II. Monticello, NY: Lubrecht and Kramer, 1979.
Peterson, R. T. and M. McKenny. *A Field Guide to Wildflowers of Northeastern and North Central North America.* Boston: Houghton-Mifflin, 1974.
Proctor, M., and P. Yeo. *The Pollination of Flowers.* Glasgow: William Collins Sons, 1973.
See also Topic 29.

Scientific American Articles

Echlin, Patrick. "Pollen." April 1968 (#1105).
Grant, Verne. "The fertilization of flowers." June 1951 (#12).
Koller, D. "Germination." April 1959 (no offprint).
Pettitt, John, Sophie Ducker and Bruce Knox. "Submarine pollination." March 1981.

Student Name ______________________________ **Date** ______________

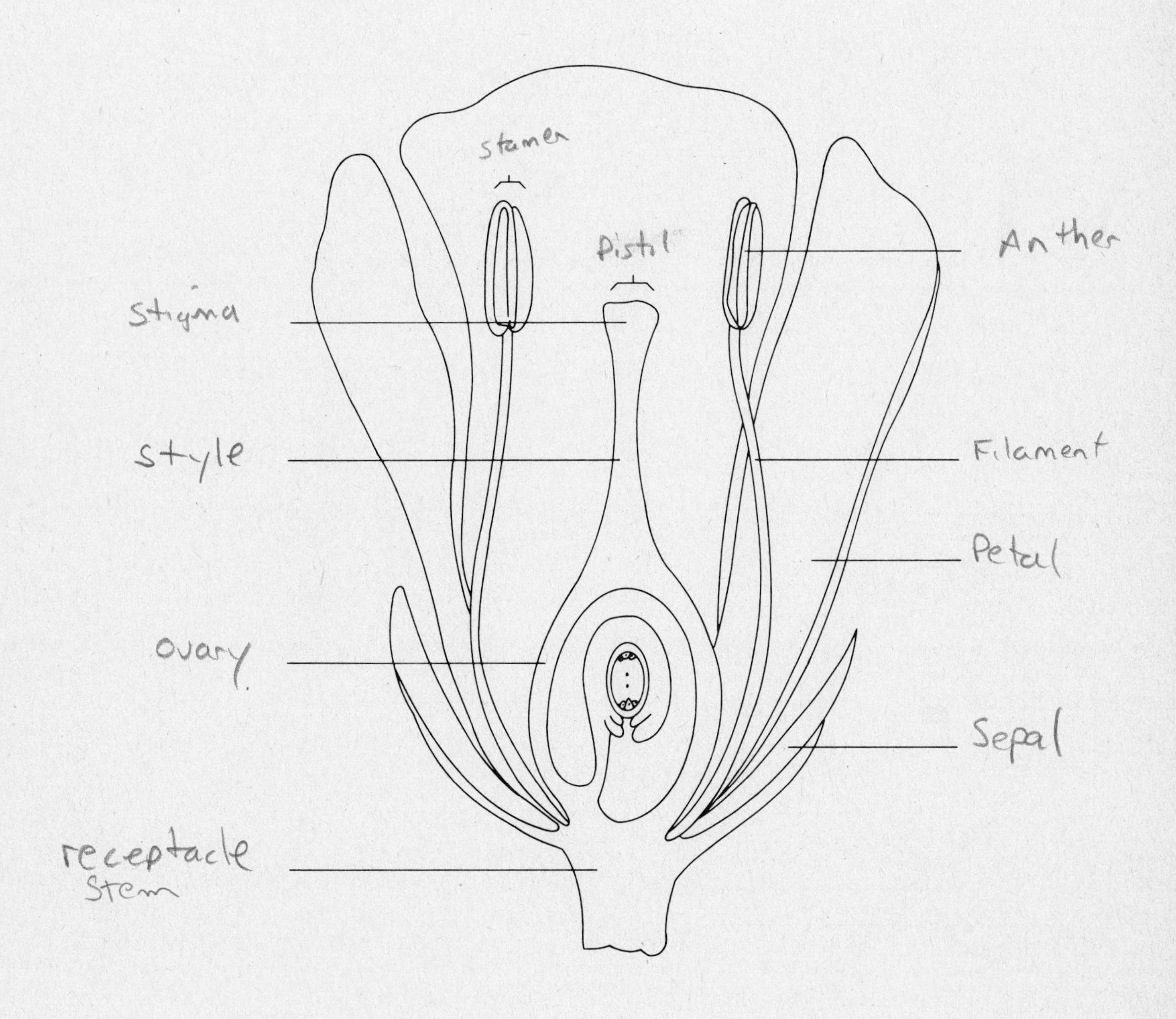

Figure 30–5. A simple, single flower. Label its parts. (Redrawn from M. K. Balbach, L. Bliss, and H. J. Fuller. *A Laboratory Manual for General Botany,* 5th ed. New York: Holt, Rinehart and Winston, 1977.)

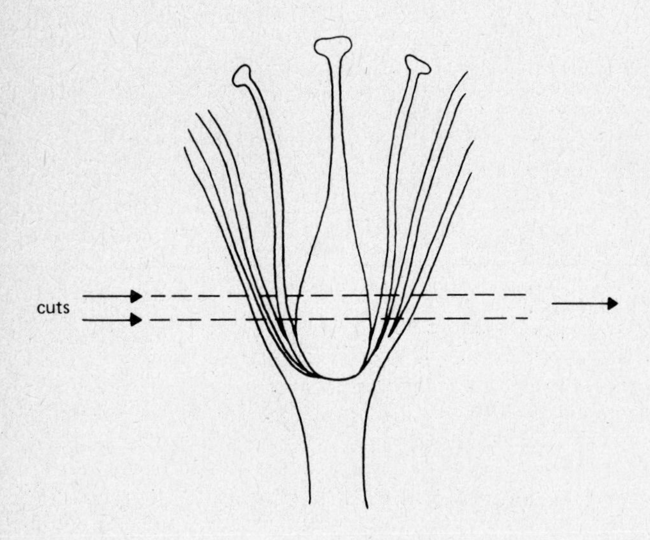

Figure 30–6. Cutting the ovary. Make a cross section by cutting at the position shown.

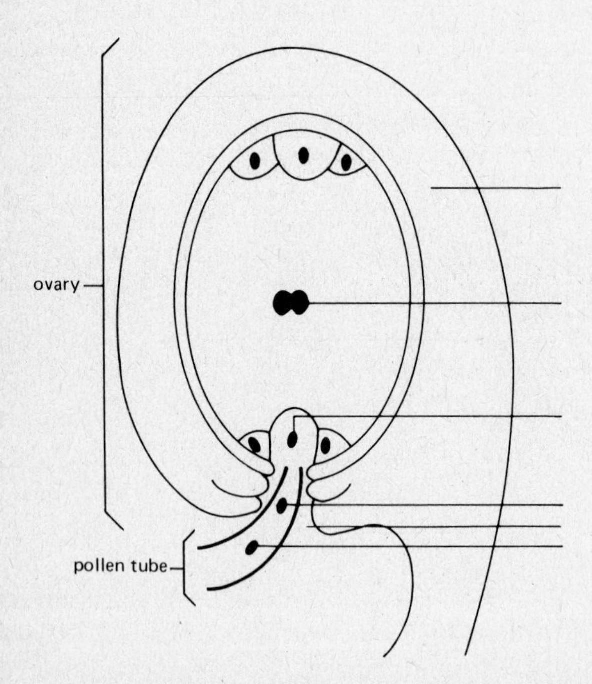

Figure 30–7. Fertilization. Label the parts indicated.

Student Name ______________________ **Date** ______________

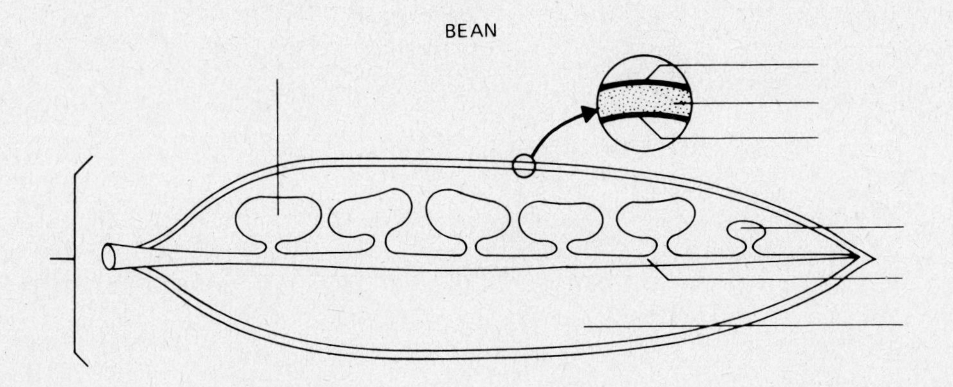

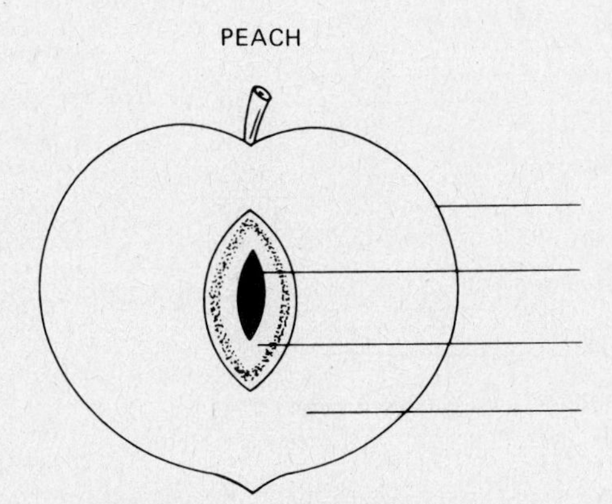

Figure 30–8. Fruits. Label the parts indicated with their correct botanical names.

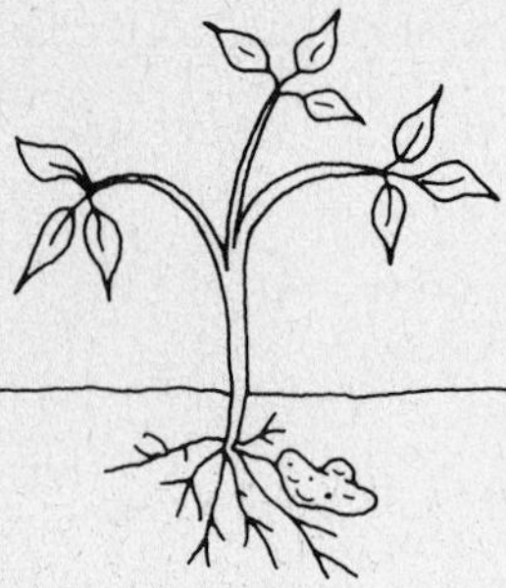

type: ____________________
example: ____________________

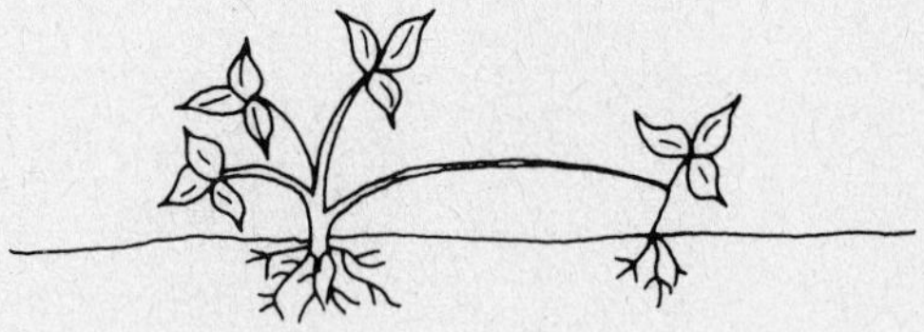

type: ____________________
example: ____________________

type: ____________________
example: ____________________

type: ____________________
example: ____________________

type: ____________________
example: ____________________

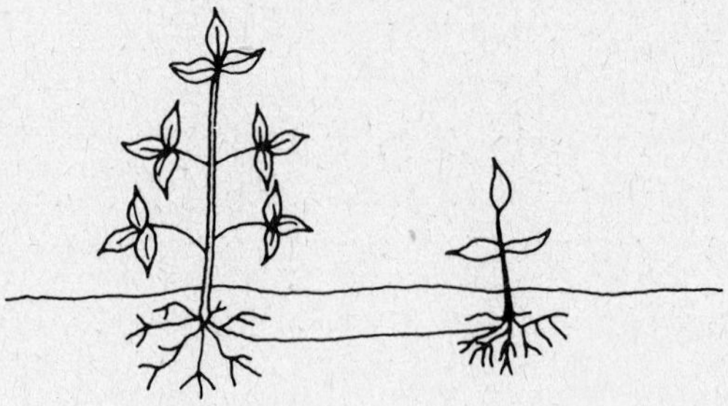

type: ____________________
example: ____________________

Figure 30–9. Vegetative or asexual reproduction. Label each type with its name or description, and note an example.

Introduction to the Projects

The following projects require more than a single laboratory period. Your instructor will give you specific directions as to how the project will be carried out. Because you will be working more independently than usual, it is important that you understand the rationale behind the experiments and exactly how they are to be done before you begin work.

You and your partner are about to become researchers in plant physiology or ecology and will carry out some experiments in order to answer certain questions. Because you may be asked to design the experiments yourself, you should read over the discussion of the scientific method given below. Then, because the first step in any research is a review of the current knowledge, you should read over the sections in your textbook relevant to your project and refer to any other sources that you can find. You will then be ready to prepare a written description of the planned work. You should bring the written plan with you to the laboratory, or hand it in ahead of time. Your instructor will check it over before you actually start your experimental work.

During the project, be careful to *record all of the manipulations and observations at the time*. Do not rely on your memory alone, or you may end up in serious difficulty when you attempt to write up the results in the laboratory report. Label all records with your name and the date they were made, since your instructor may ask you to include the raw data in your report.

When the project is finished, write up the results in a written report according to the directions given in Appendix A.

THE SCIENTIFIC METHOD

In your study of the scientific method you learned that there are two basic ways that modern science progresses. When scientists use **inductive logic,** they gather observations on a phenomenon and then formulate a generalization or scientific law from these observations. The generalization represents new knowledge and will be used to make predictions until it is shown to be invalid by a new observation. The more observations that have been made, the more likely the generalization is to be valid.

The generalization can be used to generate a **hypothesis** according to the process of **deductive logic.** A hypothesis is an unproved proposition put forward as a possible explanation of observed events to provide a basis for further study. The hypothesis itself is stated as the **major premise,** its application to a particular case is the **minor premise,** and the result of applying it to a particular case is the **conclusion.** Here is an example:

Major premise: All green plants contain chlorophyll.
Minor premise: This mint plant is green.
Conclusion: This mint plant contains chlorophyll.

Deduction does not generate new knowledge, but it proposes **experiments** that will support or disprove the major premise. In this example, if a single plant were found to be green but not to contain chlorophyll, the major premise would be discarded because it had been proved to be false. If, on the other hand, many mint plants were tested and were all found to contain chlorophyll, the major premise would be supported. The major premise would not be proved because it is not deductively valid to argue it is proved by one or more supporting observations. (There might be exceptions yet to be discovered.) When many supporting experiments have been done, the hypothesis is regarded as very probably true, and it will gain the status of a **theory** and may eventually become a **scientific law,** or "fact."

In summary, one observation that is inconsistent with the conclusion is sufficient to disprove the major premise. Many observations consistent with the conclusion will support but not prove the major premise.

When hypotheses are formulated, they often take the form:

If: A is true
Then: B is true

A is the generalization or hypothesis to be tested and B is the experimental observation to be made. According to the preceding discussion, if B is false, A must be false. If B is true, however, A may or may not be true. The example just given can be restated:

If: all green plants contain chlorophyll
Then: this green mint plant contains chlorophyll

An experiment would be done to test for the presence of chlorophyll. If it were present, the hypothesis would be supported but not proved, and if it were absent, the hypothesis would be disproved, according to deductive logic. In the latter case, the scientist would have to turn to a new hypothesis or develop a more sensitive method of testing for chlorophyll.

PROJECT A

Plant Hormones

OBJECTIVES

When you have completed this project, you should be able to:

1. Define hormone and describe how plant hormones tend to differ from animal hormones.
2. Explain the role of the plant hormone auxin in cell elongation, lateral bud development, ethylene production, root formation, and fruit formation.
3. Given a living plant or a diagram of a plant, show where each of the effects of auxin occurs.
4. Describe five criteria for establishing that a certain process is under the control of a plant hormone.
5. Given the laboratory facilities and equipment at your disposal, design and carry out an experiment to test the hypothesis that apical dominance is due to production of auxin by the apical meristem.
6. Write up your results in a formal laboratory report.
7. Describe several other experimental approaches that might be used to test this hypothesis.

TEXT REFERENCES

Chapter 1 (B); Chapter 47.
Auxin, apical dominance.

INTRODUCTION

The observations for this experiment must be made over a period of at least 2 weeks. The experimental procedures and measurements themselves do not take much time.

Hormones are chemical messengers synthesized in one part of an organism which are transported to a different part of the organism's body. There they have an effect on its development or physiology. Because animals spend a large portion of their lives in a stable adult form without much growth or differentiation, many animal hormones are directed toward maintaining the status quo, or **homeostasis.** In most plants, on the other hand, growth and differentiation tend to continue throughout the life of the organism, so plant hormones tend to be directed toward controlling growth and differentiation in the developing tissues. In some plants, aging and senescence are also closely regulated by hormones. As in animals, certain hormones regulate the events leading up to the reproduction of a mature plant. In animals, hormones are produced in specific organs, the endocrine organs, and they usually act on specific organs called the target organs. Plants, on the other hand, do not have organs with the specific function of producing hormones. Their hormones are produced by parts of the plant that have other functions as well. The hormones may then act on several different organs of the plant.

IDENTIFICATION OF PLANT HORMONES

If a certain substance is a plant hormone, you must be able to show that:

1. The substance is produced naturally by the plant.
2. It exerts a physiological effect on another part of the plant in low concentrations.

3. It is normally transported between the two parts of the plant.

The generally accepted experimental criteria for establishing that a process is under hormonal control were first articulated by Jacobs in 1959.* They can be summarized as follows:

1. **Parallel variation.** The substance must be present in the intact plant that shows the response or process, and the degree of response must vary with the concentration of the substance.
2. **Excision.** Removal of the substance from the plant should result in a loss of the regulatory effect. This may be done by any of these procedures:
   a. Surgically removing the part of the plant producing the hormone
   b. Blocking the production of the hormone by poisoning the metabolic pathway
   c. Blocking the effect of the hormone by using competitive antagonists
   d. Developing a mutant strain that lacks the suspected hormone
3. **Substitution.** Replacement of the suspected hormone should reinstate the regulatory effect. For a new hormone whose structure may not be known, this can be done by using an extract from or homogenate of the tissue thought to produce the hormone. If such a substitution experiment is successful, the next step is to isolate and identify the chemicals in the preparation that are responsible for its biological activity.
4. **Specificity.** It should be shown that other naturally occurring substances do not have the effect shown by the substance isolated in step 3. Testing of other compounds might identify factors that contribute to the observed effect and will help clarify the exact role that the isolated substance plays in the regulatory process.
5. **Generality.** Once hormonal control of a process has been demonstrated in one species, it is desirable to test whether the substance tentatively identified as the controlling hormone will exert similar control on the same process in other species.

AUXIN

Auxins are a group of structurally related small molecules derived from the amino acid tryptophan. The most commonly studied auxin is indoleacetic acid. Auxin is a stimulatory hormone that causes plant cells to elongate if it is present in the right concentration. It acts partly by stimulating the activity of **cellulases,** enzymes that soften the material between the cell walls and allow the cells to expand. If there is a greater concentration of auxin on one side of a plant stem, that side will elongate faster, causing the plant to bend toward the *other* side. Sunlight causes auxin to accumulate on the shady side of a plant stem so that the plant bends *toward* the source of the sunlight and shows **positive phototropism.**

Auxin is stimulatory only when it is present in a certain concentration range and the concentrations of other hormones are appropriate. This range is different for different tissues and, if the concentration is above the stimulatory range, the auxin will act to *inhibit* elongation and growth. In the plant, auxin is produced by the tissue of the **apical meristem** at the tips of the branches of the shoot. It is transported from the meristematic tissues to the rest of the plant. Buds are stimulated by a lower range of auxin concentrations than are stems, so the concentrations of auxin that stimulate growth of the stem will tend to inhibit the development of adjacent buds. This is the basis of **apical dominance.** As long as the apical meristem is growing, it will produce auxin in sufficient concentration to inhibit the development of the lateral buds next to it. If the meristem is removed, the auxin concentration reaching the buds will decrease, and the buds will start to develop and will now be much less sensitive to inhibition. Once a bud has started to grow into a new stem it will be stimulated by the same concentration of auxin that inhibited the bud.

Auxin is also involved to some extent in the response of roots to gravity. If a root happens to be in a horizontal position, high concentrations of auxin will accumulate on its lower side and inhibit those cells. As a result, the root will bend downward and grow towards the center of the earth. The root of a plant thus shows **positive geotropism.** Abscisic acid also accumulates on the lower side and contributes to geotropism by inhibition.

In addition to stimulating elongation, inhibiting lateral bud development, and controlling the response of plant stems to light and gravity, auxin is important in inducing the formation of lateral and adventitious roots, stimulating cell division in the cambium, stimulating cytoplasmic streaming, inducing development of the fleshy parts of some fruits, and promoting the development of female flower parts. Auxin stimulates production of another hormone, ethylene.

Study a diagram of a plant, or better yet, a living plant to review the sites of production and sites of action of auxin and the other plant hormones.

EXPERIMENTS ON APICAL DOMINANCE

Apply your understanding of the scientific method to the problem of apical dominance. Read over the following instructions before you begin to plan your project. You must then prepare a detailed written description of your experimental plan before coming to laboratory to start the experiment, and obtain the instructor's approval of the plan before starting the experiment.

Hypothesis

The hypothesis of this project is that the growth of lateral buds in bean plants is normally inhibited by auxin produced in the apical meristem.

Implications

Use your knowledge of the effects of hormones on plant development to make a list of testable or measurable

*W. P. Jacobs, "What substance normally controls a given biological process?" *Developmental Biology* 1:6, 1959.

phenomena you could study to help support or disprove the hypothesis:

1. ____

____

2. ____

____

3. ____

____

4. ____

____

5. ____

____

Which of these implications could be most easily tested experimentally? ______________

If you wished to demonstrate **parallel variation,** what kinds of experimental problems would you have to overcome?

____

____

Specificity might be difficult to demonstrate; in terms of what you know about the complexities of plant physiology, why might this criterion be hard to meet?

____

____

How would you demonstrate **generality**?

____

____

Why is it important to do so?

____

____

Materials and Methods

In this exercise you will do only **excision** and **substitution** experiments using the following materials and any others that may be available to you:

Young bean plants
Auxin in lanolin paste
Plain lanolin paste
Razor blades
Plant tags

For each of the experiments to be done, describe in your written plan the manipulations that you must do to the experimental plants, the proper controls that you must include for comparison, and the measurements that you will make.

Because you are studying the effect of the manipulation that you will perform on the development of the plant, all other factors should be kept constant.

List in your written plan all the factors that might affect the growth of your plants.

In addition, because you are going to compare experimental and control plants during the experiment, you will need to start out with plants that are as nearly identical as possible and certainly the same species, age, and size. The lateral buds on the plants must be dormant.

Why?

____

____

____

Do not choose plants in which the dormancy of these buds has been broken. Carry out your treatments quickly, so that their dormancy will not be broken by the experimental procedure. Your plants should be about 10–14 days old and should look like the young plant shown in Figure P–1.

The anatomy of the apical meristem and lateral buds of a young bean plant is shown in Figure P–2. The apical shoot of the plant should be about 2–3 cm

Figure P–1. A bean seedling. The lateral buds on either side of the shoot must be dormant at the beginning of the experiment. The shoot must be long enough so that lanolin paste can be applied to it after excision of the apical bud, without having the paste slide down directly onto the lateral buds as well. (Photo: C. Eberhard.)

long, and the lateral buds must be dormant in order for the experimental procedure to be valid.

Once the experimental manipulations have been performed, follow the growth of the lateral buds for at least a 2-week period, taking measurements as often as necessary.

How often will you be able to measure your plants?

____

Design a format for recording your data in the same way each time you take measurements. Include all measurements in your laboratory report.

BE SURE EACH ENTRY INCLUDES YOUR NAME AND THE DATE.

When you are finished with your experimental work, dispose of the plants according to the directions given by your instructor.

Results

The number of plants in your individual experiment is too small to yield statistically valid results. When a large number of such results are combined, however, the results may become statistically meaningful. Read the information on statistics given in Appendix B so that you will be able to understand the analysis of the class data. Your instructor may ask you to calculate the standard deviation for the class results. After the analysis, answer these questions:

Are the class's results statistically significant?

____

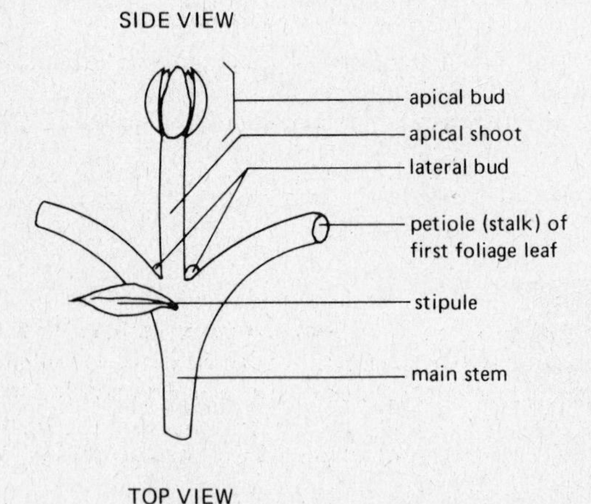

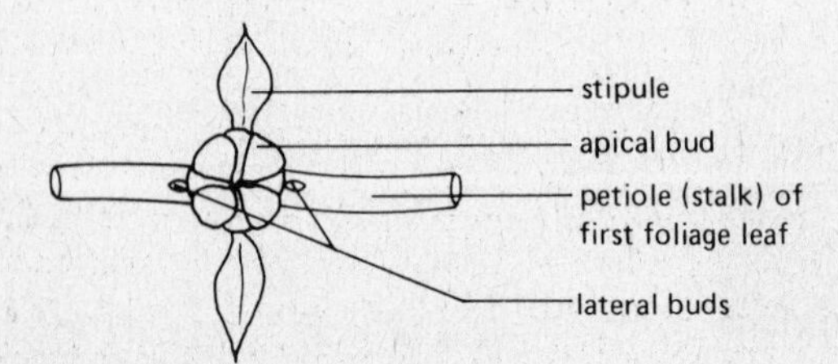

Figure P–2. Apical shoot. The shoot of a 12-day-old bean plant will show apical dominance of the lateral buds by the apical meristem located within the apical bud.

What does the statistical analysis tell us about the experiment?

____

____

____

____

Discussion and Conclusions

Is the hypothesis supported? _______________

Is the hypothesis proved? _______________

Is there an alternative hypothesis that could explain the results?

____

____

What experiments could you do to test the alternative hypothesis?

____

____

If an externally applied hormone can be shown to affect a given phenomenon, does this necessarily mean that the hormone controls the process in the intact plant?

____

What else would you want to know in support of the conclusion?

____

____

____

Write up this experiment according to the instructions given in Appendix A, including the class analysis and an explanation of what the analysis means. Hand in your report to your instructor.

EXPLORING FURTHER

Bidwell, R. G. S. *Plant Physiology*. 2nd ed. New York: Macmillan, 1979.

Flagg, R. O. "Plant growth regulators." Carolina Tips 26 (3, 9), 1963 and 27 (9, 10). Burlington, S.C.: Carolina Biological Laboratory, 1964 (reprint 1965).

Galston, Arthur W., and Peter J. Davies. *Control Mechanisms in Plant Development.* Englewood Cliffs, N.J.: Prentice-Hall, 1970.

Galston, Arthur W., Peter J. Davies and Ruth L. Satter. *The Life of the Green Plant.* 3rd ed. Englewood Cliffs, N.J.: Prentice-Hall, 1980.

Jacobs, W. P. "What substance normally controls a given biological process?" *Devel. Biol.* 1: 6, 1959.

Newcomb, E. H., G. C. Gerloff and W. F. Whittingham. *Plants in Perspective*. San Francisco: W. H. Freeman, 1965.

Salisbury, Frank B., and Cleon W. Ross. *Plant Physiology*. 2nd ed. New York: Harper and Row, 1978.

Wareing, P. F., and I. D. J. Phillips. *Control of Growth and Differentiation in Plants.* 2nd ed. Elmsford, N.Y.: Pergamon Press, 1978.

PROJECT B

Plant Nutrition

OBJECTIVES

1. List the seven elements that are macronutrients needed by all plants.
2. Give at least one example of how each macronutrient is used.
3. Give some examples of micronutrients and describe why each is needed.
4. Describe how a plant could show a deficiency of a nutrient known to be present in its soil.
5. Describe two ways in which a plant might obtain a nutrient other than from its soil or water supply.
6. Give two ways in which a plant can compensate for a limited supply of a nutrient; how would such compensation affect nutritional experiments?
7. Identify the deficient nutrient for a seedling by comparing its appearance with that of seedlings grown under conditions of known nutrient deficiencies.

TEXT REFERENCE

Chapter 45 (A–C).
Macronutrients, micronutrients, soil solution, absorption.

INTRODUCTION

The laboratory work for this project continues for 3 to 4 weeks. Once the experiment has been set up, brief measurements (15 min) must be made every few days.

Plants obtain energy reserves and the bulk of their structural material from the carbon dioxide that they take in from the air. In order to carry out the metabolic processes necessary to convert CO_2 to carbohydrates and cellulose and to simply maintain the components of their cytoplasm, however, they need many other substances in lesser amounts.

Macronutrients: Nitrogen and phosphorus are elements needed for synthesis of amino acids and nucleotides. Potassium is the main cation (ion with a + charge) in plant tissue. These elements are needed in the greatest amount. Sulfur is needed in lesser amounts for amino acids. Calcium and iron are required for many enzyme reactions (as are manganese and magnesium), while magnesium forms part of the structure of the chlorophyll molecule.

Micronutrients: Many elements are needed in tiny amounts by all plants and others are especially needed by certain plants for specialized functions. For example, when plants are associated with nitrogen-fixing bacteria, molybdenum is required because it is necessary for that reaction. Cobalt, a component of vitamin B_{12}, is also needed for nitrogen fixation. Zinc is necessary for the synthesis of auxins. Silicon, boron, copper, sodium, chlorine, and selenium are all important to at least some plants but are needed in only tiny amounts.

Plants that are deficient in a mineral will grow poorly and also show signs of deficiency characteristic of that mineral. In fact, if a plant is greatly deficient in one particular mineral, it is possible to identify which mineral from the appearance of the plant. An experienced person can describe the deficiencies found in a certain sample of soil by studying the appearance of the plants growing in it.

The characteristic effects that certain mineral deficiencies have on growing plants are shown in Table P–1. In nature, it is possible for the plant to show deficiency symptoms even though the nutrient is actually present in the soil around the plant. The nutrient may be in a form held tightly to the soil particles or insoluble at the soil pH. Even if it enters the plant, it may not be in a form that can be transported to the site where it is needed. Finally, it may be available in the needed location, but the combination of other nutrients may not be adequate for the nutrient to serve its necessary function. In this laboratory exercise, you will provide the plants with the nutrients in a readily available form and do not need to consider problems of availability, uptake, or transport.

In some cases a plant will actually be deficient for a certain nutrient but will compensate so that the effect of the deficiency is minimal. If, for example, calcium is required for a certain reaction but is present only in small amounts, magnesium or manganese, also divalent cations, will be able to substitute. Sometimes minerals can be removed from an older part of a plant and relocated to the new, growing part where they are more urgently needed. This type of recycling can prolong the active life of a plant growing under poor conditions. Compensation will complicate nutritional studies because the differences between a healthy plant and one with a deficiency will tend to be masked. Since most nutrients are needed for many processes, however, a deficiency will usually at least result in a slower growth rate, even if no overt deficiency symptoms appear in the experiment.

EFFECT OF NUTRITIONAL DEFICIENCIES

Experiment

In this exercise you will work with a partner and test the effect of two different deficiencies on young seedlings. You will measure the growth rate and observe the appearance of the experimental seedlings and compare your observations with similar ones on seedlings deficient for an unknown nutrient. The control seedlings will be given all the necessary nutrients. All of the seedlings will be grown under identical conditions of light, temperature, and moisture.

Procedure

Obtain 12 seedlings that have been grown in washed sand. Obtain 4 growth jars and label them properly with the labels provided:

A = experimental solution deficient in ____________

B = experimental solution deficient in ____________

C = experimental solution with unknown deficiency

____

D = control solution with no deficiency

TABLE P–1. SYMPTOMS OF MINERAL DEFICIENCIES IN PLANTS

Mineral that is deficient:	Symptoms that result:
Nitrogen	Leaves change from a green-yellow at the top to yellow to brown, dead leaves at the bottom. In many species a red or purpling occurs along the veins. Plant shows stunting.
Potassium	Limited growth, weak stems, and a yellow mottling of the leaves; ultimately necrotic areas on leaf tips and edges; a general overall yellowing appearance of the leaves.
Phosphorus	Plant stunted; leaves dark green; occasionally production of anthocyanins cause a red or purple color. Dead areas develop on leaves, petioles, and (in older plants) fruits, causing some dropping. Roots are poorly developed.
Sulfur	Yellowing of the younger leaves in the early stages; ultimately an overall pale green may dominate (chlorosis). Roots are poorly developed.
Calcium	First symptom usually deformation of younger leaves, and then a disintegration of terminal growing areas, due to the death of the meristem tissue.
Magnesium	Lower leaves show symptoms first, yellowing (chlorosis) from the tip and eventually falling. Veins remain green longer than the rest of the leaf.
Micronutrients	*Iron:* Young leaves light green or almost white (chlorosis), while older leaves are green. Unlike most other elements, iron cannot be withdrawn from the older leaves, so they retain a normal appearance. The yellowing of the younger leaves is most obvious between the veins. *Manganese:* Leaves become spotted with dead areas and fall. *Copper:* Tips of young leaves wither; plant may wilt even when it is watered. *Zinc:* Yellowing of lower leaves at tips and margins; leaves deformed. *Molybdenum:* Required for nitrogen metabolism, so symptoms resemble those of nitrogen deficiency.

The components of the experimental solutions are given in Table P–2.

Follow these instructions exactly in filling your bottles:

1. Add 200 mL of distilled water to each bottle.
2. Locate the column in the table for the first mineral deficiency that you have been assigned and mark it A.
3. Add each of the ingredients called for in turn, *MIXING THOROUGHLY AFTER EACH ADDITION.*
4. Fill bottle A with distilled water.
5. Locate the column in the table for the second mineral deficiency that you have been assigned and mark it B.
6. Add the ingredients called for as before and fill bottle B with distilled water.
7. Add 15 mL of an unknown solution to bottle C and fill it with distilled water. Record the unknown number in your procedure.
8. Add the components called for under "complete" in the table to bottle D and fill it with distilled water.

Measure the pH of one of your solutions by adding a drop of it to a piece of pH paper:

pH = ______________

Iron is insoluble at neutral or basic pH values. Why is the iron provided in the form of FeNaEDTA?

____

What kind of compound is EDTA? ______________

Obtain four aluminum lids for your bottles, punch four holes in each, and cover your bottles as shown in Figure P–3. One hole is for adding distilled water to your plants and the other three are for the plants themselves. Label three holes in each lid 1, 2, and 3, so you will be able to identify the individual plants.

When your bottles are ready, obtain twelve 10-day-old corn or sunflower seedlings that have been grown in washed sand and a few pieces of cotton. The plants should be about 8 cm long. Rinse the roots of the plants carefully in distilled water to remove the sand. Wrap a piece of cotton around the stem of each plant and insert three plants into the lid of each bottle. Arrange the bottles neatly in the greenhouse space assigned to your class. All bottles must be labeled with your name and the date.

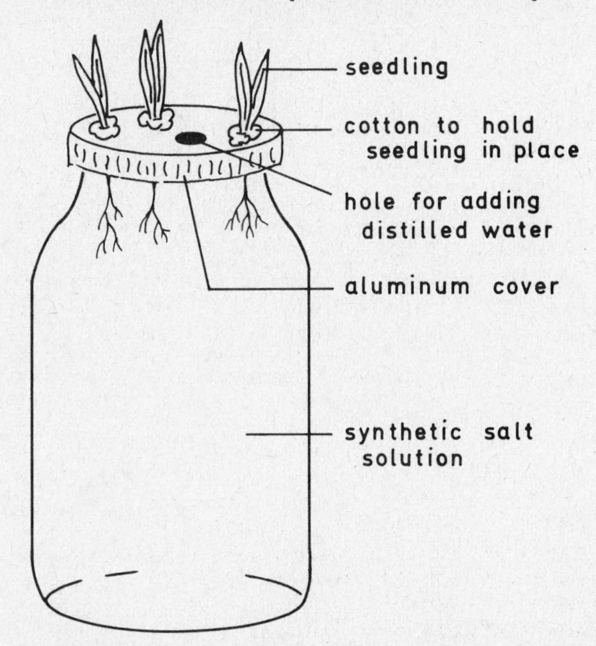

Figure P–3. Nutritional testing. Seedlings growing in a bottle to test for nutritional requirements. The bottle must be kept completely filled by adding distilled water to replace water lost.

Results

When the plants are in place in the greenhouse measure the height of each plant and record the data in Table P–3. Repeat the measurements once or twice each week for four weeks and note the appearance of the plants in the table. Notice the appearance of the experimental seedlings of the other students as well.

At the end of the experiment summarize your results and obtain the results for the other deficiencies that you did not test yourself. Summarize the data in Table P–4.

Conclusions

a. Which single mineral deficiency had the greatest effect on the seedlings? ______________

b. Did the seedlings in plain distilled water show any growth at all? ______________

c. If there was growth in these seedlings, how could you explain it? ______________

____

____

____

d. If your class studied both corn seedlings and sunflower seedlings, which type was more sensitive to iron deficiency? ______________

e. Were there other differences in corn and sunflower seedling responses to different deficiencies?

____

____

____

____

Compare the appearance and growth rate of the seedlings in the solution with an unknown deficiency (bottle C) to those of the other seedlings studied by the class.

f. Was the growth of these seedlings similar to that of the control (no deficiency) or did the unknown solution appear to be deficient in one or more minerals?

____

____

____

____

When the experiment is over, dispose of the plant remains properly and wash the bottles very carefully in hot soapy water. Rinse them six times in tap water and three times in distilled water so that all traces of the mineral solutions have been removed.

EXPLORING FURTHER

Bowling, D. J. F. *Uptake of Ions by Plant Roots.* London: Chapman and Hall, 1976.

Epstein, E. *Mineral Nutrition of Plants: Principles and Perspectives.* New York: John Wiley and Sons, 1972.

Galston, Arthur W., Peter J. Davies and Ruth L. Satter. *The Life of the Green Plant.* 3rd ed. Englewood Cliffs, N.J.: Prentice-Hall, 1980.

Galston, Arthur W., and Peter J. Davies. *Control Mechanisms in Plant Development.* Englewood Cliffs, N.J.: Prentice-Hall, 1970.

Kaufman, Peter B., John Labavitch and Anne Anderson. *Laboratory Experiments in Plant Physiology.* New York: Macmillan, 1975.

Student Name ____________________ **Date** ____________

TABLE P–2. PREPARATION OF EXPERIMENTAL SOLUTIONS*

Stock Solution		Solutions Deficient In:								
	No minerals = COMPLETE	***Ca***	***S***	***Mg***	***K***	***N***	***P***	***Fe***	***Micro-nutrients***	***All minerals = DISTILLED WATER***
		Add (mL)	*Add (mL)*	*Add (mL)*	*Add (mL)*	*Add (mL)*	*Add (mL)*	*Add (mL)*	*Add (mL)*	*No additions*
0.5 M $CaNO_3$	5	0	5	5	5	0	5	5	5	0
0.5 M KNO_3	5	5	5	5	0	0	5	5	5	0
0.5 M $MgSO_4$	2	2	0	0	2	2	2	2	2	0
0.5 M KH_2PO_4	1	1	1	1	0	1	0	1	1	0
FeNa EDTA	1	1	1	1	1	1	1	0	1	0
Microelements	1	1	1	1	1	1	1	1	0	0
0.5 M $NaNO_3$	0	5	0	0	5	0	0	0	0	0
0.5 M $MgCl_2$	0	0	2	0	0	0	0	0	0	0
0.5 M Na_2SO_4	0	0	0	2	0	0	0	0	0	0
0.5 M NaH_2PO_4	0	0	0	0	1	0	0	0	0	0
0.5 M $CaCl_2$	0	0	0	0	0	5	0	0	0	0
0.5 M KCl	0	0	0	0	0	5	1	0	0	0
	15	15	15	15	15	15	15	14	14	0

*Reprinted with permission of Macmillan Publishing Co., Inc. from *Laboratory Experiments in Plant Physiology* by Peter B. Kaufman, John Labovitch and Anne Anderson.

Student Name ______________________________ **Date** ______________

TABLE P–3. EXPERIMENTAL DATA ON MINERAL-DEFICIENT PLANTS.
(Experiment started on (date): ____________)

		Date: ________		Date: ________		Date: ________	
Plant	**Deficiency**	**Height**	**Appearance**	**Height**	**Appearance**	**Height**	**Appearance**
A-1							
A-2							
A-3							
B-1							
B-2							
B-3							
C-1	Unknown #: ______						
C-2	Unknown #: ______						
C-3	Unknown #: ______						
D-1	Complete						
D-2	Complete						
D-3	Complete						

Student Name ______________________ **Date** ______________

TABLE P–3. (contd.)

	Date: ______		Date: ______		Date: ______		Date: ______	
Plant	**Height**	**Appearance**	**Height**	**Appearance**	**Height**	**Appearance**	**Height**	**Appearance**
A-1								
A-2								
A-3								
B-1								
B-2								
B-3								
C-1								
C-2								
C-3								
D-1								
D-2								
D-3								

Student Name ______________________ **Date** ______________

TABLE P–4. SUMMARY OF EFFECTS OF NUTRITIONAL DEFICIENCIES

Plants	Deficiency	Corn Seedlings		Sunflower Seedlings	
		Total Growth (cm)	*Final Appearance*	*Total Growth (cm)*	*Final Appearance*
Controls	none				
Expt.	Ca				
Expt.	S				
Expt.	Mg				
Expt.	K				
Expt.	N				
Expt.	P				
Expt.	Fe				
Expt.	micronutrients				
Expt.	all minerals				
Unknown #					

Unknown is probably deficient: ____________ /probably not deficient: ____________

Unknown might be deficient in: ______________________________

Unknown is probably not deficient in: ______________________________

PROJECT C

The Pond Ecosystem

OBJECTIVES

When you have completed this project, you should be able to:

1. Define ecosystem, community, and succession.
2. Explain the advantages of a microecosystem (microcosm) over an outdoor ecosystem.
3. Give examples of the following that might be found in a pond: abiotic substance, producer organism, macroconsumer organism, saprobe (decomposer), detritus, primary carnivore, secondary (top) carnivore.
4. Define species, species diversity, and climax community.
5. Describe a laboratory experiment designed to compare succession in two pond microcosms.
6. Analyze the results of such an experiment and present them in a formal laboratory report.

TEXT REFERENCES

Chapter 20 (A,B); Chapter 49; Chapter 50; Chapter 51.
Binomial nomenclature, ecosystems, populations, communities, competition, succession.

INTRODUCTION

THE LABORATORY WORK FOR THIS PROJECT EXTENDS OVER A PERIOD OF AT LEAST 3 WEEKS. ONE LABORATORY PERIOD IS REQUIRED TO SET UP THE EXPERIMENT, AND THE TIME SPENT IN FOLLOWING IT IS OPEN-ENDED (10–20 HR).

The Microecosystem

An **ecosystem** includes the organisms in a certain area and their interactions with the physical environment. The energy flowing through the ecosystem will result in a diversity of species in different trophic levels and in a cycling of materials through the system. The group of organisms living in an ecosystem is called a community. The **microcosms,** or microecosystems, that we will establish in the laboratory fit the definition of ecosystems. However, the microcosms also have properties unlike those of most natural ecosystems, which must be taken into account.

Today many ecologists are turning to laboratory and field microecosystems or microcosms, which have discrete boundaries and can be manipulated and replicated at will, because outdoor ecosystems are complex, hard to delineate, and often difficult to study by the traditional means of "experiment and control." Whereas you cannot remove the sediments from a pond to see what effects this action would have, you can set up a microcosm without sediment and observe the results. Another advantage of a microcosm is that the time scale of events is shortened. Normally, **succession,** which is the pattern of changing communities over time, is very slow. In the microcosm, however, succession to a stable system takes place in a few weeks. Microcosms also generally have fewer species than their parent ecosystem as well as simpler communities.

In what ways are microcosms unlike the environments that they are supposed to mimic? Size and isolation are two obvious differences. Organisms in the microcosms tend to diverse from those in their parent communities and each other, but this problem can be eliminated to an extent by cross-seeding the jars: we mix the water in each jar with water from other jars after they have stabilized for a few days.

The Pond

Let us first consider the natural pond ecosystem. Not only is the pond a place where organisms live, but organisms make the pond what it is. Despite the complexities of this system, we can reduce it to several basic units.

ABIOTIC SUBSTANCES

These are simply **organic** and **inorganic compounds** such as H_2O, CO_2, O_2, calcium, nitrogen and phosphorus salts, and so on. A small portion of these vital nutrients is in solution and immediately available to organisms, but a much larger portion is held in reserve in particulate matter (especially in the bottom sediments) as well as in the organisms themselves. The rate of nutrient release, the solar input, temperature, and other climatic factors are the most important processes that regulate the rate of function of the entire ecosystem on a day-to-day basis.

PRODUCERS

Generally speaking, there are two main types of producers in a pond: rooted or large floating **plants** and minute floating plants, called **phytoplankton.** Phytoplankton are usually much more important than rooted vegetation in the production of food for the ecosystem.

CONSUMERS

Unicellular protozoans and zooplankton are the **microconsumers.** They feed on bacteria and planktonic organisms smaller than themselves. The **primary macroconsumers** are herbivores that feed directly on living plants or plant remains and are of two types. Zooplankton are the small animal plankton, and benthic organisms are the bottom-dwelling forms. The **secondary consumers** are carnivores such as predacious insects and game fish, which feed on the primary consumers. **Tertiary consumers** feed in turn on the secondary consumers.

DETRITUS FEEDERS (DECOMPOSERS)

The **saprobes** are the aquatic bacteria, flagellates, and fungi. They are distributed throughout the pond but are especially abundant at the mud-water interface along the bottom, where the bodies of plants and animals accumulate. Larger detritus feeders are called **scavengers.** All these organisms are saprotrophic because they feed on dead remains. They perform the vital function of recycling nutrients in the environment.

It is important to remember that although the pond may seem to be a self-contained unit, its productivity and stability over a period of years is very much determined by the input of solar energy and especially by the rate of inflow of water and materials from the watershed. The watershed is the whole drainage basin, that is, all the land that contributes runoff to the pond.

Succession

Succession, the more or less orderly process of community change, involves replacement, in the course of time, of the dominant species within a given area by other species. The ecological succession that you will be observing in your microcosms is a temporal succession. The microcosms will develop a series of different communities as the physical environment becomes modified by the community. Therefore, succession is community-controlled even though the physical environment determines the pattern, the rate of change, and often sets the limits as to how far development will go. This is very true of your microcosms because they are limited by the size of the container, the amount of incoming solar radiation, and so on. As succession proceeds, you should notice a change in species composition and an increase in species diversity. The diversity of species will increase because rare species that were unnoticed at first may increase in abundance so that they are eventually included. The successional pattern will culminate in a **climax community,** whatever that may be for a particular ecosystem.

Aquatic microecosystems derived from ponds will maintain themselves in a climax state indefinitely with only light input. They will usually reach this level in less than 100 days.

□ Be sure to examine the microcosms that have been growing for 1 or 2 years if they are on display in the laboratory.

PLANNING YOUR EXPERIMENT

Lab Groups

Work in groups of four. Each group will have two microcosm jars—one to be used for an experimental treatment, the other for a control. Two students will be in charge of each jar. There is a lot of work involved in this project, but a smoothly functioning team can handle it with a reasonable expenditure of time and effort. Because you will need to exchange data among yourselves, you should choose your group with two things in mind:

1. You should all be able to come to lab at the same time to set up your experiment and to take your subsequent observations. This way, you will all be referring to one kind of organism by the same name!
2. You should be people who see each other frequently (have classes together, live in the same dormitories, and so on) if possible. Then you will be able to get hold of each other easily to exchange data.

Materials

The water you will use comes from a nearby pond. Three types of microcosms have been set up, all containing about 3 litres of water:

1. Microcosm—no sediment. There will be little exchange of nutrients, no sediment biota, and no species refuge.
2. Microcosm—sand. No exchange of nutrients, no sediment biota, but a refuge for other species.
3. Microcosm—sediment. The "natural look." There will be nutrient exchange, a refuge for species, and a good healthy sediment biota.

You may wish to use additional chemicals, lights,

aerators, heaters, and the like. If you plan in advance what you will need and notify the staff, they will let you know whether it can be obtained. You may wish to bring additional materials of your own on the day that you set up the experiment.

Experimental Design

Prepare a detailed written description of your experimental plan before coming to laboratory to start the experiment, and obtain the instructor's approval of your plan before starting the experiment. This plan should include:

1. The nature of the experimental and control microcosms.
2. The amounts of components and concentrations of added materials or chemicals.
3. A schedule for gathering data.
4. Criteria for deciding whether or not the experimental treatment had an effect.

As a group, you should discuss and decide on a semioriginal experimental design. Here are some suggestions to mull over:

Have a microcosm with sediment as your control and one with no sediment as the experimental model. Then compare and contrast trends.

Expose one microcosm to constant light and one to the natural diurnal pattern. Again, compare and contrast trends.

Enrich one microcosm with glucose as an immediately available carbon source. Use 3 g of glucose for 3 L of water. Please be aware that this treatment creates quite a stench in the experimental jar!

Be sure that both your microcosms are receiving the *same light exposure* (both in the south window, etc.) unless you are deliberately using different exposures. Also, cross-seed your jars before you begin so that they have the same organisms at the beginning of the experiment.

STARTING YOUR EXPERIMENT

□ Choose two microcosm jars.
□ Label each microcosm with your names, the treatment, the date, and a number.
□ Record the qualitative aspects of your microcosms. What organisms can you see with the naked eye? Is there anything living in the sediment? Are there snails present? Whenever you record data, label it with your name and the date on which the observations were made. Do not rely on your memory alone, or you may end up in a state of great confusion when you attempt to write up the results. If possible, use a notebook for your data.
□ Remove a 5-mL sample from the center of your microcosm using the pipets and bulbs provided. If you are not going to examine the sample immediately, transfer it to a small sample bottle and add one drop of acid Lugol's solution to preserve the sample. Cork the sample tightly, and label it completely (name, date, microcosm, number).
□ Examine your sample in the microscope by making a wet mount of one drop or by placing a drop in a depression slide and adding a cover slip. The depression slide will allow wiggly animals more freedom of movement but it will also be more difficult to keep them in focus as they move vertically in the field.
□ While viewing the sample under low power, move the slide around to locate as many different kinds of organisms as you can. Switch to high power to search for smaller organisms. If the motile organisms are swimming too fast, add a drop of Protoslo, which will slow them down.

The search you just finished was designed to familiarize you with the different organisms. Obviously classification of the organisms is the most difficult part of this lab. Using the posters and books in the lab, try to identify the organisms that you examine. But *do not become overly concerned* with scientific identification of the organisms. It is *more* important to be able to distinguish one kind from another and to be able to communicate with other members of your group so that all of you are using the same name for the same thing.

□ When you find a new organism:

1. Draw it in your data chart, which should also have space to record the date, the kind of organism, the number in each of five fields, and the total number.
2. Place it into a taxonomic group, for example, bacterium, cyanophyte (blue-green), diatom, protozoan, multicellular animal, or multicellular plant.
3. Try to find its picture in one of the posters or books in the lab, and call it by its accepted name. For bacteria, classification as rod-, sphere-, or spiral-shaped will suffice.
4. If you cannot locate a picture that looks like the organism in one of the sources provided, give it your own name, and make sure the others in your group have a good look at it in your microscope so that all of you will be calling it the same thing. At the same time, try to distinguish and name as many different kinds of organisms as possible.

□ Now that you have identified the organisms in your sample, you will try to count them.

The following procedure is to provide a random count that gives an accurate representation of the relative abundances of organisms in the sample. You should avoid the "cheating" that would result from deliberately seeking out the most interesting or abundant organisms. Always record exactly what you see.

□ Looking at the microscope from one side, center the cover slip of your slide on the microscope stage. View the slide under low power. Count the number of each kind of organism in the field, and record it in the data table. If there are too many of some kinds of organisms to count, estimate the order of magnitude (10^2, 10^3, and so on) of individuals in each field.
□ Switch to high power to count bacteria, and be sure to make adjustments for the change in field size at the higher magnification when you record the number present.
□ Move the slide in turn to the four corners of the cover slip (again watching from the side, *not* through the microscope), and record the numbers of each type of organism present in each field.
□ In the right-hand column of the data sheet, record the total number of each type of organism observed.

□ Examine a larger sample of water in a watch glass under the dissecting microscope to identify and observe zooplankton present. You may be able to see *Daphnia* (water flea) with young in their brood pouch as well as many other interesting organisms. Carefully return this sample to your microcosm.

□ Do a chlorophyll analysis to give you some idea of the overall size of the producer trophic level. (See Figure P–4.)

1. **Sampling.** Dip a 100-mL beaker into the middle of your microcosm to get a 100-mL sample of phytoplankton. Estimate what proportion of your water is occupied by green masses of algae floating or attached to the side of the jar, and then add to your 100-mL sample of water a mass of algae to take up the same proportion of space in the beaker. (For example, if your microcosm has a clump of algae that occupies 10% of the jar, you will take out a piece of the clump that will occupy 10% of the beaker.)
2. **Filtration.** Flute a piece of Whatman filter paper, place it in a funnel, and place the stem of the funnel into an Erlenmeyer flask. Pour the contents of your beaker into the filter paper; then wait till all the liquid has gone through. Return the liquid in the flask to the microcosm.

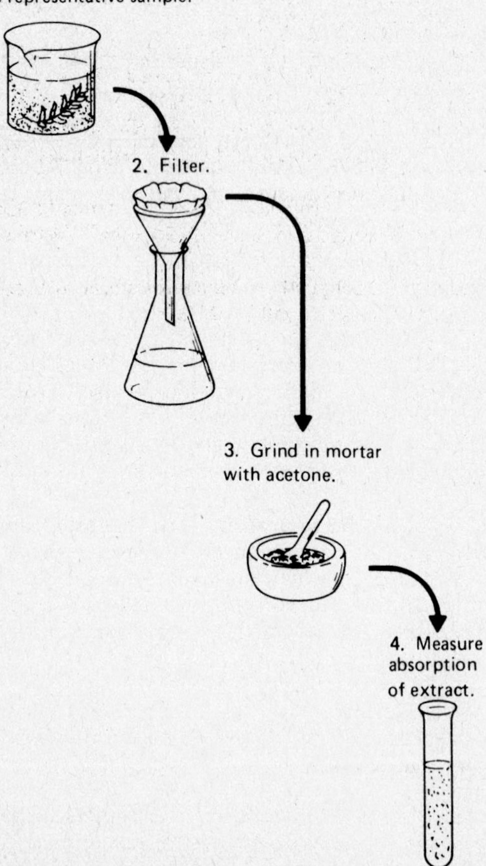

Figure P–4. Chlorophyll analysis. The concentration of chlorophyll will be proportional to the difference between the absorbance at 663 nm and 750 nm.

3. **Pigment extraction.** The algae are now clinging to the filter paper. Place the paper in the mortar and add 5 mL of acetone. Gently grind with the pestle, hard enough to break the algal cells but not hard enough to shred the paper. Remove the filter paper and let any paper fibers settle. Pour off the acetone, containing the algal pigments, into a spectrophotometer cuvette.
4. **Chlorophyll measurement.** Fill a second cuvette with acetone for your blank.

BE CAREFUL NOT TO SPILL ACETONE ONTO THE SPECTROPHOTOMETER.

Following the directions in Topic 2 for operating the spectrophotometer, take readings of both tubes at 663 nm (the absorption maximum of chlorophyll *a* in acetone solution) and at 750 nm (where chlorophyll *a* absorption is negligible). The second reading will give you some idea of the contamination of your sample by pigments other than chlorophyll and by light-scattering debris.

FOLLOWING YOUR EXPERIMENT

Sample your microcosms every 2–3 days for 3–5 weeks, making sure to keep accurate records of the distribution of large and small organisms.

Repeat the chlorophyll analysis a total of three to four times during the course of your experiment. If you expect the producers to be experiencing major population changes because of your particular experimental treatment, you should do it more often.

At the end of the experiment, follow your instructor's directions for disposing of the microcosms (some may be kept for a longer period). Clean up any equipment that you have used.

ANALYZING YOUR EXPERIMENT

Now that you have all these data, what are you going to do with them?

1. Summarize your data in tables and graphs, when possible. Graph the abundance of different kinds versus time following the example shown in Figure P–5. To simplify your data summary, you may want to lump organisms of one type (ciliated protozoan, diatom, green alga, rod-shaped bacteria, and so on) together.
2. Calculate a simple diversity index:

$$D = \frac{\text{total number of kinds}}{\log_{10} \text{ of total number of individuals}}$$

3. Classify the organisms present in your habitat as rare (10 or fewer individuals), common (11–50), or abundant (more than 50).

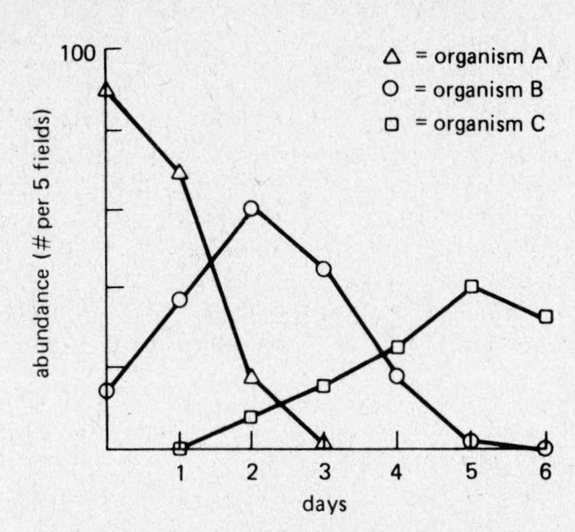

Figure P–5. Changing abundances.

4. Write a laboratory report on this experiment according to the standard form given in Appendix A of this manual. Try to answer the following questions in the course of your discussion:
   a. Compare and contrast your microcosm with the pond this water came from, or ponds in general (food chains, size, diversity, and so on). What are the properties peculiar to the microcosm?
   b. Can you draw any conclusions from your data about ponds or ecosystems in general? Did you learn anything about aquatic ecosystems or is the microcosm simply an anomaly?
   c. How could you conduct the same experiment in the field? (The answer "it is not possible" will not be accepted.)

EXPLORING FURTHER

Abramoff, P., and R. G. Thompson. "The microscope." In *Laboratory Outlines in Biology II.* San Francisco: W. H. Freeman, 1972. (Freeman laboratory separate #846.)

Bold, Harold C., and Michael J. Wynne. *Introduction to the Algae: Structure and Reproduction.* Englewood Cliffs, N.J.: Prentice-Hall, 1978.

Forbes, S. A. "The lake as microcosm." *Bulletin of the Illinois Natural Historical Survey* 15: 537, 1887.

Jahn, T., and F. Jahn. *How to Know the Protozoa.* 2nd ed. Dubuque, Iowa: W. C. Brown, 1979.

Klots, E. B. *The New Field Book of Freshwater Life.* New York: Putnam, 1966.

Krebs, C. J. *Ecology: The Experimental Analysis of Distribution and Abundance.* New York: Harper and Row, 1972.

Odum, E. P. *Fundamentals of Ecology.* 3rd ed. Philadelphia: W. B. Saunders, 1971.

Prescott, G. W. *How to Know the Freshwater Algae.* 3rd ed. Dubuque, Iowa: W. C. Brown, 1978.

Reid, George K. *Pond Life: A Guide to Common Plants and Animals of North American Ponds and Lakes.* New York: Golden Press, 1967.

Southwood, T. R. E. *Ecological Methods.* London: Chapman and Hall, 1966; New York: Barnes and Noble, 1966.

Whittaker, R. H. *Communities and Ecosystems.* 2nd ed. New York: Macmillan, 1975.

Scientific American Articles

Bormann, F. H., and G. E. Likens. "The nutrient cycles of an ecosystem." October 1970 (#1202).

Clark, John R. "Thermal pollution and aquatic life." March 1969 (#1135).

Gosz, J. R., R. T. Holmes, G. E. Likens and F. H. Bormann. "The flow of energy in a forest ecosystem." March 1978 (#1384).

Janick, J., C. H. Noller and C. L. Rhykerd. "The cycles of plant and animal nutrition." September 1976 (no offprint).

Likens, Gene E., Richard F. Wright, James N. Galloway and Thomas J. Butler. "Acid rain." October 1979 (#941).

Woodwell, George M. "Toxic substances and ecological cycles." March 1967 (#1066).

APPENDIX A

Writing a Lab Report

FORMAT

A formal lab report should follow the general format used for a research report published in a scientific journal. Although different journals require different formats, all papers have a roughly similar outline. They reflect the basic scientific method of asking a question, formulating hypotheses, conducting experiments to test the hypotheses, and interpreting the results. You may want to go to the library to look at articles in journals such as *Bioscience, Proceedings of the National Academy of Sciences,* or *Limnology* to get an idea of how to write a scientific paper. (*Scientific American* is *not* considered a scientific journal because its purpose is not communication between research scientists engaged in related work.)

Although not all journals require authors to divide their papers into clearly labeled sections, this practice will help you develop good habits in reporting your findings. Therefore, you are asked to label clearly each section in your paper except for the title.

Title

The title should consist of a few well-chosen words indicating the subject of the report. "Well-chosen" means that the title reflects the scope of your report accurately: do not call your paper "Effects of Chemicals on Animals" if you really studied how feeding dietary supplements of vitamins A and D affected the appearance of the fur of three white rats.

Abstract

The abstract is a short paragraph of 150 words or less that summarizes your experiment, including pertinent information about your experimental subjects, materials and methods, results, and conclusions. This is the part of the paper scientists read to decide whether they are interested in looking at the rest of the paper. Thus the information the abstract contains should help people decide whether your research sheds any useful light on what they are studying.

Introduction

The introduction gives the background of the experiment. It should include an explanation of the background of the general problem or area being investigated, telling why this problem is of interest and outlining what information is already known about the problem.

The problem does not necessarily have to be relevant to the health and welfare of human beings or of immediate economic importance to be of interest. Students often attempt to force a human perspective on a problem that is of interest because it is related to broad underlying principles of the structure or functioning of living things. In building up this part of your report you may want to consult outside references or, at the very least, reread the relevant parts of your text. But be sure to keep track of where you get this information and list all references used, including your textbook and laboratory guide, in a reference list at the end of the paper.

The introduction should also present the question you are trying to answer or the hypothesis you are testing. You should include what outcome you expect, and how it would help support or refute your hypothesis or answer your question.

Materials and Methods

This section should include a concise description of the materials, procedures, and equipment used. There should be enough detail so that someone else could repeat your work. Therefore, brand names of equipment, concentrations and amounts of solutions, species, size or age, sex, and other information about the experimental subjects should be included. If you follow directions from a book or paper, just say so. You need not repeat them in your paper. If, however, you change the procedure, you should explain why you did so and exactly what you did differently.

Do not include the rationale for your work in this section. Also, be sure to report your procedure as a past event rather than writing this section as a set of directions to your reader. You do not need to report attempts at the experiment with techniques that failed unless these techniques are very likely to be tried by other people in the future.

Results

Present your findings in a logical, not chronological, order. Give the results that you found, not what you think you should have found. The organism is always right! You may have to do some thinking, however, to find out why the results came out as they did. Do not explain your results in this section.

Results can often be reported more effectively in the form of one or more tables or graphs or drawings. These should have clear labels and captions. Be sure to indicate whether the data reported are single readings or averages. Statistical analyses of your data should also be included here if appropriate. In addition, a written description should summarize the results illustrated in the graph or figures. This should point out trends or inconsistencies but should not include explanations or opinions.

Discussion

Here you should give your interpretations of the data and relate them to the questions you posed in the introduction. Be careful to avoid making this section just a repetition of the introduction. If you have any data to explain away, do it here or make a new hypothesis as to why the results came out in a way you did not expect. Did the results answer your question? Did they support or disprove your hypothesis? Draw some conclusions, supporting them with your data. What is the significance of your results in the general area you studied? What are the main principles demonstrated by your results?

What further experiments should be performed to clear up discrepancies or ambiguities in your results? How might your work best be continued or extended?

Conclusion

This should be a separate section only if you have a lot of conclusions. Otherwise you should simply round off your discussion section with the conclusions that you can draw from your experiments.

References

You must refer to all the sources of information that you used in your paper. Failure to do so is a serious offense, **plagiarism,** and will result in rejection of your laboratory report or worse. All of the sources in the reference list must be cited in the paper and vice versa.

There are several possible systems for organizing your references. You may choose to list the references in the order in which you cite them in the text. In this case you place a number in your paper after a comment from the reference: for example, "The Loch Ness monster is a big worm (1)," and then list the reference from which you obtained this information as number 1 in your reference list. Alternatively, you may alphabetize your list of references and then number the list when you have finished preparing your report. In that case you can use either the number or the author's last name(s) to refer to the source.

Whichever system you use, you may sometimes want to cite the author and date of the source in your text. This has the advantage of letting your reader know something about the authority and how recent your information is without flipping to the end of the report all the time. For example, you could say, "The Loch Ness monster is a giant worm (Smith, 1970)," or "Smith (1970) has found that the Loch Ness monster is a huge worm," or "In 1970, Smith found that the Loch Ness monster is a large worm." With three or more authors, name them all the first time that you cite them and then use "Smith et al." subsequently (et al. means "and others"). You do not need to repeat the date each time unless you have more than one reference from the same author.

SUGGESTED FORMAT FOR REFERENCE LIST ENTRIES

Here are examples of three types of entries that you might make in your reference list. You may also use the format used by a prominent scientific journal or one suggested by your instructor. Once you have decided on the format, use it consistently for all of your references.

A book reference might read as follows:
Vince, Prae, D. *Photoperiodism in Plants.* New York: McGraw-Hill, 1975. pp. 104 and 194.

A reference to a journal article:
Raymond, J. A., and A. L. DeVries. (1977) "Adsorption inhibition as a mechanism of freezing resistance in polar fishes." *Proc. Nat. Acad. Sci., U.S.A.* 74: 2589–2593.

A reference to an article that appeared in a book:
Rehn, S. "Chemical defenses of ant acacias." In J. Pearson, ed., *Phytochemistry.* New York: Academic Press, 1973.

WRITING TIPS

Your report will be easier to read and understand if you follow these tips and try to conform to the accepted style of scientific writing that is required for scientific papers. Conformity can be a good thing if it increases communication!

1. Write in the past tense, not present or future: "Two rats were anesthetized and . . ."
2. Use metric units (gram, meter, litre, second) and the Celsius temperature scale.
3. Use correct abbreviations *without* periods:

10 millitres	=	10 mL
12 litres	=	12L
3 micrograms	=	3 μg
7 grams	=	7 g
24 kilograms	=	24 kg
663 nanometers	=	663 nm
8 micrometers	=	8 μm
21 centimeters	=	21 cm
9 meters	=	9 m
52 kilometers	=	52 km
37 degrees Celsius	=	37° (37°C)

4. If you start a sentence with a number, write out the number: "Twelve grams of minced toad brain were . . ."

5. For solutions, use molarity rather than normality:

   5 molar = 5 M
   .005 molar = 5×10^{-3} M or 5 mM
   .000005 molar = 5×10^{-6} M or 5 μM

   If the solution is given as a percent solution, indicate whether it was made up on the basis of weight or volume:

   5 g glycerol + 95 g H_2O = 5% glycerol (w/w)
   5 mL glycerol + 95 mL H_2O = 5% glycerol (v/v)
   5 g glycerol brought up to 100 mL with H_2O = 5% glycerol (w/v)

6. Refer to animals and plants by their scientific names. (This is not necessary when referring to humans and common laboratory animals.) Always underline or italicize the genus and species names of organisms, for example, *Stentor coeruleus.* Be as specific as you can in naming the organism, but when the name is not known completely (as in your Pond Ecosystem report), write *sp.* after the generic name, for example, *Oscillatoria sp.*
7. Avoid boring the reader with copious verbiage and excessively formal writing. It is perfectly all right to sound excited about your work if you have worked carefully and have interesting results to report.

When in doubt, ask your instructor or try to find an example in a recent journal.

APPENDIX B

Biological Experiments and Statistics

INTRODUCTION

Many scientific experiments lead to observations that take the form of numerical data. An experiment is designed to make a limited number of observations from which the scientist would like to generalize to the whole of the population that is under study. (A population is a collection of organisms or phenomena that have something in common.)

The most direct way of showing how to deal with a large body of data is to use an example to illustrate the steps in analyzing and interpreting the results. We will use a familiar case: the results of a biology exam given to 100 biology students. Suppose the biology exam consisted of 40 questions. The scores of all the students constitute the **raw data.**

If the course is graded on a "curve," the instructor and students are eager to see the plot of the **frequency distribution.** This is a graph in which the number of students receiving each score (the frequency) is plotted against the individual scores (40, 39, 38, and so on). The basis for making the graph is a table of score frequencies as is shown in Table B–1.

The table shows that 13 students received a score of 38, 4 received a score of 30, and so forth. The high score was 40 and the low score was 24, so the **range** of the scores was 24–40.

TABLE B–1. FREQUENCY DISTRIBUTION AND ***fx***

Score (x)	Frequency (f)	fx
40	6	240
39	13	507
38	13	494
37	13	481
36	9	324
35	15	525
34	8	272
33	8	264
32	3	96
31	4	124
30	4	120
29	1	29
28	1	28
27	1	27
26	0	0
25	0	0
24	1	24
		Total: 3555

ARITHMETIC MEAN

The **arithmetic mean** is the average of all scores: all scores are added together, and the total is divided by the number of scores. If the data are in the form of Table B–1, you know that 6 students received a grade of 40. You should thus multiply 40×6, 13×39 to give $f(x)$, and so on, and then add up all the products. The final total is divided by the number of students, which was 100. The results are given in Table B–1.

Mean $= \overline{x} = \Sigma x/N$, or $= \Sigma fx/N$,

where
$\overline{x}$ is the average or mean;
Σfx is the sum of all fx;
N is the number of scores.

What is the arithmetic mean for this test? ________

STANDARD DEVIATION AND VARIANCE

Variance is a measure of how far the scores tended to stray away from the average score of 35.6. One class might all score in the 30s with a mean of 35.6, and another class might include many 40s and some scores in the 20s and still have a mean of 35.6. The first class would have a lower variance than the second.

The formula for variance is:

$$\text{variance} = s^2 = \frac{\Sigma f(x - \overline{x})^2}{N - 1}$$

where

s is the standard deviation;
f is the frequency of a score x;
$\overline{x}$ is the mean of all scores;
N is the number of scores.

The difference between each score and the mean was determined, squared, and multiplied by the frequency of that score. All the products were added up, and the total was divided by the number of scores minus 1. In our example,

$$s^2 = 1068.5/99 = 10.8$$

The **standard deviation** is the square root of the variance. Thus in our example:

$$s = \sqrt{10.8} = 3.3$$

Any scores within 1 standard deviation of the mean (3.3 points, in this example) are said to be within 1 **standard deviation unit** of the mean. When there are many data points, as in our example, the mean should be expressed as mean ± standard deviation (for example, 35.6 ± 3.3). This form of presenting experimental data in a report tells the reader how much variance there was in the data gathered, and, therefore, how accurate the mean value really is. The smaller the standard deviation, the smaller the variance in the sample, and the more confident we are that the reported result is close to the true value of the mean and standard deviation for the population from which our sample was drawn.

When the class data for the plant hormone experiment (Project A) are analyzed, the results will be expressed by calculating the mean and standard deviation for each group of data. There will be a mean ± standard deviation for the lengths of the axillary buds on the control plants, another for the buds on plants treated with lanolin, and so forth.

If the frequency distribution is normal, it will form a graph that is symmetrical about the mean and approximates a bell-shaped curve. The standard deviation then takes on additional meaning. In that case, about 68% of the values, that is, the bulk of the data, will fall within 1 standard deviation unit on either side of the mean. Furthermore, 95% of the data will fall within 2 standard deviation units, and 99% will fall within 3 units of the mean. In the case of the test scores, we would expect 99% of all students taking the test to score within 9.9 points ($3s$) of the mean, or above 25.7. The standard deviation can be used in this way only if the sample was drawn from a "bell-shaped" population. (Persons who didn't study for the test or who had taken an advanced course in the subject would tend to invalidate the test results.)

SIGNIFICANT DIFFERENCE

When we examine the results of an experiment, we would like to find out whether the results are "significant" or not. There are two ways of looking at significance.

1. We predict the results that we would obtain according to a certain hypothesis. Then we see how far away from the expected our results actually were. This approach is the one you will use in testing the results from your genetics experiment to see whether they could have varied from expected due to random chance or are so far from expected that your hypothesis is doubtful (Chi square test).
2. We compare an experimental group with a control group using the **null hypothesis**: The two groups will be the same unless the experimental treatment had an effect. The more the results from the two groups differ, the less likely it is that the null hypothesis is correct.

In Project A, we would like to decide whether the axillary buds in the plant group treated by removing the apical buds grew longer than the buds in the untreated group. This can be done by using the student's t test or the Wilcoxon test to determine whether there is a significant difference between the results for the two groups.

All the plants originally came from the same population. If samples had been drawn from this population before the experiment, they might have had different means, but the difference would be due to chance alone. Such a difference would not be significant. After one group of plants has been treated in the experiment, the means may differ significantly as a result of the experimental treatment. We would like to know whether the difference is due to just random chance or to the experimental treatment. The statistical significance of the difference in the means can be determined by applying a statistical test. The test will show whether the difference in the means could be due to random chance, and is therefore not significant, or is not due to random chance but rather to the experimental treatment, and is significant.

Statistical significance is usually expressed as a decimal. A significance level of 0.05 or 5% means that we could expect to find a significant difference only about 1 out of 20 times the experiment is conducted, if there is really *no* difference between the populations. The exact significance level at which populations are assumed to be really different is fuzzy and depends on the size and type of study being conducted. As a rough guideline, however, a significance level of 0.05 is usually accepted as sufficient to assume that the two groups are likely to be from different populations. A level of 0.01 or 1% is considered highly significant.

THE CHI-SQUARE TEST

The chi-square statistic is a mathematical comparison of the data obtained with the data expected according to the rules of probability. A chi-square (χ^2) value represents the degree to which the data fit the hypothesis on which the experiment was based. If chi-square is small, the data fit the hypothesis well; a large chi-square value represents a large deviation and throws doubt on the hypothesis. The actual formula for chi-square need not concern us here. In all problems where we wish to compare the frequencies in an observed sample with an expected frequency distribution we calculate a value known as x^2, which for our purposes may be regarded as a good approximation to χ^2 for large sample sizes.

The formula for x^2 is:

$$x^2 = \sum_{i=1}^{c} \frac{(o_i - e_i)^2}{e_i}$$

where

e_i is the number expected in each class;
o_i is the number observed in each class;
c is the number of classes; and
$\sum_{i=1}^{c}$ means that the expression's value is summed for each class from 1 to c.

To calculate x^2 for a genetics experiment:

1. For each class of offspring, subtract the number expected in the class (e) from the number observed in the class (o) to find ($o - e$).
2. Square this difference $(o - e)^2$.
3. Divide the result by the number expected in the class (e).
4. Repeat for all classes of offspring.
5. Add up (Σ) all the results; the sum is x^2.

Let us apply this to a concrete example.

Problem: A genetics student crosses two fruit flies and obtains 150 offspring. He or she sorts the offspring according to sex and finds 85 females and 65 males. Is this close enough to the expected 75 of each sex so that the observed sex distribution can be regarded as due to chance? (An alternative hypothesis might be that males die as embryos more often than females.)

Solution: The experimenter obtained 150 offspring of which 85 were female and 65 male where you would have expected 75 of each sex. To calculate the x^2 value for these results we can conveniently make a table like Table B–2. The observed and expected numbers of each sex of offspring are entered in the two left columns of the table. Now we subtract the expected from the observed offspring in each class to get the deviation for each class ($o - e$). In the next column we square each deviation. Finally we divide the squared deviation for each class by the expected number in each class and add up all the results to obtain x^2, which in this case is 2.6 ($\Sigma = x^2 = 2.6$).

To determine whether this x^2 value, which we treat as chi-square, is significant, we look at a table of chi-square values (Table B–3). You will see that there is one item in the table that we have not yet mentioned and that we need before we can read off the probability: **degrees of freedom.** The degrees of freedom in a system are usually defined as the number of classes minus one. In this case, with two classes of offspring (male and female), there is only 1 degree of freedom. One way to look at degrees of freedom is to see that it represents the number of choices you can make in assigning an individual to a class. In this case, you have only one decision to make. If you decide that a particular fruit fly is not male, it must be female. If degrees of freedom are thought of as a number of decisions to be made, you can make one less decision than the number of classes, because after that number of decisions, the last class has been determined automatically. In our fruit fly problem we have an x^2 value of 2.6 and 1 degree of freedom. From the table we see that this value falls between 1.6 in the 0.2 (20%) column and 2.7 in the 0.10 column (10%). This means that we would obtain a chi-square value of 2.6 slightly more than 10% of the time if we performed numerous crosses and counted the sex of the offspring. We say that the chi-square value is **not significant.** The hypothesis that half the offspring of a fruit fly are male and half are female need not be questioned. If we obtained a chi-square value which fell in the 5% or 1% columns, we would suspect that another hypothesis would explain our results better than the one we used. A value in the 5% column would be **significant,** and a value in the 1% column would be **highly significant.**

Here is another example of the application of the chi-square statistic to a genetics problem:

Problem: A polled (hornless) bull with a straight coat (both of which are traits determined by dominant characteristics) is mated with 48 cows that show the recessive alleles of horns and curly coat. The following offspring are produced:

17 polled straight
15 horned curly
8 polled curly } recombinants?
8 horned straight }

The third and fourth phenotypes might represent recombinants.

Do these results provide convincing evidence of linkage between the two loci involved?

Solution: This problem presents us with the following two alternative hypotheses, of which only the second can be tested statistically.

1. The loci are linked and $\frac{16}{48}$ = 33% of the offspring are recombinants. This hypothesis can be tested di-

TABLE B–2. GENETICS EXPERIMENT

Class	Observed (o)	Expected (e)	Deviation ($o - e$)	$(o - e)^2$	$\frac{(o - e)^2}{e}$
Females	85	75	10	100	100/75
Males	65	75	−10	100	100/75

TABLE B–3. CHI-SQUARE VALUES

Degrees of Freedom (n)	Probability Values (P) Deviation from Hypothesis Not Significant								Deviation Significant	Deviation Highly Significant
	0.95	*0.90*	*0.80*	*0.70*	*0.50*	*0.30*	*0.20*	*0.10*	*0.05*	*0.01*
1	0.004	0.016	0.06	0.15	0.46	1.1	1.6	2.7	3.8	6.6
2	0.1	0.2	0.4	0.7	1.4	2.4	3.2	4.6	6.0	9.2
3	0.4	0.6	1.0	1.4	2.4	3.7	4.6	6.3	7.8	11.3
	Chi-square value consistent with hypothesis								Not consistent	

(Modified from Table IV in R. A. Fisher and F. Yates, *Statistical Tables for Biological, Agricultural, and Medical Research* (6th ed., 1974). Published by Longman Group Ltd., London. Previously published by Oliver and Boyd, Ltd., Edinburgh. Reprinted by permission of the authors and publishers.)

rectly only by carrying out additional breeding experiments.

2. The loci are unlinked, in which case we would expect equal numbers of the four classes of offspring. We can test this hypothesis statistically.

If we can eliminate one of these alternate hypotheses, we shall have provided strong evidence for the other. Since the only hypothesis we can test at the moment is hypothesis 2, we test it. If it must be rejected, we are left with only one alternative, that the first hypothesis is correct. To test hypothesis 2, we assume that 12 offspring were expected in each of the four classes and calculate a value of x^2 equal to 5.49. In this example there are 3 degrees of freedom. From the chi-square table, we find that we would obtain a x^2 value of 5.49 in a system with 3 degrees of freedom between 10% and 20% of the time. We have said that by convention scientists have agreed that observed results cast doubt on a hypothesis if they differ from the results predicted by the hypothesis to such an extent that the deviation observed would occur by chance less than 5% of the time. In this case, the observed deviation would occur by chance more than 10% of the time so we cannot reject hypothesis 2. In this particular case, this means that even if the loci are unlinked, the random fusion of the gametes at fertilization has resulted in offspring whose numbers differ from the 1:1:1:1 ratio of offspring expected for unlinked loci. But the chi-square test tells us that this deviation is not statistically significant. Because in science we can disprove, but never prove, a hypothesis, the answer to the problem is that these results do not provide convincing evidence of linkage between the loci. Only a larger sample size (that is, more calves) would provide the data that would permit us to make the determination whether the loci are linked or not.

Sample size has an important effect on the observed and expected results of an experiment. Intuitively, the larger the sample size taken, the closer you would come to the expected ratio. On the other hand, when sample sizes are small, you would not be surprised that results appear to deviate considerably (but *not significantly!*) from the expected.

There are two limitations on using the chi-square statistic to test closeness of fit between observed and expected results:

TABLE B–4. VALUES FOR t AT A SIGNIFICANCE LEVEL OF 0.05 (one-tailed test)*

Degrees of Freedom	Value of t (0.05 Significance Level)
1	6.314
2	2.920
3	2.353
4	2.132
5	2.015
6	1.943
7	1.895
8	1.860
9	1.833
10	1.812
15	1.753
20	1.725
25	1.708
30	1.697
40	1.684
60	1.671
120	1.658
∞	1.645

*In student's t test, the calculated value of t is compared with that in the table. If it is higher than the value in the table, the groups being compared are probably different.

(Modified from Table III in R. A. Fisher and F. Yates, *Statistical Tables for Biological, Agricultural, and Medical Research* (6th ed., 1974). Published by Longman Group Ltd., London. Previously published by Oliver and Boyd, Ltd., Edinburgh. Reprinted by permission of the authors and publishers.)

1. The chi-square method can be applied only to actual frequencies, not to percentages.
2. Chi-square is not reliable if applied to distributions in which the expected value is less than 5 for any class.

STUDENT'S *t* TEST

This test assumes that the populations under study each fit a normal distribution (bell-shaped curve), that the two populations have common variance, and that the samples within the group and between groups are independent of each other. It may be used to analyze the results from Project A.

The values of t for a significance level of 0.05 are given in Table B–4.

The degrees of freedom, N, is the sum of the degrees of freedom for each class. The degrees of freedom for a class is the number of plants in the class minus 1. If an experiment included 50 experimental and 49 control plants, N would equal $(50 - 1) + (49 - 1) = 97$. The value closest to this in the table would be 100. If your instructor provides you with the value of t for each experiment, you can interpret the results by following these steps:

1. Calculate the degrees of freedom for the experiment from the number of plants involved.
2. Note the value of t that you find in the table for that value of N at the 0.05 significance level.
3. Compare the value of t calculated from the experimental results with the value of t in the table.

If the magnitude of the observed t is *less* than the value in the table, there is *no significant difference* between the groups. We conclude that the experimental treatment of the plants had no effect. If the magnitude of the observed t is *greater* than the value in the table, there is a *significant difference* between the groups. We conclude that the experimental treatment did have an effect.

If you, rather than the instructor, are going to calculate the value of t, use the following formula which applies to cases where the two groups being compared are equal in size so that $N_1 = N_2 = N$:

$$t = \frac{(\overline{x}_1 - \overline{x}_2)\sqrt{N}}{\sqrt{(s_1)^2 + (s_2)^2}}$$

where
N = number of plants in *EACH* group;
$\overline{x}_1$ = mean of one group of plants;
$\overline{x}_2$ = mean of the second group;
s_1 = standard deviation of the first group;
s_2 = standard deviation of the second group.

Once you have the value for the magnitude of t, follow the steps given above to analyze the results of your experiment.

THE WILCOXON TEST

If your class data for Project A consisted of only a few values, or if the results did not fit a bell-shaped curve, you should use another test, the Wilcoxon (or Mann-Whitney) test, which does not depend on the frequency distribution. This test uses a ranking system to compare data from two samples.

TABLE B–5. EXCISION EXPERIMENT

Bud Length in Excised Plants (mm)	Bud Length in Control Plants (mm)
1.0	0.8
2.5	0.9
3.6	1.1
4.2	1.4
7.1	1.5
9.2	1.9

TABLE B–6. RANKED DATA

Rank	Length (mm)	C = Control Group
1	0.8	C
2	0.9	C
3	1.0	
4	1.1	C
5	1.4	C
6	1.5	C
7	1.9	C
8	2.5	
9	3.6	
10	4.2	
11	7.1	
12	9.2	

Suppose that an excision experiment were carried out with 12 plants, 6 excised and 6 control. The first step in applying the test is to arrange the data from each group in increasing order of magnitude (Table B–6).

Next, all the data are ranked, keeping track of the data from one of the groups (the control plants) (Table B–6). The rank numbers of the control and excised groups are then added:

control ranks = 1 + 2 + 4 + 5 + 6 + 7 = 25
excised ranks = 3 + 8 + 9 + 10 + 11 + 12 = 53

The lower value, in this case 25 (sum of the control group ranks), is designated the Wilcoxon statistic, W. The value calculated for W is compared with the critical values given in Table B–7. If the calculated value is equal to or less than the critical value, there is probably a significant difference between the two groups with a significance level of 0.05.

In this example, read down the column of N_1 values (for the group with the smaller mean) to $N_1 = 6$. Then read across the rows of N_2, the second group, to $N_2 = 6$. The critical value of W is 28. Our value of $W = 25$ is less than 28, so the two groups of plants are probably different within our significance level of 0.05.

If W had been 29 or greater, we would have concluded that there was no significant difference between them.

If N is greater than 25, calculate W', which can be compared to normal distribution tables:

$$W' = \frac{W - [N_1(N_1 + N_2 + 1)/2]}{\sqrt{N_1N_2(N_1 + N_2 + 1)/12}}$$

TABLE B–7. CRITICAL VALUES OF W*

N_1 / N_2	1	2	3	4	5	6	7	8	9	10	11	12	13	14	15	16	17	18	19	20	21	22	23	24	25
1	—	—	—	—	—	—	—	—	—	—	—	—	—	—	—	—	—	—	1	1	1	1	1	1	1
2	—	—	—	—	3	3	3	4	4	4	4	5	5	6	6	6	6	7	7	7	8	8	8	9	9
3	—	—	6	6	7	8	8	9	10	10	11	11	12	13	13	14	15	15	16	17	17	18	19	19	20
4	—	—	—	11	12	13	14	15	16	17	18	19	20	21	22	24	25	26	27	28	29	30	31	32	33
5	—	—	—	—	19	20	21	23	24	26	27	28	30	31	33	34	35	37	38	40	41	43	44	45	47
6	—	—	—	—	—	28	29	31	33	35	37	38	40	42	44	46	47	49	51	53	55	57	58	60	62
7	—	—	—	—	—	—	39	41	43	45	47	49	52	54	56	58	61	63	65	67	69	72	74	76	78
8	—	—	—	—	—	—	—	51	54	56	59	62	64	67	69	72	75	77	80	83	85	88	90	93	96
9	—	—	—	—	—	—	—	—	66	69	72	75	78	81	84	87	90	93	96	99	102	105	108	111	114
10	—	—	—	—	—	—	—	—	—	82	86	89	92	96	99	103	106	110	113	117	120	123	127	130	134
11	—	—	—	—	—	—	—	—	—	—	100	104	108	112	116	120	123	127	131	135	139	143	147	151	155
12	—	—	—	—	—	—	—	—	—	—	—	120	125	129	133	138	142	146	150	155	159	163	168	172	176
13	—	—	—	—	—	—	—	—	—	—	—	—	142	147	152	156	161	166	171	175	180	185	189	194	199
14	—	—	—	—	—	—	—	—	—	—	—	—	—	166	171	176	182	187	192	197	202	207	212	218	223
15	—	—	—	—	—	—	—	—	—	—	—	—	—	—	192	197	203	208	214	220	225	231	236	242	248
16	—	—	—	—	—	—	—	—	—	—	—	—	—	—	—	219	225	231	237	243	249	255	261	267	273
17	—	—	—	—	—	—	—	—	—	—	—	—	—	—	—	—	249	255	262	268	274	281	287	294	300
18	—	—	—	—	—	—	—	—	—	—	—	—	—	—	—	—	—	280	287	294	301	307	314	321	328
19	—	—	—	—	—	—	—	—	—	—	—	—	—	—	—	—	—	—	313	320	328	335	342	350	357
20	—	—	—	—	—	—	—	—	—	—	—	—	—	—	—	—	—	—	—	348	356	364	371	379	387
21	—	—	—	—	—	—	—	—	—	—	—	—	—	—	—	—	—	—	—	—	385	393	401	410	418
22	—	—	—	—	—	—	—	—	—	—	—	—	—	—	—	—	—	—	—	—	—	424	432	441	450
23	—	—	—	—	—	—	—	—	—	—	—	—	—	—	—	—	—	—	—	—	—	—	465	474	483
24	—	—	—	—	—	—	—	—	—	—	—	—	—	—	—	—	—	—	—	—	—	—	—	507	517
25	—	—	—	—	—	—	—	—	—	—	—	—	—	—	—	—	—	—	—	—	—	—	—	—	552

*If the calculated value of W is equal to or less than the value shown in the table, the two groups are statistically different at a significance level of 0.05. The values of N_1 are the numbers of entries in the first group and the values of N_2 are the numbers in the second group.

(Adapted from Table 1 in L. R. Verdooren, "Extended tables of critical values for Wilcoxon's test statistic," *Biometrika* **50:** 177, 1963. Used with the permission of the Biometrika Trust.)

APPENDIX C

Laboratory Equipment

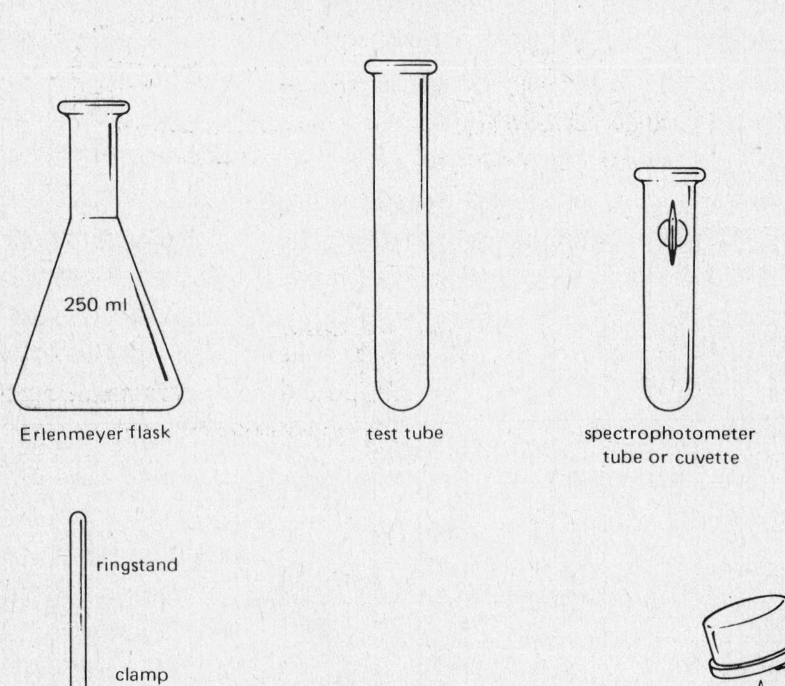

Erlenmeyer flask

test tube

spectrophotometer tube or cuvette

beaker

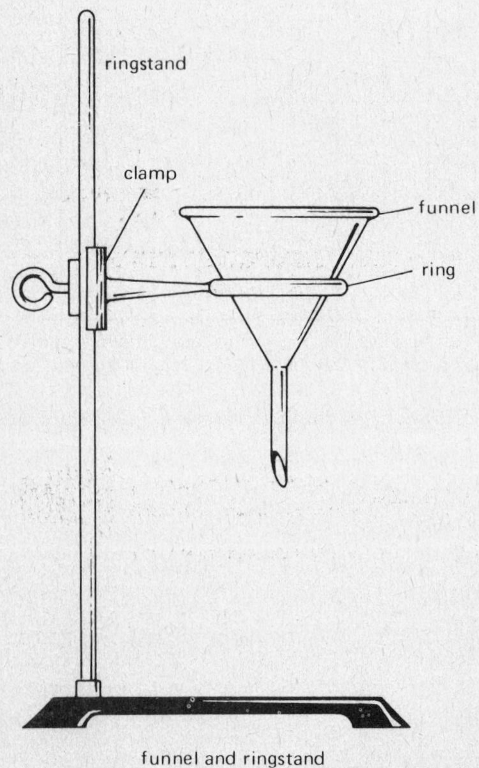

funnel and ringstand

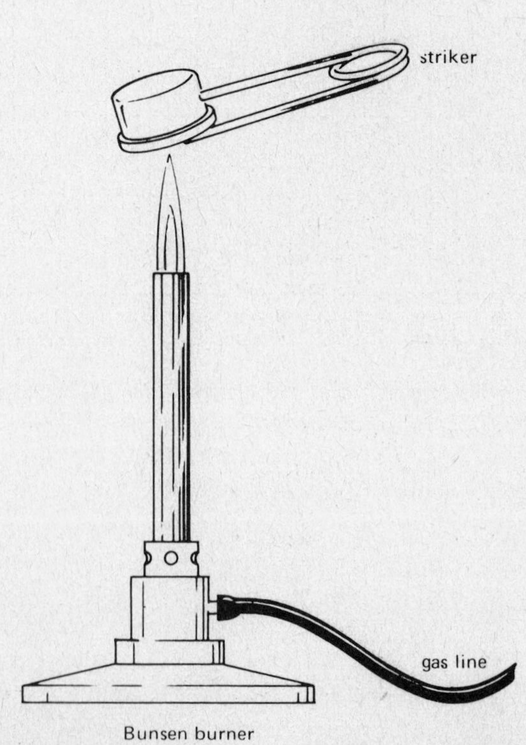

Bunsen burner

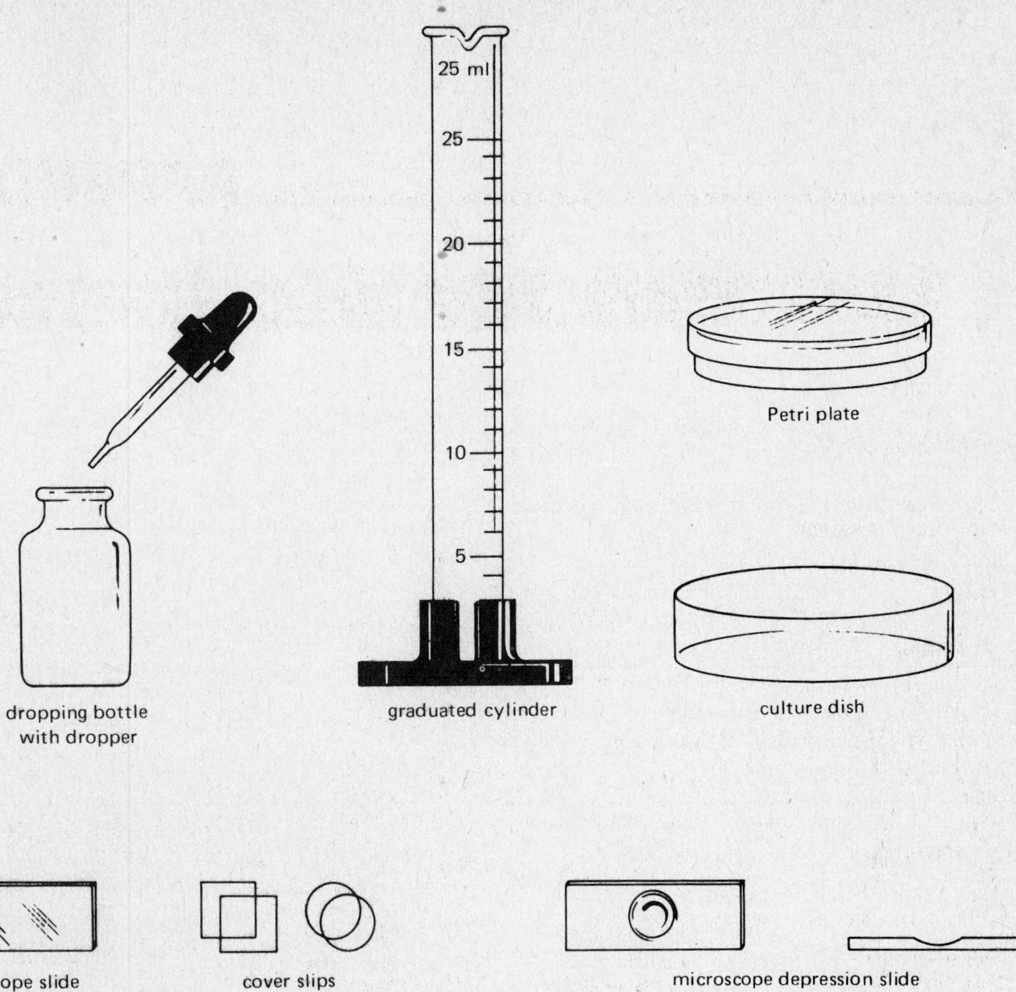

microscope slide

cover slips

microscope depression slide

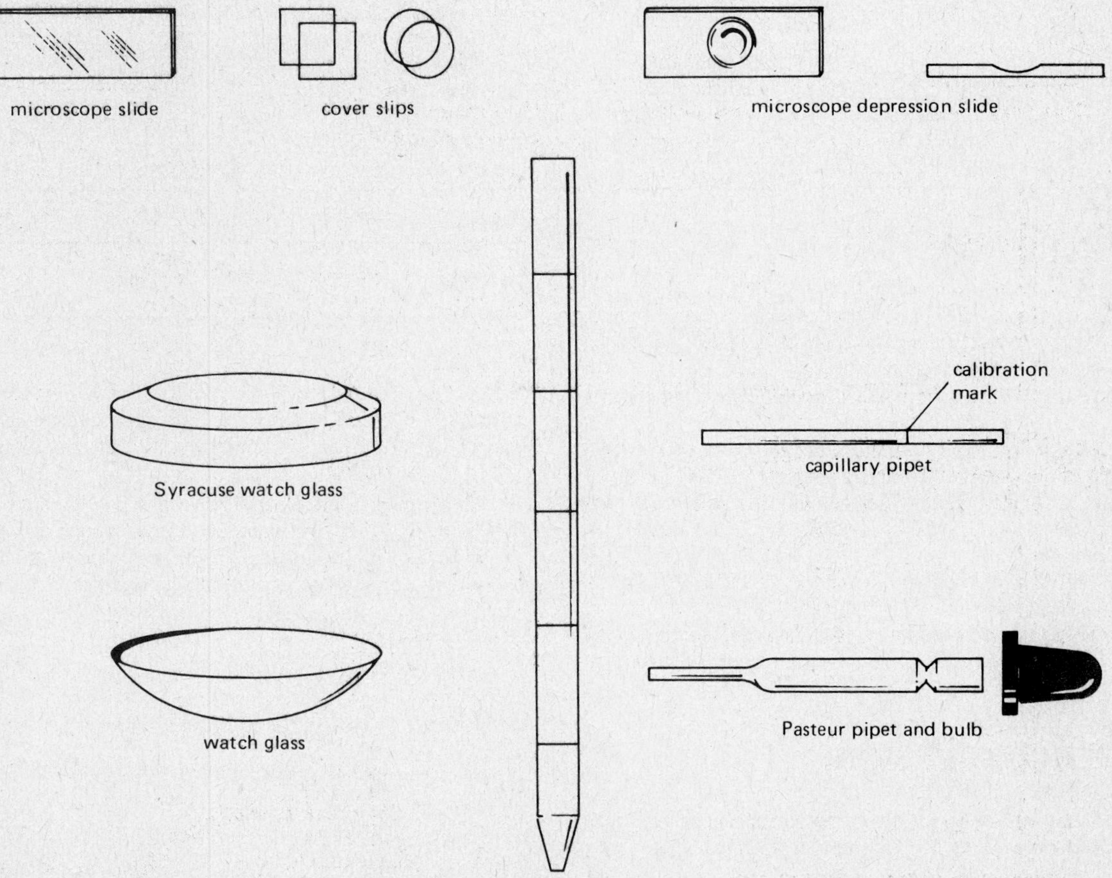

APPENDIX D

The Metric System

INTRODUCTION

These are the basic units of the metric system (abbreviations are in parentheses):

GRAM	METER	LITRE
(g)*	(m)	(L)
mass or weight	length	volume

All other units are powers of 10 smaller or larger than the basic unit. The same prefixes are used to name these units regardless of the basic unit:

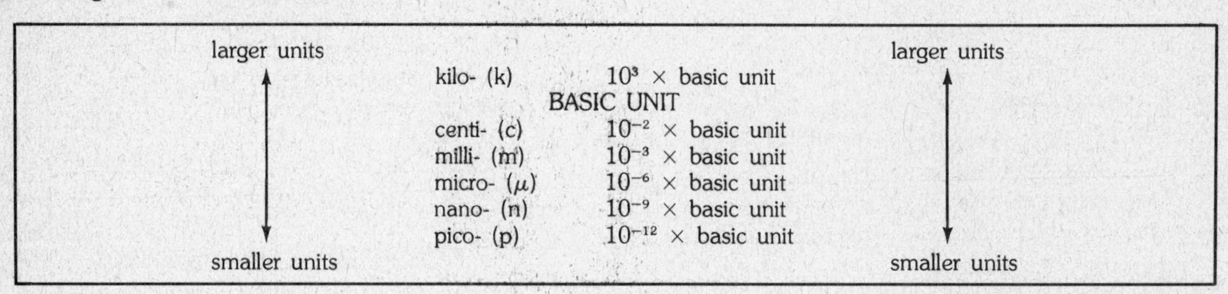

larger units

kilo- (k)	$10^3 \times$ basic unit
BASIC UNIT	
centi- (c)	$10^{-2} \times$ basic unit
milli- (m)	$10^{-3} \times$ basic unit
micro- (μ)	$10^{-6} \times$ basic unit
nano- (n)	$10^{-9} \times$ basic unit
pico- (p)	$10^{-12} \times$ basic unit

smaller units

CONVERSION

Within one type of measurement, such as weight, it is possible to convert between larger and smaller units by simply moving the decimal:

Example: Convert 4729 grams (g) to kilograms (kg).

Solution: Working within weight we change into a *larger* unit so our value will be *smaller*. We will *divide* to obtain a smaller value. We will use the *factor 10^3* because there are 10^3 meters/1 kilometer

$$4{,}729 \text{ g} \div 10^3 = 4\ 7\ 2\ 9. = 4.729 \text{ kg}$$

Ease of conversion between units is a big advantage of the metric system. Anyone who can change dollars to cents can do it!

INTERCONVERSION

The metric system is set up so that certain units are identical *with respect to the standard reference, water:*

1g	=	**1 cm^3**	=	**1 mL**
H_2O		H_2O		H_2O

So, 27 mL of H_2O will occupy a cube 3 cm on each edge, 27 cm^3, and weigh 27 g. For other substances, the density (g/mL) must be known to interconvert units.

Example: Your water bed measures 2 meters (m) $\times$ 1.8 m $\times$ .25 m. How much does it weigh in kg?

Solution: The volume of the bed needs to be expressed in cm^3 before converting to weight:

$$200 \text{ cm} \times 180 \text{ cm} \times 25 \text{ cm} = 9 \times 10^5 \text{ cm}^3$$

Then 9×10^5 cm^3 is equivalent directly to 9×10^5 g

$$9 \times 10^5 \text{ g} = 9\ 0\ 0\ 0\ 0\ 0. = 900 \text{ kg}$$

Interconversion is a powerful tool that allows you to make "back of the envelope" calculations amazingly quick.

*By convention, metric abbreviations never have periods or use "s" for plural: 1 gram (1 g) and 187 grams (187 g).

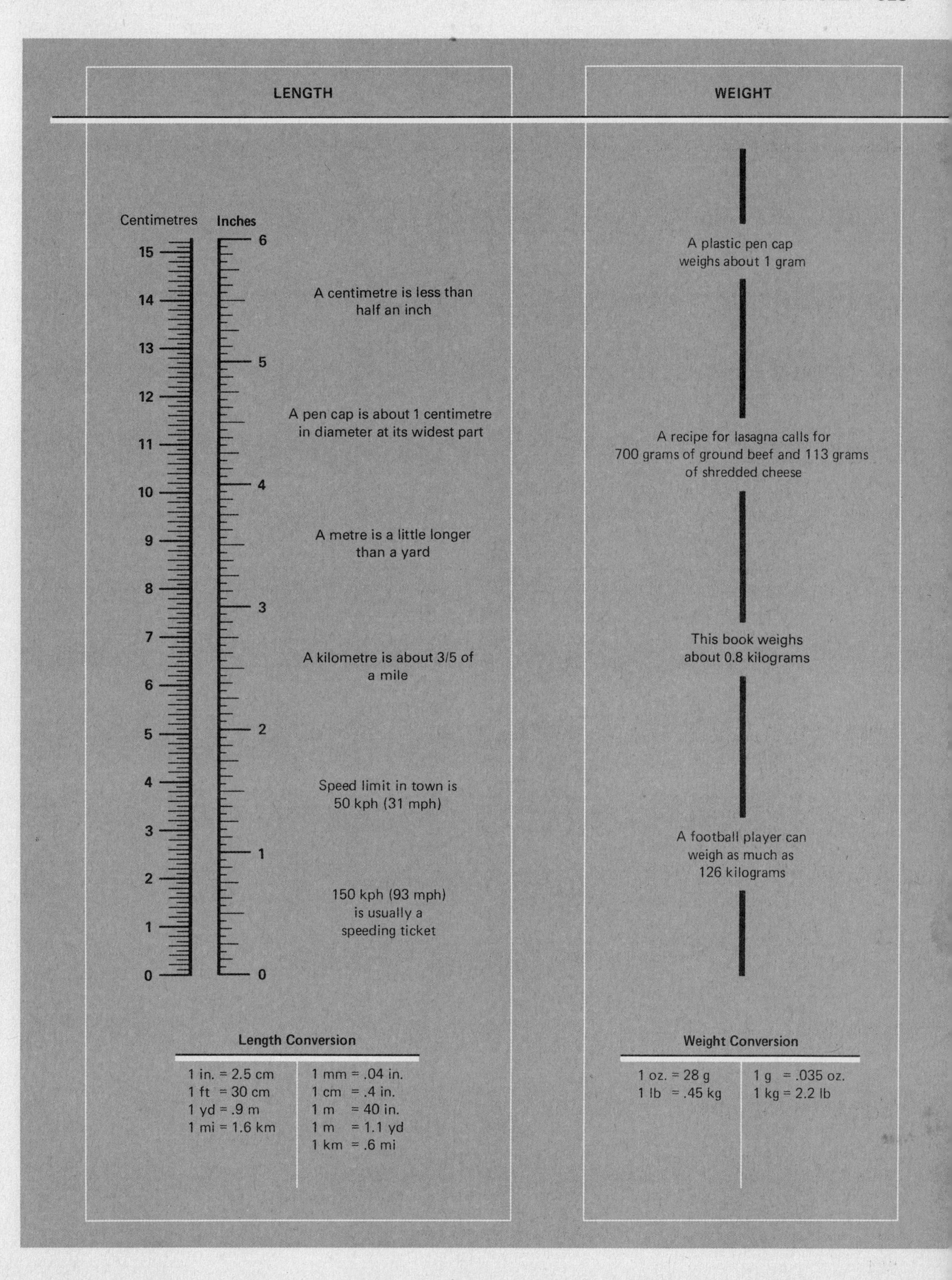
LENGTH
WEIGHT
Centimetres
Inches
15
14
13
12
11
10
9
8
7
6
5
4
3
2
1
0
6
5
4
3
2
1
0
A centimetre is less than half an inch
A pen cap is about 1 centimetre in diameter at its widest part
A metre is a little longer than a yard
A kilometre is about 3/5 of a mile
Speed limit in town is 50 kph (31 mph)
150 kph (93 mph) is usually a speeding ticket
A plastic pen cap weighs about 1 gram
A recipe for lasagna calls for 700 grams of ground beef and 113 grams of shredded cheese
This book weighs about 0.8 kilograms
A football player can weigh as much as 126 kilograms
Length Conversion
1 in. = 2.5 cm
1 ft = 30 cm
1 yd = .9 m
1 mi = 1.6 km
1 mm = .04 in.
1 cm = .4 in.
1 m = 40 in.
1 m = 1.1 yd
1 km = .6 mi
Weight Conversion
1 oz. = 28 g
1 lb = .45 kg
1 g = .035 oz.
1 kg = 2.2 lb

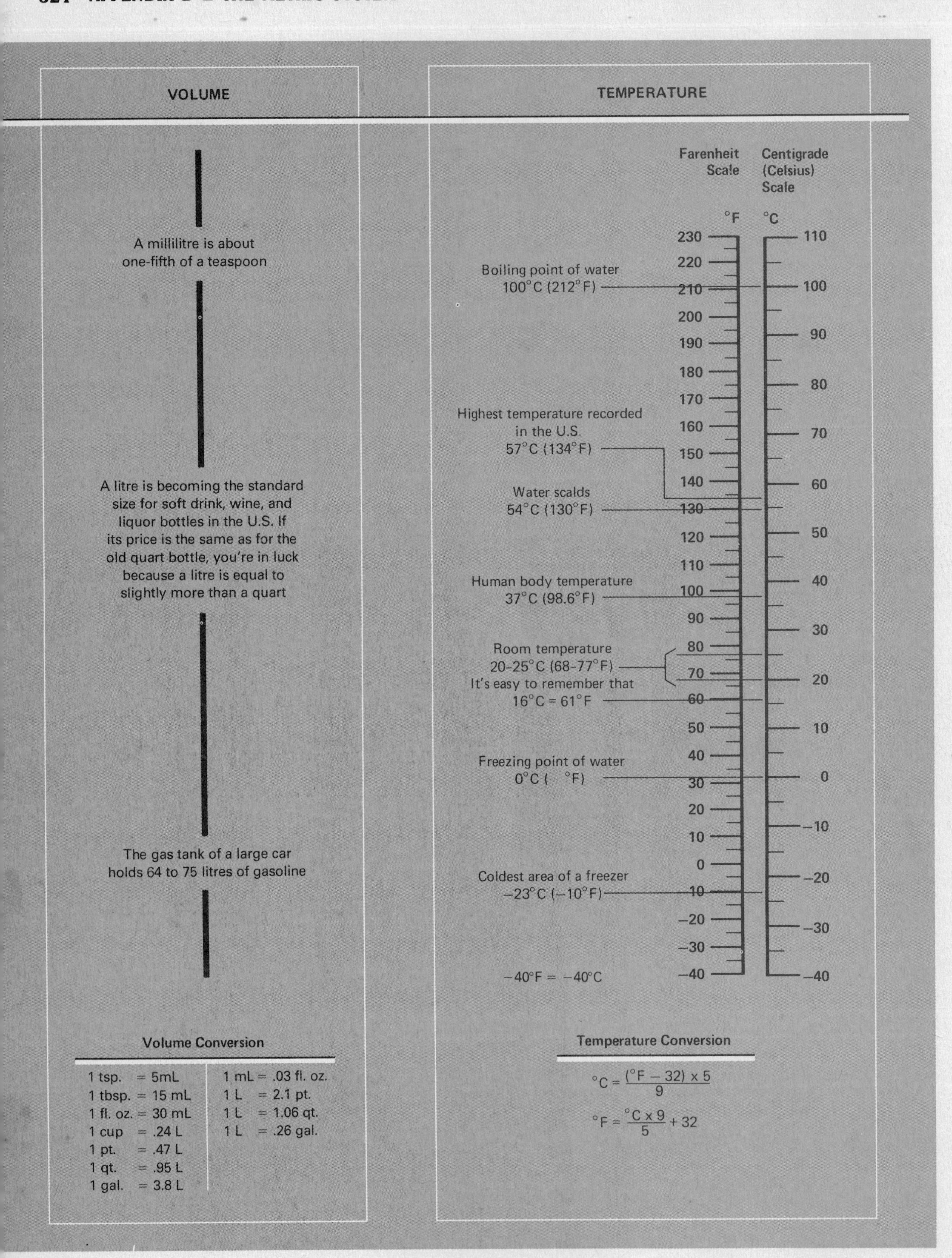

Volume Conversion

1 tsp. = 5mL	1 mL = .03 fl. oz.
1 tbsp. = 15 mL	1 L = 2.1 pt.
1 fl. oz. = 30 mL	1 L = 1.06 qt.
1 cup = .24 L	1 L = .26 gal.
1 pt. = .47 L	
1 qt. = .95 L	
1 gal. = 3.8 L	

Temperature Conversion

$$°C = \frac{(°F - 32) \times 5}{9}$$

$$°F = \frac{°C \times 9}{5} + 32$$

INDEX

Abbreviations, 312
Abiotic substances, 306
Abortion, 210
Abscisic acid, 284, 294
Absorbance, 10–11, 52–53, 58
Absorption, 121
  maximum, 52
  spectrum, 51, 56
Abstract, 311
Acetocarmine, 26
Acetone, 4, 49, 52, 308
Acetylcholine, 194–195, 225
Acid, 1, 9
Acoelomate, 138
Active site, 9
Active transport, 35–36, 38
Adaptation, 232
Adrenal gland, 204
Adrenalin 194–195
Afterimages, 232
Agamospermy, 284–285
Agar, 36, 114
Agaricus campestris, 123
Agonistic displays, 252, 256–257
Albumin, 191
Aldehyde group, 1–2
Alfalfa sprouts, 283–284
Algae, 121, 125–127
  brown (Division Phaeophyta), 125
  green (Division Chlorophyta), 125–127
  red (Division Rhodophyta), 125
Allantoic duct, 204
Allantois, 218, 220–221
Alternation of generations,
  algae, 121, 129
  cnidarians, 138
Alumina, 3
Alveoli, 179
Amino acids, 1–2, 4, 8–9, 203
  names of, 8
Amino group, 1
Ammonia, 203, 218
Amoeba, 23, 27
Amniocentesis, 86, 89
Amniotes, 163–167
Amniotic egg, 163
Amphibia, Class, 157, 161–163
Amyloplasts, 268–269
Anabaena, 115
Anaphase, 62–64, 67–69
Anemia, 88, 194
Aneurans, 161
Angiosperms, 265–274, 279–285
Animalia, Kingdom, 111
Annelida, Phylum 143–145, 149
Annuals, 266
Annulus, 132
Antenna, 76, 142, 150
Anther, 280–281
Antheridium, 132–133
Antibodies, 86–87
Antigen, 86–87
Anus, 208
Aorta, 184–186
Apical dominance, 269, 294–296
Apical meristem, 268–269, 294
Appendix, 175
Apple, 232
Arachnida, Class, 150
Arachnoidiscus, 118
Arbacia punctulata, 215
Archegonium, 132–133
Arista, 76
Aristotle's lantern, 214
Arteries, 170, 183, 185–186
Arterioles, 186
Arthropoda, Phylum, 149–154
Artiodactyla, Class, 169
Ascaris, (roundworm), 66–68, 140–141
Ascospores, 123
Ascus, 122
Astral rays, 62, 64
ATP, 41, 44, 49–50, 54, 62
Atrium, 184–185
Atropine, 195–196
Autonomic nervous system, 194
Autosomes, 78, 91
Autotroph, 29, 113
Auxin, 284, 294
Aves, Class, 157, 161, 163–165
Axial skeleton, 240
Axolotl, 180

Bacillus, 114
Bacteria, 27, 33, 113–115, 297
Banana, 285
Barbiturate, 164
Bark, 133, 270–271, 274
Barr body, 89, 93
Base, 9
Basidia, 123–124
Basidiospores, 124
Bean, 44, 97, 282–283, 288, 294–296
Beer, 3, 42
Beer's Law, 10
Behavior, 225, 251–257
Benedict's test, 2
Bent little finger, 86–87, 93
Betta splendens, 256–257
Bile, 173
Biofeedback, 194
Bioluminescence, 115, 117
Biston betularia, 96
Biuret test, 2
Bivalvia, Class, 146–147
Bladder, urinary, 204–207
Blade, 125, 273
Blastula, 216
Blind spot, 229–231
Blood, 149, 154, 159, 162, 182–189
  cells, 88, 93, 192–193
  groups, 86–88
  islands, 218
  pressure, 194
  type, 85–87
Bone, 240–241, 248
Book lungs, 150
Brain, 144–145, 150–152, 154, 157, 160–161, 163, 165, 167, 226–227
Bread, 42
Breast feeding, 208–210
Brine shrimp, 138
Bronchi(-oles), 179
Bud, 269, 273, 294–296
Buffer, 10, 53, 192
Butanol, 4

Caecum, 175
Calcium, 297–298, 306
Calvin cycle, 49–50, 53
Calyx, 280
Cambium, 268, 271–272, 274
Caminalcules, 106–107
Cap (fungus), 123–124
Capillaries, 183, 186–187, 194, 196
Capsule (moss), 130
Carbohydrate, 1, 42, 54, 174
Carbon dioxide (CO_2), 41–43, 47–50, 52–53, 57, 182, 203, 251, 306
Carbon fixation, 49, 52–54
Carbonic acid (H_2CO_3), 43, 182
Carbonyl group, 2
Carboxyl group, 1
Cardiac muscle, 241
Cardiac stomach, 151
Carnivora, Class, 169
Carotenes (carotenoids), 50, 55, 125
Carpel, 280
Carrier, 88
Carrot, 268
Cartilage, 240–241
Catalysis, 9
Catalysts (biological), 9
Cell(-s), 22–23, 25–31, 37–38, 175, 192–193, 267–274
  cycle, 61–64
  division, 62, 267
  poles, 62
  reproduction, 61–71
  respiration, 36, 41–44
  walls, 27, 31, 38, 121, 294
Cellulase, 116, 294
Cellulose, 2, 3, 31, 64, 121
Central nervous system, 225–228
Centrioles, 62, 64
Centromere, 61–62, 65–66, 68, 89
Cephalization, 143, 149
Cephalothorax, 150–151
Ceratium, 117
Cerebellum, 227
Cerebrum, 226–227
Cestoda, Class, 139–140
Chaetae, abdominal, 76
Cheek cells, 26
Chelicerae, 150–151
Chemical gradient, 35–36
Chemicals, industrial, 42
Chemoreceptors, 229, 232, 251
Chemotaxis, 251, 253
Chiasmata, 67
Chick embryo, 217–221
Chi-square, values of, 318
Chi-square test, 79, 315–318
Chitin, 149
Chlamydomonas, 23, 25, 116, 125–126
Chloragen cells, 145
Chlorophyll, 49–51, 55, 115, 125, 129, 284
  extract, 52, 308
Chloroplasts, 38, 51, 53, 117
Chlorosis, 298
Chondrichthyes, Class, 157–160
Chorion, 218, 221
Chromatids, 61, 65, 67
Chromatin, 62
Chromatogram, 4, 49–51, 55
  measuring, 5
Chromatography, 3–4, 8, 49–51, 55
Chromosome(-s), 61–64, 66–68, 73
  homologous, 64
  X, 86, 88–89, 91
  Y, 88, 91
Chrysophyta, Phylum, 117–118
Cilia, 27, 62
Circulatory system, 143, 159, 161–165, 167, 183–189, 191–196
Citric acid cycle, 41
Clam, 146–147
Claspers, 76–77, 159
Classification, 103–106, 307
Clay, 42
Cleavage, 68, 216–217
Cleavage furrow, 63
Clones, 284
Club mosses (Subdivision Lycopsida), 131, 136
Cnidaria, Phylum, 137–139
Coccus, 114
Cochlea, 230–231
Cockroach, 180
Coconut, 3
Cocoon, 144
Coelom(-ate), 137–138, 142–143, 149
Coenocyte, 61, 122
Coffee, 3
Coleoptile, 283–284
Coleus, 54, 269
Collagen, 29
Collenchyma, 269
Colon, 175
Colonies (bacteria), 114–115, 119
Color, 51, 55
Columnar epithelium, 29
Communication, 252, 254
Community, 305–306
Compact bone, 241
Companion cells, 269
Concentration gradient, 35–36
Conception, 208
Conclusion, 291, 296, 299, 309, 312
Condom, 208–209
Cone, 133
Congo red, 28
Conidiospores, 123
Conifer (Class Coniferae), 104–106, 133–134
Connective tissue, 29
Consumers, 306
Contraception, 204, 208–211
Contractile vacuole, 27, 36, 117
Contraction, muscle, 244–245

Convection currents, 36
Copper, 297–298
Copper sulfate, 2
Coprinus, 124
Cork, 271
Corn, 283–284, 299
Corn starch, 3
Cornea, 229–230
Corolla, 280
Corpus callosum, 226–227
Cortex, cerebral, 226
Cotyledon, 283–284
Covalent bonds, 1
Cowper's glands, 206–207
Coxa, 76
Cranial nerves, 226, 228–229
Crayfish, 150–152
Cream (contraceptive), 209
DCIP, 53, 58–59
Crop milk, 164
Crossing over, 67
Crustacea, Class, 150–152
Crystal violet, 27
Ctenoid scales, 160
Cuticle, 30, 130, 269, 274
Cutin, 269
Cyanobacteria, 115
Cycads, 133
Cyst (nematode), 141
Cytochrome b, 44
Cytochrome c, 44
Cytochrome oxidase, 41, 44, 48
Cytokinesis, 61–64, 67–68
Cytoplasm, 26

Dandelion, 283–284
Data (handling), 291–292, 296, 299, 307–309, 312, 314–319
DCIP, 53, 58–59
Deamination, 203
Deciduous trees, 133
Decomposers, 121, 306
Deductive logic, 291
Deficiency symptoms, 298
Degrees of freedom, 316
Denaturation, 9, 12
Detritus, 121
Detritus feeders, 306
Development,
  animal, 213–221
  plant, 266–274, 283–284
Dialysis, 36–37
Diaphragm,
  contraceptive, 208–209
  vertebrate, 173–174, 181
Diastole, 194
Diatomaceous earth, 118
Diatoms, 117–118
Dichotomous key, 104
Dicots, 266, 274, 283
Dictyostelium, 124
Differentiation, 293
Diffusion, 35–37
Digestion, 4, 169–175
Digestive tract, 137, 139–147, 149–154, 159–166, 171–175
Dikaryon, 123
Dinoflagellates, 117
Diploid (2N), 64–67, 121
Diptera, Order, 74
Dipylidium caninum, 140
Disaccharide, 2
Discussion, 296, 308, 312
Dissection,
  fetal pig, 169–175, 177–179, 184–189, 204–208
  procedures, 169–170
  tools, 170
Distribution (frequency), 314–315
Disulfide bonds, 1
Diversity index, 308
Division (cell), 61–68
DNA replication, 61–62, 65
Dogfish shark, 158–160
Dormancy, 283–284, 296
Douches, 209
Down's syndrome, 89, 91
Drosophila, 73–79
  anesthetizing, 74–75
  genetics experiment, 77–79
  life history, 74
  metamorphosis, 154
  morphology, 75–77
  taxis, 252–253
Drumstick, 89
Ductus arteriosus, 185, 189
Ductus venosus, 188, 189

Ear, 229–231
Ear lobes (attached), 86–87, 93
Earthworm, 143–145
Eclosion, 74
Ecosystem (pond), 305–309
Ectoderm, 138, 217
EDTA, 299
Edward's syndrome, 91
Effectors, 225, 239–245
Egestion, 203
Egg (hen's), 218–219
Egg cell (ovum), 66, 68, 134, 205, 208–210, 215
Egg white, 2–3, 7
Elaters, 131
Electron transport, 41, 44, 49–50, 52–53
Electrons, 41, 44
Electroreceptors, 229
Elodea, 37–38, 43, 48, 52–53, 57
Embryo, 68, 215–222, 283–284
Embryo sac, 281, 282
Embryophytes, 129–136
Endoderm, 138, 217
Endodermis, 268
Endopodite, 150, 152
Endosperm, 134, 281, 283–284
Endothermic, 163
Entropy, 35
Enzymes, 9–12, 44
Enzyme-substrate complex, 9
Ephedra, 133
Epicotyl, 283–284
Epidermis, 26, 30, 268, 274
Epiglottis, 173, 179
Epithelium, 29, 33–34
Equipment, 321–322
Equisetum, 131
Erythroblastosis fetalis, 88
Esophagus, 174, 179, 186
Ethanol (alcohol), 3, 20, 41, 54
Ether, 74–75
Etiolated, 284
Eukaryotes, 113
Euglena, 116–117, 120
Euglenoid movement, 116
Euglenophyta, Phylum, 116
Eukaryotic cells, 25
Euplotes, 23
Eustachian tubes, 162, 165, 179, 230–231
Evergreen leaf, 131
Evolution, 96, 106
Excretion, 203–205
Excretory system, 204–205
Exopodite, 150, 152
Exoskeleton, 149
Experimental design, 294–295, 307
Eye, 229–230
  compound, 76, 153
  simple, 76, 153
Eyespot, 117

F_1, 78, 83
FAD/$FADH_2$, 41
Fat, 3, 174
Fatigue (muscle), 245
Fatty acid, 1, 3
Feathers, 163
Femur, 76
Fermentation, 41–42
Ferns (Class Filiaceae), 131–133
Fertilization, 61, 66, 134
  double, 134, 281–282, 288
  external, 161, 215–217
  internal, 149, 159, 205, 208, 213
Fertilizer, 268
Fetal pig, 169–175, 177–179, 184–189, 204–208
Fetus, 188
Fibers, 269, 273
Fibrinogen, 191
Fibrous root, 267
Fiddlehead (fern), 132
Filter feeders, 146
Filter paper, 3, 38, 49, 308
Fire coral, 138
Fish, 43, 48, 253–257
Five Kingdoms, 111
Fixed action pattern, 252
Flagellum(-a), 30, 62, 114, 117
Flame cells, 139
Flatworms (Phylum Platyhelminthes), 138–140
Flavin mononucleotide (FMN), 115
Flexure, 220
Floodlamp, 52
Flour (white, soy), 3
Flower, 266, 279–280, 287
Fluorescence, 50
FMN, 115
Foam (contraceptive), 209
Foods, 2–3, 7
Foot (molluscan), 146
Foramen ovale, 184, 189
Formalin, 169
Fossils, 131
Fovea, 229
Fragmentation (lichen), 125
Frog, 161–163, 179, 181, 187
  muscles, 243–245
  pithing, 195
Fronds, 131
Fruit, 282–283, 289
Fucus, 125
Functional groups, 1
Fungi, Kingdom, 111, 121–125
Fungus, 24, 38, 41, 111, 121–125

G_1 (cell cycle), 61–62
G_2 (cell cycle), 61–62
Galactose, 42
Gall bladder, 173–174
Gametes, 61, 66–67, 121, 279
Gametophyte, 121, 126, 130, 133, 134
Gamma globulin, 191
Gas, 42
Gas exchange, 177–182
Gastrovascular cavity, 137
Gastrula, 216
Gelatin, 3
Gene fixation, 98
Gene,
  flow, 95
  pool, 97–98
Genes, 67
Genetic drift, 98, 101
  variation, 65
Genetics, human, 85–92
  Mendelian, 73–79
  population, 95–98
Genotype, 86–87, 95–96, 97, 99
Genus, 103, 313
Geotaxis, 251, 253
Geotropism, 267
Germ cells, 61
Germinal disc, 217
Germination, 283–284
Gibberellin, 283–284
Gill rakers, 161
Gills, 123, 146–147, 152, 159, 161, 179–180
Ginkgo, 133
Girdled, 272
Gladiolus, 280
Glottis, 173, 179
Glucose, 2–3, 7, 37, 49–50, 54, 307
Glycerol, 3
Glycolysis, 41, 44
Goblet cells, 29, 33
Gonads, 66
Gonionemus, 138
Gonium, 125–126
Gonyaulax, 117
Gradient, chemical, 35–36
  concentration, 35–36
  hydrogen ion (H^+), 50
Grasshopper, 152–154, 180
Green gland, 152
Grex (pseudoplasmodium), 124
Guaiacol, 10–12
Guano, 164
Guard cells, 30
Gymnosperms, 133, 272–273

Habituation, 233
Hair cells, 230
Halteres, 76
Haploid (N), 64–66, 121
Hardy-Weinberg, 95–96
Heart, adult, 183–184
  fetal, pig, 184–185
  frog, 194–196
  human, 196
Heartwood, 272, 278
Heat sink, 52
Hemocoel, 154
Hemoglobin, 88, 193–194
Hemophilia, 86
Hepatic portal system, 188–189
Herbs, 266
Heterocysts, 115
Heterotrophs, 41, 113, 116
Heterozygous, 88, 96
Hickory, 270
Hill reaction, 53, 58–59
Hitchhiker's thumb, 86, 93
Holdfast, 125–126
Homeostasis, 225, 293
Homologous chromosomes, 64–67
Homozygous, 77, 96
Honey, 2–3, 7
Hormones, 204–206, 209–210, 293–296
Horsetails (Subdivision Sphenopsida), 131–136
Human reproduction, 208–211
Hydra, 24, 137–139
Hydrochloric acid (HCl), 43, 48, 173, 192
Hydrogen bonds, 1, 9
Hydrogen ion, 9–10, 50
Hydrogen peroxide, 10–12
Hydrolysis, 4
Hydrophobic interactions, 3, 9
Hydroxylamine, 12
Hydroxyl group, 1
Hypertension, 194
Hyperventilation, 182
Hypha(-ae), 122
Hypocotyl, 283–284
Hypodermic impregnation, 142
Hypothalamus, 227
Hypothesis, 291, 295, 315–318
  Null, 315
Hypotonic, 37

Imbibition, 283–284
Immersion oil, 21
Immune system, 191, 193
Implant (contraceptive), 209
Indeterminate growth, 267
Indoleacetic acid, 294
Inductive logic, 291
Inguinal canal, 206–207
Inhibitors, 42
Insect, 24, 73–77, 116, 152–154
Insecta, Class, 152–154
Inspiration, 181
Instar, 74, 154
Intercalary meristem, 274
Intercourse, 208
Interkinesis, 68
Internode, 269
Interphase, 62–64, 69
Intestine,
  cells 27–28
  large 175
  small 174
Introduction, 311
Iodine test, 2, 54, 283
Ionic bonds, 1, 9
Iris, pigmented, 86, 93
  structure, 229–230
Iron, 297–298
IUD (intrauterine device), 208–209

Jacobs, W.P., 294
Jelly (contraceptive), 209
Joint, 239, 248
Journal references, 312–313

Karyotype, 86, 89–91, 94
Ketone group, 1–2
Kidney, 159–164, 166, 203–206
  stones, 204
Kinesis, 251
Kinetochore, 61
Klinefelter's syndrome, 91
Kymograph, 242–244, 255–256

Laboratory equipment, 320–321
Lactobacillus, 27, 114
Lactose, 2, 27
Lanolin, 295
Larva, 74, 152, 154, 213, 216
Larynx, 178
Lateral line organ, 158–159
Lateral root, 267
Law, scientific, 291
Leaf, 24, 34, 54, 131, 133
  arrangement, 273–274
  cells, 30–31, 268, 274
  venation, 273
Leaflets, 273
Learning, 251–252
Leeuwenhoek, Antony van, 19
Length (meter), 322–324
Lenses, cleaning, 20
Lenticels, 270–271
Lichens (Division Mycophycophyta), 97, 123–125
Ligament, 239–240, 248
Light, 10, 51–54, 58–60
Lilac, 269
Liver, 173–174
Liverworts, 130
Locomotion, 239
Locule, 280–282
Luciferase, 115
Lugol's solution, 2, 307
Lumen, 28
Lungs, 179, 181–182, 203
Lycopodium, 131
Lymphatic system, 175, 187, 191
Lysis, 37

Macronucleus, 28
Macronutrients, 174, 297
Magnesium, 297–298
Magnification, 21–22
Malpighian tubules, 152, 154
Maltose, 2
Mammalia, Class, 157, 161, 165–167, 169
Mammary glands, 165, 171
Mandibles, 150, 152–154
Manganese, 297–298
Mantle, 146
Margarine, 3, 7
Marrow, 240
Maxillae, 150
Mean, arithmetic, 314
Mechanoreceptors, 229–230, 232–233
Medulla, 226–227
Medusa, 137–138
Megasporangium, 134, 281–282
Meiosis, 61, 121, 127, 213
  animal, 66–68
  mechanics of, 64–66
Membrane, cell, 35, 117
  semipermeable, 35, 37
  thylakoid, 50
  vesicles, 64
Menstrual cycle, 208, 210
Meristem, 267–269, 298
Mesenteries, 173
Mesoderm, 138, 217
Mesophyll, 30, 274
Metameric segmentation, 143, 149
Metamorphosis,
  arthropod, 149
  insect, 74, 154
  sea urchin, 215–217
Metaphase, 62–65, 67–69
Materials, 295, 306–307, 311
Methods, 295, 311
Methyl cellulose, 28
Methyl orange, 192
Methylene blue, 26
Metric system, 312, 322–324
Molecules, 1–5
Microecosystem, 305–309
Micronutrients, 297–298
Micropyle, 281–283, 288
Microscope, 19–24
  compound, 20–23
  stereo, 23–24
Microsporangium, 133, 281
Microtubules, 62
Middle lamella, 64
Milk, 3, 27
Mineral deficiency, 298
Mitochondria, 62
Mitosis, 61–62
  animal, 62–66
  mechanics of, 64–66
  onion, 64
  plant, 64, 69
  whitefish, 62–63
Mitotic spindle (apparatus), 62, 64
Models (fish), 257
Molarity, 313
Molasses, 42
Molds, slime (Division Myxomycophyta), 124
Molecules, 1–5
Mollusca, Phylum, 146–147, 149
Molt, 74, 149, 152, 154
Molybdenum, 297–298
Monera, Kingdom, 111, 113–115, 119
Monocots, 266, 274, 284
Monosaccharide, 1–2
Morula, 216
Mosses (Class Musci), 130, 136
  sphagnum, 130
Motivation, 252
Mucosa, 28–29
Multicellularity, 137
Multiple alleles, 87
Muscle(-s), 28–29, 232, 239, 240–245
  fiber, 241
  frog, 243
  human, 241–242
  physiology, 242–245
Mushroom (Class Basidiomycetes), 123–124
Mussel, 146–147
Mutants, 77–78, 82–83
Mutation, 95
Mycelium, 122
Myelin, 226

Nadi reagent A, B, 44
NAD^+/NADH, 41
NADHP, 49–50, 53–54
Naiads, 154
Nares, 177
Nasopharynx, 178
Natural flora, 114
Natural selection, 95, 96–98, 100
Nauplius larva, 152
Navel, 171
Navicula, 118
Necrosis, 298
217–218
Necturus, 180
Negative pressure breathing, 181
Nematocysts, 137–138
Nematoda, Class, 140–141
Nephridium, 144–147
Nephron, 204
Nervous system, 225–233
Neural tube, 217
Neurons, 225–226, 228
Neurulation, 216
Neutral red, 38
Nictitating membrane, 162, 164
Ninhydrin, 4–5
Nitrogen, 297–298, 306
Noctiluca, 117
Node (plant), 131, 269
Non-disjunction, 89, 91
Noradrenalin, 225
Normality, 313
Nose, 177–178
Notochord, 157–158, 217–218
Nuclear membrane, 25, 62–63, 68
Nucleolus, 26, 62–64
Nucleus, 25–26, 62–64, 117
Numerical taxonomy, 106
Nutrition (plant), 297–299
Nymphs, 154

Ocelli, 76, 153
Oil, 3, 7
Oil immersion, 21
Oligochaetae, Class, 143–145
Onion, juice, 3
  cells, 25, 33, 64
Oocyte, 66–67
Oogenesis, 66–68
Operculum, 160–161, 180
Optimum (enzyme), 9–10
Optomotor response, 255–256
Oral groove, 28
Organ, 137
Organogenesis, 216
Orientation, 251–257
Osmosis, 36–37
Osmotic potential, 36
Osteichthyes, Class, 157, 160–161
Ovary, animal, 66, 205
  flower, 280–282, 288
Oviduct, 66, 205
Oviparous, 161, 163
Ovipositor, 77
Ovoviviparous, 159
Ovulation, 208, 210
Ovule, 134, 280
Ovum (egg cell), 66, 68
Oxidation, 41
Oxidative phosphorylation, 44
Oxygen (O_2), 44, 50, 52–53, 57, 115, 188, 306

P_1, 78, 83
Palate, 171, 178
Palisade cells, 30, 274
Pancreas, 174
Pandorina, 125
Paper, 2, 7
Parafilm, 42, 53
Paramecium, 23, 27–28, 33, 251
P. bursaria, 23
Paramylum bodies, 117
Parasite, 121
Patau's syndrome, 91
Peach, 282, 288
Peanuts, 3
Pedigree, 85–86
Pedipalps, 150–151
Pellicle, 117
Penicillium, 123
Penis, 142, 206–207
Peppered moth, 97
Pepsin, 173
Peptide bonds, 2
Perception, 230–233
Perch, 160–161, 180
Perennials, 266
Pericarp, 282–283
Pericycle, 268
Peripheral nervous system, 225, 228–229
Peritoneum, 28, 172–173, 175
Peroxidase, 10, 15–17
Petals, 280
Petiole, 132, 273
Petroleum ether, 49
Peziza, 123
pH, 9–11, 43, 82, 192, 251, 298–299
Pharyngeal gill slits, 157, 220
Pharynx, 173, 178
Phenolphthalein, 43
Phenotype, 86–87, 95–97, 99
Phloem, 30–31, 130, 268, 270–271, 274
Phosphorus, 297–298, 306
Photobacterium fischeri, 115
Photoreceptors, 229–231, 251
Photosynthesis, 49–54, 121, 284
Photosynthetic pigments, 49, 55
Photosystems I, II, 50–51
Phototaxis, 251–253
Phototropism, 294
Phycocyanin, 125
Phycoerythrin, 125
Physarum, 124
Phytochrome, 284
Pigeon, 163–165
Pigments 49–50
Pill (contraceptive), 209
Pineapple, 285
Pistil, 280
Pith, 269
Pithing (frog), 195
PKU (phenylketonuria), 86
Placenta, 165, 188, 189
Placoid scales, 158
Plagiarism, 312
Planarian, 24, 138–139
Plankton, 306
Plant, cells, 29–31, 267–274
  function, 265–274, 279–285
  hormones, 293–296
  nutrition, 297–299
  structure, 265–274, 279–285
Plantae, Kingdom, 111, 125–127, 129–136
Plasma, 191–192
Plasma membrane, 26
Plasmodium, 124
Platelets, 192–193
Pluteus, 215–217
Poisons, 44
Polar body, 66–67
Pollen, 133, 280–281
  tube, 134, 281
Pollination, 279–280
Polychaetes, 147
Polymers, 1
Polyp, 137–138
Polysaccharide, 2
Pond ecosystem, 305–309
Portuguese Man-of-War, 138
Positive pressure breathing, 181
Potassium, 297–298
  bromide, 36–37
  cyanide, 44
  ferricyanide, 36–37
Potato, 2, 7, 232, 271
  starch, 3
Pregnancy, 208, 210
Premise, 291
Primary,
  growth, 266, 271
  phloem, 271–272
  root, 267
  xylem, 271–272
Primata, Class, 169
Proboscis, 76
Producers, 306
Proglottid, 139–140
Prokaryotic cells, 25, 113
Pronuclei, 68
Prophase, 62–64, 69
Proprioceptors, 232–233
Prostate, 207
Prosthetic group, 9
Protein, 2, 4, 9, 174, 203
  test for, 2
Prothallus, 133
Protista, Kingdom, 111, 116–118
Protonema, 130
Protopodite, 150, 152
Protoslo, 28
Pseudocoelom(-ate), 137–138, 140
Pseudopods, 27
PTC (phenylthiocarbamide), 96, 99
Pteropsida, Subdivision, 131–134
Pulmonary
  artery, 184–185
  circuit, 162
  veins, 187
Pupa, 74, 154
Pyloric valve, 173–174
Pyrrophyta, Phylum, 117
Pyruvic acid, 41

Quadrupeds, 161–167
Quiescent center, 64, 267
Quinine sulfate, 232

Radicle, 267, 283
Rat, 165–167
Receptacles, 125
Receptor cells, 225
Reciprocal crosses, 77–79, 81
Reciprocal inhibition, 241
Rectum, 208
Reducing sugar, 2
Reference list, 312
Report (laboratory), 79, 296, 308–309, 311–313
Reproduction, asexual (vegetative), 126, 284–285, 290
  cell, 61–71
  sexual, 122, 126, 205–211, 285
Reptilia, Class, 161, 163
Reservoir, 117
Resin, 273
Resolution, 21
Respiration, animal, 177–182
  aerobic, 41–42
  anaerobic, 44
  cell, 36, 41–44
Results, 296, 299, 312
Reticular activating system, 227
Reticular formation, 227
Retina, 229–230
R_f (chromatogram), 4, 55
Rh blood group, 87–88
Rhizoids, 122, 130
Rhizome, 131
Rhizopus (Class Zygomycetes), 121
Rhythm (contraception), 210
Rituals, 252
RNA, 62
Root, 64, 267–268
  cap, 64, 267
  hairs, 267–268
Rotifera, Phylum, 142
Roundworm *(Ascaris)*, 66–68, 140–141

S (synthesis, cell cycle), 61–62
Salivary, amylase, 171
  glands, 178
Salt (table), 3, 9
Saprobes, 121
Sapwood, 272, 278
Scallop, 146
Scars, 270
Schooling, 253–255
Scientific method, 291–292, 294
Sclerotium, 124
Scolex, 139–140
Scutellum, 76
Sea urchin, 213–216
Secondary, growth, 266, 271
  phloem, 271–272
  xylem, 271–272
Seed coat, 134
Seeds, 133–134, 279, 283–284
Semicircular canals, 157–158, 230–231
Seminal vesicles, 207
Seminiferous tubules, 68
Semipermeable membrane, 35, 37
Senescence, 293
Sense organs, 229–230, 251
Sepals, 280
Septa, 143, 145
Serum, 192
Sex, chromosomes, 78, 88–89, 91
  comb, 76
Sex-linkage, 78
Shark, 158–160
Shoot, 268–269, 296
Siamese fighting fish, 256–257
Sickle cell anemia, 86, 88
  hemoglobin, 194
Sieve tube members, 270
Sign stimulus, 252
Significance (statistical), 315
Silica, 131
  gel, 3, 4
Silk glands, 150
Silver, bromide, 39
  chloride, 39
  ferricyanide, 39
  nitrate, 36–37
Sino-atrial node, 194
Siphons, 146–147
Sister chromatids, 61–62, 65, 67
Skeleton, 239–241, 248
Smooth muscle, 241
Snails, 43, 48
Snyder's medium, 114–115, 119
Sodium, azide, 44
  benzoate, 96, 99
  bicarbonate, 53, 57
  carbonate, 38, 40
  chloride, 36–37, 191, 232
  hydroxide, 2, 192
Soil, 298
Solute, 35–36
Solvent, 35
  front, 4–5, 50, 55
Somite, 217, 220
Sorus(-i), 132
Sowbug, 251
Soy flour, 3
Species, 192, 306, 313
Spectrophotometer, 10–11, 52–53, 308
  instructions, 10–11
Spectrum, absorption, 51
  visible, 51, 56
Sperm cells, animal, 66–68, 207–210, 215
  plant, 130, 279
Sperm nucleus, 281–282
Spermatocyte, 66, 68
Spermatogenesis, 66
Spermopsida, Subdivision, 104–106
Sphagnum moss, 130
Spider, 150
Spinal cord, 227–228
Spinal nerves, 228
Spindle fiber, 62, 64–65
Spinnerets, 150–151
Spiracles, 152, 158, 180
Spirillum, 114
Spirogyra, 23
Spleen, 173–174
Sporangia, 122, 131
Spores, 121
Sporophyll, 131, 133
Sporophyte, 121, 126, 133
Squamous epithelium, 29
Stalk, fungus, 123–124
  moss, 130
Stamen, 280
Standard deviation, 296, 315–316
Standards (chromatography), 4
Starch, 2–
Statistical tests,
  Chi square, 315–318
  student's *t*, 318
  Wilcoxon, 318–319
Statistics, 79, 296, 315–320
Stems, 131, 266–273
Stentor, 23, 28, 33
Sterile jacket cells, 130
Sterilization, 204, 210
Sternites, 77
Stethoscope, 194
Stimuli, 251–252, 257
Stipe, 125
Stolons, 122
Stomata, 30, 130, 266, 274
Stretch receptors, 241
Striated muscle, 241
Strobilus(-i), 131
Stroma, 50
Strongylocentrotus purpuratus, 215
Student's t test, 318
Stunting, 298
Suberin, 271
Submucosa, 28–29
Substrate, 9
Subunits, 1, 4
Succession, 305–306
Sucrose, 3, 42, 44, 53, 232
Sudan III test (fat), 3
Sugar, 1–3
  Benedict's test, 2
Sulfhydryl group, 1
Sulfur, 297–298
Sunflower, 299
Sweat, 203
Swim bladder, 160
Swimmerets, 150–151
Symbiosis, 114, 116
Symmetry, 214
Sympathetic nervous system, 195
Synapses, 225
Synapsis, 65, 67
Syncytium, 61
Synedra, 118
Syrinx, 165
Syrup, 3
Systemic circuit, 162
Systole, 194

t (values of), 317
Tangelo, 285
Tap root, 267
Tarsus, 76
Taste test, 96
Taxis, 251–253
Taxon(-a), 103–104, 307
Taxonomy, 103
Teeth, 158, 162, 166–167, 171
Telophase, 62–64, 67–69
Temperature scales, 324
Temporal summation, 245
Tendon, 241
Tentacles, 137–138
Tergites, 77
Termite, 116
Testes, 68, 206–207
Tetany, 245
Tetrad, 65, 67
Tetrapods, 161–167
Tetrazolium, 283
Thalamus, 227
Theory, 291
Thermoreceptors, 229, 233
Thumb crossing, 86, 93
Thylakoid membrane, 50–51
Thymus, 167, 178
Thyroid, 167, 179
Tibia, 76
Tissues, 25–26, 28–31, 34, 137, 217, 267–274
Title, 311
Toad, 161
Tongue, 232–233
  rolling, 86–87, 93
Tonoplast, 38
Tooth decay, 114, 119
Torsion, 218, 220
Trachea, 178–179
Tracheal system, 152–153, 180
Tracheids, 272–273
Tradescantia, 280
Transpiration, 266–267
Transport, active, 35–36, 38
Transport system, 183–189, 191–196
Tricarboxylic acid cycle, 41
Trichinella spiralis, 141
Trichinosis, 141
Trichonympha, 116
Trinacria, 118
Trochanter, 76
Trochophore larva, 144
Tropotaxis, 252
Tube feet, 213–214
Tube nucleus, 281
Turbellaria, Class, 138–139
Turbinate bones, 178
Turner's syndrome, 89, 91
Turnip peroxidase, 10–12, 15–17,
  extract, 11–12
Twig, 270–271
Twitch (muscle), 244–245
Tympanic membrane, 162
Typhlosole, 145, 147

Ulothrix, 126
Ultraviolet light, 51, 55
Umbilical arteries, 186, 189
  cord, 171, 188
  vein, 188–189
Unguligrade locomotion, 171
Urea, 203
Urethra, 204–207
Uric acid, 218
Urine, 203
Uterus, 66, 205
Uvula, 171

Vacuoles, 26
  central, 38
  contractile, 27, 36
  food, 27
Vagina, 206
Vagus nerve, 194–196
Valves, heart, 184
  veins, 187, 194
Variance, 314–315
Variation, genetic, 65, 96
Vasa deferentia, 206–207
Vascular, bundles, 30–31, 269
  cambium, 268, 271–272
  tissue, 30, 130–131, 265–274
Vegetative propagation, 122–123, 126, 131–132, 284–285, 290
Veins, 170, 183, 187–188
Venae cavae, 184, 185, 187–188
Venation, 273
Ventricle, 184–185
Venules, 186
Vertebra, 157–158, 218
Vertebrata, Subphylum, 157–167, 169
Vessels, 273
Vibrissae, 165
Villi, 174–175
Visceral mass, 146
Vitamin K, 175
Vitelline membrane, 215
  vessels, 218, 220–222
Viviparous, 165
Vocal cords, 178
Volume (litre), 322–324
Volvox, 23–24, 30, 125
Vomerine teeth, 162

W values, 319
Water, 37, 266–267
  metabolic, 41, 44
Water vascular system, 214
Wavelength (light), 51–53, 56
Weight (gram), 323–326
Wet mount, 22–23, 307
Whatman #1 filter paper, 3, 38, 49, 308
Wheat grain, 283
White blood cells, 27
Whitefish (mitosis), 62
Whittaker, R.H., 111
Widow's peak, 86–87, 93
Wilcoxon test, 318–319
Wine, 42
Wood, 272
Woody, plants, 266
  tissue, 270–273
Writing tips, 312

Xanthophylls, 50, 55
X chromosome, 86, 88–89, 91
Xylem, 30–31, 130, 266–268, 270–271, 274
Xylene, 20

Yeast (Class Ascomycetes), 23, 28, 38, 41–42, 47, 122–123
Yolk sac, 217–218
Yogurt, 26

Zebra danio (fish), 255
Zinc, 297–298
Zoospores, 126
Zygospores, 122, 126
Zygote, 61, 66, 68, 213